Crustacea

Frontispiece. Above left: Carl Claus (courtesy of J. Sieg, University of Osnabrück, Vechta).
Above right: William Thomas Calman (photograph by M. D. Burkenroad). Below left:
H. Graham Cannon (courtesy of J. Dalingwater and L. Lockey, University of Manchester).
Below right: Sidnie M. Manton (photograph by R. Siewing).

CRUSTACEA

Frederick R. Schram
San Diego Natural History Museum

New York   Oxford
OXFORD UNIVERSITY PRESS
1986

Oxford University Press

Oxford   New York   Toronto
Delhi   Bombay   Calcutta   Madras   Karachi
Petaling Jaya   Singapore   Hong Kong   Tokyo
Nairobi   Dar es Salaam   Cape Town
Melbourne   Auckland

and associated companies in
Beirut   Berlin   Ibadan   Nicosia

Published by Oxford University Press, Inc.,
200 Madison Avenue, New York, New York 10016

Oxford is a registered trademark of Oxford University Press.

Library of Congress Cataloging-in-Publication Data

Schram, Frederick R.
Crustacea.

Includes index.
1. Crustacea.   I. Title.
QL435.S37   1986   595.3   85–11526
ISBN 0–19–503742–1

Printing (last digit):   9 8 7 6 5 4 3 2 1

Printed in the United States of America
on acid-free paper

To those who have gone on before

PREFACE

This book was originally undertaken to assuage my own sense of frustration in using the currently available single-volume references on crustacean form and function to obtain an overview of the natural history and evolution of Crustacea. The classic volume in this regard has been W. T. Calman's 1909 treatise *Crustacea*. Despite all that has happened in the field since 1909, that book is still a standard reference, even though quite out of date. Calman provided a remarkably concise synthesis of what was known about crustacean biology up to that time, and so effective was he at doing this that he established the framework upon which the discipline has developed even to the present. Another excellent reference is Kukenthal and Krumbach's (1927) *Handbuch der Zoologie*. Aside from being out of date (and in German), this compendium provides rather complete treatment of the groups it covers. However, it is somewhat uneven given the variety of authors for the various chapters. The multi authored *Traité de Zoologie* suffers from this same uneven treatment and, in addition, is not nearly so comprehensive as Calman or the *Handbuch*. Finally, there is the English edition of Kaestner's (1970) *Textbook of Zoology* which, despite its being a single-author work, is quite uneven and incomplete regarding many aspects of anatomy and development, though, in spite of these shortcomings, it frequently serves as a textbook in carcinology courses.

In quite another category is McLaughlin's *Comparative Morphology of Recent Crustacea* (1980). This book is excellent to the task it was designed to do, that is, as a guide or manual on strictly anatomical matters. I have found it useful in offering access to the literature, since it has reference lists for each group it deals with. However, it covers only recent forms and does not deal with fossil groups. It provides a fine glossary of crustacean anatomical terms.

So, given this background, I originally set out to treat all known groups of crustaceans in as even a manner as I could manage, covering basic external and internal anatomy, modes of locomotion, feeding mechanisms, reproductive biology, habitat and biogeographic distributions, embryonic and larval development, fossil record, taxonomy, phylogeny, and evolution—much in the manner of Calman. My object has not been to cover all aspects of crustacean biology, especially those dealing with physiology, histology, and ultrastructure. Other compendia are currently available that do so much more effective a job than I could do. Rather I wished to present an overview of those aspects of crustacean studies that have been useful in elucidating the interrelationships of constituent groups and the evolution of Crustacea as a whole. Some disciplines within the biology of crustaceans serve rather poorly in this regard. For example, several of my chapter reviewers suggested that I cover more subjects like neurophysiology and field biology. Praiseworthy as these suggestions were, the fact remains that little can be gleaned about crustacean evolution from such subjects, not because they cannot potentially make such contributions, but rather because information in such fields to date has not been accumulated and organized in a manner designed to elucidate crustacean relationships. To have produced a book with this kind of information in it would have strained my capacities and simply was not the kind of book I wanted to write.

However, as preparation of this book progressed, I was struck by how much is still to be learned about basic structure and function, natural history, and evolutionary biology of crustaceans! For example, though ostracodes are the most numerous of crustaceans and quite diverse, living in many different habitats and obviously 'munching' on many different things, we know relatively little about their feeding. Certainly our knowledge is not on a par with what we do know about feeding strategies in cladocerans or thoracican barnacles. Or, despite their ready availability the world over and ease of handling in the laboratory, we lack knowledge about the basic biology of some branchiopod groups, like the notostracans and conchostracans, though we have been making significant strides in understanding other branchiopod groups, like the anostracans and cladocerans. Or, in very diverse and successful groups like isopods, amphipods, and decapods, though we eagerly describe numerous new taxa each year, we still don't know enough about the function of structures in these groups to begin to understand their adaptive radiations—for example, there is no agreement on the higher taxonomy of such large groups.

Nor have students of crustaceans been particularly consistent in their view of the science. Carcinologists seem to continue to be so impressed by differences among some taxa, that they have totally disregarded the significance of their similarities, for example, isopods and amphipods, or leptostracans and 'cephalocarids' and branchiopods. Conversely, in other taxa they frequently focus on similarities in order to minimize differences, for example, the natantian decapods, which include such diverse groups as dendrobranchs, euzygids, and eukyphids, or in malacostracans *sensu lato*, which encompass the phyllopodous leptostracans and the stenopodous hoplocaridans and eumalacostracans.

Such shortcomings and inconsistencies are not to be taken as condemnations of carcinology as a whole; rather, they are merely noted in passing as manifestations of what has been a characteristic phenomenon in the history of many disciplines. Carcinologists have got to become more critical in evaluating their work. They must cut to the heart of what has been published in carcinology to date and not accept things merely because they are 'in the literature.' More than in many other animal groups, carcinology has been dominated by certain key ideas, generated by prominent figures of the past, revered for their capacities for synthesis. This has had bad and good effects on the development of the science: bad in that a too rigid adherence to the pronouncements from the past has often stifled the development of fresh approaches to problem areas; good in that well-done synthetic overviews by single individuals can be quite effective toward channeling work along productive lines of enquiry. So for better or worse, carcinology has been strongly influenced by its traditions.

Four people really came to stand out in this regard as I prepared this volume. Carl Claus (1835-1899), the first of these, was one of the most prolific of the nineteenth century carcinologists. His interests extended from copepods to malacostracans; from species descriptions, through studies of development, to pioneering analyses of basic form and function. It was Claus who established the separate status of tanaidaceans from isopods, who allied leptostracans with the malacostracans, and who was among the first to advance a scheme of the phylogenetic relationships among the higher categories of crustaceans.

William T. Calman (1871-1952), perhaps the most preeminent worker in this century, is of course best known for his classic volumes *Crustacea* (1909) and *Life of the Crustacea* (1911). The former was especially crucial for giving basic overviews of all groups and providing us with the higher taxonomy of the crustaceans used up to this time. However, Calman was also a fine taxonomist at lower taxonomic levels; for example, of this, his careful work on both fossil and recent syncarids has never been found wanting in accuracy. His syncarid papers also illustrate another strength of Calman's work, effective combination of both fossil and recent lines of evidence in order to arrive at a complete overview whenever possible.

H. Graham Cannon (1897-1963) was noteworthy for two reasons. He was responsible for a whole series of papers on a wide range of groups from branchiopods through malacostracans that examined limb functional morphology and ontogenetic development. These have recently been found wanting in some respects, especially in regard to some biases induced by his experimental approaches. However, this has only come with the hindsight born of newer techniques; Cannon's works must be judged in the context of their time, of which they were exemplary efforts. In addition, because of his position as professor at the universities of Sheffield and Manchester, Cannon also produced a whole coterie of his own students and influenced others at sister institutions. Around Cannon was centered a British school that worked on problems of arthropod functional morphology and development.

Chief among Cannon's protégés was Sidnie M. Manton (1902-1979). Her range of interests extended beyond the crustaceans to also encompass onychophorans, myriapods, and cheliceriforms. Her controversial conclusions about arthropod polyphyly, based on her exacting studies of locomotory functional morphology and limb development, have in turn influenced a new generation of carcinologists to more critically examine old assumptions about the meaning of structure and function and to reevaluate long-entrenched ideas about crustacean relationships. Manton, along with Cannon, developed the biramous theory of limb evolution that stood in contrast to Borradaile's mixopodial theory—warring concepts that still battle today within the pages of this book.

These people were not always correct in their conclusions, nor were their methods entirely without flaw. Nor were they the only people to make major contributions to crustacean studies. To try and name all of the major contributors would risk leaving some out. However, these four were the people who in large part did provide the major paradigms within which carcinology has developed and who did produce and/or inspire a significant percentage of the corpus of knowledge upon which our current understanding of crustacean evolution is based. However, we should not accept this corpus of knowledge uncritically. Not to question our predecessors and our peers is not to do real science.

Much work remains to be done, as this book should serve to point out. We still lack knowledge on the internal anatomy of groups like spelaeogriphaceans, mictaceans, some aspects of the brachypodans (cephalocarids), amphionidaceans, conchostracans, and tantulocarids. Knowledge of feeding mechanisms is at best incomplete in all groups; and for some we know currently virtually nothing about how they feed, for example, conchostracans and notostracans. Complete data on ontogeny and larval development are extant for only a very few groups. Data on breeding and reproductive biology are scattered in the literature and have never been subjected to coherent analysis and interpretation. And biogeographic analysis in some taxa still awaits the stabilization of taxonomy in those groups.

It was also my hope that this book might serve to standardize the terminology and orthography used in the discipline. For example, thoracopods not modified as maxillipeds are variously referred to in the literature as pereopods, peraeopods, or pereiopods (the last is used herein), or first antennae and first maxillae are also known as antennules and maxillules (preferred here). I also tend to name limbs in the thoracic series by their ordinal numbers as well as by any specialized names they may hold; for example, there is never any doubt about where a first thoracopod is, but one may not readily recognize where something like a pyllopod is located (the first thoracopod of gnathiidean isopods). In addition, my choice of the anglicized versions of formal taxa will probably not please everyone, for example, ostracodes for Ostracoda, or cirripedes for Cirripedia, or peneids for Penaeidae. My researches on matters of orthography have led me to the choices I made, but I realize that whatever choices I could have made would have left some people unhappy.

So while I hope this book will serve as a reference text in crustacean evolutionary biology, I trust it will also serve as a guidebook and to point the way toward productive lines of research. I would feel my efforts well justified if in a few years this book were quite out of date. For that would indicate my vision was not misdirected.

Several years ago it was remarked to me that the day of single-author compendia was over, that no one person could hope to encompass and comprehend all the available knowledge on a subject. In a sense this is true, and certainly among recently published books, edited multi-author volumes are the norm [see, for example, *The Biology of Crustacea* (Academic Press) or *Crustacean Issues* (Balkema)]. Maybe the days of a Libbie Hyman are gone, but there is still much to be gained from one person trying to develop an overview of a subject. Coherence, unity, or pattern can more easily emerge with the overview of a single person in a way not possible with the view of a 'committee.' This was the strength of synthesizers like Claus, Calman, Cannon, and Manton. Their visions helped shape a science. If their work subsequently constrained thought, that was the fault of their exegetes, not of the visionaries.

Of course, mistakes will be made and inaccuracies will creep in. This book is no exception. I am sure these will be brought quickly to my attention. In the end these imperfections in the book are mine alone, but I have tried to mitigate them by seeking reviews of almost all chapters by relevant authorities. Without exception, people have responded generously in this regard. I am immensely indebted to everyone who have given of their valuable time and had input into developing this book. These include Drs. D. T. Anderson, D. Belk, E. L. Bousfield, G. A. Boxshall, D. E. G. Briggs, R. C. Brusca, M. D. Burkenroad, D. L. Felder, B. E. Felgenhauer, A. Fleminger, G. Fryer, L. F. Gardiner, M. J. Grygier, R. R. Hessler, R. F. Maddocks, R. B. Manning, J. Mauchline, P. A. McLaughlin, W. A. Newman, M. L. Reaka, W. D. I. Rolfe, H. K. Schminke, J. Sieg, I. G. Sohn, J. Stock, R. Swain, L. Watling, D. I. Williamson, and G. D. Wilson. The artwork was done by Bryan Burnett, Robert Chandler, and in particular Michael Emerson. The typing of the basic manuscript and various revisions could not have occurred without the yeoman service of Deanne Demere and Marjorie Rea. And nothing at all would have been forthcoming without the insistent encouragement of my wife Joan.

However, I especially want to 'thank' the Crustacea, for being such a compelling, fascinating group that some days I can hardly wait to get to work in the morning to find out more about them. Geoffrey Smith expressed it well of syncarids in *A Naturalist in Tasmania* (Clarendon Press, 1909):

> Goethe somewhere remarks that the most insignificant natural object is, as it were, a window through which we look into infinity. And certainly when I first saw the Mountain Shrimp walking quietly about in its crystal-clear habitations, as if nothing of any great consequence had happened since its ancestors walked in a sea peopled with strange reptiles, by a shore on which none but cold-blooded creatures plashed among the rank forests

of fern-like trees, before ever bird flew or youngling was suckled with milk, time for me was annihilated and the imposing kingdom of man shrunk indeed to a little measure.

San Diego   F.R.S.
May 1985

CONTENTS

Crustacea

1

WHAT ARE CRUSTACEANS?

Crustacea is an arthropodous phylum of animals whose members can be characterized by possessing a five-segment cephalon with two sets of antennae, a pair of mandibles, and two sets of maxillae; which display a tendency to fuse the segments of the head to form a cephalic shield and to develop from the posterior aspect of the cephalon a posteriorly directed shield over the body (a carapace); which exhibit a tendency to regionalize the body segments into distinct tagmata and to specialize the associated appendages; and which utilize anamorphic development that typically commences with a unique larva or ontogenetic stage termed a nauplius.

Crustaceans have a definite preference for marine conditions. However, there is hardly a habitat type on the planet that does not have crustaceans in it, albeit fresh water and terrestrial crustaceans usually have rather strict requirements associated with their ability to be present. Unlike other arthropodous groups, which exploit essentially one basic body plan in a sense, when you've seen one insect or arachnid, you have seen them all, the crustaceans exhibit a greater degree of diversity in form than that seen in any other animal phylum. There may be more species of insects than any other group in the world and more individual nematodes, but the inherent capacity of crustaceans to tagmatize and specialize body segments and appendages insures that there are more basic kinds within the crustaceans than any other group in the world. For this reason, aside from the features of the head, it is impossible to characterize crustaceans except by noting *tendencies* toward certain conditions or states. There are certain common anatomical themes noted among crustaceans, and as an orientation to the phylum these will be reviewed here.

APPENDAGES   Some groups, such as remipedes and many maxillopodans (especially larvae), have a curious set of frontal filaments in the vicinity of the antennules. Frontal filaments are not thought to be true limbs but are rather considered to be singularly developed sensory organs.

The first set of true appendages is the antennules (frequently referred to as first antennae). Until the discovery of the remipede *Speleonectes*, it was generally assumed that only the Malacostraca had multiramous antennules. Thus it would have appeared that, except for malacostracans, crustaceans conformed to the general arthropodan state wherein the primary preoral antenniform appendages are uniramous. However, there may be some basis to postulate that in fact biramous antennules are a diagnostic feature of adult crustaceans. Those forms with biramous antennules are those that have the most primitive body type with either complete lack of trunk tagmosis or at least possession of limbs on all trunk segments. Thus, it would appear that the biramous condition is secondarily lost in various groups (yielding to triramy in hoplocaridans or uniramy in maxillopodans, phyllopodans, and some eumalacostracans). The antennules typically appear as uniramous anlagen in the course of development; so the convergent appearance of uniramous antennules in many groups of crustaceans could be another manifestation of paedomorphosis in the phylum (see Chapter 44). By the same token, however, one could just as easily conceive of multiramous antennules as being convergently developed in different groups of adult crustaceans. Either one of these alternatives possesses interesting issues for phylogeny within the Crustacea (see Chapter 43). Antennular innervation is deutocerebral.

The antennae (also known as second antennae) are primarily postoral in origin. This is made evident in the course of development, where the typical adult preoral condition is arrived at late in ontogeny. The antennae frequently serve a locomotory and sometimes a food-gathering function in the larval stages but rarely function in this regard in adults where they are primarily sensory. The development of these appendages as sensory antennae in crustacean adults would thus seem to be a distinctive feature for the phylum. The antennal innervation is tritocerebral.

The food processing appendages of the crustacean head are the mandibles, maxillules (= first maxillae), and maxillae (= second maxillae). In all of these, the distal elements of the limb are typically reduced to 'palps' while the proximal protopodal elements develop specialized endites to handle the food. In addition, the mouth is usually marked anteriorly by a labrum, or upper lip, and often by posterior elements called paragnaths, or lower lips.

The first trunk appendages are sometimes modified as maxillipedes to assist in food processing as in nectiopodan remipedes, mystacocarids, some copepods, cirripedes, hoplocaridans, and most eumalacostracans. One or several such pairs of maxillipedes may be developed. However, such limb specializations may or may not be accompanied by fusion of the relevant segments to the head. For example, the nectiopodan first trunk segment is fused to the cephalon, but the maxillipede-bearing segment in mystacocarids is free; those eumalacostracans with maxillipedes generally fuse the relevant segments that bear them to the cephalon, but in mysidans the maxillipede bearing segment is actually free. Though classic theory states the primitive cephalon of crustaceans ends with the maxillary segment, one should be aware that in many instances one or more of the trunk segments can fuse to the head and that this fusion occurs sometimes in some otherwise apparently very primitive groups.

The basic crustacean trunk limb (Fig. 1-1) consists of a protopod, to which can be attached one or more branches. Distally these branches are

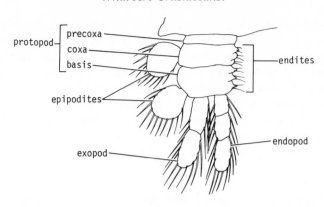

Fig.1-1. Diagrammatic hypothetical crustacean limb illustrating all possible component parts.

termed the exopod (outer branch) and the endopod (inner branch); proximally the protopod may bear lateral exites or epipodites and medial endites. The protopod generally is composed of two segments or joints, the coxa (proximal) and basis (distal). However, Hanson (1925) was among the first to maintain that the protopod is primitively three-segmented, with a proximalmost precoxa. Some groups do in fact clearly possess a precoxa, for example, hoplocaridans. In other taxa the issue is not at all clear, for example, bathynellaceans or branchiurans. The problem is: when is a 'precoxa' an extension of the body wall or true limb joint? Some crustacean groups have no apparent separate protopodal joints whatsoever, such as branchiopods and remipedes. Though much has been made of this issue of protopodal joints in terms of crustacean phylogeny (most of it confusingly so), a greater significance of single segment or multisegment protopods probably lies in relating the morphology of crustacean limb parts to particular functional requirements.

SHIELDS   The tendency to develop head and body shields is one of the characteristic phenomena noted in the course of crustacean evolution. The segments of the cephalon are typically fused dorsally into a single shield; only anostracans lack this fusion. This distinctive phenomenon caused Secretan (1980), in an interesting analysis, to treat the formation of the head shield as an integral stage in an evolutionary path which ultimately leads to a complete carapace. Thus she viewed only anostracans as truly lacking a carapace and considered that all other crustaceans possessed a carapace to some greater or lesser degree. This somewhat unusual approach may have some merit; however, its utility is closely tied to an understanding of what the carapace might be.

   The body shield, or carapace, can take a variety of forms (see e.g., Secretan, 1964). It need not always arise from the maxillary segment. Boxshall (1983) stated misophrioid copepods have a carapace extension from the maxillipedal segment, and in ostracodes the carapace arises from the naupliar segments. Müller (1983) described fossil forms with carapacelike shields apparently arising from trunk segments. Nor does the carapace always develop in the same way (Dahl, 1983). Thus, various carapaces noted within Crustacea may not be homologous. Add to this the fact that there are a variety of Cambrian arthropods that are definitely not crustaceans, but that have body shields (e.g., Briggs, 1983). Clearly, the issues involved with defining a carapace are not ones that are likely to lend themselves to concise diagnoses and solutions for some time to come.

TAGMATA   The tendency toward body tagmatization, or somite specialization, is also diagnostic of crustaceans. As with the issue of 'shields,' a great deal of variation between crustacean groups is manifested, but some generalizations are possible. The trunk is typically divided into an anterior thorax and posterior abdomen. Only the remipedes and few branchiopods have an undifferentiated trunk. The number of segments in these two regions is typically specific to a group and varies in degree of development. For example, cephalocarids and mystacocarids have an 'abdomen' which is essentially differentiated from the 'thorax' only because it lacks appendages, while malacostracans have an abdomen with structural and functional differences from that of the thorax.

The thorax, because of its universal importance as a functional unit—no matter what the group—has received much attention. Unfortunately, there has been a corresponding tendency to dismiss the abdomen as merely the 'tail end' of the body. Makarov (1978) declares that this neglect has led to lack of knowledge about this region which must be remedied to arrive at an adequate understanding of crustacean evolution. He believes we should delineate a third trunk region separate from the abdomen *sensu stricto*, the urosome. This would designate the posterior terminus of the body, which in many cases is as important in locomotion as the more generally acknowledged thorax. It was consideration of similar issues that brought Bowman (1971) to his controversial position (disputed by Schminke, 1976) that attempted to distinguish between true telsons and anal segments. The telson is classically defined as the terminal unit of the body whose mesoderm is derived directly from the teloblasts after metamere budding has ceased. The implication here is that the telson is not a 'true segment' (see, e.g., Kaestner, 1970; or McLaughlin, 1980); however, it is a confusing enough issue that Moore and McCormick (1969) said it was a somite. In actual practice, *telson* has come to mean the terminal unit or 'segment' of the body, without any reference to the fine points of mesodermal origins. Indeed, one can in fact quibble about whether in the course of development the teloblasts anteriorly bud off the segmental units, or whether the terminal 'somite' simply keeps dividing into two segments (the more anterior of which loses the capacity to further divide). In fact Oishi (1959, p. 307) made some observations in this regard that 'strongly suggests that the development of the telson should not be considered separately from the differentiation of teloblasts.' Such discussions really are academic, however, since to categorically dispose of the terminal unit by a catchall term—*telson*—masks the very real anatomical differences that do exist in the crustacean body termini and that can have meaning in terms of functional morphology and ultimately in understanding the evolution of crustaceans.

By not having a 'name for it' we lose the power to 'understand it.' This confusion extends to what name is applied to the 'extensions' of the telson [variously caudal lobes, caudal rami, furcal lobes, furcal rami, caudal furca(e), or simply the furca] that sometimes act like and even look like, but are not, true appendages. Yet the only way we can come to an adequate understanding of the urosome, in the sense of Makarov (1978), is to make clear what we are speaking about. For this reason I think Bowman (1971) has done a service in trying to outline terms that are defined by specific anatomical conditions. I have tried to follow his lead in this book. Thus the following definitions are relevant. The *telson* (Fig. 1-2B, E) is the last body unit (= segment, if you prefer) in which the anus is not terminal. The *anal segment* (Fig. 1-2A, C) is the last body unit in which the anus is terminal.

Fig. 1-2. Urosome variations among crustaceans. (A) *Hutchinsoniella macracantha*, illustrating an anal segment with relatively simple caudal rami; (B) *Lepidurus lemmoni*, telson with proximal portions of caudal rami; (C) *Derocheilocaris typicus*, anal segment with moderately complex caudal rami; (D) *Anthracaris gracilis*, uropods flanking telson with distal furcae and a terminal lobe; (E) *Euphausia pacifica*, telson with furcae; (F) *Schisturella pulchra*, a complex urosome with three sets of uropods and a deeply cleft telson.

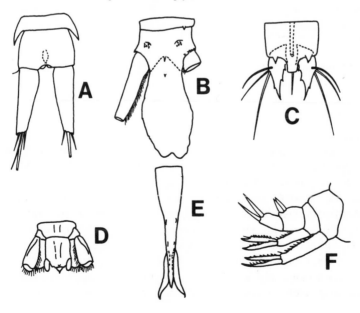

(In this respect the proximal portions of the telson are homologous to the anal segment.) The *rami* (= caudal rami) are appendagelike structures that arise on an anal segment (Fig. 1-2A, C) or near the base of a telson (Fig. 1-2B) that serve to assist in locomotion. The rami are analogs of *uropods* (Fig. 1-2D, F), which are specialized appendages of the segment just anterior to the anal segment or telson. The *furcae* (Fig. 1-2D, E) are small lobes or spines that are located near the terminus of the telson.

GUT   The crustacean digestive system consists of ectodermally derived fore- and hindgut elements lined with cuticle and an endodermal midgut. The presence of large paired midgut caeca is a distinctive crustacean feature. The midgut hardly ever provides evidence of metamerism, though in stomatopods the caeca have metameric outpocketings and nectiopodans have paired metameric diverticula along the length of the entire midgut. The midgut tube is generally quite long, though isopods lack a midgut tube and have only a fore- and hindgut with the digestive caeca.

CIRCULATION   The crustacean circulatory system is generally thought to be open, though in barnacles where the system has been examined closely (Burnett, 1972, 1977) it has proven to be virtually closed. The heart sits in a pericardial chamber. The chamber acts as an atrium to collect the blood which then passes through openings or ostia into the heart. The number of ostia varies. The heart is segmentally derived, and the ostia are the remnants of the intersegmental spaces between the primordial segmental cardiac anlagen. The respiratory pigment is generally hemocyanin, but other pigments have been noted in various groups.

EXCRETION   Excretion in crustaceans can be achieved in several ways, including nephrocytes, phagocytes, caecal and gill epithelia, and nephridia. The last are the most common and are generally found in association with the antennae and/or maxillae. Very rarely are these glands supplemented by segmental organs in the body, but such can occur as in bathynellaceans and mysidaceans.

REPRODUCTION   Sexes in crustaceans are typically separate. Hermaphroditism is widespread and is generally either protandric or protogynic, the latter under the control of an androgenic gland in the last thoracomere that degenerates through the life cycle. However, remipedes, cephalocarids, and cirripedes have functional organs of both sexes. Some branchiopods and ostracodes exhibit parthenogenesis, and such is also suspected in some tanaids. Gonads can have a variety of forms (sometimes compact, sometimes tubular), and their locations in the body vary (sometimes thoracic, sometimes abdominal). The location of gonopores also varies (Table 1-1) and has been useful in the classification of some groups.

Table 1–1. Body plan of various crustacean groups. (Modified from Boxshall, 1983)

	Total body somites	Somite of gonopore ♂	♀	
Remipedia	37	20	20	
Most Branchiopoda	up to 42	16	16	
Anostraca	25–32	17	17	25 in *Polyartemia*
Cephalocarida	25	11	11	
Copepoda	16	12	12	
Cirripedia	16	12	6	
Mystacocarida	16	9	9	
Branchiura	9(+?)	9	9	
Ostracoda	8	8	8	
Tantulocarida	17	?	10	
Malacostraca	20 or 21	13	11	

NERVOUS SYSTEM   A primitive arrangement of the crustacean nervous system consists of a brain with proto-, deuto-, and tritocerebrum (elements of this last are sometimes included in the circumesophageal connectives), and a postoral sequence of paired ganglia in each segment joined by transverse commissures. The trend in central nervous system evolution is to fuse the postoral elements, an extreme example of which is seen in brachyurans with a single large postoral ganglionic mass.

EYES   Among the most distinctive of sensory organs of crustaceans are the various photoreceptors, the frontal eyes or nauplius eye complex, Gicklhorn's organ, and the compound eyes. Elofsson (1965, 1966) has provided the most comprehensive review of the frontal eyes. Within this complex of photoreceptors Elofsson distinguished between the nauplius eyes *sensu stricto* and the affiliated dorsal and ventral frontal organs. He recognized four basic arangements of structures within the photoreceptor. In the Maxillopoda the nauplius eye has three pigmented cups, lacks any associated frontal organs, and has tapetal and lens cells not seen in other types. In the Malacostraca the nauplius eye, when present, is combined in a variety of ways with the dorsal and ventral frontal organs (Table 1-2), and has an everse arrangement of sensory cells in the three cups combining rhabdomeres and rhabdomes. In Anostraca (Fig. 1-3A) only paired ventral frontal organs occur near the nauplius eye of three cups. In the other phyllopods there are paired distal and single posterior median frontal organs associated with a nauplius eye with four sensory cups. Elofsson (1966) felt that this frontal eye or nauplius eye complex is mistakenly considered a primitive structure. Rather it shows a wide range of specializations. Although four basic types can be recognized, none should be treated as more primitive or ancestral in any way to the others. The entire frontal eye complex is frequently associated with the X-organ, but this structure is

Table 1–2. Variation in frontal eye types among malacostracans. (Modified from Elofsson, 1965)

	Ventral frontal organ	Dorsal frontal organ	Nauplius eye
Leptostraca			
Stomatopoda	+	+	+
Anaspidacea		+	+
Euphausiacea	+		+
Caridea	+	+	+
Sergestoidea	+	+	+
Astacidea	+	+	+
Anomura	+	+	+
Brachyura			
Mysidacea	+		
Amphipoda			
Isopoda			
Cumacea			
Tanaidacea			

endocrine in function and has nothing to do with photoreception (Fig. 1-3A, B).

Elofsson (1970) recognized Gicklhorn's organ as an additional photoreceptor, found only in copepods. These are paired structures, each member having two cells that closely resemble retinula cells. Although these are loosely placed in some proximity to the frontal eye, they are quite separate from the latter (Fig. 1-3B).

The most prominent photoreceptors in crustaceans are the compound eyes. (Fig. 1-4). Unlike the simple frontal eyes, the compound eyes are composed of ommatidia, repeated units quite complex in structure (Shaw and Stowe, 1982). The faceted cornea is a transparent portion of the body cuticle secreted by the immediately underlying corneagen cells. Below the cornea of each ommatidium is the crystalline cone. There are variously two to five cells here, though generally there are only two or four. Deficiencies below four are typically marked by some complementary number of accessory cone cells. Internal to the crystalline cone are the retinular cells, which range in number from five in anostracans up to 17 in oniscoid isopods, with eight as the norm. The light-sensitive organelles of the retinular cells are the central microvilli, which form the rhabdome. The rhabdome can take a variety of forms depending on functional need (Elofsson, 1976), with particular adaptations to perceive polarized light. Flanking the cone and retinular cells are various pigment cells. These latter either screen the optic units of the ommatidium so that only direct oncoming light passes directly down to the rhabdome, the appositional eye, or they act as mirrors to reflect light from several ommatidia onto a single point in the field of rhab-

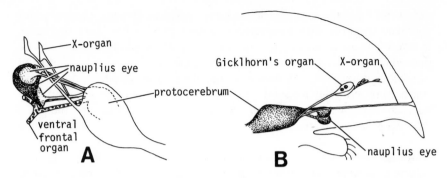

Fig. 1-3. Representative frontal complexes in crustaceans. (A) the anostracan *Branchinecta paludosa*; (B) anterior head of the copepod *Pareuchaeta norvegica*. (From Elofsson, 1966)

domes, the superpositional eye (Land, 1978). The latter mirror optics of the malacostracan eye are quite different from the lens optics seen in the superpositional eyes of insects. As is the case with the frontal eyes, Elofsson (1976) felt the compound eyes of crustaceans, as well as all arthropods, display such a variety of form that convergences among and within arthropodous phyla are rampant, and phylogenetic speculations strongly based on eyes are to be suspect, the misinterpretations of Paulus (1979) notwithstanding.

Fig. 1-4. Appositional ommatidium. Cell R8 functions as an accessory rhabdomere distal to the principle rhabdomere formed by the microvilli of cells R1–7. (Modified from Shaw and Stowe, 1982)

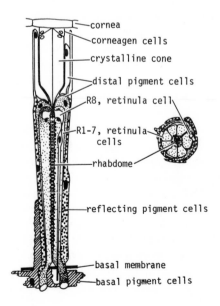

RECEPTORS   Crustaceans also have a wide array of chemo-, thermo-, and mechanoreceptors. These sensors are associated with pits, hairs, and setae in or on the cuticle. Details of structure and function of these receptors are limited (Ache, 1982; Bush and Laverack, 1982). Although some cataloging of setal types exists for particular groups (e.g., Factor, 1978), there is no standard terminology for setal types applicable to all crustaceans, and little concrete work has been done on function of particular setae and the possible receptors associated with them.

ONTOGENY   In regard to early ontogeny, it was Anderson (1973) who clearly documented the differences in blastomere fate maps among various arthropods. That of the Crustacea is very distinctive and unlike any other arthropodous group (see Chapter 2), and the pattern is very consistent within crustaceans (Anderson, 1982) notwithstanding criticisms to the contrary by Weygoldt (1979).

Cleavage in crustaceans is a modified spiral type. Variations in presumptive areas on the blastula are clearly related to yolk content. This is particularly evident in the degree of separation noted between presumptive endoderm and mesoderm. The former is typically posterior to the latter, though the difficulty encountered here prompted Weygoldt (1958) to suggest use of the term 'mesendoderm' and not worry about the differentiation. This may have some merit for many malacostracans but ignores the distinctions evident in other groups.

LARVAE   All crustaceans either go through a free nauplius larval stage or at least evidence an egg-nauplius phase in early ontogeny. Oddly, only the ostracodes, cirripedes, copepods, and among malacostracans, euphausiaceans, dendrobranchiates, and possibly procaridideans have free naupliar larvae. Subsequent to the nauplius, a series of metanaupliar stages are passed through, again either as free larvae or within the egg. These metanaupliar stages are entered with the proliferation of postnaupliar segments by the telocells. Subsequent to the metanaupliar stages, postlarval or juvenile stages may be passed through before a definitive adult morphology is reached.

A variant terminology for larvae can be used (see Williamson, 1982), which would define larvae on the basis of limb function rather than the ontogenetic factors used above. In such a system, the nauplius is a stage where the antennules, antennae, and mandibles are used for propulsion. If other appendages are present but are not used, then the larva can be termed a metanauplius though still treated as part of the naupliar phase. The zoea is a larval phase in which the thoracopods are used for propulsion. The megalopa or decapodid (unique to Malacostraca) is a larval phase in which the pleopods are used for propulsion. This functional system seems to have greatest use among malacostracans, wherein the zoeal phases are sometimes subdivided into protozoeal and mysis phases. However, this has limited application to other groups, where other names have been applied (Table 1-3), and even variant terms have been used within malacostracans based on particular structural peculiarities within groups. Clearly, the reader of crustacean ontogenetic literature must be aware of such variant terminologies that connote developmental, functional, or structural concepts. The resultant confusion, however, is more annoying to the uninitiated rather than betraying any real defect in our understanding of the subject matter.

PREANTENNAL SEGMENT   One final problem concerning crustacean development deals with the existence of a 'preantennal segment.' The naupliar mesoderm in crustaceans differentiates from growth of mesodermal bands anteriad from the blastopore. However, all crustaceans develop another mass of mesodermal tissues anterior to the antennular mesoderm which typically gives rise to the musculature of the labrum and parts of the foregut, and in malacostracans it also contributes to the formation of the anterior aorta. Knowledge of the derivation of this preantennular mesoderm is lacking for some groups. Sometimes this tissue differentiates from the

Table 1–3. Variant terminology used among groups of crustaceans for various stages in larval development.

	Nauplius	Metanauplius	Juvenile or postlarval
Cephalocarida	−	"nauplius"	cephalopodid
Branchiopoda	+	metanauplius	
Mystacocarida	?	metanauplius	
Ostracoda	+	instars	
Copepoda	+	copopodid	
Cirripedia	+	cyprid	
Lysiosquilloidea		antizoea	erichthus
Gonodactyloidea			erichthus
Squilloidea			erichthus
Bathynellacea		parazoea + bathynellid	
Euphausiacea	+	calyptopis + furcilia	cyrtopia
Amphionidacea		amphion	megalopa
Penaeoidea	+	protozoea + mysis	postlarva
Sergestoidea	+	elephocaris + acanthosoma	mastigopus, decapodid
Eukyphida		zoea	megalopa, decapodid
Euzygida		zoea	megalopa, decapodid
Eryonoidea		eryoneicus	eryoneicus, decapodid
Palinuroidea		phyllosoma	puerulus, decapodid
Astacidea		zoea	megalopa, decapodid
Anomura		zoea	megalopa, decapodid
Brachyura		zoea	megalopa, decapodid
Peracarida			manca

anterior ends of the naupliar mesodermal bands, as in cirripedes, anaspidaceans, and tanaidaceans; other times it arises as a distinct set of mesodermal inpocketings of the protocerebral ectoderm before developing a set of coelomic cavities, as in most malacostracans. The exact meaning of this preantennular mesoderm is not at all clear. Speculation has raged for decades whether this represents a true segment in the anteriormost regions of the cephalon or whether this merely arises through variant modes of developing a mesodermal supply to the anterior nauplius.

The above presentation is only meant to be a general overview. Subsequent chapters will take up specific groups and cover these matters in such detail as is available. Although it is difficult to concisely and specifically define Crustacea in a manner applicable to all groups, except for structure of the head, the overall facies of these animals is such that crustaceans are nonetheless easy to recognize.

REFERENCES

Ache, B. W. 1982. Chemoreception and thermoreception. In *Biology of Crustacea*, Vol. 3 (H. L. Atwood and D. C. Sandeman, eds.), pp. 369–98. Academic Press, New York.

Anderson, D. T. 1973. *Embryology and Phylogeny in Annelids and Arthropods*. Pergamon Press, Oxford.

Anderson, D. T. 1982. Embryology. In *Biology of Crustacea*, Vol. 2 (L. G. Abele, ed.), pp. 1–41. Academic Press, New York.

Bowman, T. E. 1971. The case of the nonubiquitous telson and the fradulent furca. *Crustaceana* **21**:165–75.

Boxshall, G. A. 1983. A comparative functional analysis of the major maxillopodan groups. *Crust. Issues* **1**:121–44.

Briggs, D. E. G. 1983. Affinities and early evolution of Crustacea: the evidence of the Cambrian fossils. *Crust. Issues* **1**:1–22.

Burnett, B. R. 1972. Aspects of the circulatory system of *Pollicipes polymerus*. *J. Morph.* **136**:79–180.

Burnett, B. R. 1977. Blood circulation in the balanomorph barnacle *Megabalanus californicus*. *J. Morph.* **153**:299–306.

Bush, B. M. H., and M. S. Laverack. 1982. Mechanoreception. In *Biology of Crustacea*, Vol. 3 (H. L. Atwood and D. C. Sandeman, eds.), pp. 399–468. Academic Press, New York.

Dahl, E. 1983. Malacostracan phylogeny and evolution. *Crust. Issues* **1**:189–212.

Elofsson, R. 1965. The nauplius eye and frontal organs in Malacostraca. *Sarsia* **19**:1–54.

Elofsson, R. 1966. The nauplius eye and frontal organs of the non-Malacostraca. *Sarsia* **25**:1–128.

Elofsson, R. 1970. A presumed new photoreceptor in copepod crustaceans. *Zeit. Zellforsch.* **109**:316–26.

Elofsson, R. 1976. Rhabdom adaptation and its phylogenetic significance. *Zool. Scripta* **5**:97–101.

Factor, J. R. 1978. Morphology of the mouthparts of larval lobsters *Homarus americanus*, with special emphasis on their setae. *Biol. Bull.* **154**:383–408.

Hansen, H. J. 1925. On the comparative morphology of the appendages in the Arthropoda. *A. Crustacea*. Gyldendalske, Copenhagen.

Kaestner, A. 1970. *Invertebrate Zoology.* Vol. III. *Crustacea.* Interscience, New York.

Land, M. F. 1978. Animal eyes with mirror optics. *Sci. Am.* **239**(6):126–34.

Makarov, R. R. 1978. Kaudalnaya tagma visshikh rakoobraznikh, yeye biologichyeskaya spyetsifika i proiskhozheniye. *Zh. Obshchyey Biologii* **39**:927–39.

McLaughlin, P. A. 1980. *Comparative Morphology of Recent Crustacea.* Freeman, San Francisco.

Moore, R. C. and L. McCormick. 1969. General features of Crustacea. In *Treatise on Invertebrate Paleontology*, Part R, *Arthropoda* **4**(1) (R. C. Moore, ed.), pp. R57–120. Geol. Soc. Am. and Univ. Kansas, Lawrence.

Müller, K. J. 1983. Crustacea with preserved soft parts from the Cambrian of Sweden. *Lethaia* **16**:93–109.

Oishi, S. 1959. Studies on the teloblasts in the decapod embryo. 1. Origin of teloblasts in *Heptacarpus rectirostris*. *Embryologia* **4**:283–309.

Paulus, H. F. 1979. Eye structure and monophyly of Arthropoda. In *Arthropod Phylogeny* (A. P. Gupta, ed.), pp. 299–383. Van Nostrand Reinhold, New York.

Schminke, H. K. 1976. The ubiquitous telson and the deceptive furca. *Crustaceana* **30**:292–300.

Secretan, S. 1964. La carapace des Crustacés différents modes d'adaptation aux segments du corps. *Ann. Paléontol.* **50**:191–208, 2 pls.

Secretan, S. 1980. Comparaison entre des Crustacés à céphalon isolé, à propos d'un beau matérial de Syncarides du Paléozoique, implications phylogéniques. *Geobios* **13**:411–33, 4 pls.

Shaw, S. R., and S. Stowe. 1982. Photoreception. In *Biology of Crustacea*, Vol. 3 (H. L. Atwood and D. C. Sandeman, eds.), pp. 291–367. Academic Press, New York.

Weygoldt, P. 1958. Die Embronalentwicklung des Amphipoden *Gammarus pulex pulex*. *Zool. Jahrb. Abt. Anat.* **77**:51–110.

Weygoldt, P. 1979. Significance of later embryonic stages and head development in arthropod phylogeny. In *Arthropod Phylogeny* (A. P. Gupta, ed.), pp. 107–35. Van Nostrand Reinhold, New York.

Williamson, D. I. 1982. Larval morphology and diversity. In *Biology of Crustacea*, Vol. 2 (L. G. Abele, ed.), pp. 43–110. Academic Press, New York.

2

CRUSTACEA AND
OTHER ARTHROPODS

A great deal of debate has taken place over the controversial contentions of Anderson (1973) and Manton (1977) that Arthropoda is a polyphyletic taxon. That is, although the various arthropod groups share some features in common, these are not in fact unique to arthropods, and that fundamental aspects of early development and the functional morphology of locomotion force the conclusion that arthropodlike organisms form at least four distinct phyla (Uniramia, Cheliceriformes, Trilobitomorpha, and Crustacea). This is not a particularly new idea, having its origins in Tiegs and Manton (1958). However, it is one that has infuriated many people, particularly entomologists (see, e.g., many of the contributions in Gupta, 1979). Anderson and Manton dealt almost exclusively with living forms (Uniramia, Chelicerata, and Crustacea). Schram (1978) integrated into their scheme knowledge about pycnogonid locomotion (Schram and Hedgpeth, 1978) and development (Dogiel, 1913; Morgan, 1891; Sanchez, 1959) and concluded that Chelicerata and Pycnogonida were sister groups within a clade Cheliceriformes. Schram (1978) also felt it best to retain for the time being a separate phylum status for Trilobitomorpha. Sawyer (1984) established the continuum of 'arthropodization' from annelids through clitellates to the uniramians.

The central theme in these investigations of arthropod relationships is the multiplicity of convergent development. *No* phyletic arrangement of arthropods can avoid this. For example, in the late 1800s the arthropods were viewed as closely related. The merostomes were considered to be Crustacea. The arachnids were aligned with the myriapod–hexapods in the Tracheata, based on the possession of tracheae in many of these animals. The presence of Malphigian tubules in insects and many arachnids served to strengthen the supposed links between these two groups. However, Lankester (1881) effectively demonstrated the affinities of *Limulus* with scorpions and thus all arachnids. The Chelicerata were established as a group separate and distinct from all others. The arachnids, with their tracheae and Malphigian tubules, were then realized to be derived from marine merostomes and not myriapods. Later, study of Malphigian tubule development revealed that these structures were evaginations of the proctodeum or hindgut, that is, ectodermally derived, in the insects, and outgrowths of the midgut, that is, endodermally derived, in arachnids. So the initial conclusion that excretory structures were convergent, derived from taxonomic considerations of chelicerate affinities, were eventually seconded by embryonic studies.

JAWS Subsequent to the elucidation of the status of the Chelicerata, the crustaceans and myriapod–hexapod groups were allied together as a sub-phylum Mandibulata (Snodgrass, 1938), but the work of Manton (1964) on the functional morphology of arthropod jaws altered our understanding of mandible evolution (Fig. 2-1). There are two basic jaw forms in arthropods: a gnathobasic type, in which only the modified coxa is used for biting, and the whole-limb type, in which the entire appendage is employed, the biting surface being the tip of the distalmost segment or distalmost part of the whole-limb jaw. The gnathobasic form is found in the crustacean mandible (and on the prosomal appendages 2 to 6 in *Limulus*). The whole-limb jaw is found in the onychophorans, myriapods, and hexapods. Manton (1964) concluded that these jaw types are so distinctive that neither one could have given rise to the other.

Within each basic jaw type there are different modes of action possible. The gnathobasic jaw of Crustacea primitively employs the coxal promotor–remotor muscles to produce the anteroposterior rolling action of the molar process around a dorsoventral axis. In some eumalacostracans, for example, stomatopods, the specialized development of an incisor process posterior to the molar process produces a secondary transverse action of this incisor process. Powerful remotor muscles produce the grinding (rolling) action of the molar process and the biting (transverse) action of the incisor. The weaker promotor muscles part the molar and abduct the incisor processes.

The living merostomes use prosomal appendages 2 through 6 as both walking limbs *and* biting limbs. The coxae on these appendages are capable of two different actions. A series of promotor-remotor muscles move the coxae anteroposteriorly when the animal walks. Separate sets of special abductor–adductor muscles move the coxae transversely when the coxae are used for biting. It appears that the muscles used and the modes of action of the crustacean and limulid gnathobasic jaws are so different that there is little possibility that they are related in any way to each other and are therefore only analogous.

The whole-limb jaw can also exhibit different modes of action. The onychophoran mandible has an anteroposterior slicing action, each jaw frequently moving in opposite phase to the other, just like the onychophoran walking legs. Some hexapods have a rolling, grinding action similar to crustaceans. The basic promotor–remotor action in hexapods can also be converted in some forms to a transverse action with the development of special incisor processes. As occasionally happens in the eumalacostracans, the hexapods, especially pterygote insects, greatly reduce the anterior molar process and completely convert the mandible to a transverse type that still employs the promotor–remotor muscles. The segmented

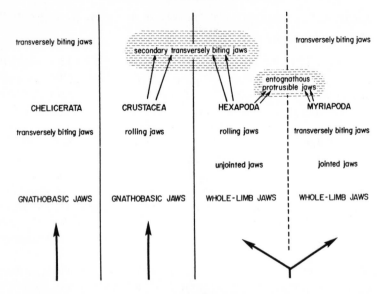

Fig. 2-1. The distribution of the principal forms of mandibles or jaws and the derivation of the jaw mechanism. Vertical lines indicate separate (heavy) or common (broken) origins, while shaded areas indicate convergence. (From Manton, 1964)

mandibles of myriapods have a transverse action with the musculature being largely adductor in function. The abduction is achieved indirectly by the endoskeletal tentorium, which swings downward and forward from the inner surface of the cranium and pushes the jaws apart. This is a basic arrangement for all myriapod groups and is so different from that of the hexapods that neither can have directly given rise to the other, disproving the view that insects may have been derived from symphylans as has been commonly suggested in the past (e.g., Snodgrass, 1938).

The taxon Mandibulata is thus seen to have no basis (Fig. 2-1). Similarities between crustacean and myriapod–hexapod jaws are merely convergent. Though the variations in jaw function between Myriapoda and Hexapoda are striking, there remain a great enough number of shared derived characters in all classes, such as the form of the labrum, derivation of the jaw, tracheae, uniramous lobopodial limb, and manner of potential entognathy (though this has arisen several times within the myriapod–hexapod group), that it appears that the onychophoran–myriapod–hexapod line forms a fairly coherent unit.

COMPOUND EYES   When the concept Mandibulata is discarded, another striking convergence is revealed: that of compound eyes. Recent work has indicated differences in the pigment screen between crustacean and insect compound eyes (Struwe et al., 1975), and there is a demonstrated lack of homology between insect primary pigment cells and crustacean corneagen cells (Elofsson, 1970) contrary to what had been previously supposed. Elofsson (personal communication) believes insect and crustacean compound eyes to be as different from each other as are vertebrate and cephalopod eyes, that is, though a striking gross superficial similarity exists, constituent parts have quite different embryonic histories. Compound eyes are also possessed by merostome chelicerates and trilobites. The compound eyes of *Limulus* have a thick cornea covering over all the ommatidial units, and this is a pattern not only in the living limulids but also is found in the fossil merostomes. Crustaceans and insects have a typically distinct cornea over each ommatidium in the compound eye. Little is known about the eyes of trilobites that can be effectively compared with the detail available in the eyes of living arthropods. Though trilobite compound eyes are composed of several optical units (Clarkson, 1973) some may not have been as closely aligned and coordinated as those of living arthropods (Clarkson, 1966). Several different types of trilobite eyes are recognized (Clarkson, 1975, 1979; Jell, 1975).

Though striking, this convergence of arthropod eyes is no more profound than that which occurs between the eyes of some cephalopods and vertebrates. Arthropods being what they are, there are only certain optimal ways to solve the problem of visual perception. The ocellus as the basic optimal unit is common to all groups of arthropods and, given this simple structure as a foundation, natural selection has produced similar compound eyes in various sorts of crustaceans, insects, trilobites, and merostomes. This problem will be returned to below.

LIMB MORPHOLOGY   The most characteristic shared derived feature of the onychophorans–myriapods–hexapods is the lobopodial uniramous appendage. This appendage evolved from a condition where a long series of identical limbs had to be precisely coordinated to achieve locomotion. A lack of coordination would lead to interference of one limb with an adjacent one and stumbling (Manton, 1969).

The distinct derived body shapes of the various groups within this assemblage—named the Uniramia by Manton (1973a)—are correlated with habit. The unique onychophoran unsclerotized deformable body, connective tissue endoskeleton, and unstriated muscle enables these animals to squeeze through narrow openings and spaces, allowing them to escape predators (though some authorities use these same characters as criteria for maintaining Onychophora as a separate phylum). The distinctive diplosegments of the Diplopoda are related to a need to develop motive force in burrowing or pushing through leaf litter, soil, and decomposing wood. In the Chilopoda the lengthening of the legs, variation in body segment size, and special muscle insertions enable these animals to exploit a running, carnivorous habit. The Symphyla have divided tergites that allow them to twist and flex their bodies in climbing under, over, around, and in between obstructions and in executing sharp-angle turns to escape and elude predators. The Hexapoda have reduced the number of legs, lengthened the appendage, and spread out the field of movement of each limb (with result-ant increase of mechanical advantage) in order to allow them to exploit the resultant versatility of movement and speed.

The cheliceriformes developed from long-legged animals with few appendages on the body and typically no more than five postoral limbs on the adult prosoma (Manton, 1973a, b). There is no necessity to rigidly coordinate movements among such a small number of legs and so this group consequently executes rather inaccurate stepping movements in con-trast to the Uniramia. In arachnids the stepping movement does not typi-cally involve a promotor–remotor swing of the coxa. Rather, the cheliceriformes employ a 'rocking' action in lengthening the stride of the leg to greater effect than any other arthropods. This motion is so termed from the position of the dorsum of the appendage during movement: on the propulsive backstroke, the dorsum is directed forward; on the recovery stroke, when the leg is brought forward, the dorsum is directed posteriorly. The appearance of a single isolated leg would then describe a rolling or rocking motion. Manton (1973b) is not clear on just how this rocking is achieved, but Schram and Hedgpeth (1978) noted a similar movement in pycnogonids and attributed it in part to sets of adjacent but separate exten-sor muscles in the basalmost segments of the legs.

The primitive living crustaceans have flat, paddlelike appendages (Hessler and Newman, 1975; Schram, 1983a) that are used in a metachro-nal, swimming pattern. The legs were apparently directed ventrolaterally and were not involved in food getting (see Chapter 44).

EMBRYOLOGY   More convincing data for recognizing three arthropodous phyla come from the comparative embryological studies of Anderson (1973). His findings (Fig. 2-2) for Annelida, Uniramia, Crustacea, and Chelicerata were combined with data on the Pycnogonida and summarized in a review by Schram (1978).

The uniramians and annelids are seen to have a basic similarity of development. The annelids have spiral cleavage in eggs with little to moderate amounts of yolk (while uniramians do not exhibit spiral cleavage because of a modification induced by large amounts of yolk in the egg). In annelids the presumptive endoderm (midgut) arises from the 3A, 3B, 3C, and 4D cells located along the ventral part of the blastoderm. This midgut area is enclosed by an overgrowth (epiboly) of cells from the dorsal blasto-derm. The stomodeum arises from the 2b cell at the time of gastrulation as a solid mass of cells that subsequently hollows out and forms a mouth. The presumptive mesoderm arises from the 4d lineage, is located *posterior* to the presumptive endoderm and becomes internalized during the epiboly of the ectoderm. The mesoderm then grows forward as a pair of bands from which the somites bud. The presumptive ectoderm of the embryo develops from the 2d lineage.

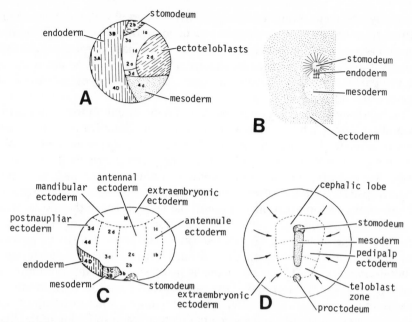

Fig. 2-2. Blastoderm fate maps of various types of annulates. (A) annelid, *Tubifex*; (B) ony-chophoran, *Peripatus*; (C) cirripede crustacean; (D) xiphosuran chelicerate. Note relation-ships of stomodeum, endoderm, and mesoderm. (Derived from Anderson, 1973)

The uniramians have a developmental pattern like that of the annelids, although cell lineages cannot be traced because of the derived loss of spiral cleavage. The presumptive endoderm is a group of cells on the ventral sur-face of the blastomeres. (Frequently large anterior and smaller posterior midgut sections can be delineated, as in the Onychophora.) Gastrulation typically occurs with the presumptive endoderm migrating inward and becoming vitellophagic. The stomodeum forms a solid mass of cells, which then hollows out as a tube. The presumptive mesoderm in the onychophor-ans and chilopods arises from an area posterior to the presumptive endo-derm (in the same position as in the annelids) and after involution, grows forward as a pair of bands. The annelid and uniramian development can thus be seen to conform to the same basic plan; however, uniramians are generally epimorphic in their development, that is, usually hatching with the adult complement of segments.

The crustacean pattern is entirely and strikingly different from that of the uniramians. The cleavage of crustaceans is basically spiral and can in some instances allow the tracing of cell lineages. The presumptive endo-derm arises from the 4D cell only. The presumptive mesoderm arises from the 3A, 3B, and 3C lineages and is thus *anterior* to the presumptive endo-derm, rather than posterior as in the uniramians and annelids. The pre-sumptive ectoderm arises from the 3d and 4d cells and is very early zoned into regions conforming to the nauplius topography. The stomodeum arises from the 2b cell but does so independent of gastrulation. Thus the presumptive area relationships in the crustacean blastoderm are quite different from that seen in the uniramians. The Crustacea have anamorphic development, typically passing through a more or less extensive larval series after hatching.

Chelicerate embryology has not received the attention that other arthropod groups have had. From what is known, the chelicerates do not have spiral cleavage. The cleavage is total, however, despite the amount of yolk present. The presumptive endoderm is segregated internally during cleavage. Typically the egg divides completely. Then the nuclei and surrounding cytoplasm 'float' to the surface of the developing embryo and begin to divide and develop a cell layer around the yolky cells on the inside. The presumptive mesoderm exists as a small, elongate area along the ventral midline of the blastodisc. The mesoderm sinks inward as a gastric groove and proliferates cells that spread outward between the ectodermal and endodermal layers. The stomodeum is associated with the presumptive mesoderm at the anterior end of the gastric groove and forms the mouth at the time of mesoderm initiation. The embryonic ectoderm occupies the rest of the germinal disc and is zoned in patterns corresponding to the first four postoral embryonic segments. Nothing can be determined concerning cell lineages in chelicerates (or if they indeed ever had any). Certain features of their development allow them to be separated from other arthropods, namely, the simultaneous separation of presumptive endoderm from other regions with cleavage, and the initiation and sinking of mesoderm by delamination along the ventral midline of the embryo. Chelicerates have epimorphic development.

It is unclear where Pycnogonida fit into this embryonic scheme; Anderson (1973) did not deal with them. Little reliable work has been done on pycnogonid development. A short review of what is known from the literature is given in Schram (1978). Their cleavage is total and nonspiral. Sanchez (1959) claimed to detect a spiral arrangement in the eight-cell stage of *Callipallene*. Morgan (1891), however, indicated that cell lineages in *Callipallene* could not be traced and that the 'micromeres' in such an arrangement are the only cells that eventually form the germinal disc. The yolk in pycnogonids can range from small to large amounts and the patterns of the presumptive areas in the blastoderm vary among forms with different amounts of yolk. In the eggs with little yolk, the presumptive midgut forms from a surface cell that sinks to the interior and proliferates to form a syncytium (Dogiel, 1913). The mesoderm in such forms develops from cells that surround the presumptive midgut cell and delaminate mesodermal cells at the time the endoderm differentiates. Morgan (1891) reported multipolar delamination of the endoderm in the pycnogonids with little yolk that he studied, *Phoxichilidium maxillare* and *Tanystylum orbiculare*, and is unclear as to how the mesoderm developed in these forms.

In pycnogonids with moderate amounts of yolk, such as *Nymphon stromii*, the presumptive midgut develops from dorsal 'macromere cells' of the embryo, which come to be enveloped by an overgrowth of ventral 'micromere cells.' The presumptive mesoderm arises from some cells around the edge of the micromere 'cap' which, when the 'blastopore' reaches the equator of the embryo, migrate under the micromere cap and proliferate mesoderm. Dogiel (1913) is unclear as to whether this is a true migration of cells or a marginal delamination from the micromere cap.

In forms with large amounts of yolk, like *Chaetonymphon spinosum* or *Callipallene empusa*, the division of the 'macromere cells' stops at an advanced stage (cytokinesis is typically incomplete in many of these cells). Only the micromeres continue to divide, spread over the yolky macromeres, and form a germinal disk (Morgan, 1891). Cells that form the endodermal tissue arise by multipolar delamination from the disc. The mesoderm arises by multipolar delamination of cells from around the region of the involuting stomodeum.

The presumptive ectoderm of pycnogonids is zoned into five regions and soon gives rise to the ventral organs corresponding to the protocerebral brain and first four postoral embryonic ganglia. Many pycnogonids have epimorphic development, though some of them are anamorphic with a protonymphon larva.

The precise affinity of pycnogonids based on what is known of their embryology is inconclusive. The early separation of the endoderm during cleavage, the association in at least some of the pycnogonids of mesoderm formation with developing stomodeum, and the zonation of the ectoderm suggests distant relationship with the chelicerates. However, a great deal more conclusive information is needed, especially on the forms with large amounts of yolk in the egg, before pycnogonid embryology can be effectively related to that of other arthropods. Sister group status of pycnogonids and chelicerates, however, is also effectively corroborated by features of the endoskeletal system (Firstman, 1973), neural anatomy (Schram, 1978), and locomotory morphology (Schram and Hedgpeth, 1978).

SOFT ANATOMY   Additional confirmation concerning the separateness of uniramians, cheliceriforms, and crustaceans has been made. Clarke (1979) reviewed arthropod internal anatomy, especially guts, and observed three basic patterns corresponding to these groups. Schaller (1979) in his examination of sexual behavior concluded there was no basis to unite crustaceans and uniramians into a taxon Mandibulata. Neither of these workers, however, took any position as to whether arthropods were mono- or polyphyletic.

DIPHYLY   Counterarguments to the polyphyletic interpretation of arthropods have been put forth under various guises. There are two approaches: monophyly (see, e.g., several, but not all, contributions in Gupta, 1979), and those who believe in a diphyletic scheme that opposes the uniramians as one taxon to all the other arthropods, sometimes referred to as *schizora-mians*. Both these approaches present some problems.

The attempted union of crustaceans with trilobites and cheliceriforms has been advanced on several grounds, one of the main ones being developmental patterns. Trilobites are unusual as fossils in that a great deal is known about their larval sequences (Whittington, 1957, 1959). The earliest stage in the series is the protaspis, which has the appearance of a reduced trilobite cephalon lacking the typical elaborations of glabella and free cheeks. Some workers have questioned whether the protaspis is the earliest trilobite larval stage. Fortey and Morris (1978) have described what they term a 'preprotaspis,' naupliuslike, phaselus larva, but Schram (1982) felt that such a claim for what are nothing other than rather nondescript, cap-like microfossils was perhaps unjustified. The protaspis remains the only securely identified earliest trilobite larva.

Apparently not all protaspids were fossilized; the protaspids of the primitive olenellid trilobites have never been found. Sclerotization of the earliest stages in trilobite development may have been a relatively late evolutionary event. Many of the larger types of protaspids, those in excess of 0.4 mm, have the axis divided into five rings. (Smaller protaspids show no sign of this division.) The most anterior ring is the largest, is associated with the eyes, and possibly represents an acron. The remaining four segments, ending in the posterior occipital, have been interpreted as the basic four segments of the adult trilobite cephalon (Hendricksen, 1926; Beklemishev, 1969). This axial segmentation of the protaspid is frequently lost in the higher trilobites, for example, in the Lichidae (Whittington, 1956) the protaspids have no segmental grooves but have five sets of spines thought to correspond to the five basic axis segments.

Trilobites have been frequently linked phyletically to merostomes. Iwannoff (1933) was one of the first to suggest that the development of *Tachypleus* indicated a phyletic relationship with the trilobites. Iwanoff rejected the so-called 'trilobite larva' of the limulids as indicating a relationship of the two groups (he suggested the larva was more like the Carboniferous merostome *Euproops* than a trilobite). However, his chief reason for linking the two groups was that in *Tachypleus* the initial delineation of body segments by the mesoderm is into four somites, and he compared these to the four 'postacronal' segments of the trilobite protaspis. This arrangement of larval segments is similar to that of the pycnogonids, where the ectoderm produces five ventral organs that eventually give rise to the protocerebral (acronal) part of the brain and the first four embryonically postoral ganglia. In those pycnogonids that have a protonyphon larva, the first three of these ganglia are associated with the chelifore, pedipalp, and oviger segments respectively—the last ganglion not being associated with any appendages at this stage. All these ontogenetic similarities may serve to link trilobites and cheliceriforms into the Arachnomorpha (sensu Størmer, 1944).

The crustaceans, however, have an altogether different embryonic segment pattern. In the initial nauplius stage typically three ganglia and three sets of appendages appear: the antennules, innervated by the deutocerebral portion of the brain, and the antennae and mandibles, innervated by the first two postoral ganglia. It would seem that trilobites, chelicerates, and pycnogonids all share early developmental stages with four postoral segments, while the Crustacea seem to be quite distinctive with only two.

Considerable discussion has been published through the decades on the comparative anatomy of arthropod head segments, and like the larval debates, these have been used to try and unite various groups to each other. Manton (1949) felt that in the end most of these arguments were probably not to be taken too seriously.

The homology of head segments is necessarily related to the homology of brain regions. Bullock and Horridge (1965), after reviewing the literature, opted for the simplest, most parsimonious arrangement they could come up with, one that would permit a correspondence of nerve roots to similar regions in all arthropod groups and the annelids. This sort of approach was first developed and is extensively used in vertebrate morphology; it is certainly more pragmatic and may be more logical than any other. If we are, in fact, dealing with separate phyla, it may be unreasonable to demand the homology of all head structures. Using this scheme, the anteriormost appendages of arthropods can be compared (Table 2-1).

Bullock and Horridge considered the protocerebrum and deutocerebrum to be two parts of an asegmental anterior neural mass (acronal). [There seems to be some embryonic indication that there may be true somites in the preoral region (Manton, 1960). Thus the term 'acronal' here may be misleading.] The tritocerebrum is the first in the postoral series of ganglia. Pycnogonids and chelicerates do not have a deutocerebrum. The Crustacea and Uniramia are the only groups with true preoral (possibly acronal) appendages.

The trilobite cephalic condition has been the subject of some controversy. Cisne (1974, 1982) records four somites in the cephalon of *Triarthrus*, an antenna-bearing segment and three leg-bearing segments. He claims that the antennal segment is preoral in derivation. Cisne homologizes this segment with the cheliceral segment of the chelicerates, but he mistakenly claims that this segment is preorally derived in all groups of arthropods. However, Bullock and Horridge clearly point out the chelicerate chelae are tritocerebral in affinity, that is, postoral in embryonic derivation. In all arthropod embryos the region forming the mouth and labrum moves posteriorly, while the lateral segmental tissue migrates forward. In addition, Cisne points out that, while he does not have any nervous system preservation in his fossils, the gut in his specimen travels some distance anteriad from the mouth and loops around *in front of all* the cephalic musculature, including that of the antennal segment (Fig. 2-3). This structural

Table 2–1. Topological comparisons of appendages in various arthropodal groups based on neural innervation scheme of Bullock and Horridge (1965). Trilobites based on data of Cisne (1974). Only the anteriormost segments of the body are dealt with. The proto- and deutocerebra are usually considered to be "acronal," though that term implies a subjective judgment concerning head metamerism. The tritocerebrum is derived from the first embryonic postoral segment. These comparisons are to imply analogies, though they may in part also produce homologies in some instances. (From Schram, 1978)

		Chelicerata	Pycnogonida	Crustacea	Onychophora	Myriapoda-Hexapoda	Trilobita
Preoral							
	Protocerebral	—	—	—	—	—	—
	Deutocerebral	—	—	antennule	antenna	antenna	antenna
	Tritocerebral I	chelicerae	chelifores	antenna	("lips")	(labrum)	first leg
	II	pedipalps	pedipalps	mandible	mandible	mandible	second leg
	III	first leg	ovigers	maxilla I	slime papilla	maxilla I	third leg
	IV	second leg	first leg	maxilla II	first leg	maxilla II	fourth leg
Postoral	V	third leg	second leg	first leg	second leg	first leg	fifth leg
	VI	fourth leg	third leg	second leg	third leg	second leg	sixth leg
	VII		fourth leg	third leg	fourth leg	third leg	seventh leg
	VIII			fourth leg	fifth leg	(fourth leg)	

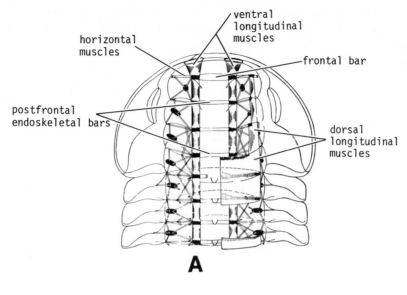

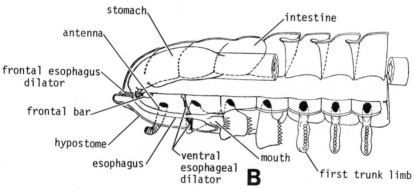

Fig. 2-3. (A) dorsal and (B) lateral views of head and first three trunk segments in *Triarthrus eatoni*. Note relationship of skeletomusculature and digestive systems with esophagus passing anteriad of the antennal frontal bar and other unmodified segmented structures related to it. (Modified from Cisne, 1974)

arrangement would really seem to preserve the most primitive of arthropod states with little or no movement of any of the cephalic segments anteriad during development, with only the mouth migrating posteriad, the antenna and legs all being originally postoral in origin. Linking this adult trilobite cephalic anatomy to the protaspid evidence presented above would indicate that the four segments of the late protaspid *are* postoral in position. It is very important to remember that the adult position of somite derivatives in relation to the arthropod mouth does not correspond to the relative position in the embryos.

Analysis of the limb morphology in fossil arthropods has been difficult to amalgamate with the functional studies of living forms. This has been especially problematical for the aglaspids and trilobites. Aglaspids are now recognized not to have chelae and thus are not Chelicerata (Briggs et al., 1979). Relatively little is known concerning trilobite limbs. Of the hundreds of trilobite genera, only six are known at all well. On this limited basis trilobites have often been characterized as having similar append-ages. Though trilobite appendages are alike in general form, having a strong telopod and a 'filamentous branch' arising from the coxa, an exam-ination of these few species reveals a potential for a wide range of func-tional variations (Fig. 2-4).

Olenoides (Whittington, 1975b), *Triarthrus* (Cisne, 1975), *Naraoia* (Whittington, 1977), and possibly *Cryptolithus* (Bergström, 1972; Camp-bell, 1975) possess medially directed spines of setae on the coxa that may have functioned like *Limulus* coxae, or possibly utilized a pushing action more like crustaceans. *Phacops* (Stürmer and Bergström, 1973), *Naraoia*, and *Cryptolithus* have spines and 'gnathic' structures on more distal seg-ments of the telopod. *Ceraurus* has a completely unadorned telopod, no 'gnathic' structures at all. The morphology of the filamentous branch exhi-bits even more diversity. (The use of terms like 'preepipodite' or 'exite' leads to phyletic and anatomical conclusions that are not necessarily justi-fied.) Bergström (1969) gives some convincing arguments against the fila-mentous branch being able to function in respiration. Bergström points out that the cuticle of the filaments is relatively thick. Though the filaments are flexible, they do not collapse or fold as a structure with a thin cuticle would do, and the filaments are preserved equally as well as the trilobite telopod. This would seem to imply a cuticle on these filaments too thick to sustain a respiratory exchange of gases, though Whittington (1975b) disagrees. Bergström also raises a question about the mechanics of getting body fluid out into the filaments and back and the ability of the trilobite body plan to sustain the necessary high body fluid pressure to achieve it. Bergström feels that the apparently thin cuticle on the underside of the trilobite pleura is a better candidate for the respiratory surface.

Schram (1983a) raised the possibility that in light of the very primitive body plan of the remipede *Speleonectes lucayensis*, the phyletic theories of crustacean origins based on cephalocarids (Hessler and Newman, 1975) may have to be rejected. Cephalocarids have gnathobasic trunk limbs apparently utilized for both swimming and feeding. If remipedes, with their biramous paddles used only for locomotion, intervene as the most primitive crustacean type, then the cephalocarids can be seen as very specialized crustaceans (Schram, 1982). Consequently, the derived gnatho-basic form of trilobite limbs may allow a further connection to be made to the gnathobasic form of merostomes. This would again suggest some con-sideration to be given to the concept of Arachnomorpha and would further argue for the isolated phyletic position of the crustaceans. This separate-

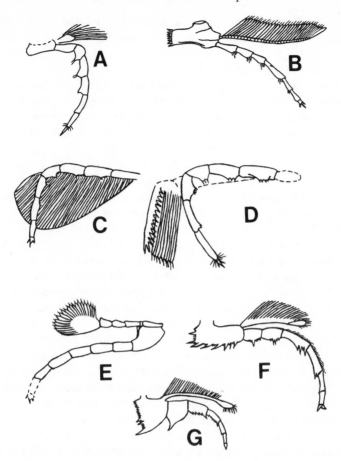

Fig. 2-4. Trilobite appendages. (A) *Phacops*, anterior view (modified from Stürmer and Bergström, 1973); (B) *Triarthrus eatoni*, anterior view (from Cisne, 1975); (C) *Cryptolithus tesselatus*, ventral view (from Campbell, 1975); (D) *C. tesselatus*, ventral view (from Bergström, 1972); (E) *Ceraurus*, anterior view (modified from Størmer, 1939); (F) *Olenoides serratus*, anterior view (modified from Whittington, 1975b); (G) *Naraoia compacta* (from Whittington, 1977). Filamentous branches of A, B, E, F, and G actually extending into the plane of the figure.

ness is further reinforced when one considers that the primitive crustacean condition (verified again in remipedes) might be one of cephalophagy (Cannon and Manton, 1927; or Chapter 44), feeding only with the perioral limbs with the more posterior limbs being used only for locomotion. This would stand in contrast to the mixed function seen in merostome and trilobite limbs.

Thus while it would seem that a possible argument for trilobitomorphs and cheliceriforms being sister groups could possibly be made, relating crustaceans to these is much more difficult without recourse to elaborate sequences of hypothetical 'paper animals' and all the problems those engender when dealing with the Phylogenetic Uncertainty Principle (Schram, 1983b).

MONOPHYLY Another group of investigators seeks to maintain arthropod monophyly. For the most part, many of these people are insect workers, strongly devoted to the cladistic methodology of the Hennig school and even more strongly committed to maintaining the status of Mandibulata. Their arguments, however, possess major defects.

Weygoldt (1979) has presented the best rebuttal to the embryological arguments of Anderson. Although Weygoldt concedes that evolutionary systematic analysis, such as those evidenced in ontologic and functional morphologic studies, can elucidate the pathways evolution has taken, he maintains that such an approach cannot show relationships between groups. Such an assertion, if it is granted, would imply that the strength of evolutionary studies like those of Anderson and Manton are most effective in falsifying suggested possible relationships. Weygoldt (1979, p. 109) concludes, 'Therefore it cannot replace the approach of establishing homology.' I think that even the proponents of the polyphyletic nature of arthropods would not disagree with this contention.

Weygoldt goes on to question Anderson's use of fate maps in phyletic analysis. He wonders about the accuracy of such maps, but insists, without documentation, that (even if fate maps are true) presumptive areas can shift or disappear. To grant this, however, is to make the concept of homology totally meaningless. Weygoldt is not alone in this, since Clarke (1979) says that even the germ layer origin of organs is 'irrelevant' in determining homology. Weygoldt goes on to reject the consistent similarity of presumptive patterns within Anderson's major arthropod groups, while in effect telling us to use Anderson's technique. 'If their [adult arthropods] ontogeny is followed backward, the developmental stages become increasingly similar up to the stage that shows the basic body pattern' (p. 126). Of course this is exactly what Anderson did and discovered instead distinctly different ontogenetic organization evident in basic arthropod body plans. In view of these inconsistencies of logic, the arguments Weygoldt makes for homology of head parts between various arthropod groups are unconvincing. Presumably, he would reject arguments of homology based on cephalic topology and innervation as outlined above in Table 2-1.

Paulus (1979) seeks to establish arthropod monophyly based on shared possession of faceted or compound eyes. He advances three possible interpretations of compound eyes: (1) that they are convergent between various arthropod groups, but that ommatidia are homologous, (2) that ommatidia are convergent, and (3) that the first arthropods had compound eyes that were then further modified in each group. It is this third alternative that he favors. The question then resolves itself into one of polarity, that is, as to whether compound eyes are a shared primitive or derived character. One possible outgroup analysis to an actual animal would seem to imply the former. The Cambrian wormlike creature *Opabinia* is seen to have stalked compound eyes (Whittington, 1975a). Since *Opabinia* is an annulate, but obviously not an arthropod, this would seem to indicate that if compound eyes are shared among all the first arthropods and other annulates, then such commonality is primitive and thus not useful in establishing arthropod relationships. Indeed the problem with using eyes in establishing arthropod relationships is concisely summarized by Paulus himself when commenting on the case of trying to establish the monophyly of such a tightly related group as the Arachnida, which he concluded 'as a whole is not provable on the basis of the lateral eyes. The same is true with the other characters . . . which are very evident but not provable because of the strong possibility of convergences' (p. 312).

Like Weygoldt, Paulus interprets his evidence toward preconceived ends. For example, Paulus concedes that myriapod eyes are quite different from any seen among hexapods. However, rather than taking this as congruent with the conditions that Manton observed in regard to myriapod locomotory morphology (see above), he rejects the anatomy of both as indicating separate myriapod and hexapod origins in favor of an interpretation that myriapod eyes were secondarily modified from a more 'typical' hexapod type.

Such twisting of facts to fit a preconceived end is all too typical in the monophyletic arguments. Boudreaux (1979) maintains that Manton based her arguments purely on suppositions; however, he himself offers us only another series of suppositions based on hypothetical ancestral 'paper animals.' No justification is offered for why one set of suppositions may be better than what are supposed to be another set. Baccetti (1979) quite openly informs us that, in regard to sperm characters, morphologic features must be chosen carefully to achieve the correct relationships of taxa. In Baccetti's case, the achievement of arthropod monophyly based on sperm requires a *Limulus*-like ancestor.

What then is the status of characters that have been said to be autapomorphies of Arthropoda? In addition to compound eyes, other characters usually put forward are: a chitinous cuticle (and with this a hormonal system necessary to achieve periodic molting), a segmented body, and jointed appendages. It has already been pointed out that with the nonarthropodous annulate *Opabinia* as an outgroup, compound eyes might just as well be considered a plesiomorphy for arthropods, that is, characteristics not of arthropods *per se* but possibly of some even higher category of annulates. The pertinent features of the cuticle are not unique to arthropods either. Chitin is noted in Annelida (another annulate candidate for an arthropod outgroup), Mollusca, Bryozoa, and Brachiopoda. Nor is the molt chemistry unique to arthropods. Ecdysone has been identified in nematodes (Horn et al., 1974) and coelenterates (Sturaro et al., 1982), and phytoecdysones have been identified in vascular plants (Tombes, 1979). (The former is especially interesting in that Clarke, 1979 suggested the possibility of aschelminth origins for arthropods.) The feature of body segmentation is characteristic of a whole group of 'phyla' that may or may not be related to each other, including Annelida and several diverse annulate Middle Cambrian and Carboniferous forms of uncertain affinities. Finally, the 'definitive' arthropod character, jointed limbs, have themselves come to be seen as not unique to this group. The problematic Cambrian 'arthropod limb' *Anomalocaris* has now been identified as a cephalic structure on an animal from the Burgess Shale that is clearly annulate but not an arthropod (Whittington and Briggs, 1982).

What are we left with in the final analysis? Arthropods are a group of animals with a segmental chitinous cuticle, jointed limbs, and, frequently, compound eyes, who share these characters, not uniquely with each other, but with a wide array of annulate and other phyla. The concept of arthropods is thus akin to that of lophophorates or aschelminths—incidentally groups that have also been proposed for separate phylum status. It would appear to this author that the various lines of arthropods are separate phyla. However, if tradition must prevail, there are at least three, possibly four or more, distinctly separate groups, of which Crustacea are among the most isolated.

Just exactly how many arthropodous phyla there might be was not a great concern of Anderson or Manton, restricted as they were to the consideration of living forms; three or four seemed sufficient. The growing knowledge about 'trilobitoids' from the Middle Cambrian Burgess Shale now appears to indicate many more very distinctly different arthropodous clades (Whittington, 1979), so many in fact as to cause Briggs (1983) to balk at characterizing them all as 'phyla.' Briggs tabulated cephalic appendages in Burgess Shale arthropods divided on the basis of whether limbs were anterior or posterior to the mouth. The 'formula' he arrived at gave cephalic configurations (Table 2-2) that were similar in principle to those seen in Table 2-1 but not equivalent, since they were based strictly on external considerations with no reference to homologies based on internal form and nervous innervation. Uncertainties were encountered by Briggs due to preservation, but Table 2-2 reveals a very diverse array of what appear to be very distinct clades based solely on preliminary analysis of external cephalic architecture (with little or no reference as yet to limb morphology). However, this kind of analysis has its limits. That animals may have different head configurations does not preclude a relationship. An example is the trilobite *Rhenops* (Bergström and Brassel, 1984), which has one extra head segment from that usually assumed for the Trilobita.

Are *each* of these cephalic *Baupläne* to be interpreted as a potential 'phylum'? If so, does this strain the credibility of 'arthropody' as a polyphyletic phenomenon? If not, are all these 'arthropods' to be drawn together by a series of 'paper animals'? You pays your money and takes your choice.

Table 2–2. Configuration of head appendages in Cambrian arthropods and the Devonian trilobite *Rhenops*. The ratio of preoral and postoral appendages is expressed as a "formula", the preoral number given first. Assuming that no more than two appendages will be preoral (the case in living arthropods), there are only three possible configurations for each total number of cephalic appendages. Not all the possible combinations ("formulas") actually occur. In many of the genera, the arrangement is uncertain due to poor preservation. (Modified from Briggs, 1983)

Appendage number	Formula	Arthropods
1	0 + 1	
	1 + 0	?*Plenocaris, Sidneyia*
2	0 + 2	
	1 + 1	*Branchiocaris, Sarotrocercus*
	2 + 0	*Marrella*
3	0 + 3	
	1 + 2	*Habelia, Leanchoilia*
	2 + 1	
4	0 + 4	*Yohoia*
	1 + 3	*Olenoides, Naraoia, Burgessia, Molaria,* ?*Aglaspis, Actaeus, Alalcomenaeus*
	2 + 2	?Phosphatocopina
5	0 + 5	
	1 + 4	*Dala, Rhenops*
	2 + 3	?*Perspicaris, Canadaspis,* ?*Waptia,* ?*Odaraia*
6	0 + 6	
	1 + 5	*Emeraldella*
	2 + 4	

REFERENCES

Anderson, D. T. 1973. *Embryology and Phylogeny in Annelids and Arthropods*. Pergamon Press, New York.

Baccetti, B. 1979. Ultrastructure of sperm and its bearing on arthropod phylogeny. In *Arthropod Phylogeny* (A. P. Gupta, ed.), pp. 609–64. Van Nostrand Reinhold, New York.

Beklemishev, W. N. 1969. *Principles of Comparative Anatomy of Invertebrates*. Univ. of Chicago Press, Chicago.

Bergström, J. 1969. Remarks on the appendages of trilobites. *Lethaia* 2:395–414.

Bergström, J. 1972. Appendage morphology of the trilobite *Cryptolithus* and its implications. *Lethaia* 5:85–94.

Bergström, J., and G. Brassel. 1984. Legs in the trilobite *Rhenops* from the Lower Devonian Hunsrück Slate. *Lethaia* 17:67–72.

Boudreaux, H. B. 1979. Significance of intersegmental tendon system in arthropod phylogeny and a monophyletic classification of Arthropoda. In *Arthropod Phylogeny* (A. P. Gupta, ed.), pp. 551–86. Van Nostrand Reinhold, New York.

Briggs, D. E. G. 1983. Affinities and early evolution of the Crustacea: the evidence of the Cambrian fossils. *Crust. Issues* 1:1–22.

Briggs, D. E. G., D. L. Bruton, and H. B. Whittington. 1979. Appendages of the arthropod *Aglaspis spinifer* and their significance. *Paleontol.* 22:167–80.

Bullock, T. H., and G. A. Horridge. 1965. *Structure and Function in the Nervous System of Invertebrates*, Vols. I and II. Freeman, San Francisco.

Campbell, K. S. W. 1975. The functional morphology of *Cryptolithus*. *Fossils and Strata* **4**:65–86.

Cannon, H. G., and S. M. Manton. 1927. On the feeding mechanism of a mysid crustacean, *Hemimysis lamornae*. *Trans. Roy. Soc., Edinb.* **55**:219–52.

Cisne, J. S. 1974. Trilobites and the origin of arthropods. *Science* **186**:13–18.

Cisne, J. S. 1975. The anatomy of *Triarthrus* and the relationships of the Trilobita. *Fossils and Strata* **4**:45–64.

Cisne, J. S. 1982. Origin of the Crustacea. In *Biology of Crustacea*, Vol. I (L. G. Abele, ed.), pp. 65–92. Academic Press, New York.

Clarke, K. U. 1979. Visceral anatomy and arthropod phylogeny. In *Arthropod Phylogeny* (A. P. Gupta, ed.), pp. 467–549. Van Nostrand Reinhold, New York.

Clarkson, E. N. K. 1966. Schizochroal eyes and vision in some phacopid trilobites. *Paleontol.* **9**:464–87.

Clarkson, E. N. K. 1973. The eyes of *Asaphus raniceps*. *Paleontol.* **16**:425–44.

Clarkson, E. N. K. 1975. The evolution of the eye in trilobites. *Fossils and Strata* **4**:7–32.

Clarkson, E. N. K. 1979. The visual system of trilobites. *Paleontol.* **22**:1–22.

Dogiel, V. 1913. Embryologische Studien an Pantopoden. *Zeit. wiss. Zool. Abt. A* **107**:575–756.

Elofsson, R. 1970. Brain and eyes of *Zygentoma*. *Entomol. Scand.* **1**:1–20.

Firstman, B. L. 1973. The relationships of the chelicerate arterial system to the evolution of the endosternite. *J. Arachnol.* **1**:1–54.

Fortey, R. A., and S. F. Morris. 1978. Discovery of nauplius-like trilobite larvae. *Paleontol.* **21**:823–33.

Gupta, A. P. 1979. *Arthropod Phylogeny*. Van Nostrand Reinhold, New York.

Hendricksen, K. L. 1926. The segmentation of the trilobite head. *Medlemsbla. Dan. Geol. Fosen., Copenhagen* **7**:1–32.

Hessler, R. R., and W. A. Newman. 1975. A trilobitomorph origin for Crustacea. *Fossils and Strata* **4**:437–59.

Horn, D. H. S., J. S. Wilkie, and J. A. Thomson. 1974. Isolation of β-ecdysone from the parasitic nematode *Ascaris lumbricoides*. *Experientia* **15**:1109–10.

Iwanoff, P. P. 1933. Die embryonale Entwicklung an *Limulus moluccanus*. *Zool. Jahrb., Abt. Anat.* **56**:163–348.

Jell, P. A. 1975. The abathochroal eye of *Pagetia*, a new type of trilobite eye. *Fossils and Strata* **4**:33–44.

Lankester, E. R. 1881. *Limulus*, an arachnid. *Quart. J. Micro. Sci.* **21**:504–48.

Manton, S. M. 1949. Studies on the Onychophora VII. *Phil. Trans. Roy. Soc. Lond.* (B)**233**:483–580.

Manton, S. M. 1960. Concerning head development in the arthropods. *Biol. Rev. Cambridge Phil. Soc.* **35**:265–82.

Manton, S. M. 1964. Mandibular mechanisms and the evolution of arthropoda. *Phil. Trans. Roy. Soc. Lond.* (B)**247**:1–183.

Manton, S. M. 1969. Evolution and affinities of Onychophora, Myriapoda, Hexapoda, and Crustacea. In *Treatise on Invertebrate Paleontology*, Part R, *Arthropoda* **4**(1) (R. C. Moore, ed.), pp. R15–56. Geol. Soc. Am. and Univ. Kansas Press, Lawrence.

Manton, S. M. 1973a. Arthropod phylogeny—a modern synthesis. *J. Zool.* **171**:111–30.

Manton, S. M. 1973b. The evolution of arthropodan locomotory mechanisms, Part II. *J. Linn. Soc. Zool.* **53**:257–375.

Manton, S. M. 1977. *The Arthropoda*. Oxford Univ. Press, Oxford.

Morgan, T. H. 1891. A contribution to the embryology and phylogeny of the pycnogonids. *Stud. Biol. Lab.* **5**:1–76.

Paulus, H. F. 1979. Eye structure and the monophyly of the Arthropoda. In *Arthropod Phylogeny* (A. P. Gupta, ed.), pp. 299–383. Van Nostrand Reinhold, New York.

Sanchez, S. 1959. Le développement des Pycnogonides et leur affinities avec les Arachnides. *Arch. Zool. Exp. Gen.* **98**:1–101.

Sawyer, R. T. 1984. Arthropodization in the Hirudinea: evidence for a phylogenetic link with insects and other Uniramia. *Zool. J. Linn. Soc.* **80**:303–22.

Schaller, F. 1979. Significance of sperm transfer and formation of spermatophores in arthropod phylogeny. In *Arthropod Phylogeny* (A. P. Gupta, ed.), pp. 587–608. Van Nostrand Reinhold, New York.

Schram, F. R. 1978. Arthropods: a convergent phenomenon. *Fieldiana: Geol.* **39**:61–108.

Schram, F. R. 1982. The fossil record and evolution of Crustacea. In *The Biology of Crustacea*, Vol. I (L. G. Abele, ed.), pp. 93–147. Academic Press, New York.

Schram, F. R. 1983a. Remipedia and crustacean phylogeny. *Crust. Issues* **1**:23–28.

Schram. F. R. 1983b. Method and madness in phylogeny. *Crust. Issues* **1**:312–50.

Schram, R. F., and J. W. Hedgpeth. 1978. Locomotory mechanisms in Antarctic pycnogonids. *Zool. J. Linn. Soc.* **63**:145–69.

Snodgrass, R. E. 1938. Evolution of the Annelida, Onychophora, and Arthropoda. *Smith. Misc. Coll.* **97**(6):1–159.

Størmer, L. 1939. Studies on trilobite morphology. I. The thoracic appendages and their phylogenetic significance. *Norsk. Geol. Tidsskr.* **19**:143–273.

Størmer, L. 1944. On the relationships of the fossil and recent Arachnomorpha. *Skr. Nov. Vidensk. Akad., Oslo* **5**:1–158.

Sturaro, A., A. Guerriero, R. DeClauser, and F. Pietra. 1982. A new, unexpected marine source of a molting hormone. Isolation of ecdysterone in large amounts from the zoanthid *Gerardia savaglia. Experimentia* **38**:1184–85.

Stürmer, W., and J. Bergström. 1973. New discoveries on trilobites by x-rays. *Paleontol. Zeit.* **47**:104–41.

Struwe, G. E., E. Hallberg, and R. Elofsson. 1975. The physical and morphological properties of the pigment screen in the compound eye of shrimp. *J. Comp. Physiol.* **97**:257–70.

Tiegs, O. W., and S. M. Manton. 1958. The evolution of the Arthropoda. *Biol. Rev. Cambridge Phil. Soc.* **33**:255–337.

Tombes, A. S. 1979. Comparison of arthropod neuroendocrine structures and their evolutionary significance. In *Arthropod Phylogeny* (A. P. Gupta, ed.), pp. 645–67. Van Nostrand Reinhold, New York.

Weygoldt, P. 1979. Significance of later embryonic stages and head development in arthropod phylogeny. In *Arthropod Phylogeny* (A. P. Gupta, ed.), pp. 107–36. Van Nostrand Reinhold, New York.

Whittington, H. B. 1956. Beecher's lichid protaspis and *Acanthopyge consanguina. J. Paleo.* **30**:104–9.

Whittington, H. B. 1957. The ontogeny of trilobites. *Biol. Rev. Cambridge Phil. Soc.* **32**:421–69.

Whittington, H. B. 1959. Ontogeny of Trilobita. In *Treatise on Invertebrate Paleontology*, Part O, *Arthropoda 1* (R. C. Moore, ed.), pp. O127–44. Geol. Soc. Am. and Univ. Kansas Press, Lawrence.

Whittington, H. B. 1975a. The enigmatic animal *Opabinia regalis. Phil. Trans. Roy. Soc. Lond.* (B)**271**:1–43.

Whittington, H. B. 1975b. Trilobites with appendages from the Middle Cambrian, Burgess Shale. *Fossils and Strata* **4**:97–136.

Whittington, H. B. 1977. The Middle Cambrian trilobite *Naraoia. Phil. Trans. Roy. Soc. Lond.* (B)**280**:409–43.

Whittington, H. B. 1979. Early arthropods, their appendages and relationships. In *The Origin of Major Invertebrate Groups* (M. R. House, ed.), pp. 253–68. Academic Press, London.

Whittington, H. B., and D. E. G. Briggs. 1982. A new conundrum from the Middle Cambrian Burgess Shale. *3rd N. Am. Paleo. Conv. Proceed.* **2**:573–75.

3

NECTIOPODA
AND ENANTIOPODA

The Carboniferous species *Tesnusocaris goldichi* (Brooks, 1955) was always a difficult form to attempt to place within the arthropods. Its unusual morphology and limited preservation only allowed it to be placed within the Crustacea with trepidation. However, the discovery of *Speleonectes lucayensis* (Yager, 1981), allowed both these forms to be allied into a previously unsuspected line of primitive crustaceans, the class Remipedia.

NECTIOPODA

DEFINITION   No carapace, only a cephalic shield; no eyes; ventral cephalon with frontal processes, biramous antennules, paddlelike biramous antennae; mandibles 'internalized' into atrium oris; maxillules, maxillae, and maxillipede as mouthparts, well developed and prehensile; basal maxillulary endites functioning in place of external 'mandibles'; first trunk segment fused to cephalon; trunk not regionalized, each segment with biramous appendages ventrolaterally directed; anal segment with simple oval caudal rami.

HISTORY   The first species, *Speleonectes lucayensis*, was described by Yager (1981) from deep within an anchialine cave on Grand Bahama Island. Since then, related species have been found on other islands in the Bahamas, Turks and Caicos, and the Canary Islands. Schram (1983) initially pointed out the phylogenetic significance of this group. The recognition of their sister-group status to the Carboniferous Enantiopoda (*Tesnusocaris goldichi*) causes me herein to name a distinct order for them, the Nectiopoda.

MORPHOLOGY   The cephalon is relatively short, about one-twelfth the total body length. The head is covered by a cephalic shield that tapers anteriorly and folds over the front of the cephalon. The head shield is generally marked by a faint transverse groove. There are no eyes (Fig. 3-1). The anterior part of the cephalon bears a pair of small rodlike frontal processes (Fig. 3-2A), each process bearing a stout spine distinctly smaller than the main branch.

36

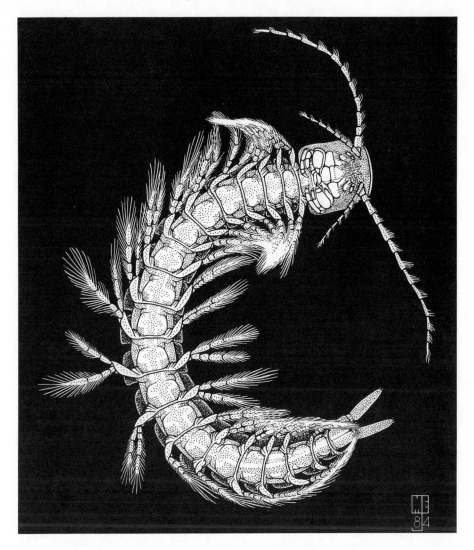

Fig. 3-1. Ventral view of body of *Speleonectes ondinae*, from the Canary Islands.

The antennules (Fig. 3-2B) are biramous. The two-segmented peduncle has a battery of long, ribbonlike aesthetascs on the proximal segment. The dorsal flagellum has numerous large joints and is about twice as long as the cephalon; in contrast, the ventral flagellum has fewer joints and is only about half as long as the head. One species has the segments of the ventral flagellum fused into a slender, bladelike element. Each joint of the flagella bears a tuft of setae distally.

The antennae (Fig. 3-2C) are biramous paddles, moderate in size. The

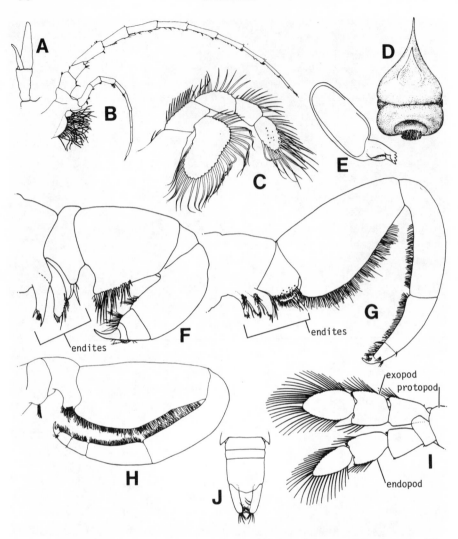

Fig. 3-2. Appendages of *Lasionectes entrichomus*. (A) frontal process; (B) antennule; (C) antenna; (D) mandible; (E) maxillule; (F) maxilla; (G) maxillipede; (H) tenth trunk limb; (I) anal segment and caudal rami.

protopod is two-segmented. The three-segment endopod is curved laterally around the single-segment oval exopod. Plumose setae mark the margins of all the antennal joints.

The mandibles (Fig. 3-2E) are symmetrical and bear distinct incisor and molar processes separated by large laciniae mobili. The mandibles are 'internalized' under the large fleshy labrum and lie in an atrium oris. There they form a mandibular mill effective in triturating food as it passes into the foregut. The large labrum (Fig. 3-2D) is more broad posteriorly than anteriorly and is frequently equipped posteromedially with a dense array of ribbon setae. The paragnaths are rounded lobes with dense batteries of ribbon setae, and the intermaxillulary space is also generally covered with ribbon setae. The entire perioral area is quite setose.

The maxillules (Fig. 3-2F) are robust, seven-segmented, uniramous, prehensile, or subchelate appendages. The proximalmost segment has a flat endite with long terminal spines, some of them barbed. The second segment has a paddlelike endite with numerous stout spines and setae along the margin. These two endites immediately flank the mouth and apparently function as 'mandibles.' The third joint has a generally conical endite, terminally mounted with spines and subterminal setae. The fourth joint has a broad and elongate endite variously armed with spines and dense rows of setae. The distal joints are all robust and variously equipped with either tufts or rows of setae. Flexure in the limb occurs between joints 4 and 5 to achieve either a prehensile or subchelate condition. The terminal talonlike claw is distally marked by a large pore.

The maxillae (Fig. 3-2G) and maxillipedes (Fig. 3-2H) are quite similar but vary in relative size depending on species and in the development of the basal endites. The terminal claws are complex arrangements of spines, setal pads, and pores, that seem to be genus specific in their structure.

The trunk appendages (Fig. 3-2I) are biramous and directed ventrolaterally. The endopod is four-segmented, the exopod three-segmented. All trunk appendages are homonomous, except for a tendency for the first limb to be somewhat slender in form and the posteriormost limbs to be small.

The terminal anal segment is elongate and bears long and lobelike terminal caudal rami (Fig. 3-2J). The anus proper seems to be covered with a small flap.

The foregut is composed of a muscular anterior portion capable of considerable dilation and a posterior, narrow, tubular portion. The foregut extends the length of the head and enters the midgut at about the level of the cephalon–trunk border. The entire length of the midgut is marked by a series of paired diverticula in each segment extending into the lateral pleural lobes of the trunk segments. These outpocketings grow smaller toward the posterior part of the body. A short hindgut occupies the anal segment.

The circulatory system consists of an elongate middorsal vessel running the length of the body. It appears, however, that the muscular portion of the heart proper is restricted only to the cephalon.

Excretion is facilitated by a set of well-developed maxillary glands occupying the posterolateral quarters of the cephalon.

Knowledge of the reproductive system is presently still rather limited. The ovaries occur in the posterior portion of the cephalon and are continuous with the oviducts, which extend dorsal of the gut to the fourteenth trunk segment. The gonopores are located on the medial surface of the

protopod of the fourteenth trunk limb. The animals appear to be hermaphrodites, with the male system extending from the head posteriad lying ventral to the midgut.

The nervous system is very well developed. The brain is a large organ in
the anterior part of the cephalon. The circumesophageal commissures connect to a simple ladderlike ventral nerve cord. The cephalic ganglia are
rather large, but the trunk segment ganglia are only modestly developed.
The entire cuticle is equipped with pores and tiny sensillia. Though some of
the pores may be secretory in nature, it seems that many of them
(especially on the cephalic limbs) are the sites of chemo- and mechanoreceptors.

NATURAL HISTORY   These creatures are currently known only to occur in
anchialine limestone caves and flooded lava tubes. They live below a distinct halocline in brackish layers of waters generally deep within the caves.
Dennis Williams (personal communication) relates that oxygen in the
remipede habitat is very low, around 0.5 parts per billion—virtually
anoxic. Yet the animals are moderately active, good swimmers. When collected and maintained in aquaria, the animals take to ceaseless, rather
frenetic swimming and literally burn themselves out within a few days.

Swimming at any speed is achieved with regular metachronal beats. If
the animal strives to escape or swim against gravity toward the surface, the
metachronal pattern can be shifted to a brief, singular, power stroke with
all limbs pushing in unison. However, once under way the nectiopodans
then quickly return to a metachronal pattern. From films taken of living
nectiopodans, the cycle of beat seems to repeat every 8 to 10 segments.

The robust, prehensile to subchelate mouthparts would seem to imply a
carnivorous mode of feeding. Indeed, Williams (personal communication)
reports once seeing a speleonectid feeding on *Typhlatya garciai*, a caridean
commonly associated with the West Indian nectiopodans. The prey was
grasped in the flexed mouthparts and pressed tightly to the mouth. When
feeding was completed, an empty cuticle was set afloat. Distinct food particles are never seen in sections of the gut, so that apparently the action of
the mandibular mill and muscular esophagus thoroughly macerates food.

Nothing is currently known concerning nectiopodan breeding habits
nor details of development. Several of the known species have been found
in association with juveniles. In general form these resemble the adults,
but they are smaller, lack gut diverticula and gonopores, and only have
from 12 to 17 segments.

The distribution of nectiopodans is a classic Tethyan pattern (Fig. 3-3).
They have been found in caves in the Bahamas, and Turks and Caicos, as
well as a drowned lava tube in the Canary Islands. The Canary species is in
the same genus, *Speleonectes*, as one of the Bahamian forms.

TAXONOMY   As of this writing six species in four genera are known. These
are actively being studied.

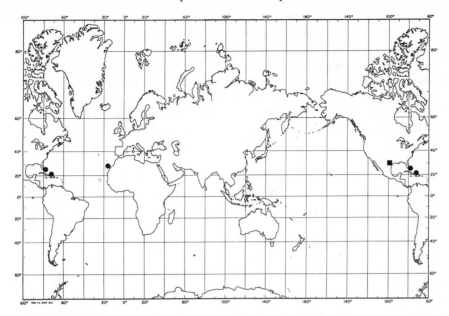

Fig. 3-3. Biogeographic distribution of Nectiopoda ●, and Enantiopoda ■.

ENANTIOPODA

DEFINITION   Large sessile compound eyes; biramous antennules, large and paddlelike; biramous antennae, small and thin; trunk segments with pairs of homonomous paddlelike appendages, seven-segmented 'exopod' and flaplike 'endopod.' Tail end of body is unknown.

HISTORY   The single known species, *Tesnusocaris goldichi* Brooks, 1955, was described from the Lower Pennsylvanian, Tesnus Formation, Brewster County, Texas. The animal was always something of a problematic form. Brooks (1955) placed *Tesnusocaris* within the cephalocarids, and Birshtein (1960) made this the basis of his higher classification of orders for the Cephalocarida. *Tesnusocaris goldichi* is the only species in the Order Enantiopoda. Hessler (1969) rightly rejected *Tesnusocaris* as a cephalocarid, but no suggestion as to what its exact affinities might be was advanced. Schram (1983) linked *Tesnusocaris* with the living remipedes.

MORPHOLOGY   The compound eyes are large, oval, sessile and consist of some 800 facets each. The cephalon has a subrectangular head shield with the anterior width somewhat narrower than that of the posterior.

The biramous antennules have a large ventral ramus and a small flaplike dorsal branch. They are setate along their posterior margins. The total length of the dorsal ramus is about 5.5 mm, the ventral is 53 mm.

The antennae are small, thin, and biramous. The basal portion is unknown, as are the distal elements, due to lack of preservation of other than the medial portions of the rami (Schram et al., 1986).

Mediad of the antennae is a large wedge-shaped labrum, wider posteriorly than anteriorly. Beneath the labrum is a pair of tooth-bearing mandibles, whose exact size is indeterminate.

The maxillules and maxillae are robust limbs and are directly posterior to the mandibles. They lie about one-third the length of the cephalon from the posterior margin, that is, the head appendages occupy only the anterior two-thirds of the cephalon.

The trunk is posteriorly tapered, becoming gradually narrower, but its exact length is undetermined. At least the first 15 segments bear a set of homonomous appendages of a rather peculiar form (Fig. 3-4C). The appendages articulate on the sternites rather close to the midline. The outer branch or 'exopod' has seven flattened segments, which increase in width distally. The inner branch or 'endopod' is flaplike, but its exact morphology and relationship to the outer branch is uncertain. The margins of both of these branches are setose.

NATURAL HISTORY   Little can be discerned about the paleobiology of this species. It was apparently a coastal shallow-water form. Of biogeographic interest is that the fossil is from Texas, a region contiguous to the known distribution of the living order Nectiopoda.

Fig. 3-4. Reconstruction of *Tesnusocaris goldichi*. (A, B) dorsal and ventral views of body; (C) detail of trunk limbs. (From Brooks, 1955; but ventral reconstruction of head does not accord with description in this book's text based on restudy of holotype.)

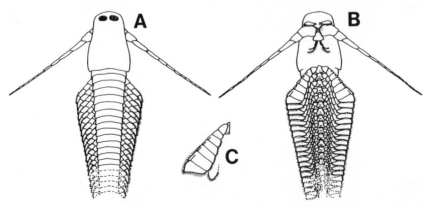

REFERENCES

Birshtein, Ya. A. 1960. Podklass Cephalocarida: Chlenistongie, Trilobitoobraznie, Rakoo-braznie. In *Osnovy Paleontologii* (Ya. A. Orlov, ed.), pp. 421–22. Akademii Nauk, Moskva.

Brooks, H. K. 1955. A crustacean from the Tesnus Formation of Texas. *J. Paleo.* **29**:852–56.

Hessler, R. R. 1969. Cephalocarida. In *Treatise on Invertebrate Paleontology*, Part R, Arthropoda **4**(1) (R. C. Moore, ed.), pp. R120–28. Geol. Soc. Am. and Univ. Kansas Press, Lawrence.

Schram, F. R. 1983. Remipedia and crustacean phylogeny. *Crust. Issues* **1**:23–8.

Schram, F. R., M. J. Emerson, and J. Yager. 1986. Remipedia, Part 1, Systematics. *Memoirs San Diego Nat. Hist. Mus.* **15**.

Yager, J. 1981. Remipedia, a new class of Crustacea from a marine cave in the Bahamas. *J. Crust. Biol.* **1**:328–33.

4

AESCHRONECTIDA
AND PALAEOSTOMATOPODA

These two Paleozoic orders illustrate the diversity of the hoplocaridan body plan. The aeschronectidans were relatively unspecialized with regard to feeding limbs. The palaeostomatopods were an early experiment on a rapacious carnivore pattern.

AESCHRONECTIDA

DEFINITION Hoplocaridans in which carapace covers entire thorax, with well-developed lateral wings; maxillae sometimes as long as thoracic limbs; thoracopods achelate, endopod as four-jointed finely setose stenopod, exopods flaplike; pleopods with branching gill tufts on protopod; telson short, subrectangular, and flat; uropods (at least the exopods) as long thin blades, extending well beyond the telson.

HISTORY The order was first recognized by Schram (1969). However, some of the species of *Crangopsis*, originally described by Salter (1861) and Peach (1882) and originally thought to be decapods, have proven to be aeschronectidans as well (Schram, 1979a). It was the identification of hoplocarid features (viz., the enlarged abdomen with pleopodal gills, triflagellate antennules, articulated rostrum over a probable cephalic kinesis, thoracic protopods of three segments and endopods of four segments) on the aeschronectidans that initially led Schram to view the hoplocarid morphotype as very ancient and completely distinct from that of eumalacostracans and to justify their treatment as separate subclasses of Malacostraca.

MORPHOLOGY The aeschronectidans were moderate to large in size (Fig. 4-1). They suffer, as many fossil forms do, from the vagaries of preservation; however, the genus *Kallidecthes* is the most completely known. The carapace ranged from subtriangular to subrectangular in lateral view with no traces of furrows or ridges. The compound eyes were stalked.

The antennules (Fig. 4-2A) had long peduncles of three segments (which can be extremely large, e.g., in the genus *Aratidecthes*). The three flagella were prominent and coequal in length.

The antennal peduncles (Fig. 4-2B) had at least two moderate distal

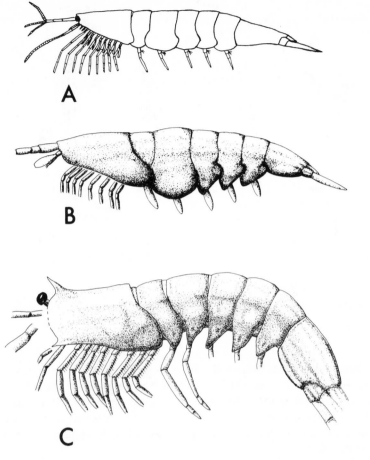

Fig. 4-1. Types of Aeschronectida. (A) *Kallidecthes richardsoni*; (B) *Crangopsis socialis* (from Schram, 1979a); (C) *Aenigmacaris cornigerum*. (From Schram and Horner, 1978)

segments. Presumably there were one or more proximal joints, since a small oval scaphocerite partially covers the two distal segments. *Kallidecthes eageri* has at least three peduncular segments but no preserved scaphocerite. The flagellum was long and thin.

The mandibles (Fig. 4-2C) generally were rather small and not well preserved. They could be quite prominent (*K. eageri*), with a slight incisive process.

The maxillules (Fig. 4-2D) were rather diminutive with a short three-segment palp.

The maxillae (Fig. 4-2E) were apparently very small in the aratidecthids, but the kallidecthids and aenigmacarids had long stenopodous four-segment endopods mounted on a short three-segment protopod. No exopod is noted on such maxillae.

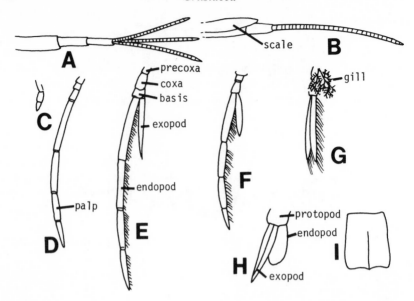

Fig. 4-2. Appendages of *K. richardsoni*. (A) antennules; (B) antennae; (C) maxillule; (D) maxilla; (E) first thoracopod; (F) eighth thoracopod; (G) first pleopod; (H) uropod; (I) telson.

The thoracopods (Fig. 4-2F) were all similar in form, though they may have been either subequal in size to each other, or decreased in length towards the posterior. The protopod was composed of three short segments and the endopod of four segments (frequently with short setae along their posterior margins). The exopod was either developed as a thin blade or as thin flaps.

The pleopods (Fig. 4-2G) had robust two-segment protopods with two simple flaplike rami. The proximal protopodal segment bore the tufts of short branching gills. However, the pleopods in the aenigmacarids were very distinctive (Fig. 4-1C); their protopods had a long, apparently uniramous stenopod having four segments (the proximal two were long, the distal two rather short).

The uropods (Fig. 4-2H) were bladelike. The exopods were long and bladelike, with the inner margin sometimes setose. The endopod was a short, oval flap. The telson was a short, broad, thin, blunt plate, about equal in length to the uropodal endopods and much shorter than the exopods.

Nothing is known concerning internal anatomy in this order.

NATURAL HISTORY The nature of the thoracopods can allow us to infer some aspects of aeschronectid biology. The simple achelate setose limbs might seem to indicate a filter feeding mode. The possible function of the stenopodous maxillae seen in some families pose something of a problem, though such long maxillary palps are not entirely unknown; for example, they occur in some decapod types. However, Kunze (1983), in her consideration of the origin of the stomatopod feeding morphology, felt that all hoplocaridans had to have come from a common carnivorous habit.

Only one species, *Crangopsis socialis*, is associated with a fauna, which seems to indicate it was probably an open water swimming form. The other species are found affiliated with near-shore shallow water communities that might imply at least occasional epibenthic residence on their part. The well-developed stenopodous pleopods on the aenigmacarids would seem to suggest that these forms were probably epibenthic. Modestly successful in the Late Paleozoic these hoplocaridans were most effective at whatever they did since they dominate the faunas in which they are found in terms of absolute numbers of individuals.

As in the case with so many Late Paleozoic malacostracan groups, aeschronectidans were endemic to the Carboniferous equatorial island continent Laurentia (Schram, 1977), being found today as fossils only in Europe and North America.

DEVELOPMENT   Nothing is as yet known about aeschronectidan development. Schram (1969) reported a size frequency distribution for a population of *K. richardsoni*, but this contained nothing that could be considered juvenile stages.

TAXONOMY   Three families are recognized at present. The Carboniferous Aenigmacarididae have the peculiar stenopodous pleopods and the well-developed maxillary palp. The Pennsylvanian Kallidecthidae have normal paddlelike pleopods but also carry the specialized maxillae. The Carboniferous Aratidecthidae possessed both unspecialized maxillae and pleopods but bore enlarged antennae. Not all species are known equally well. *Kallidecthes richardsoni* and *Aratidecthes johnsoni* are known in great morphologic detail, while on the other hand species of the aenigmacarid genus *Joanellia* are rather poorly preserved and thus incompletely understood at present.

Family Kallidecthidae Schram, 1969
Family Aenigmacarididae Schram and Horner, 1978
Family Aratidecthidae Schram, 1979

PALAEOSTOMATOPODA

DEFINITION   Hoplocaridans with the carapace covering the thorax, or reduced to expose posterior thoracic segments along the dorsal midline; thoracomeres not sharply tagmatized; thoracopods 2 to 5 subequal and subchelate, ischiomerus relatively shorter than the carpus, long propodus sometimes serrate; telson frequently with a terminal spike, usually (if not always) bearing terminal small caudal furcae.

HISTORY Though species assignable to this group have been known since the last quarter of the nineteenth century (Meek, 1872; Peach, 1882, 1908), it was not until Brooks (1962, 1969) that the distinctive and characteristic form of the thoracopods was recognized and thus allowed a separate order to be erected to accommodate them. Schram (1969) more clearly delineated palaeostomatopods as compared to the Stomatopoda, the living order that they in general resemble and assembled a review of the group (Schram, 1979b).

MORPHOLOGY Due to the vagaries of preservation, no species of palaeostomatopods are completely known (Fig. 4-3). In general, species in the genus *Perimecturus* afford us the most information.

The carapace covered the whole of the thorax, with a tendency to expose the posterior thoracomeres middorsally in *Archaeocaris* (Fig. 4-3F). The carapace was smooth, without grooves, though posterior longitudinal ridges occurred in the genus *Perimecturus* (Fig. 4-3B, C).

The antennules had a peduncle of at least two segments. The three flagella, at least in *P. parki*, were well developed, though their exact length is not known (Fig. 4-3B).

The antennae had at least a two-segment peduncle. The scaphocerite was variously developed, but was known to be really large only in *Bairdops beargulchensis* (Fig. 4-3E). Again, the flagellum was well developed but of unknown length.

Mandibles were known only from species of *Archaeocaris*, where it was a relatively massive appendage.

Nothing is known concerning the maxillules, maxillae, and first thoracopod.

It appears the second through fifth thoracopods were subchelate. However, assignment of position and even exact number of subchelipedes is not absolutely certain. All subchelipedes were apparently equal in size. The precoxa and coxa were short, and the basis is very long. The ischiomerus was relatively short and attached distally to a relatively longer carpus. However, exact ratios of these limb segments differ in the various genera. The propodus and dactylus were equally long, with the propodus frequently armed with a series of well-developed teeth while the dactylus was a thin smooth blade. The nature of the posterior thoracopods is completely unknown.

The pleopods were well preserved only in *Bairdops elegans*. The protopod was robust and the rami rather delicate. The pleopods are also known to carry tufty gills.

The uropods were rather bladelike in *Perimecturus* and *Bairdops* and

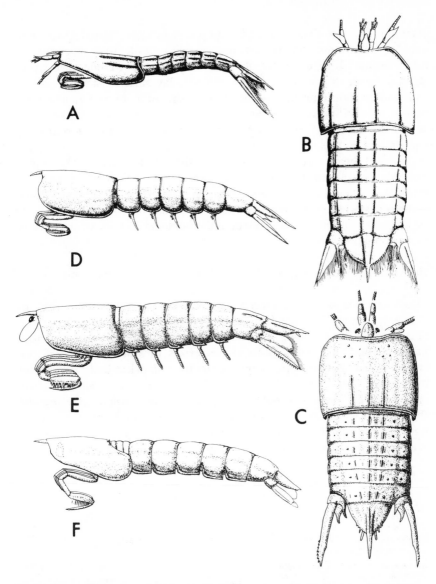

Fig. 4-3. Types of Palaeostomatopoda. (A, B) *Perimecturus parki*; (C) *P. rapax*; (D) *Bairdops elegans*; (E) *B. beargulchensis*; (F) *Archaeocaris vermiformis*. (A, B, D from Schram, 1979a; C, E from Schram and Horner, 1978; F from Schram, 1979b.)

more lobelike in *Archaeocaris*. The telsons in the former two genera were also clearly developed with a terminal spike. It appears that all forms have small furcal lobes distally.

NATURAL HISTORY   The palaeostomatopod subchelate thoracopods indicate with certainty that we are dealing with carnivorous creatures. In addition, the rather distinctive flattened body form of *Perimecturus* would seem to imply this particular genus probably lay buried just at the surface in muddy bottoms awaiting the hapless passing of its prey. The living stomatopods are known for their rather elaborate behavior patterns. It is tempting to speculate whether palaeostomatopods may have developed in the same way (Schram, 1979b). However, the relatively short temporal duration of the order (Late Devonian through late Mississippian) and their undoubted replacement by primitive stomatopod types (forms whose morphology is otherwise so very similar) might indicate the palaeostomatopods were not nearly so well adapted in terms of behavior. They thus appear to be an early but aborted experiment in the hoplocaridan line of active rapacious carnivores.

DEVELOPMENT   Nothing of course is known concerning development.

TAXONOMY   At present six species in three genera are recognized. They are currently placed in a single family, the Perimecturidae, though their differences are such that each genus could have its own family. The specialized flattened body of *Perimecturus* characterizes it, whereas *Bairdops* and *Archaeocaris* share a more primitive similar cylindrical body type. However, the elaborate tailfan unites *Perimecturus* and *Bairdops* from *Archaeocaris*. The dorsal reduction of the carapace isolates *Archaeocaris* from the other genera. Furthermore, the degrees of difference expressed between the palaeostomatopod species are of the order of magnitude that are typically used within the living stomatopods to justify genera.

REFERENCES

Brooks, H. K. 1962. Paleozoic Eumalacostraca of North America. *Bull. Am. Paleo.* **44**:163–338.
Brooks, H. K. 1969. Palaeostomatopoda. In *Treatise on Invertebrate Paleontology*, Part R, Arthropoda 4(2) (R. C. Moore, ed.), pp. R533–35. Geol. Soc. Am. and Univ. Kansas Press, Lawrence.
Kunze, J. C. 1983. Stomatopoda and the evolution of the Hoplocarida. *Crust. Issues* **1**:165–88.
Meek, F. B. 1872. Descriptions of new western Paleozoic fossils, mainly from the Cincinnati group of the Lower Silurian Series of Ohio. *Proc. Acad. Nat. Sci. Philadel.* **24**:335–36.
Peach, B. N. 1882. On some new Crustacea from the Lower Carboniferous rocks of Eskdale and Liddesdale. *Proc. Roy. Soc. Edinb.* **30**:73–91.

Peach, B. N. 1908. A monograph on the higher Crustacea of the Carboniferous rocks of Scotland. *Geol. Surv. Great Britain, Paleontol. Mem.* **1908**:1–82.

Salter, J. W. 1861. In T. Brown, Notes on the Mountain Limestone and Lower Carboniferous rocks of the Fifeshire coast from Burntisland to St. Andrews. *Trans. Roy. Soc. Edinb.* **22**:394.

Schram, F. R. 1969. Some Middle Pennsylvanian Hoplocarids and their phylogenetic significance. *Fieldiana: Geol.* **12**:235–89.

Schram, F. R. 1977. Paleozoogeography of Late Paleozoic and Triassic Malacostraca. *Syst. Zool.* **26**:367–79.

Schram, F. R. 1979a. British Carboniferous Malacostraca. *Fieldiana: Geol.* **40**:1–129.

Schram, F. R. 1979b. The genus *Archaeocaris*, and a general review of the Palaeostomatopoda. *Trans. San Diego Soc. Nat. Hist.* **19**:57–66.

Schram, F. R., and J. Horner. 1978. Crustacea of the Mississippian Bear Gulch limestone of central Montana. *J. Paleo.* **52**:394–406.

5

STOMATOPODA

DEFINITION   Carapace reduced, exposing part of the fifth and all of the sixth through eighth thoracomeres; thorax highly tagmatized, anterior five segments with subchelate appendages of varying sizes with the ischiomerus often equal to or longer than the carpus, posterior three segments with stiltlike stenopods; pleopods biramous and flaplike, with gills carried on the proximal part of the exopods; tailfan typically elaborate, specialized in connection with uropods to support abdomen off the substrate.

HISTORY   The stomatopods have long been recognized as a distinct group of crustaceans, since Latreille defined the group in 1817. Its exact affinities as well as constituents have not been so easily delineated, since they have been allied at various times with decapods, 'schizopods,' and leptostracans and enlarged to include various larval types as well as some of the more peculiar decapods. The larvae are so distinctive that they were originally described as separate genera before their true affinities were discerned (see, e.g., Gurney, 1946). Calman (1904) finally formalized the separate position of stomatopods when he established the superorder Hoplocarida. At that time they were generally conceded to be an early and distinct branch of the Eumalacostraca. With the recognition of the distinctive but obviously related fossil orders Palaeostomatopoda and Aeschronectida, Schram (1969a, b) postulated a distinct origin for the hoplocaridans separate from the eumalacostracans. A distinct Paleozoic suborder, Archaeostomatopodea, was recognized by Schram (1969b) intermediate in many respects between palaeostomatopods and the living suborder, Unipeltata. An extensive and ongoing taxonomic revision of the living stomatopods has been under way for the past several years focused around the extensive work of Manning (see, e.g., Manning, 1980).

MORPHOLOGY   In the living stomatopods (Fig. 5-1), the body is somewhat dorsoventrally flattened, though not always clearly so. The eyes are stalked and quite large and, together with the antennules, occupy a separate anterior kinetic region of the cephalon. The compound eyes are often clearly divided into upper and lower units by a middle band of ommatidia and seems to be related to perfection of binocular vision on each side of the animal (Manning, et al., 1984a, b; Schiff and Manning 1984; Abbott et al., 1984). There is a large epistome in front of the mouth, placing the mandibles and mouth opening quite far back from the anterior end of the ani-

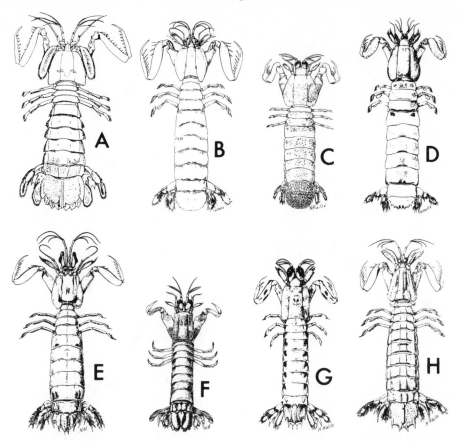

Fig. 5-1. Various family types of unipeltate stomatopods illustrating the four superfamilies within the group. (A) *Bathysquilla crassispinosa*; (B, C, D) Lysiosquilloidea, (B) *Lysiosquilla monodi*, (C) *Coronida bradyi*, (D) *Acanthosquilla septemspinosa*; (E, F, G) Gonodactyloidea, (E) *Eurysquilla galatheae*, (F) *Protosquilla folini*, (G) *Pseudosquilla ciliata*; (H) *Squilla aculeata*, a Squilloidea. (A from Holthuis and Manning, 1969; B–H, from Manning, 1977)

mal. The posture of the stomatopods is very distinctive, the pleopods and 'maxillipedes' are held off the bottom by raising the body up on 'stilts' formed by the posterior thoracopods and the downwardly directed telson and uropods.

The antennules (Fig. 5-2A) have an elongated but thin peduncle of three segments. This bears three short to moderate sized subequal flagella.

The antennae (Fig. 5-2A) have a well-developed two-segment protopod. The moderate to large setose oval scaphocerite has two segments: a small basal element and the distal scale proper. The flagellum is more weakly developed with two thin, long, proximal, peduncularlike segments and a distal relatively short annulated portion.

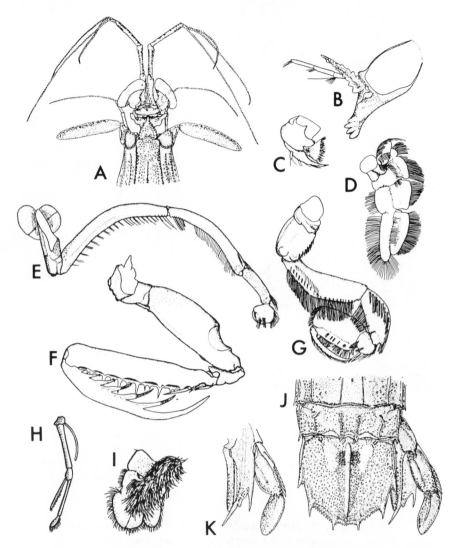

Fig. 5-2. Appendages of stomatopods. (A) anterior cephalon with antennules, antennae, stalked eyes, and articulated rostrum displayed; (B) mandible; (C) maxillule; (D) maxilla; (E) first maxillipede; (F) large second maxillipede; (G) form typical of the third through fifth maxillipedes; (H) last thoracopod of a male (note genital papilla); (I) pleopod; (J) dorsal view of telson and uropod; (K) ventral view of uropod. [(A, J, K) *Harpiosquilla harpax* (from Manning, 1969); (B, C, D, E, F, G) *H. stephensoni* (from Kunze, 1981); (H) *Squilla mantis* (modified from Holthuis and Manning, 1969); (I) *S. mantis* (From Calman, 1909).]

The mandibles (Fig. 5-2B) are very well developed. There is generally a three-segment palp. The incisor process has a single row of sharp teeth and is used merely to grip food. The molar process is large, armed with two rows of teeth, extends up into the proventricular stomach, and is used for grinding. The labrum is rather small and continuous with the epistome. There is a set of small fringed paragnaths near the incisor processes: these press tightly against the labrum when the mandibles are abducted laterally.

The maxillules (Fig. 5-2C) have two well-developed basal endites; the distal one is developed terminally as a large tooth with some adjacent setae. Attached to the distal endite is a small one-segment palp.

The distinctive maxillae (Fig. 5-2D) consist of four segments, which all seem to bear more or less well-developed endites. Both the medial and lateral margins are setose.

The first five thoracopods in stomatopods, termed maxillipedes, are subchelate and usually carry epipodites. The first maxillipedes (Fig. 5-2E) are most distinctive. They are long and lie along the side of the head and carapace. These rather flexible appendages are decorated with long setae along the ventral surface and shorter setae dorsally. The terminal subchelipede is very small. The first maxillipede generally is used in cleaning the antennae, eyes, and mouthparts, as well as grooming the entire body, including the pleopods and tailfan.

The second thoracopods (Fig. 5-2F) are large, massive, and very distinctive from all the posterior ones. The coxa, basis, and carpus are short, the ischiomerus huge. The propodus and dactylus form a deadly subchelipede often armed with spines and teeth. This distal subchelate part of the maxillipede is typically modified to facilitate smashing or spearing action (Fig. 5-5).

The third, fourth, and fifth thoracopods (Fig. 5-2G) are all similar. Subchelate in form, they are also rather setose. The basis is very long. The propodus is also short and rather wide. They bear respiratory epipodites.

The sixth, seventh, and eighth thoracopods (Fig. 5-2H) are slender, biramous, and without epipodites. The protopod has a short precoxa and basis and a longer coxa. The outer branch is a slender single segment; the inner branch is a two-segment stenopod. The terms exopod and endopod cannot apparently be used here without qualification, since Claus (1871) claimed that the outer branch is an endopod and the inner stenopod is an exopod, the positions rotating during the course of development. The last thoracopod in the male carries a long 'papilla' extending from the precoxa to act as a penis.

The pleopods (Fig. 5-2I) are rather broad, biramous flaps. The endopods bear an appendix interna on their medial margins that hook to their counterpart, allowing a pleopod pair to move back and forth as a unit. The branchial stems arise near the base of the exopods and bear the branching tufts of the gills.

The tailfan (Fig. 5-2J) is a broad complex structure. The uropod rami are rather robust; the protopod is variously developed as a posteriorly projecting spine or blade between the rami (Fig. 5-2K). The telson is broad and deep. It is characteristically (depending on family) decorated with spines, ridges, grooves, processes, and denticles, or sometimes is smooth and rather inflated.

One of the more distinctive aspects of the internal anatomy of the stomatopods is the digestive system (Fig. 5-3), analyzed in great detail by Kunze (1981, 1983). There is no esophagus; the mouth opens directly into the stomach or proventriculus, which occupies the whole of the cephalon. The cardiac stomach extends forward from the mouth and is all but devoid of ossicles or setae except for small pairs in the anterior and posterior areas. The cardiac stomach is connected to the posterior pyloric stomach through a small channel. This junction is blocked by a complex array of ossicles, termed the posterior cardiac plate, which act as a sieve through

Fig. 5-3. Diagram of stomatopod gut of *Harpiosquilla stephensoni*. (A) lateral view; (B) dorsal view. (From Kunze, 1981)

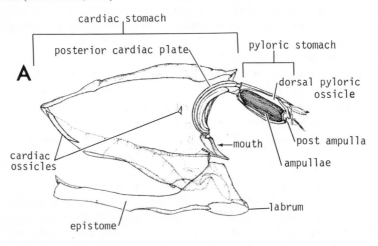

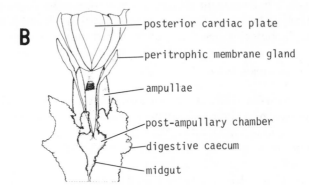

which all food passes. The pyloric stomach is relatively narrow and filled with another complex of ossicles. The pyloris opens posterolaterally into the digestive glands and posteriorly into the midgut. The digestive gland is massive, filling most of the body, ramifying into the telson, and having lateral branches into each of the four posterior thoracomeres and each pleomere. The midgut is a simple narrow tube extending from the pyloric stomach back to the hindgut at the level of the fifth pleomere.

The circulatory system of stomatopods is also rather distinctive (Fig. 5-4). The heart extends for the whole length of the body, with six pairs of ostia in the thorax and seven in the abdomen. The system is virtually closed (Burnett, 1972), with especially rich capillary networks in the brain and subneural region, but exhibits some peculiarities in the pattern of arteries in the abdomen. Komai and Tung (1931) observed that the first pleomere in *Oratosquilla oratoria* contained two sets of arteries, but the first pair ended blindly in a muscle mass in the anterior portion of the somite. The other segments of the abdomen were supplied by arteries that actually arose in the anterior part of the next most posterior segment before giving rise to branches that supply the muscles and the appendages. Siewing (1956) feels this indicates that the first two of an original seven pleomeres fused to form the stomatopod six-segment abdomen. Burnett (1972) found contradictory evidence for his pattern in *Hemisquilla ensigera*. The general arrangement of the heart, arteries, and ostia indicates a rather primitive condition (Burnett, 1984). This is made more evident by the presence of a total of eight thoracic ostia in the developing embryo, the second and third eventually disappearing.

The excretory system centers around a well-developed maxillary gland in the adult (Balss, 1926). The larvae may utilize antennal glands, but this is not as yet certain (Kaestner, 1970).

The ovaries lie above the digestive glands and also extend throughout most of the length of the body. The oviducts enter a single median pouch or seminal receptacle, which opens by a median pore on the sternite of the sixth thoracomere. The testes extend only from the fourth abdominal segment to the telson. Both gonads are fused posteriorly near the telson. The vasa deferentia extend forward to the eighth thoracomere to open on the precoxae of the eighth thoracopods by means of long processes called the genital papillae. Sperm cords, rather than spermatophores, are formed in the distal portions of the duct. In addition, in the posterior thoracomeres of the male are a pair of accessory glands, similar in appearance to testes and joined at their anterior extension (Deecaraman and Subramoniam, 1980). These glands also open on the tip of the genital papillae by ducts parallel to the vasa deferentia. A set of androgenic glands have been reported on coxal muscles in the eighth thoracopods (Charniaux–Cotton, 1960). Mature sperm are round and vesicular, and resemblances to decapod sperm are only superficial (Komai, 1920).

The nervous system has modified its form in relation to the unique

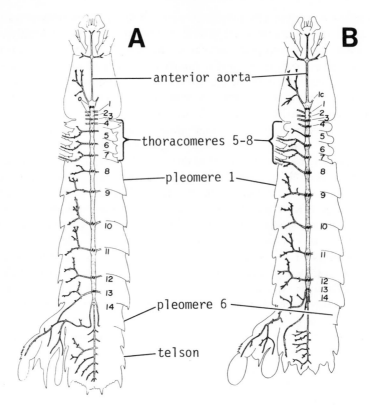

Fig. 5-4. Stomatopod circulatory system. Numbers label lateral arteries. (A) *Squilla oratoria*; (B) *Hemisquilla ensigera*. (From Burnett, 1972)

stomatopod body plan (Bullock and Horridge, 1965). The supraesophageal ganglion is located in the cephalon in association with the region anterior of the cephalic kinesis. As a result, the circumoral connectives are extremely long, in response to the great development of the epistome. The ganglia from the mandibles back to that of the fifth thoracomere are fused into a single, long, postoral ganglion, again in response to the compact arrangement of segments in the mouth and maxillipede region.

NATURAL HISTORY   The stomatopods are entirely marine and mark one of the very few radiations of obligate carnivores within the Crustacea. Most crustacean carnivores are actually low-level or facultative carnivores, subsisting mostly as scavengers. The mantis shrimp, however, eat only live prey. They lie buried and hidden in muds, they dig burrows, or they occupy dens in coral or rock rubble, striking at prey as it passes by. They kill typically by either smashing their victims (usually hard-bodied prey like mollusks and crabs), or spearing them (typically fish, polychates, and shrimplike crustaceans). To do this they use the large subchelipede of the second maxillipedes (Fig. 5-5). They also selectively pick up some prey, such as various small crustaceans, by scooping them up with their maxillipedes. Although specializations for particular types of food procurement differ in different families, all species are capable of using all three modes. The great variety of organisms typically found in stomatopod stomach contents testifies to their eclectic feeding habits (Camp, 1973; Kunze, 1981).

Once prey is secured, the mouthparts and maxillipedes are used to shred the victim. The mandibles in this process serve largely to bite or grasp the food, while other appendages stuff food into the mouth. Once inside the cardiac stomach, a combination of mandibular molar grinding, muscular action of the walls, and the injection of digestive enzymes through the pyloric stomach from the digestive glands combine to break up the food (Kunze, 1981). Material then passes through the various ossicular sieves in the cardiac and pyloric stomachs into the digestive glands. From the digestive gland remnant material is periodically injected into the midgut and surrounded by a peritrophic membrane. The undigestible shell and cuticular fragments in the cardiac stomach are regurgitated once the digestibles have been moved through the pyloric stomach (Fig. 5-3).

Stomatopods are capable of executing complex behavior sequences (Dingle, 1969a), study of which indicates there is both inter- and intraspecific communication. The amount of information transmitted is equivalent to that seen among hermit crabs; the amount of information transmitted per unit time is twice that of honey bees and three times that of fire ants.

This communicative ability finds use in resource partitioning between coexisting individuals and species (Caldwell and Dingle, 1975; Dingle and Caldwell, 1972, 1975, 1978; Reaka and Manning, 1981). Stomatopods that exhibit more specialized social, courtship, and agonistic behaviors and more extensive parental care of young occupy preformed dens in coral or

Fig. 5-5. Comparison of types of second maxillipedes. (A) *Mesacturoides crinitus*, a smasher, and (B) *Harpiosquilla harpax*, a spearer. (From Manning, 1978; Dingle and Caldwell, 1978)

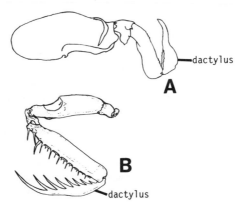

rock habitats. Species with less specialized and less aggressive behaviors excavate their own burrows in soft bottom habitats. Larger taxa typically display more complex agonistic behavior patterns than small taxa. However, despite the greater complexity of behavior, hard-bottom dwelling and large forms do not show greater taxonomic diversity between regions, greater species radiations or higher levels of extinction than forms with less complex behaviors or that are smaller in size (Reaka and Manning, 1981). The intensity and effectiveness of identical agonistic behaviors within populations of a species can depend upon other associated species (Caldwell and Dingle, 1977). For example, *Haptosquilla glyptocercus* in Enewetak is more successful in defending itself against and evicts from its own dens individuals of *Gonodactylus incipiens* and *G. falcatus*; but in Phuket, Thailand, *H. glyptocercus* is considerably less effective in defense when matched against *G. viridis* and *G. chiragra*.

Though the evolution of such complex behavior capabilities apparently has had little effect on rates of evolution within stomatopods (Reaka and Manning, 1981), it probably has had some some effect on the evolutionary success of stomatopods as compared to other hoplocaridan types (Schram, 1979) and in competition with eumalacostracan forms. The mantis shrimp cannot be characterized in any way as 'shy or retiring.' They do not possess the eumalacostracan caridoid escape behavior, that is, escape tailfirst. Rather, they have very distinct behaviors under stress (Reaka, personal communication). Squilloids will stidulate with the reflexed spiney uropods and telson. Gonodactyloids will coil up and block the entry of their dens with the armored telson. However, when the attacker presses his challenge and the stomatopod is forced to retreat, then it will double up or turn around and then retreat head first—quite unlike most eumalacostracans.

Needless to say, mating in such naturally aggressive creatures can have its little anxieties. Serène (1954), Dingle and Caldwell (1972), and Reaka and Manning (1981) have analyzed courtship. Males generally are always ready to mate except just after a previous liaison. The female is very particular in her receptivity, and it is not always evident to a male whether he will be received with indifference (no motion while the male grasps her abdomen with his maxillipedes, creeps up her back, all the while stroking her with quivering antennules and the first maxillipede) or receive a chop to the head. After complete mounting and more male stroking of the carapace, this time involving some of the maxillipedes, the male curls his abdomen around the female (Fig. 5-6), causing her to roll over. With ventral surfaces opposed, the male inserts his genital papillae into her seminal receptable. For 10 to 50 seconds the male thrusts his abdomen up and down injecting the sperm cords into the seminal receptacle while the relaxed female raises her head and tail up and down. The male then either moves slowly off while the female rights herself, or she finally manages to throw him off. The secretions of the male accessory gland appear to break

Fig. 5-6. Stomatopod copulation. (From Dingle and Caldwell, 1972)

down the cement of the sperm cords (Deecaraman and Subramoniam, 1980), and thus may act as an activation agent inside the female.

The male and female may share the same den, repeatedly copulating for as long as a week (Reaka and Manning, 1981), in a behavior probably designed to insure that no other male mates with the female. The female becomes gradually more aggressive until the male is evicted. The eggs are then laid and remain with the female in the chamber through hatching and beyond until the early larval stages are passed through. The female guards, cleans, and aerates the developing eggs and embryos. The females during this time would only rarely make short excursions outside the den, returning immediately to guard the eggs if danger threatened.

Mantis shrimp seem capable of some elementary learning responses in the laboratory (Reaka, 1980). However, it is not clear as yet whether stomatopods can be induced to undergo extensive behavior modification. Nor is it currently clear what role color response, if any, plays in mantis shrimp behavior, though many species are brightly marked and seem to use markings in inter- and intraspecific reactions. Hazlett (1979) has used colored models to elicit threat displays in *Gonodactylus oerstedii*, though he was uncertain whether it was true color or merely light reflectivity that triggered the reactions.

Reaka (1975, 1976, 1979) has carried out some interesting research on molting in stomatopods. Mantis shrimp, especially the coral living forms, can be very long-lived and exhibit very long periods between molts. Most interesting is the manner of molt; essentially a median suture occurs only along the middorsal line of the last three thoracomeres. Lateral sutures occur along the sides of the carapace, abdomen, and posterior portions of the eighth thoracomere that connect with the middorsal suture, the primary point of exit in the molt. The peculiar arrangement on the eighth thoracomere (with both medial and lateral sutures) indicated to Reaka (1975) that there may be some evidence to suggest that the first of the primitive seven abdominal segments in stomatopods fused with the eighth thoracomere to result in a six-segment abdomen.

Zoogeographic distribution in mantis shrimp is just now coming to be understood as the taxonomy of the group is reaching revision. Previously, when it was thought there was only a few genera and species, stomatopods were viewed for the most part to be rather ubiquitous. Now, though the issue is still clouded, several interesting points can be mentioned.

Stomatopods are essentially tropical and warm temperature water forms. Though some species can be found extending into cooler waters, they are the exception, and to date there are no known polar forms. The bathysquillids seem to be ancient relicts possibly distantly related in their morphology to the extinct Jurassic sculdids (Manning, 1980; Manning and Struhsaker, 1976). With a mere handful of species, their disjunct distribution would confirm this relict status (Fig. 5-7). A relict status could also be assigned to the single genus *Hemisquilla*. Other families undergoing rapid radiation, such as the squillids (Fig. 5-8) or gonodactylids (with many

Fig. 5-7. Biogeographic distibution of bathysquillids.

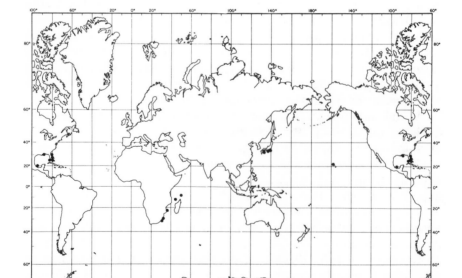

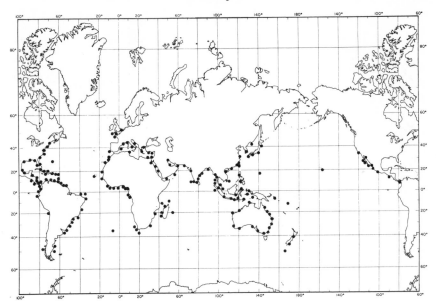

Fig. 5-8. Biogeographic distribution of squillids.

genera, each with large numbers of species) are found essentially every-where.

Other families, for example, *Harpiosquilla* in the squillitoids (Fig. 5-9) and the pseudosquillids (Fig. 5-10) and eurysquillids in the gonodactyl-loids, are essentially Indo-Pacific with tropical Atlantic elements (Reaka and Manning, 1981). These latter could be relatively recent migrants out of an Indo-West Pacific center of origin by way of the Banguilla current, or they could be remnants of an ancient Tethyan distribution. Of special interest in this regard is the truly circumtropical species *Heterosquillites mccullochae*. Manning (personal communication) is of the opinion prob-ably that this species may truly represent an ancient relict.

It would seem that the biogeography of stomatopods indicates a moder-ately old group, of Mesozoic origins, with a few relict and Tethyan pat-terns, but containing a large contingent undergoing active radiation within the confines of the tropic and temperate belts. This is a perfectly genera-lized statement, and more meaningful statements await a detailed review of stomatopod distribution patterns.

DEVELOPMENT   Komai (1924) and Shiino (1942) investigated the embryo-nic development of *Oratosquilla oratoria*, while Manning and Provenzano (1963) compared some features of the development of *Gonodactylus oer-stedii* with these.

The stomatopod egg is very rich in yolk. The nuclei make their way to

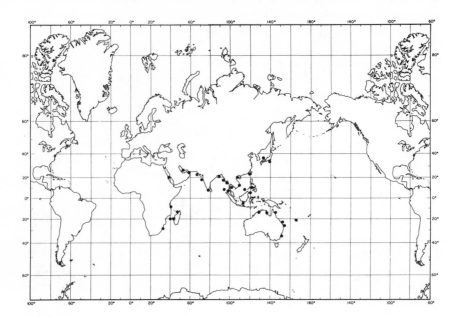

Fig. 5-9. Biogeographic distribution of harpiosquillids.

Fig. 5-10. Biogeographic distribution of pseudosquillids.

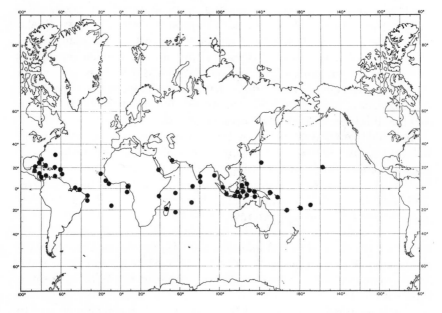

the surface of the egg and undergo only partial cleavage. Eventually an area on the surface emerges where the nuclei are slightly more crowded, the germinal disc (Fig. 5-11A). The earliest differentiation on the disc (Fig. 5-11C) results in the formation of the anlagen of the optic lobes and a posterior area known as the ventral plate. The plate develops (Fig. 5-11B) the blastopore, from whose anterior end the primary mesoderm delaminates. The primary mesoderm extends forward to join a secondary preantennular mesoderm (Fig. 5-11F) and thus forms a mesodermal ring around the edge of the germinal disc. The preantennulary mesoderm later disintegrates without forming or contributing to any structures and represents in some minds the vestiges of the preantennulary or acronal segment. Later, the posterior region of the blastopore gives rise to the endoderm, which in turn comes to form a compact mass of yolk-laden cells (Fig. 5-12B).

Eventually an egg-nauplius stage is formed (Figs. 5-11E, G). The three primary appendages arise while the ventral plate develops into a protruding thoracoabdominal rudiment (Fig. 5-11H). The naupliar appendage anlagen thicken and rise up above the germinal disc (Figs. 5-11H, 12A). A teloblast ring develops around the blastopore prior to the initiation of the egg-metanauplius stage (Fig. 5-12E), which then begins to bud off the maxillulary and maxillary segments in sequence. These cephalic segments, though arising on the rudiment, associate immediately with the germinal disc and act to extend the disc posteriorly. Subsequent appendage and segment anlagen remain with the thoracoabdominal rudiment (Fig. 5-12F).

Several features of stomatopod development are shared with decapods. These include the mode of early cleavage, mesodermal yolk cells, method of blood vessel formation, and maintenance of the dorsal curvature of the embryo. All seem related to the shared presence of large amounts of yolk in the egg. Several other features of mantis shrimp development are shared with leptostracans. These are the constitution of the teloblast ring, endodermal formation of a large midgut and digestive gland lobes, segmental heart and arteries, separation of the first thoracomere from the cephalon, and the possible partial indications of a seventh somite at the end of the abdomen. However, these are all plesiomorphic characters and, though interesting in themselves, are no basis to establish taxonomic links between the groups. Several points of mantis shrimp development are unique to the group, namely, the complete lack of coelomic development, the presence of only a rudimentary antennal gland without any lumen, and the development of peculiar anal glands.

One embryonic feature deserves further mention. Shiino (1942) observed in the last pleomere an apparent double ganglionic mass, one anterior, the other posterior (Fig. 5-12G). He felt this matched the condition seen in other groups, indicating a seventh ganglionic mass in the abdomen. However, the interganglionic cell groups are only rudimentary (Fig. 5-12H), and no epithelial furrows develop. The masses do remain until hatching, when they shrink and merge.

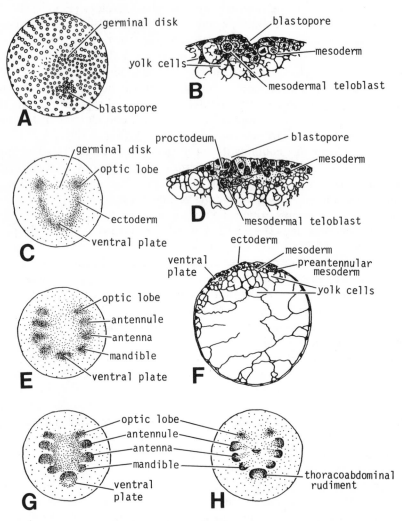

Fig. 5-11. Early embryology of stomatopods. (A) late blastula showing the initiation of the germinal disc and blastopore; (B) longitudinal section through stage in A; (C) early differentiation of germinal disc; (D) longitudinal section through C; (E) beginning of egg-nauplius stage; (F) diagrammatic longitudinal section through D; (G) late egg-nauplius stage; (H) development of raised limb buds and the thoracoabdominal rudiment. (Modified from Shiino, 1942)

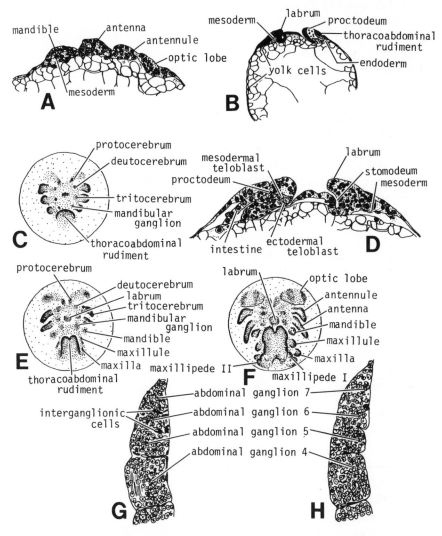

Fig. 5-12. (A) lateral longitudinal section through H showing raised limb buds; (B) diagrammatic medial longitudinal section through H showing rudiment and endodermal development; (C) just prior to beginning of egg-metanauplius stage; (D) median longitudinal section through K; (E) early egg-metanauplius; (F) later egg-metanauplius; (G) medial sagittal section of posterior nerve cord showing partial separation of ganglia by rudimentary interganglionic cells; (H) lateral sagittal section displaying the incomplete nature of supposed separation of sixth and seventh ganglionic anlagen. (Modified from Shiino, 1942)

Precious little has been done to outline complete sequences of larval development in stomatopods since Giesbrecht (1910), and only a few species have been completely worked out: for example, *Gonodactylus oerstedii* (Manning and Provenzano, 1963; Provenzano and Manning, 1978), *Squilla armata* (Pyne, 1972), and *Squilla empusa* (Morgan and Provenzano, 1979). The first three larval stages pass within the female den with thigmokinetic and negative phototaxic behavioral adaptations to facilitate larval protection (Dingle, 1969b). These three stages are characterized by a lack of developed mandibles, incipient maxillary and maxillipede development, and a great deal of yolk. At the fourth stage the yolk is exhausted, appendage development completed, and behavioral modifications induced. At this time the larvae leave the den and become pelagic. A total of four pelagic larval stages occur in *G. oerstedii* (Fig. 5-13) and nine such stages in *S. empusa*. The changes through the entire larval sequence generally are gradual. However, a recognizable metamorphosis takes place at the end of the larval series into a postlarva stage. Suddenly the carapace is reduced, carapace spines lost, rostrum reduced and articulated, and the

Fig. 5-13. Larval development in *Gonodactylus oerstedii*. (A, B, C, D, E, F, G) first through seventh stages; (H) postlarva. (From Manning and Provenzano, 1963; Provenzano and Manning, 1978)

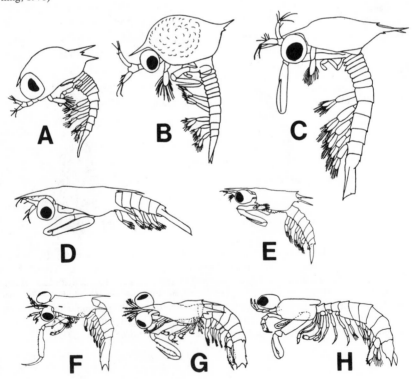

tailfan developed with the inclusion of uropodal and telsonic spines. At this metamorphosis the postlarva settles to the bottom and takes up adult life.

The traditional view of stomatopod larvology is that there are two types (Gurney, 1946). The antizoea, attributed to *Lysiosqulla* and *Coronida*, has uniramous antennules, five pairs of biramous thoracopods, and no pleopods. Other genera hatch as pseudozoeas with biramous antennules, two uniramous thoracopods, and five pairs of pleopods. All genera were then said to develop into an ericthus larva, with the exception of *Squilla* that develops into an alima. The difference between an erichthus and alima larva lies in the number of intermediate denticles on the telson. Although the antizoea types lie within the superfamily Lysiosquilloidea, the massive taxonomic revision presently in progress within the stomatopods precludes any reasonable larval classification at this time since only 10% of all known stomatopod species have identifiable larvae (Provenzano and Manning, 1978).

FOSSIL RECORD   Knowledge of the ancient forms of this group is still growing. There are essentially two types of fossils: those Mesozoic and Cenozoic types assignable to the living suborder Unipeltata and the Paleozoic Archaeostomatopodea. Of the Unipeltata, only the Jurassic Sculdidae cannot be assigned to any living families. All other fossil forms in this category are essentially assignable to recent families (though the materials frequently leave much to be desired). In the Archaestomatopodea, only one family is recognized to date, the Tyrannophontidae (Fig. 5-14); its three species in two genera form anatomical intermediates between the living Stomatopoda and the extinct order Palaeostomatopoda.

TAXONOMY   The classification of this group has been undergoing constant revision since the early 1960s. What was once thought to be a minor group with few genera and species is now a respectable radiation (Manning, 1980) with some 67 genera in 13 families and four superfamilies (Fig. 5-1). Revisionary and review work continues, and no general consensus seems near concerning the phylogeny with the group until the taxonomy is finally stabilized.

The family arrangement of the order is summarized here:

Suborder Archaeostomatopodea Schram, 1969 Middle Mississippian–
    Upper Pennsylvanian
    Family Tyrannophontidae Schram, 1969
Suborder Unipeltata Latreille, 1925 Upper Jurassic–Recent
    Superfamily Bathysquilloidea Manning, 1967 Upper Jurassic–Recent
    Family Bathysquillidae Manning, 1967 Recent
    Family Sculdidae Dames, 1886 Upper Jurassic
    Superfamily Lysiosquilloidea Giesbrecht, 1910 Recent
    Family Lysiosquillidae Giesbrecht, 1910

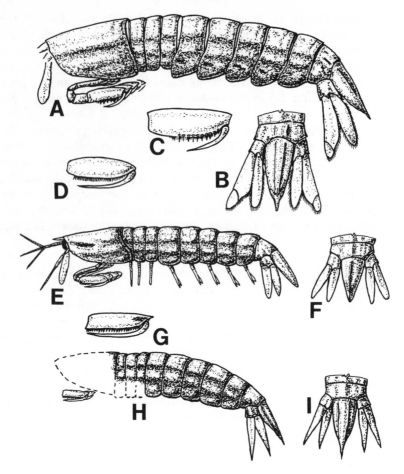

Fig. 5-14. Carboniferous tyrannophontid archaeostomatopodeans with lateral views of bodies and propodus/dactylus of subchelae, and dorsal views of tailfans. (A, B, C) *Gorgonophontes peleron*; (D, E, F) *Tyrannophontes theridion*; (G, H, I) *T. fraiponti*. (From Schram, 1984)

Family Coronididae Manning, 1980
Family Erythrosquillidae Manning and Bruce, 1984
Family Nannosquillidae Manning, 1980
Superfamily Gonodactyloidea Giesbrecht, 1910 (?Cretaceous)
    Upper Miocene–Recent
    Family Gonodactylidae Giesbrecht, 1910 (?Cretaceous) Upper
        Miocene–Recent
    Family Eurysquillidae Manning, 1977 Recent
    Family Hemisquillidae Manning, 1980 Recent
    Family Odontodactylidae Manning, 1980 Recent
    Family Protosquillidae Manning, 1980 Recent

Family Pseudosquillidae Manning, 1977 Recent
Superfamily Squilloidea Latreille, 1803 ?Cretaceous–Recent
Family Squillidae Latreille, 1803 ?Cretaceous-Recent
Family Harpiosquillidae Manning, 1980 Recent

REFERENCES

Abbott, B. C., R. B. Manning, and H. Schiff. 1984. An attempt to correlate pseudopupil sizes in stomatopod crustaceans with ambient light conditions and behavior patterns. *Comp. Biochem. Physiol.* **78A**:419–26.

Balss, H. 1926. Stomatopoda. In *Handbuch der Zoologie*, Vol. 3(1) (W. Kükenthal and T. Krumbach, eds.), pp. 1039–74. de Gruyter, Berlin.

Bullock, T. H., and G. A. Horridge. 1965. *Structure and Function in the Nervous System of Invertebrates*, Vols. 1 and 2. Freeman, San Francisco.

Burnett, B. R. 1972. Notes on the lateral arteries of two stomatopods. *Crustaceana* **23**:303–305.

Burnett, B. R. 1984. Striated muscle in the wall of the dorsal abdominal aorta of the California spiney lobster *Panulirus interruptus. J. Crust. Biol.* **4**:560–66.

Caldwell, R. L., and H. Dingle. 1975. Ecology and evolution of agonistic behavior in stomatopods. *Naturwiss.* **62**:214–22.

Caldwell, R. L., and H. Dingle. 1977. Variation in agonistic behavior between populations of the stomatopod, *Haptosquilla glyptocercus. Evolution* **31**:220–23.

Calman, W. D. 1904. On the classification of the Malacostraca. *Ann. Mag. Nat. Hist.* **13**(7):144–58.

Calman, W. T. 1909. *Crustacea*. In *A Treatise on Zoology*, Vol. 7 (E. R. Lankester, ed.). Adam and Charles Black, London.

Camp, D. K. 1973. Stomatopod Crustacea. *Mem. Hourglass Cruises* **111**:1–92.

Charniaux–Cotton, H. 1960. La glande androgène du crustacè stomatopode *Squilla mantis. Bull. Soc. Zool. Fr.* **85**:110–14.

Claus, C. 1871. Die Metamorphose der Squilliden. *Abh. K. Gesell. Wiss. Göth.* **16**:111–63, 35 pls.

Deecaraman, M., and T. Subramoniam. 1980. Male reproductive tract and accessory glands of a stomatopod *Squilla holoschista. Int. J. Invert. Reprod.* **2**:175–88.

Dingle, H. 1969a. A statistical and information analysis of aggressive communication in the mantis shrimp *Gonodactylus bredini. Anim. Behav.* **17**:561–75.

Dingle, H. 1969b. Ontogenetic changes in phototaxis and thigmokinesis in stomatopod larvae. *Crustaceana* **16**:108–10.

Dingle, H., and R. L. Caldwell. 1972. Reproduction and maternal behavior of the mantis shrimp *Gonodactylus bredini. Biol. Bull.* **142**:417–26.

Dingle, H., and R. L. Caldwell. 1975. Distribution, abundance, and interspecific agonistic behavior of two mudflat stomatopods. *Oecologia* **20**:167–78.

Dingle, H., and R. L. Caldwell. 1978. Ecology and morphology of feeding and agonistic behavior in mudflat stomatopods. *Biol. Bull.* **155**:134–49.

Dingle, H., R. C. Highsmith, K. E. Evans, and R. L. Caldwell. 1973. Interspecific aggressive behavior in tropical reef stomatopods and its possible ecological significance. *Oecologia* **13**:55–64.

Giesbrecht, W. 1910. Stomatopoda. *Fauna Flora Neapel Monogr.* **33**:1–239.

Gurney, R. 1946. Notes on stomatopod larvae. *Proc. Zool. Soc. Lond.* **116**:133–75.

Hazlett, B. A. 1979. The meral spot of *Gonodactylus oerstedii* as a visual stimulus. *Crustaceana* **36**:196–98.

Holthuis, L. B., and R. B. Manning. 1969. Stomatopoda. In *Treatise on Invertebrate Paleontology*, Part R, *Arthropoda* 4(2) (R. C. Moore, ed.), pp. R535–52. Geol. Soc. Am. and Univ. Kansas Press, Lawrence.

Kaestner, A. 1970. *Invertebrate Zoology*, Vol. 3. Interscience, New York.

Komai, T. 1920. Spermatogenesis of *Squilla oratoria*. *J. Morph*. **34**:307–33.

Komai, T. 1924. Development of *Squilla oratoria* I. Change in external form. *Mem. Coll. Sci. Kyoto Imp. Univ*. **1**(B):273–83.

Komai, T., and Y. M. Tung. 1931. On some points of the internal structure of *Squilla oratoria*. *Mem. Coll. Sci. Kyoto Imp. Univ*. **6**(B):1–15.

Kunze, J. C. 1981. The functional morphology of stomatopod Crustacea. *Phil. Trans. Roy. Soc. Lond*. (B)**292**:255–328.

Kunze, J. C. 1983. Stomatopoda and the evolution of the Hoplocarida. *Crust. Issues* **1**:165–88.

Manning, R. B. 1969. A review of the genus *Harpiosquilla*, with description of three new species. *Smith. Cont. Zool*. **36**:1–41.

Manning, R. B. 1977. A monograph of the West African stomatopod Crustacea. *Atlantide Repts*. **12**:25–181.

Manning, R. B. 1978. A new genus of stomatopod crustacean from the Indo-West Pacific region. *Proc. Biol. Soc. Wash*. **91**:1–4.

Manning, R. B. 1980. The superfamilies, families, and genera of recent stomatopod Crustacea, with diagnoses of six new families. *Proc. Biol. Soc. Wash*. **93**:362–72.

Manning, R. B., and A. J. Provenzano. 1963. Studies on development of stomatopod Crustacea I. Early larval stages of *Gonodactylus oerstedii*. *Bull. Mar. Sci. Gulf and Carib*. **13**:467–87.

Manning, R. B., H. Schiff, and B. C. Abbott. 1984a. Cornea shape and surface structure in some stomatopod Crustacea. *J. Crust. Biol*. **4**:502–13.

Manning, R. B., H. Schiff, and B. C. Abbott. 1984b. Eye structure and the classification of stomatopod Crustacea. *Zool. Scripta* **13**:41–44.

Manning, R. B., and P. Struhsaker. 1976. Occurrence of the Caribbean stomatopod *Bathysquilla microps* off Hawaii, with additional records for *B. microps* and *B. crassispinosa*. *Proc. Biol. Soc. Wash*. **89**:439–50.

Morgan, S. G., and A. J. Provenzano. 1979. Development of pelagic larvae and postlarva of *Squilla empusa*, with an assessment of larval characters within the Squillidae. *Fish. Bull*. **77**:61–90.

Provenzano, A. J., and R. B. Manning. 1978. Studies on development of stomatopod Crustacea II. The later larval stages of *Gonodactylus oerstedii* reared in the laboratory. *Bull. Mar. Sci*. **28**:297–315.

Pyne, R. P. 1972. Larval development and behavior of the mantis shrimp, *Squilla armata*. *J. Roy. Soc. N. Zeal*. **2**:121–46.

Reaka, M. L. 1975. Molting in stomatopod crustaceans I. Stages of the molt cycle, setagenesis, and morphology. *J. Morph*. **146**:55–80.

Reaka, M. L. 1976. Lunar and tidal periodicity of molting and reproduction in stomatopod Crustacea: a selfish herd hypothesis. *Biol. Bull*. **150**:468–90.

Reaka, M. L. 1979. Patterns of molting frequencies in coral dwelling stomatopod Crustacea. *Biol. Bull*. **156**:328–42.

Reaka, M. L. 1980. On learning and living in holes by mantis shrimp. *Anim. Behav*. **28**:111–15.

Reaka, M. L., and R. B. Manning. 1981. The behavior of stomatopod Crustacea, and its relationships to rates of evolution. *J. Crust. Biol*. **1**:309–27.

Schiff, H., and R. B. Manning. 1984. Description of a unique crustacean eye. *J. Crust. Biol*. **4**:604–14.

Schram, F. R. 1969a. Polyphyly in the Eumalacostraca? *Crustaceana* **16**:243–50.

Schram, F. R. 1969b. Some middle Pennsylvanian Hoplocarida and their phylogenetic significance. *Fieldiana: Geol*. **12**:235–89.

Schram, F. R. 1979. The genus *Archaeocaris* and a general review of the Palaeostomatopoda. *Trans. San Diego Soc. Nat. Hist.* **19**:57–66.

Schram, F. R. 1984. Upper Pennsylvanian arthropods from black shales of Iowa and Nebraska. *J. Paleo.* **58**:197–209.

Serène, R. 1954. Observations biologiques sur les stomatopodes. *Mem. Inst. Oceangr. Nhatrang* **8**:1-93.

Shiino, S. M. 1942. Studies on the embryology of *Squilla oratoria*. *Mem. Coll. Sci. Kyoto Imp. Univ.* **17**(B):11–174.

Siewing, R. 1956. Untersuchungen zur Morphologie der Malacostraca. *Zool. Jahrb. Anat.* **75**:39–176.

6

ANASPIDACEA

DEFINITION   Carapace absent; eyes stalked, sessile, or absent; first thoracomere fused with cephalon, second through eighth free; mandible with accessory incisor process; first thoracopod generally as a maxillipede with spinose gnathobase; pereiopod endopods uniform, exopods typically well developed though occasionally absent posteriorly, branchial epipodites on first five or six pereiopods; pleon with six free segments; pleopods variously well developed or absent, endopods rudimentary or absent, except in males where they contribute to a petasma on the first two pairs.

HISTORY   Living anaspidaceans were not described until 1893, from Mt. Wellington outside Hobart, Tasmania, though the distinctive nature of the fossil relatives had been known by that time for almost a decade. The living syncarids were assigned to the order Anaspidacea, and subsequent species described to the present seemed to indicate a restriction of the order to Tasmania and southernmost Victoria (Sayce, 1908; Smith, 1908; Knott and Lake, 1980; Nicholls, 1931; Swain et al., 1970, 1971; Williams, 1965b).

Noodt (1963) described some groundwater syncarids from South America, and subsequently (Noodt, 1965) designated them a separate order of their own, the Stygocaridacea. This arrangement prevailed until the discovery of stygocarids in Australia and New Zealand (Schminke, 1980), the description of the psammaspid anaspidaceans similar to stygocarids (Schminke, 1974; Knott and Lake, 1980), and reassessments of syncarid morphology (Schminke, 1978; Knott and Lake, 1980). These resulted in the assignment of stygocarids to family status within the Anaspidacea.

MORPHOLOGY   The Anaspididae contain the most primitive genera within the order (*Anaspides, Paranaspides, Allanaspides*). These taxa exhibit the basic structural plan upon which the other forms are based (Fig. 6-1).

The body is long and cylindrical. The cephalon is marked with a cervical groove. In *Anaspides* the body segments have parallel margins, but in *Paranaspides* the first pleomere is wedge-shaped, forcing the body into a permanent dorsoventral flexure. *Allanaspides* resembles *Anaspides* except that it possesses a peculiar fenestra dorsalis ('a transparent oval') on the anterior dorsal part of the body. The eyes are stalked.

The antennules (Fig. 6-2A) are biramous, with long flagella of unequal length. The peduncle has three segments, the most proximal of which contains a statocyst.

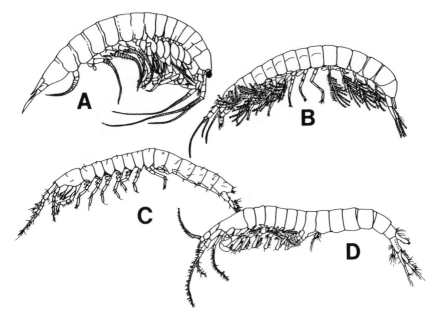

Fig. 6-1. Anaspidacean body forms. (A) *Anaspides tasmaniae* (from Schminke, 1978); (B) *Koonunga cursor* (from Sayce, 1908); (C) *Stygocarella pleotelson* (from Schminke, 1980); (D) *Psammaspides williamsi* (from Schminke, 1974).

The antennae have (Fig. 6-2B) a four-segment peduncle. The modest oval scaphocerite arises from the second segment, and the flagellum is long.

The mandibles (Fig. 6-2C) have a prominent three-segment palp, a serrated incisor process, and a molar process flanked with a setal lobe. The paragnaths are a pair of setose lobes.

The maxillules (Fig. 6-2D) have a set of two endites and a small exite. These are marginally setose or spinose. The palp is quite tiny.

The maxillae (Fig. 6-2E) have a large protopod with three densely setose endite lobes and a short but stout one-segment palp.

The thoracopods are all subequal in development, with a tendency for the first one to carry enlarged gnathobases and reduce the exopod (Fig. 6-2F). The endopod is robust, the exopod annulate and prominently setose, the epipods typically two in number. The knee occurs between the carpus and propodus (Fig. 6-2G). The eighth thoracopods lack epipods and exopods and are not directed parallel to the other thoracopods but rather have a more posterior orientation. *Allanaspides* lacks exopods on the seventh thoracopods as well.

The pleopods are composed of the annulate exopods (Fig. 6-2H), the endopods being typically reduced. In males the anterior endopods form a copulatory organ. On the first pair (Fig. 6-2I) they are thick, inwardly

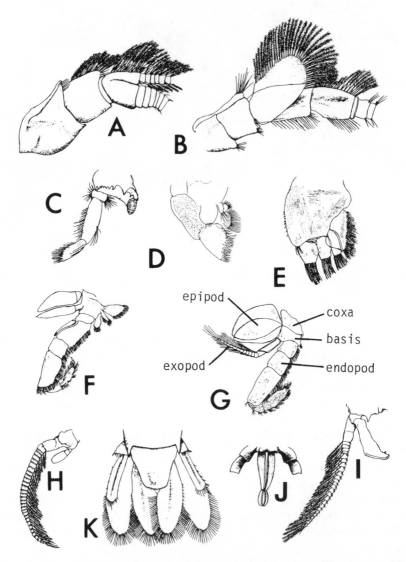

Fig. 6-2. Appendages of *Anaspides tasmaniae*. (A) antennule; (B) antenna; (C) mandible; (D) maxillule; (E) maxilla; (F) first thoracopod; (G) second thoracopod; (H) first pleopod of female; (I) first pleopods of male; (J) second pleopods of male; (K) tailfan. (From Thomson, 1894)

curved lobes linked together by retinacula. On the second pair (Fig. 6-2J) the endopods are two-segment spoon-shaped structures. Both pairs of endopods are directed forward. The second lie in a trough formed by the first and are positioned between the endopods of the first pleopod and the thoracic sternites.

The uropods are biramous (Fig. 6-2K). The rami are oval flaps. The exopod bears an incomplete diaeresis.

There are variations in appendages from the scheme outlined above among the anaspidacean families. Some of the more obvious are presented here. Only the anaspids possess a scaphocerite; the other families generally lack it (Fig. 6-3A). The mandibular palp is present in all families except the stygocarids (Fig. 6-3B). The palp on the maxillule is present only in the anaspidines and is lacking in the stygocaridines (Fig. 6-3C). The first and last two pairs of thoracopods can exhibit considerable variation. The thoracopods may lack exopods and/or epipodites, and the first may be

Fig. 6-3. Appendage variations in anaspidaceans. (A, B, C, D) *Stygocarella* pleotelson (from Schminke, 1980). (A) antenna; (B) mandible; (C) maxillule; (D) fourth thoracopod. (E) *Parastygocaris goerssi*, first thoracopod or maxillipede (from Noodt, 1970). (F, G) *Eucrenonaspides oinotheke* (from Knott and Lake, 1980). (F) first pleopod of female; (G) second pleopod of female. (H) *Psammaspides williamsi*, uropod (from Schminke, 1974).

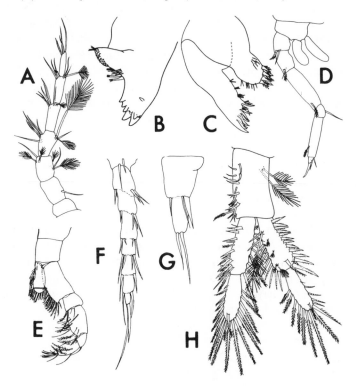

developed as a robust maxillipede (Fig. 6-3E). The exopods on the seventh thoracopod may be reduced (psammaspids) or absent (koonungids and stygocarids); stygocarids (Fig. 6-3D) sometimes lack exopods on all thoracopods. The pleopods in the nonanaspid families exhibit various degrees of reduction, being uniramous in koonungids and on the first pair in psammaspids (Fig. 6-3F), reduced in psammaspids (Fig. 6-3G), or absent in stygocaridines. The uropods, when not part of a tailfan, are quite distinctively spinose (Fig. 6-3H). The anus is ventral and proximal in all families except stygocarids where it is terminal.

Our knowledge of syncarid internal anatomy is based largely on *Anaspides* (Fig. 6-4). The digestive system has an esophagus entering a cardiac stomach, which has long ridges of chitin bearing setae. Siewing (1956, 1963) homologized the pars ampullaris between the caeca and pyloric stomach with that of stomatopods; however, these latter are also structurally and functionally similar to what is found in decapods, and the similarity is probably only due to function. The caeca arise as a mass of long strands at the anterior end of the midgut and extend the length of the thorax. There are also single short median dorsal caeca in the first and fifth pleomeres. The proctodeum extends to the level of the sixth pleomere.

Excretion is achieved with maxillary glands. These are large convoluted tubular structures in the posterior region of the head.

The circulatory system is very elaborate. The heart extends from the first thoracomere to the fourth pleomere. A single pair of ostia are found in the third thoracomere, and anterior and posterior dorsal aortae extend out from the heart. There are seven pairs of lateral arteries that have visceral and podial branches. The first arises from the anterior end of the heart and extends into the head. The second pair is located in the last thoracomere, with one of the branches forming a ventral aorta to supply the thoracopods. The third pair arises in the first pleomere to supply the thoracic viscera. The other four pairs supply the viscera and in the abdomen the pleopods. Even the ultrastructure of the heart is quite elaborate (Tjønneland et al., 1984), which suggests that though syncarids as a group are sometimes considered primitive, their circulatory system is quite advanced.

The nervous system centers on a large supraesophageal ganglion, but oddly there is no subesophageal ganglion since the postoral ganglia are all separate and distinct. The nerve cord is developed as a band of fibers connecting the segmental ganglia. Large lateral giant fibers in the nerve cord have been demonstrated as effective in the control of the caridoid escape reaction in *Anaspides* (Silvey and Wilson, 1979). The antennules carry statocysts. The eyes may be stalked (anaspids), sessile (*Koonunga*), or completely absent (all other forms).

The sexes are separate. The testes extend from the fifth thoracomere back through the abdomen. The vas deferens arise anteriorly and extend to the gonopores on the eighth sternite; however, in *Allanaspides* they fuse and open by a single median gonopore. The ovaries are similar in position

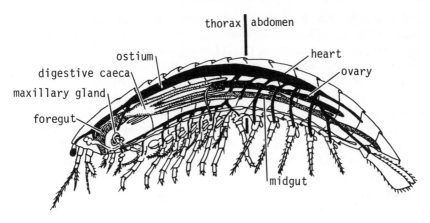

Fig. 6-4. Internal anatomy of an anaspidacean. (Modified from Siewing, 1956)

to the testes, except they may extend forward to the level of the second thoracomere. The female gonopores open near the base of the sixth thoracopods. The eighth sternite in females is modified anteriorly into a spermathecum.

NATURAL HISTORY   Manton (1930) made some general observations on feeding and locomotory behavior in *Anaspides* and *Paranaspides*. Despite the phylogenetic import of syncarids, little has been done in regard to field observations besides Manton's report. In 1980, (Schram, 1984) collected and observed a number of Tasmanian syncarids of the above genera as well as *Allanaspides* and *Micraspides* in the field and laboratory. Manton's observations were generally confirmed and extended.

Anaspides is in constant motion under laboratory conditions, but is typically more quiescent in the field. The thoracic exopods constantly vibrate back and forth in order to ventilate the epipods. The animals maintain a constant random patrol of their home pools; the flagella of the antennules and antennae orient in different directions and constantly sweep about. The first thoracopod (= maxillipede) seems to repeatedly test the substrate in a searching process. Manton recorded that *Anaspides* is rather inefficient in finding food and in holding onto it once secured. It appears omnivorous, seeming to prefer to scavenge, but it can be observed to pick up sand grains and manipulate them with its mouthparts. The pleopods are directed ventrolaterally and are used to push the body along in metachronal coordination with the thoracopods. When startled *Anaspides* is rather clumsy in its escape attempts; it does not swim as well as other anaspidids. It executes a single caridoid flip and throws itself up into the water column, whereupon it then floats effortlessly until it drifts back to the bottom. This reaction affords the best way to catch *Anaspides*, but also makes it vulnerable to many kinds of predator, particularly fish. It does move quite well on land if accidentally dropped there and can be rarely found naturally exploring around outside its pools (Swain and Reid, 1983). *Anaspides* is among the most widely dispersed of all the syncarids, being found in streams, lakes, and caves over much of central and western Tasmania with two currently recognized species (Williams, 1965a, b; Lake and Coleman, 1977). It would seem to be undergoing active diversification.

Paranaspides lives in algal and macrophyte mats on lake bottoms (Fulton, 1982). It seems to be more effective in its ability to move than *Anaspides*. It can execute the single escape flip, dart in different directions on the bottom, lay still for long periods, or be an excellent swimmer. When disturbed, *Paranaspides* executes the caridoid escape reaction but afterward does not drift helplessly; rather, individuals orient themselves in the water and swim to the bottom. The pleopods, either in swimming or walking along the bottom, are well coordinated with the thoracopods. *Paranaspides* does not do well out of water, unlike *Anaspides*; instead the body will totally collapse. While at rest its tailfans are held up at an angle off the bottom. It also constantly beats its exopods back and forth to ventilate the epipods.

Allanaspides lives in crayfish burrows and small ponds in grass swamps on the sides of gentle slopes. It is found only in certain remote areas of southwestern Tasmania. Adults beat their exopods more or less constantly (juveniles less so); however, they seem to move the exopods in a somewhat swirling or rotary manner and so create a vortex. This beating sets up a current that moves water under the head and back toward the tail. In this regard the front thoracopods are typically oriented anteriorly, under the head, and might conceivably assist in feeding. The pleopods are held stiffly and directed laterally to form a subcircular field around the abdomen and assist in locomotion like oars pushing along the bottom. When at rest, the first two pleopods beat most vigorously to ventilate the thoracic epipods. The animal swims very well, even upside down. (This was something Smith, 1908 and Hickman, 1937 also observed in *Anaspides*.) *Allanaspides* is apparently a detritus feeder; in the laboratory it was seen to pick up fecal pellets and fondle them in its mouthparts.

Micraspides is the most infaunal of the anaspidaceans yet observed. Like *Allanaspides*, it also lives in the burrows of parastacids in waterlogged grass swamps in western Tasmania, but the animals seldom come to the surface unless a burrow collapses and the resultant pool fills with water (Roy Swain, personal communication). They are the most flexible of anaspidids, bending the body dorsoventrally as well as laterally. They are thus ideally suited to climbing in, around, over, and under obstacles, and as a result are easy to spot when searching in waterplants for them. Their occurrences in the field are erratic, though when a population is located it tends to be relatively dense. *Micraspides* does not beat or move anything when it is stationary. The pleopods are rather stiff (like *Allanaspides*) and cooperate entirely within the metachronal sequence with the thoracopod to move the animal along the substrate. The pleopods thus differ only in form, but not function, from the thoracopods. When disturbed, they take evasive action by turning at some angle or flexing the body to change direction 180°.

The observations on limb function above confirm the work of Macmillan, Silvey, and Wilson (1981) on *Anaspides*. In addition, they observed that the only exception to the metachronal pattern of the limbs occurs in the rhythm of the last thoracopod and that the beat of the respiratory thoracic exopods is independent from that of the locomotory portions of the limbs. What all this may mean in terms of evolution of locomotory mechanisms in the Eumalacostraca will depend on further work among groups outside the Syncarida. Cannon and Manton (1929) advanced some ideas concerning feeding mechanisms in *Anaspides tasmaniae, Paranaspides lacustris*, and *Koonunga cursor*. They felt that *Anaspides* and *Paranaspides* are probably raptors as well as maxillary filter feeders like *Hemimysis lamornae* but that *Koonunga* is a raptor only. However, intriguing as their analysis reads, it was based solely on the examination of preserved material. A detailed analysis of anaspidacean feeding based on living material still remains to be done.

Modes of mating have not yet been determined for syncarids. Hickman (1937), in the course of his embryological studies, made some observations relevant to reproduction in *Anaspides*. The eggs are attached to bits of bark or plant, or laid in crevices. Development is slow. Eggs are laid in spring or early summer (Swain and Reid, 1983) and take from 32 to 35 weeks to hatch, those laid in the fall take 60 weeks to hatch because of a winter dormancy. Animals hatch as juveniles. Collections of *Allanaspides* made in May by Roy Swain and myself yielded populations with large proportions of juveniles.

The distribution of anaspidaceans (Fig. 6-5) is a classic Gondwana relict pattern, though variations within that range are striking. The anaspidine families are restricted to Tasmania and southern Victoria, but typically each of these species is known from only a handful of localities at best. The stygocaridines, however, seem to have been more successful; this is perhaps related to their groundwater habits. The psammaspids are, so far, known only from a locality in New South Wales (*Psammaspides williamsi*) and from one site in northwest Tasmania (*Eucrenonaspides oinotheke*— from a ground spring in a house wine cellar at 6 Payton Place, Devonport, certainly one of the most restricted and unusual distributions on record). The stygocarids are widely dispersed (Noodt, 1970; Schminke, 1980): *Stygocaris*, with a species in Victoria, one in New Zealand, and another in Chile; *Parastygocaris*, restricted to Argentina; *Oncostygocaris*, in southern Chile; and *Stygocarella*, from New Zealand. The only fossil anaspidacean fits this same pattern; *Anaspidites antiquus* occurs in the Triassic of Sydney, New South Wales.

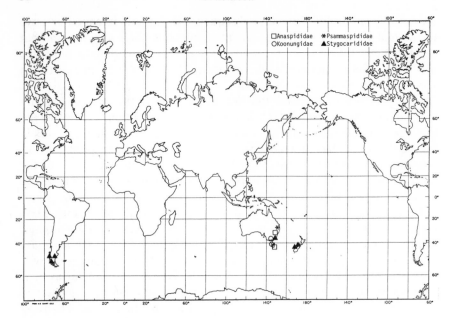

Fig. 6-5. Biogeography of Anaspidacea.

DEVELOPMENT   The single most comprehensive work on syncarid ontogeny still remains that of Hickman (1937). Segmentation is total (Fig. 6-6A, B) and leads to a large blastocoel surrounded by large yolk-laden cells (Fig. 6-6C). The primary mesoderm arises from cells near the poles that migrate inward to form a central mass of cells (Fig. 6-6D, E, F). The embryo then elongates and forms a ventral oval furrow (Fig. 6-6G, H, I) that invaginates to form endoderm (Fig. 6-6H). The archenteron is then quickly obliterated. The blastopore virtually closes (Fig. 6-6J) and all but completely disappears. The blastocoel is also filled in by the development of the endoderm, and the primary mesoderm mass is squeezed against the ectoderm on either side of the embryo. It takes about eight weeks to achieve gastrulation. At this point a dormant period ensues; for spring and summer clutches this rest period is short, while for fall and winter clutches the pause lasts several weeks.

When development resumes, an egg-nauplius results (Fig. 6-6K) when the ventral germinal disc is formed. As a recognized nauplius is assumed (Fig. 6-6L, M), a terminal mesodermal teloblast is established (Fig. 6-6N, O) which begins to bud off segments (Fig. 6-6P). After the elapse of at least eight months from the time of laying, the hatchling stage is achieved (Figs. 6-6Q, 6-7).

The internal development of *Anaspides* seems to represent a typical malacostracan type and shares a number of similarities with patterns seen in both *Nebalia* and *Hemimysis*: the two rows of seven ectodermal telo-

blasts initially form midventrally just anterior to the blastopore; the ectodermal teloblasts become centered in this midventral position; there are
four pairs of mesodermal teloblasts; there is development of postmandibular somites from anterior to posterior; the development of the heart is in
stages, first the side walls, then the floor, and finally the roof; a median
dorsal organ is present; and there is lack of a free-swimming larva.

Certain features are shared only with *Nebalia*: a ring of 19 ectodermal
teloblasts is formed around a caudal papilla; the caecal lobes arise as

Fig. 6-6. *Anaspides tasmaniae* development. (A) two-celled stage; (B) four-celled stage; (C)
16-celled stage; (D, E) primary mesoderm formation; (F) just prior to gastrulation; (G, I)
initiation of gastrulation; (H, J) formation of archenteron and closing of blastopore; (K) ventral egg-nauplius at week 9 (summer sequence); (L) week 10; (M) week 11; (N) week 13, with
initiation of teloblast formation; (O) week 15; (P) week 16, with somite formation well
advanced; (Q) just prior to hatching. (Modified from Hickman, 1937)

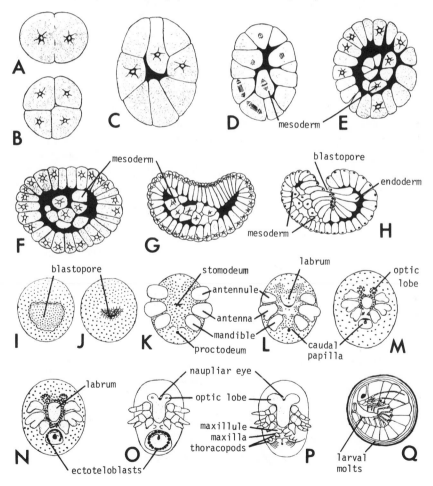

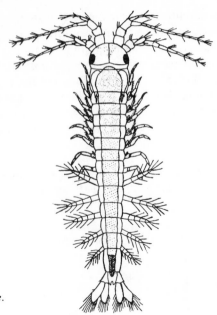

Fig. 6-7. Juvenile *Anaspides tasmaniae*.
(From Hickman, 1937)

outpocketings from the endodermal yolk sac; and genital primordia arise in the coelomic pouches very late in development.

Several ontogenetic features are similar to other groups. The holoblastic cleavage leading to a large blastocoel compares well with events seen in copepods, branchiopods, and some decapods. Other branchiopod features in *Anaspides* are the early appearance of coiled and end-sac mesodermal primordia leading to eventual maxillary glands, the persistence of yolk into late stages of development, and a prolonged dormant period.

Taken together, these developmental features do not seem to bear much direct phylogenetic import but rather emphasize the central position and general nature of anaspidacean development.

FOSSIL RECORD   Only a single fossil anaspidacean (Brooks, 1962) is currently recognized (Schram, 1984), *Anaspidites antiquus* (Fig. 6-8). Though

Fig. 6-8. *Anaspidites antiquus*, Triassic of New South Wales. (From Schram, 1984)

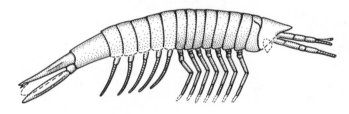

its short thoracic segments present a very distinctive form, this Triassic fossil from New South Wales is otherwise indistinguishable from the living Anaspididae and is best left in that family.

TAXONOMY   The taxonomic relationship of the order has been in flux since the early 1960s. Until that time only a few isolated forms in Tasmania and Victoria constituted the known types. With the addition of new species, genera, and families, our understanding has been clarified. Knott and Lake (1980) recognized two suborders. The Anaspidinea are characterized by a simple lobelike rostrum, maxillulary palp, natatory and annulate pleopods, a prominent seminal receptacle in females, and a flattened tailfan. The suborder Stygocaridinea are characterized by a lack of a maxillulary palp, reduced or absent pleopods, a small seminal receptacle if present, and a telson not forming a tailfan with the uropods. Schminke (1975) clearly outlines the differences between families with these suborders. A current summary taxonomy for the group is therefore presented here:

Suborder Anaspidinea Calman, 1904 Triassic–Recent
   Family Anaspididae Thomson, 1893 Triassic–Recent
   Family Koonungidae Sayce, 1908 Recent
Suborder Stygocaridinea Noodt, 1965 Recent
   Family Stygocarididae Noodt, 1962 Recent
   Family Psammaspididae Schminke, 1974 Recent

REFERENCES

Brooks, H. K. 1962. On the fossil Anaspidacea, with a revision of the classification of the Syncarida. *Crustaceana* **4**:229–42.
Cannon, H. G., and S. M. Manton. 1929. On the feeding mechanism of the syncarid Crustacea. *Trans. Roy. Soc. Edinb*. **56**:175–89.
Fulton, W. 1982. Notes on the distribution and life cycle of *Paranaspides lacustris. Bull. Aust. Soc. Limnol*. **8**:23–25.
Hickman, V. V. 1937. The embryology of the syncarid crustacean, *Anaspides tasmaniae. Pap. & Proc. Roy. Soc. Tasm*. **1936**:1–35.
Knott, B., and P. S. Lake. 1980. *Eucrenonaspides oinotheke* from Tasmania, and a new taxonomic scheme for Anaspidacea. *Zool. Scripta*. **9**:25–33.
Lake, P. S., and D. J. Coleman. 1977. On the subterranean syncarids of Tasmania. *Helistite* **15**:12–17.
Macmillan, D. L., G. Silvey, and I. S. Wilson. 1981. Coordination of the movements of the appendages in the Tasmanian mountain shrimp *Anaspides tasmaniae. Proc. Roy. Soc. Lond*. (B)**212**:213–31.
Manton, S. M. 1930. Notes on the habits and feeding mechanisms of *Anaspides* and *Paranaspides. Proc. Zool. Soc. Lond*. **1930**:791–800.
Nicholls, G. E. 1931. *Micraspides calmani*, a new syncaridan from the west coast of Tasmania. *J. Linn. Soc. Lond. (Zool.)* **37**:473–88.
Noodt, W. 1963. Anaspidacea in der südlichen Neotropis. *Verh. Dtsch. Zool. Ges. Wien*. **1962**:568–78.

Noodt, W. 1965. Natürliches System und Biogeographie der Syncarida. *Gewässer Abwässer.* **37/38**:77–186.

Noodt, W. 1970. Zur Eidonomie der Stygocaridacea, einer Gruppe interstitieller Syncarida (Malacostraca). *Crustaceana* **19**:227–44.

Sayce, O. A. 1908. On *Koonunga cursor*, a remarkable new type of malacostracous crustacean. *Trans. Linn. Soc. Lond. (Zool.)* **11**:1–16.

Schminke, H. K. 1974. *Psammaspides williamsi*, ein Vertreter einer neuen Familie mesopsammaler Anaspidacea. *Zool. Scripta.* **3**:177–83.

Schminke, H. K. 1975. Phylogenie und Verbreitungsgeschichte der Syncarida. *Verh. Dtsch. Zool. Ges. Bochum.* **1974**:384–88.

Schminke, H. K. 1978. Die phylogenetische Stellung der Stygocarididae— unter besonderer Berücksichtigung morphologischer Ähnlichkeiten mit Larvenformen der Eucarida. *Zeit. f. Zool. Syst. u. Evol.-forsh.* **16**:225–39.

Schminke, H. K. 1980. Zur Systematik der Stygocarididae und Beschreibung zweier neuer Arten. *Beaufortia* **30**:139–54.

Schram, F. R. 1984. Fossil Syncarida. *Trans. San Diego Soc. Nat. Hist.* **20**:189–246.

Siewing, R. 1956. Untersuchungen zur Morphologie der Malacostraca. *Zool. Jahrb. Abt. Anat.* **75**:39–176.

Siewing, R. 1963. Studies in malacostracan morphology: Results and problems. In *Phylogeny and Evolution of Crustacea* (H. B. Whittington and W. D. I. Rolfr, eds.), pp. 85–103. Mus. Comp. Zool., Cambridge.

Silvey, G. E., I. S. Wilson. 1929. Structure and function of the lateral giant neurone of the primitive crustacean *Anaspides tasmaniae*. *J. Exp. Biol.* **78**:121–36.

Smith, G. W. 1908. Preliminary account of the habits and structure of the Anaspididae, with some remarks on some other freshwater Crustacea from Tasmania. *Proc. Roy. Soc. Lond.* (B)**80**:465–73.

Swain, R., and C. I. Reid. 1983. Observations of the life history and ecology of *Anaspides tasmaniae*. *J. Crust. Biol.* **3**:163–72.

Swain, R., I. S. Wilson, J. E. Ong, and J. L. Hickman. 1970. *Allanaspides helonomus* from Tasmania. *Rec. Queen Vict. Mus.* **35**:1–13.

Swain, R., I. S. Wilson, and J. E. Ong. 1971. A new species of *Allanaspides* from southwestern Tasmania. *Crustaceana* **21**:196–202.

Thomson, G. M. 1894. On a freshwater schizopod from Tasmania. *Trans. Linn. Soc. Lond. (Zool.)* **6**:285–303.

Tjønneland, A., S. Økland, A. Bruserud, and A. Nylund. 1984. Heart ultrastructure of *Anaspides tasmaniae*. *J. Crust. Biol.* **4**:226–32.

Williams, W. D. 1965a. Subterranean occurrence of *Anaspides tasmaniae*. *Int. J. Speleol.* **1**:333–37.

Williams, W. D. 1965b. Ecological notes on Tasmanian Syncarida, with a description of a new species of *Anaspides*. *Int. Rev. ges. Hydrobiol.* **50**:95–126

7

PALAEOCARIDACEA

DEFINITION First thoracomere not fused to cephalon, although frequently reduced; first thoracopod typically reduced in form; eighth thoracopod parallel in alignment to the other thoracopods; thoracic exopods and pleopods variously flaplike or annulate; tailfan well developed, uropodal exopods usually with diaeresis.

HISTORY Species of fossil syncarids have been entering the literature since the middle 1800s. *Uronectes fimbriatus* (Jordan) 1847 was the first and has been the object of a long series of papers by various authors over generations that have yielded a confusing array of name changes and morphological interpretations. Meek and Worthen (1865) described *Acanthotelson stimpsoni* (Fig. 7-1A) and *Palaeocaris typus* (Fig. 7-1B), which have become the best known and most clearly understood forms (Brooks, 1962b). However, it was not until 1885 and 1886 that Packard recognized the separate status of these fossils and erected the taxon, Syncarida—this a decade before the first living syncarids were ever described, from Tasmania. The fossil syncarids generally were considered, however, to occupy merely a separate family within the Anaspidacea. Finally, it was Brooks (1962a, b) who clearly recognized the separate status of the Paleozoic fossils from the modern types, and erected a separate order for them. Schminke (1975) defined the palaeocaridaceans most clearly in regard to other major syncarid groups, and Schram (1984) has reviewed all known fossil syncarids and revised their taxonomy.

MORPHOLOGY The most completely known palaeocaridacean is *Acanthotelson stimpsoni* from the Pennsylvanian of Illinois. Though it was specialized in regard to certain aspects of its anatomy, it affords a sound basis for understanding the entire group.

With the exception of the slightly shorter first thoracomere, all the body segments were subequal. The anterior rim of the cephalon was developed as a small rostrum. The head shield was marked by a set of cephalic grooves. The eyes were stalked and fit into an optic notch on the anterior margin of the head shield.

The antennules (Fig. 7-2A) had long three-segment peduncles. The flagella were moderate in size, but unequal in length.

The antennae (Fig. 7-2B) had long four-segment peduncles. The flagellum was approximately as long as the body. The scaphocerite was

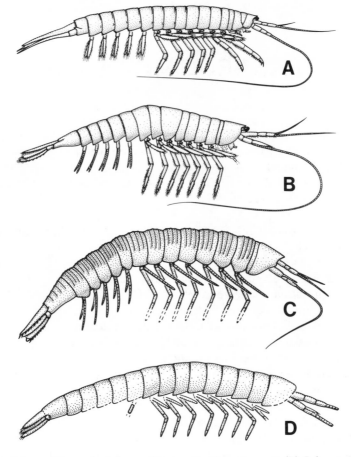

Fig. 7-1. Palaeocaridacean body types. (A) *Acanthotelson stimpsoni*; (B) *Palaeocaris retrac-tata*; (C) *Praeanaspides praecursor*; (D) *Minicaris brandi*. (From Schram, 1984)

apparently absent in *Acanthotelson*, but in other genera of the order it was typically developed as an oval structure arising on the second segment of the peduncle.

The mandibles (Fig. 7-2C) were massive and bore a prominent incisor process, and a prehensile palp of at least two and probably three segments. The mandibles were located posterior to a small triangular labrum, and were apparently posteriorly flanked by a set of lobelike paragnaths.

The maxillule had a prominent three-joint palp. The maxilla is not completely known in *Acanthotelson* but those of *Palaeocaris typus* are known to have at least two segments. There were no indications of any gnathobases on these appendages, and this fact, combined with the great development of the endopod palp, is rather noteworthy.

The first thoracopod (Fig. 7-2D) was reduced over that of any other thoracic appendage. There was a five-segment endopod; the nature of an exopod, if any, is not known. The entire appendage was about half the length of succeeding thoracopods, and although the first thoracomere was free, this appendage was more like a maxillipede in terms of form and probable function.

The second and third thoracopods (Fig. 7-2E, F) in *Acanthotelson* were rather specialized as robust, spinose raptorial legs. The third pair was biramous with a flaplike exopod. These legs in other palaeocaridaceans were variously developed. For example, *Uronectes* had only the second pair as raptors, while *Palaeocaris* (Fig. 7-1B) displays a more generalized and more typical array with none of these anterior thoracopods modified.

Fig. 7-2. Appendages of *Acanthotelson stimpsoni*. (A) antennule; (B) antenna; (C) mandible; (D) first thoracopod; (E) second thoracopod; (F) third thoracopod; (G) eighth thoracopod; (H) first pleopod; (I) telson and uropods. (Modified from Brooks, 1962b)

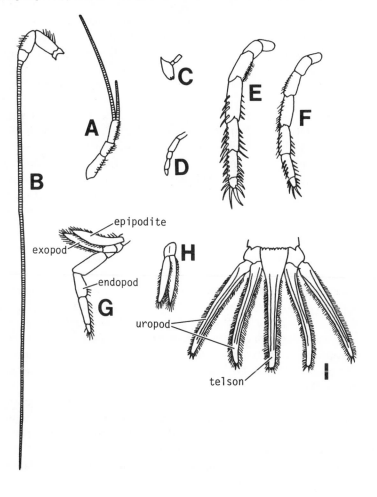

A more generalized thoracopod form (Figs. 7-1A, 2G) was seen on the posterior appendages of *Acanthotelson*. The five-segment endopod displays some slight setation; the exopod was prominent and flaplike and almost matched in size by the long flaplike epipods. The squillitids (Fig. 7-1C) had annulate exopods.

The pleopods in *Acanthotelson* (Fig. 7-2H) had a single-segment protopod bearing two long flaplike rami with setose margins. In contrast, *Squillites* had uniramous annulate pleopods and *Praeanaspides* had biramous ones (Fig. 7-1C).

The uropods (Fig. 7-2I) in *Acanthotelson* were long and spikelike, as was the telson. A more typical tailfan was seen in *Palaeocaris* (Fig. 7-1B), in which broad flaplike uropods (the exopod with a diaeresis) flanked a rounded telson.

Information on internal anatomy of palaeocaridaceans is difficult to come by, of course. However, two points can be mentioned here. The gut, frequently preserved as a detrital infilling, was apparently simple and did not appear to display any prominent stomach enlargements or specializations. Excretion was achieved by antennal glands, since specimens are available that clearly display a nephropore (e.g., Brooks, 1962a, b, Pl. 59, Fig. 6). Interestingly, sexual dimorphism has never been noted, even from those localities with thousands of specimens with exceptional preservation.

NATURAL HISTORY   Conclusions about palaeocaridacean life-style must be largely inferential. These syncarids apparently preferred brackish to fresh waters. They are found typically only in those sorts of marginal paleogeographic and faunal situations that appear to have represented deltaic, swampy, and lagoonal habitats (Schram, 1979, 1981). They are frequently found in association in brackish facies with the rather specialized pygocephalomorph mysidaceans, and in lagoonal habitats with aenigmacarid aeschronectidans.

Palaeocaridacean guts are frequently infilled with detritus, presenting a perfect gut cast. However, this does not imply necessarily that they were detritus feeders. On the contrary, by inference from what is known of the living anaspidacean syncarids and what is implied by the variety of anterior thoracopod morphologies seen in the fossils, diverse feeding habits can be inferred. For example, a form like that seen in *Palaeocaris* or *Squillites* is rather generalized and may reflect versatile feeding habits akin to those of living anaspids like *Paranaspides* or *Allanaspides*. On the other hand, forms like *Acanthotelson* and *Uronectes* may indicate a more active benthic scavenger behavior.

The lack of sexual dimorphism would seem to imply primitive copulatory behavior with both eggs and sperm probably being shed free into the water. Such behavior would agree with the general primitive condition palaeocaridaceans seem to represent as a whole.

The stratigraphic and geographic distribution of palaeocaridaceans is rather striking. They are known to date only from the Carboniferous and Permian. They are restricted in the lower part of their stratigraphic range to North America and Europe. Schram (1977) pointed out such a distribution corresponds to the existence of a tropical island continent, Laurentia, in the late Paleozoic. All eumalacostracans and hoplocaridans of that time period share this biogeographic pattern. Later Permian palaeocaridaceans occur in Brazil; and in the Triassic, fossils identified as probably related to anaspids are found in New South Wales, Australia. The suggestion is that all these forms were endemic to Laurentia, and only after the formation of Pangaea in the Permian did dispersal to other continental regions occur.

DEVELOPMENT Despite the abundance of fossils of palaeocaridacean adults of various size classes (Brooks, 1962b), nothing of their development, even as juveniles, is known.

TAXONOMY Brooks (1962a, b, 1969) proposed a family level taxonomy of the group based mostly on body form and anterior thoracopod specializations. Subsequently, Schram and Schram (1974) and Schram (1979) discovered information that indicated that the form of exopods and pleopods is much more variable among genera than had been suspected. Schram (1984) produced a generic and familial revision of the palaeocaridaceans based on this diversity of appendage form.

Family Minicarididae Schram, 1984 Lower Carboniferous–Lower Permian
Family Acanthotelsonidae Meek and Worthen, 1865 Upper Mississippian–Lower Permian
Family Palaeocarididae Meek and Worthen, 1865 Pennsylvanian (Upper Carboniferous)
Family Squillitidae Schram and Schram, 1974 Upper Mississippian–Lower Permian

REFERENCES

Brooks, H. K. 1962a. On the fossil Anaspidacea, with a revision of the classification of the Syncarida. *Crustaceana* **4**:229–42.
Brooks, H. K. 1962b. The Paleozoic Eumalacostraca of North America. *Bull. Am. Paleo.* **44**:163–338.
Brooks, H. K. 1969. Syncarida. In *Treatise on Invertebrate Paleontology*, Part R, *Arthropoda* **4**(1) (R. C. Moore, ed.), pp. R345–59. Geol. Soc. Am. and Univ. Kansas Press, Lawrence.
Jordan, H. 1847. Entdeckung fossiler Crustaceen in Saarbrückenschen Steinkohlengebirge. *Verk. natur. Ver. preuss. Rheinl. Westf.* **4**:89–92.
Meek, F. B., and A. H. Worthen. 1865. Notice of some new types of organic remains from the Coal Measures of Illinois. *Proc. Acad. Nat. Sci. Philad.* **1865**:46–51.

Packard, A. S. 1885. The Syncarida, a group of Carboniferous Crustacea. *Am. Nat.* **19**:700–703.

Packard, A. S. 1886. On the Syncarida, a hitherto undescribed synthetic group of extinct malacostracous Crustacea. *Mem. Nat. Acad. Sci.* **3**:123–28.

Schminke, H. K. 1975. Phylogenie und Verbreitungsgeschichte der Syncarida. *Verk. Dtsch. Zool. Ges.* **1974**:384–88.

Schram, F. R. 1977. Paleozoogeography of late Paleozoic and Triassic Malacostraca. *Syst. Zool.* **26**:367–79.

Schram, F. R. 1979. British Carboniferous Malacostraca. *Fieldiana: Geol.* **40**:1-129.

Schram, F. R. 1981. Late Paleozoic crustacean communities. *J. Paleo.* **55**:126–37.

Schram, F. R. 1984. Fossil Syncarida. *Trans. San Diego Soc. Nat. Hist.* **20**:189–276.

Schram, J. M., and F. R. Schram. 1974. *Squillites spinosus*, from the Mississippian Heath Shale of central Montana. *J. Paleo.* **48**:427–64.

8

BATHYNELLACEA

DEFINITION Body with eight free thoracomeres; last pleomere fused (?) as a pleotelson; eyes absent; antennules biramous; mandible with well-developed palp; first thoracopod similar to the following thoracopods; eighth thoracopod in males as a copulatory (?) organ; thoracopods biramous but endopod with only three or four segments; pleopods, if present, reduced in number and size or vestigial; well-developed uropods; anus terminal; caudal rami; no tailfan.

HISTORY *Bathynella natans* was described in 1882 by Vejdovsky from well water in Prague, but its systematic affinities were something of a problem from the very beginning. Vejdovsky left the issue unresolved; however, even with no further specimens to study, Calman (1899) suggested affinities with *Anaspides tasmaniae* and to related fossil forms from the Paleozoic. Chappuis (1915) redescribed *B. natans* in detail from new and more abundant material and placed it in a separate suborder Bathynellacea within an order Anomostraca [= Syncarida]. Calman (1917) disagreed with this separate subordinal status because of the great similarity in regard to the free first thoracomere of *Bathynella* to the fossil forms (the latter at that time were treated as a family within the anaspidaceans). Little work was subsequently done on the group until after World War II, and not until the extensive studies of Noodt and Schminke has the group come into its own. Noodt (1965) reestablished the ordinal distinction of Chappuis, and Schminke (1975) has defined the relationship of bathynellaceans to other syncarids. Bowman (1971) and Schminke (1976) differed in their interpretations of the sixth pleomere, telson, and whether there are furcae or not.

MORPHOLOGY Bathynellaceans are small interstitial groundwater forms (Fig. 8-1). They range in size from the 0.5 mm *Acanthobathynella knoepffleri* up to the relatively gigantic 3.4 mm *Bathynella magna*. The body is subcylindrical and elongate. The head is relatively long with a faint mandibular groove and lacks eyes and a rostrum. The first thoracomere is free and is as fully developed as any other thoracic somite. The individual pleomeres are equal in size to the individual thoracomeres.

The antennules (Fig. 8-2A) are biramous, a tiny setose lappet on the fifth segment represents the second branch.

The antennae (Fig. 8-2A) are biramous in the Bathynellidae. The peduncle consists of two segments. The scaphocerite (when present) is a small oval flap; the flagellum consists of five prominent segments.

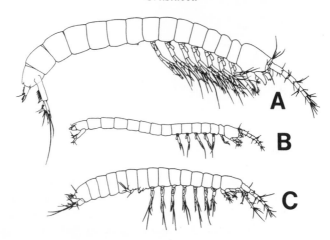

Fig. 8-1. Bathynellacean body form. (A) *Notobathynella williamsi*; (B) more paedomorphic *Hexabathynella halophila*; (C) *Bathynella* sp. (From Schminke, 1974)

The labrum is toothed, without teeth (Fig. 8-2B), or setose. The mandible (Fig. 8-2C) has a palp of up to three segments. The mandibular teeth are prominent and have an accessory incisor process fused to the tooth row. The paragnaths, if present, are deeply cleft as two lobes (Fig. 8-2D) and rather elongate with only terminal setal tufts.

The maxillule (Fig. 8-2E) has two endites with stout setae. There are three brush setae, which may be vestiges of a palp.

The maxilla (Fig. 8-2F) has four segments, the three distal segments bear heavily setose endites. The terminal segment is small with long, simple setae.

All thoracopods (when present) are essentially similar (Figs. 8-2G, H, I). There seems to be a very short precoxal segment, but whether this is in fact a separate true joint or just part of an articulating ring is not clear. The coxa may carry a small vesicular epipodite. In bathynellids, the exopod is a single segment and is marked by a distinctive narrowing on its distal half with long setae carried at and beyond this 'shoulder.' The endopod has three or four segments. In parabathynellids, the exopod may consist of several segments. The eighth thoracopod, if present, is highly modified. In females it is reduced to varying degrees (Fig. 8-2K), while in males the coxa is elaborated as a lobe that bears the gonopore (Fig. 8-2J).

The first pair of pleopods is usually the only one that may be present in bathynellids (Fig. 8-2L). It consists of a simple two-segment appendage with a few simple setae. Parabathynellids lack pleopods altogether.

The uropod protopods (Fig. 8-2M) are armed distally with stout spines. The exopod is short and conical with apical setae, while the endopod is cylindrical with a group of distal spines and setae. The interpretation of the 'telson' is the subject of some controversy. Bowman (1971) felt the anus was terminal on the sixth pleomere and that the telson was deeply cleft into

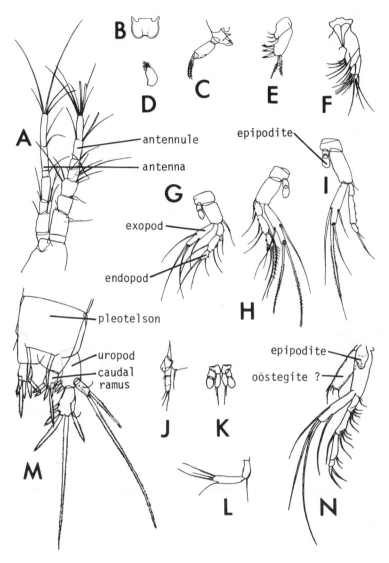

Fig. 8-2. Appendages of *Nannobathynella marcusi*. (A) antennule and antenna; (B) labrum; (C) mandible; (D) paragnath; (E) maxillule; (F) maxilla; (G) first thoracopod; (H) second thoracopod; (I) seventh thoracopod; (J) eighth thoracopod of male; (K) eighth thoracopod of female; (L) pleopod; (M) tail end of animal (from Noodt, 1972). (N) second thoracopod, *Bathynella inlandica* (from Uéno, 1954).

two distinct lobes as occurs in some amphipods. However, Schminke (1976) disagrees with this interpretation and propounds the traditional view that the telson is fused with the sixth pleomere as a pleotelson and adorned with a pair of broad furcae, each armed distally with a row of spines. In view of the definitions in Chapter 1, I interpret the terminal anus to be flanked by a pair of caudal rami and these in turn are flanked by a pair of uropods—a unique condition among crustaceans.

Knowledge of the internal anatomy (Fig. 8-3) has not essentially been improved upon since Chappuis (1915) and is largely based on what Chappuis called *Bathynella natans*. The digestive system is most unusual for a malacostracan. There is no 'stomach' as such, but the esophagus extends to the level of the sixth thoracomere. This is followed by a 'stomach' with longitudinal glandular ridges that extend to the eighth thoracomere. This 'stomach' opens into a wide 'midgut,' which possesses a thick glandular dorsal wall and thin ventral walls and which extends to the fourth pleomere. The hindgut also appears to have glandular walls. There are no digestive diverticula of any kind.

The heart lies in the fourth thoracomere; no ostia are known. A dorsal artery extends forward to the head. Chappuis interpreted the posterior vessel as a vein bringing blood from the tail region to the heart. If true, this would be a strange condition for crustaceans, and Calman (1917) suggested this was probably better considered to be a posterior artery.

The maxillary gland is also most unusual. There is a coelomic sac in the posterior region of the head. This opens up into a long duct that extends back to the fourth thoracomere and that then turns and runs anteriorly again to the maxillae to terminate in a vesicle opening by a muscle-operated slit on the maxillary surface. In addition to the maxillary gland, accessory excretory structures are located throughout the body. There are masses of paired 'nephrocytes' in the body somites and a massive 'uropodal gland' in the last pleomere that opens on the uropods.

The nervous system occupies a relatively large amount of space. The brain is a large undifferentiated lobe in the supraesophageal space. The ventral nerve cord is a continuous unit with little differentiation between ganglia and connectives.

The gonads are abdominal, with ducts extending forward to open on the respective sixth or eighth thoracomeres depending on the sex.

NATURAL HISTORY Bathynellaceans typically live in interstitial groundwater habitats. They crawl over and around sediment grains and browse on detritus, protozoans, fungi, and bacterial films found there. The only exceptions to this general groundwater habitation are *Bathynella baicalensis* and *B. magna* that live at depths of 20 to 1440 m in Lake Baikal. Bathynellaceans generally prefer fresh water, though some species of *Hexabathynella* are collected in brackish conditions from several localities around the world. *Thermobathynella adami* is found in waters of 55°C.

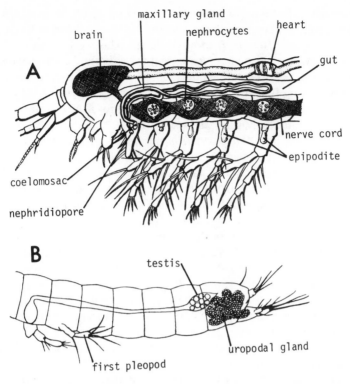

Fig. 8-3. Internal anatomy of *Bathynella natans*. (A) anterior body region; (B) posterior region displaying testes and excretory uropodal gland. (From Chappuis, 1915)

Modes of reproduction are not completely known with certainty. Females lay relatively large eggs (0.1 mm) in reference to the size of the body, and only one or two eggs at a time. Copulatory behavior is not known, though the male eighth thoracopod is said to be copulatory in function (which may or may not be so).

Bathynellaceans are found worldwide and, with limited dispersal capabilities, are obviously a very ancient group. Schminke (1974) and Schminke and Wells (1974), based largely on the better known parabathynellids, recorded groups of genera aligned along strong Eurasian and Gondwanan tracks, and originally postulated a possible dispersal from a center of origin out of east Asia. Since, Schminke (1981a) has come to agree with Schram (1977) who felt, based on fossil evidence, a more likely interpretation involved a tropical Laurentian origin in the late Paleozoic from a primitive syncarid stock, with a dispersal outward to northern and southern land areas with the formation of Pangaea in the Permian. The Eurasian and Gondwanan tracks seen in bathynellaceans then record a dispersal from an ancient core. The subsequent breakup of Pangaea resulted in a vicariance of a formerly contiguous fauna into separate north and south components.

DEVELOPMENT   Bathynellaceans are the only syncarids for which a larval developmental sequence is known. Jakobi (1954) studied the development of *Antrobathynella stammeri* in great detail. This work remains the best source for information on bathynellacean ontogeny and clarifies some inaccuracies in the remarks of Chappuis (1948). The eggs are prepared by the female for laying two at a time. Within 48 hours of laying, a 32-cell stage blastula is reached. The eggs are yolky and soon reach an egg-nauplius (Fig. 8-4). Even at this early stage the caudal rami appear as anlagen, though Chappuis at one time felt *Bathynella* possessed a telson and no furca or rami and that the anal segment only bears uropods. Chappuis himself and subsequent workers, including Jakobi, came to feel that there were furca. Jakobi supplies few details of development prior to hatching, but this remains a profitable area of future investigation for someone to pursue.

Segments and appendage anlagen are added until a stage (Fig. 8-5) is reached where the first pair of thoracopods are developed, the second pair exist as anlagen, the mandibles are only a simple lobe, and a fair degree of yolk is still found in the gut. The embryo hatches at this point to what Schminke (1981b) terms parazoea I, which possesses 10 postcephalic segments with the last three rather incompletely separated from each other. The animals do not move particularly well at this stage, though they manage to locomote with body gyrations. At the next stage, parazoea II, the second thoracopods are developed, the third appear as anlagen, the mandible becomes functional with teeth on it, another segment is added to the body, and yolk disappears from the gut. Presumably, the animal feeds from this stage on, but the body form is still incomplete and still termed 'larval.'

At the molt to stage 4, the regularity of development breaks down, even while the adult number of segments is attained. Four functional thoracopods appear, but there may or may not be anlagen for the fifth, and the uropods may be variously developed. Likewise, in the fifth stage through

Fig. 8-4. Egg-nauplius stage of *Antrobathynella stammeri*. (A) ventral view; (B) lateral view. (From Jakobi, 1954)

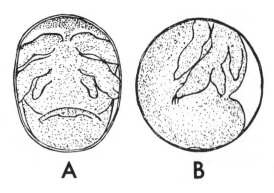

A   **B**

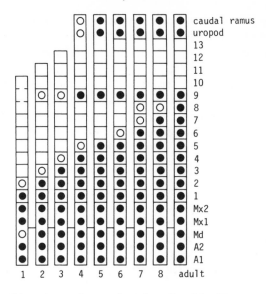

Fig. 8-5. One possible pathway of postembryonic mode of development of *Antrobathynella stammeri*. ○ indicates rudimentary limb, ● reveals a more or less definitive form. For variant pathways see Jakobi (1969), Serban (1973), or Schminke (1981b).

the eighth, thoracopods, pleopods, and uropods are developed; there may be 0, 1, 2, or 3 anlagen for the sixth through eighth thoracopods (see Schminke, 1981b). This versatility of appendage formation gives a clear indication how paedomorphic variations between bathynellacean genera could easily have arisen. Eventually an adult condition is reached, though the exact pathway can apparently vary in the sequence of appendage and anlagen appearance. A complete adult stage is attained under laboratory conditions after nine months from hatching.

TAXONOMY   The full development of a free first thoracomere in the bathynellaceans is reminiscent of the form seen in the extinct palaeocaridaceans. However, Paleozoic forms like *Minicaris brandi*, *Erythrogaulos carrizoensis*, and *Nectotelson krejcii* have a large first thoracomere more akin to that seen in the bathynellaceans and rather unlike that of the other palaeocaridaceans, who typically have the first thoracic segment free but reduced by various degrees.

Two living families are recognized. The more primitive Bathynellidae have paragnaths, scaphocerites, untoothed labrum, prehensile mandibular palps of one to three segments, thoracic exopodites always as one segment, and one or two pairs of pleopods. The advanced Parabathynellidae have no paragnaths, no scaphocerite, a toothed or setose labrum, nonprehensile mandibular palps of one segment, thoracic exopods with one or more segments, and pleopods absent or highly reduced.

The entire thrust of bathynellacean evolution is directed around increasing degrees of progenetic paedomorphosis (Fig. 8-1A, B), that is, they tend to reduce the degree of morphologic development necessary to achieve a reproductive stage. They generally abort the development of pleopods and, to varying degrees, the posterior thoracopods. Their extreme paedomorphic state, combined with a number of their rather distinctive peculiar anatomical conditions, has caused some disagreement over their exact affinities. For example, Serban (1972), in a detailed treatment of external anatomy of bathynellaceans, placed them in Podophalocarida, a taxon coequal in status with phyllocarids, other syncarids, hoplocarids, peracarids, pancarids, and eucarids.

REFERENCES

Bowman, T. E. 1971. The case of the non-ubiquitous telson and the fradulent furca. *Crustaceana* **21**:165–75.
Calman, W. T. 1899. On the characters of the crustacean genus *Bathynella*. *J. Linn. Soc. Lond.* **27**:338–45.
Calman, W. T. 1917. Notes on the morphology of Bathynella and some allied Crustacea. *Quart. J. Micr. Soc.* **62**:489–514.
Chappuis, P. A. 1915. *Bathynella natans* und ihre Stellung im System. *Zool. Jahrb. (Syst.)* **40**:147–76.
Chappuis, P. A. 1948. Le développment larvaire de *Bathynella*. *Bull. Soc. Sci. Cluj.* **10**:305–9.
Jakobi, H. 1954. Biologie, Entwicklungsgeschichte und Systematik von *Bathynella natans*. *Zool. Jahrb. (Syst.)* **83**:1–62.
Jakobi, H. 1969. Contribuicao à ontogenia de *Bathynella* e *Brasilibathynella*. *Bol. Univ. Fed. Paraná, Zool.* **3**:131–42.
Noodt, W. 1965. Natürliches System und Biogeographie der Syncarida. *Gewässer. Abwässer.* **37/38**:77–186.
Noodt, W. 1972. Brasilianische Grundwasser—Crustacea, 2. *Nannobathynella, Leptobathynella,* und *Parabathynella* aus der Serra do Mar von São Paulo. *Crustaceana* **23**:152–64.
Schminke, H. K. 1974. Mesozoic intercontinental relationships as evidenced by bathynellid Crustacea. *Syst. Zool.* **23**:157–64.
Schminke, H. K. 1975. Phylogenie und Verbreitungsgeschichte der Syndarida. *Verh. Dtsch. Zool. Ges.* **1974**:384–88.
Schminke, H. K. 1976. The ubiquitous telson and the deceptive furca. *Crustaceana* **30**:292–300.
Schminke, H. K. 1981a. Perspectives in the study of the zoogeography of interstitial Crustacea: Bathynellacea (Syncarida) and the Parastenocarididae (Copepoda). *Int. J. Speleol.* **11**:83–89.
Schminke, H. K. 1981b. Adaptation of Bathynellacea to life in the interstitial ('Zoea Theory'). *Int. Rev. ges. Hydrobiol.* **66**:575–637.
Schminke, H. K., and J. B. J. Wells. 1974. *Nannobathynella africana* and the zoogeography of the family Bathynellidae. *Arch. Hydrobiol.* **73**:122–29.
Schram, F. R. 1977. Paleozoogeography of late Paleozoic and Triassic Malacostraca. *Syst. Zool.* **26**:367–79.
Serban, E. 1972. Bathynella (Podophallocarida Bathynellacea). *Trav. Inst. Spéal. 'Emile Racovitza'* **11**:11–224.
Serban, E. 1973. Sur le processus de la pléonisation du péréion dans l'ordre des Bathynellacea. *Bijdr. Dierk.* **43**:173–201.
Uéno, M. 1954. The Bathynellidae of Japan. *Arch. Hydrobiol.* **49**:519–38.

9

BELOTELSONIDEA, WATERSTONELLIDEA, AND 'EOCARIDACEA'

These orders were a series of Late Paleozoic taxa that thrived during Carboniferous time but apparently represented lines that became extinct in the Permian. They illustrate the diversity of the earliest eumalacostracan radiation.

BELOTELSONIDEA

DEFINITION   Carapace covering but not fused to thorax; all thoracopods equally developed, uniramous; thoracic sternites wide; pleopods present, biramous flaps; uropods broadly rounded; telson subtriangular with distal furcal lobes.

HISTORY   First described by Packard (1886) on the basis of a few specimens, *Belotelson* was originally thought to be a large syncarid. Brooks (1962b) preferred to interpret these as caridoids; however, it was not until Schram (1974) that a complete reconstruction of *Belotelson magister* was available. Their position as a separate order within the Eumalacostraca was realized only as the distinct series of eumalacostracan *Baupläne* were recognized (Schram, 1981b).

MORPHOLOGY   *Belotelson magister* (Fig. 9-1A, B, C) is the best known species. The carapace was smooth but marked by a very elongate rostrum and a peculiar posterior middorsal pore of unknown function. The eyes were large, spherical, and stalked and set into an optic notch in the carapace.

The antennules had large three-segment peduncles. The proximalmost joints were the longest and notched to accommodate the stalked eyes. The flagella were moderately long and well developed.

The antennae apparently had two short protopodal segments. The scaphocerite was large with a well-sclerotized outer margin of a broad oval scale. The two proximalmost joints of the flagellum were large elongate peduncles. The flagellum itself was long and well developed.

The mandibles had at least a large pointed incisor process. They were posterior to a modest labrum and were under a pair of large maxillules.

101

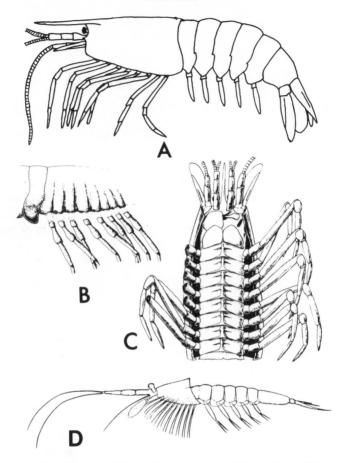

Fig. 9-1 (A, B, C) *Belotelson*. (A) lateral reconstruction of *B. magister*; (B) lateral thorax
(carapace removed) of *B. traquairi*; (C) ventral cephalic region of *B. magister*; (D) lateral
reconstruction of *Waterstonella grantonensis*. (A, B, C from Schram, 1974, 1979, D from
Briggs and Clarkson, 1983)

The maxillules and maxillae were similar and formed a set of large
platelike endites, apparently gnathic in function.

The uniramous thoracopods were all equal, none apparently modified
as maxillipedes. The bases of these legs were difficult to interpret on the
fossils. *Belotelson traquairi* seemed to have distinct short coxa and basis
followed by larger ischium and merus, with some evidence for gills on the
coxa (Schram, 1979). The distal endopods were robustly developed as
walking legs.

The large abdominal pleura all overlapped those of the next posterior
segment. The biramous pleopods were developed as swimmerettes. The

uropods were broad oval flaps. The exopod had a well-sclerotized lateral margin, while the endopods were more diaphanous lobes.

The telson was subtriangular and distally equipped with a pair of modest furcal lobes.

Little is known about the internal anatomy of *Belotelson*. The sides of the thorax are marked by prominent grooves that apparently correspond to the endoskeletal elements inside the thorax to accommodate the origins of the coxal and basal musculature. In addition, some specimens are preserved with detrital gut casts (but this does not necessarily imply belotelsonids were detritovores).

NATURAL HISTORY   The known species of *Belotelson* are found associated with a shallow near-shore marine community. They appear to have been relatively abundant since they are usually among the more prominent crustacean components of these faunas (Schram, 1981a).

Belotelson seems to have probably been a scavenger, low-level carnivore; that is, while not equipped with chelipedes of any kind to allow active capture of live prey, the gnathic mouthparts would seem to imply a meat diet of some sort. The large endites of the maxillules and maxillae formed opposed gnathobases for probable shearing action, and the large incisor process on the mandibles would have had a biting action. Like modern-day lobsters, the belotelsonids probably patrolled an area, scavenging what they could, but not neglecting any hapless living prey that they might have stumbled across and that didn't put up too stiff a resistance.

Neither oöstegites nor evidence of other brooding structures have ever been noted on the fossils.

So far, *Belotelson* is noted only from Carboniferous localities in Europe and North America, implying an endemism to the tropical island continent of Laurentia in the Coal Age.

TAXONOMY   To date only two species of belotelsonids have been recognized, and these within a single genus.

WATERSTONELLIDEA

DEFINITION   Carapace complete, unfused to thorax; body thin, elongate; cephalon long, antennular peduncles as long as thorax; thoracopods biramous, short, very setiferous; sixth pleomere and tailfan as long as first five pleomeres.

MORPHOLOGY   The small, thin, cylindrical body was divisible into four subequal parts (Fig. 9-1D): cephalon, thorax, anterior five pleomeres, and sixth pleomere and tailfan. The carapace was thin and covered the thorax, but was not fused to it.

The antennules were long. The peduncle, of three subequal segments, was as long as the thorax. The flagella were moderately long.

The antennae have short peduncles, to which were mounted distally the oval scaphocerite and short flagellum.

Nothing has yet been discerned concerning mouthparts. The thoracopods were all subequal. The protopods were short, as were the biramous branches terminally mounted with long setae. The carapace laterally covers the thoracopodal protopods.

The first five pleomeres were subequal with only poorly developed pleura, each bearing biramous, terminally setose pleopods. The last pleomere was very elongate. The spatulate telson was flanked by long uropods, each ramus of which is two-segmented and terminally setose.

NATURAL HISTORY   Nothing is known about internal anatomy of this animal.

These animals are preserved in gregarious associations that seem to represent schools (Schram, 1979). This, combined with the size and general form (reminiscent of living types like *Lucifer* or *Arachnomysis*), indicate a pelagic habit. Neither traces of brood pouches nor traces of eggs attached to appendages have ever been noted. It thus appears that the animals did not brood eggs, at least not for long (cf. *Lucifer*), and certainly had no brood pouch.

TAXONOMY   At present, only a single species is recognized, *Waterstonella grantonensis*.

'EOCARIDACEA'

This is sort of a catchall order, whose membership has slowly shrunk through the years as forms become better understood and assigned elsewhere. As a result, because the remaining unassigned taxa are too poorly known, no diagnostic characters can be given for the order. A brief overview of the included families is given, with comments on possible affinities.

EOCARIDIDAE   *Eocaris oervigi* (Brooks, 1962a) is known from one incomplete specimen (Fig. 9-2A). The large abdomen and carapace outline might indicate aeschronectid hoplocaridan affinities. The genus *Devonocaris* (Brooks, 1962a, b) (Fig. 9-2B) is known only from fragmentary material and was originally thought to be assignable to *Palaeocaris* (Wells, 1957).

ANTHRACOPHAUSIIDAE   The genus *Anthracophausia* (Fig. 9-2C) as defined by Schram (1976) presents a distinct body form, but the current incomplete knowledge concerning appendages makes definitive assignment of the

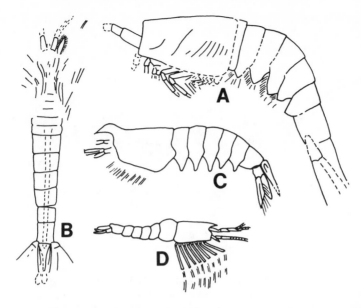

Fig. 9-2. Partial reconstructions of eocarids. (A) *Eocaris oervigi*; (B) *Devonocaris cuylerensis* (from Brooks, 1962a); (C) *Anthracophausia dunsiana*; (D) *Essoidia epiceron* (from Schram, 1974, 1979)

genus difficult. For example, discovery of almost completely intact specimens from the Carboniferous of Bearsden in Scotland reveals unusual antennae with robust and rather spinose segments (Fig. 9-3). Such limbs are similar to structures seen in isopods, amphipods, and tanaidaceans, and in those instances are associated with a burrowing or tube-dwelling life style. It is possible *Anthracophausia* may represent a fossorial euphausiacean. *Peachocaris strongi* (Brooks, 1962b), originally allied with *Anthracophausia*, is likely a mysidacean (see Chapter 29).

ESSOIDIIDAE   The single species, *Essoidia epiceron* (Schram, 1974) (Fig. 9-2D) is largely distinguished from the anthracophausiids by the abdominal flexion occurring on the first pleomere. However, incomplete knowledge concerning appendages hinders definitive assignment of this species to an order.

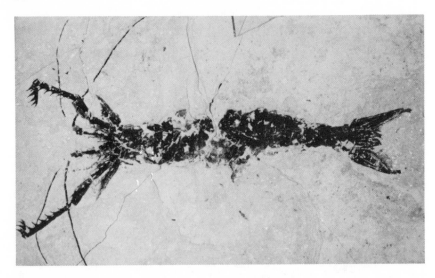

Fig. 9-3. *Anthracophausia* sp. from the Carboniferous of Bearsden, Scotland. Cephalon to the left, tailfan to the right; note long, biramous flagella of antennules and robust spinose segments of the antennae. (Photograph provided by Dr. W. D. I. Rolfe, Hunterian Museum, University of Glasgow.)

REFERENCES

Briggs, D. E. G., and E. N. K. Clarkson. 1983. The Lower Carboniferous Granton 'shrimp bed,' Edinburgh. *Sp. Pap. Paleo.* **30**:161–77.
Brooks, H. K. 1962a. Devonian Eumalacostraca. *Arkiv. f. Zool.* **(2)15**:307–17.
Brooks, H. K. 1962b. The Paleozoic Eumalacostraca of North America. *Bull. Am. Paleo.* **44**:163–338.
Packard, A. S. 1886. On the Syncarida a hitherto undescribed synthetic group of extinct malacostracous Crustacea. *Mem. Nat. Acad. Sci. Wash.* **3**:123–28.
Schram, F. R. 1974. Mazon Creek caridoid Crustacea. *Fieldiana: Geol.* **30**:9–65.
Schram, F. R. 1976. Some notes on Pennsylvanian crustaceans of the Illinois basin. *Fieldiana: Geol.* **35**:21–28.
Schram, F. R. 1979. British Carboniferous Malacostraca. *Fieldiana: Geol.* **40**:1–129.
Schram, F. R. 1981a. Late Paleozoic crustacean communities. *J. Paleo.* **55**:126–37.
Schram, F. R. 1981b. On the classification of Eumalacostraca. *J. Crust. Biol.* **1**:1–10.
Wells, J. W. 1957. An anaspid crustacean from the Middle Devonian of New York. *J. Paleo.* **31**:983–84.

10

MYSIDACEANS

DEFINITION   Carapace typically envelops thoracomeres at least laterally, unfused to all but one to three of the most anterior segments; thoracopods with annulate exopods, the first and sometimes the second pair modified as pediform maxillipedes, lamellar epipod on first thoracopod; dendrobranchiate gills sometimes on body wall near leg bases; oöstegite brood pouch; well-developed tailfan; young shed as juveniles with complete set of appendages.

HISTORY   The first species of mysidacean, *Praunus flexuosus*, was recognized in 1776 by O. F. Müller. Originally classified along with stomotopods and leptostracans, it was Latreille in 1817 who erected the taxon Schizopoda for the mysids and leptostracans. The latter were soon separated from the mysidaceans. The former then remained alone in the Schizopoda until Milne Edwards described the first euphausian in 1830. In 1883 Boas suggested separate orders for Mysidacea and Euphausiacea and recognized two suborders of mysidaceans, the Lophogastrida and Mysida. G. O. Sars, in his monograph on the Challenger Schizopoda, rejected the suggestion of Boas, but the works of Hansen (1893) and Calman (1904) ensured the general acceptance of Boas' system within a larger group, Peracarida. This arrangement, like the original suggestion of Latreille, persisted unchallenged for some time until Schram (1981b) and Watling (1981, 1982) suggested possible alternative taxonomic schemes to explain mysidacean relationships. Schram (1984) proposed separate ordinal status for all three mysidacean suborders.

MORPHOLOGY   The carapace generally covers the entire thorax (at least laterally) and can be fused to the first to third thoracomeres. Variants occur; for example, in the genus *Longithorax* (Fig. 10-1F) the last thoracomere is completely exposed and twice as long as any of the anterior pleomeres. The carapace is typically undecorated except for a single cervical groove. The rostrum is generally small (Fig. 10-1B & D) to virtually absent (Fig. 10-1C), though in a form like *Gnathophausia* (Fig. 10-1A) the rostrum is long and serrate. The posterior margin of the carapace is frequently excavated dorsally to expose the posterior thoracomeres (Fig. 10-1C). The eyes are stalked and bear hexagonal facets.

The antennules (Fig. 10-2A) have three-segment peduncles and two well-developed flagella. Males in the Mysidae, however, typically bear a

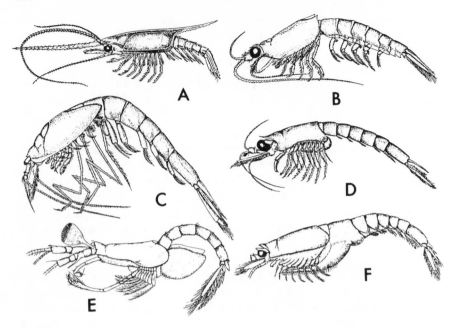

Fig. 10-1. Mysidacean body forms. (A) *Gnathophausia zoea*; (B) *Lophogaster typicus*; (C) *Eucopia sculpticauda*; (D) *Boreomysis arctica*; (E) *Arachnomysis leuckartii*; (F) *Longithorax fuscus*. (From Tattersall and Tattersall, 1951)

setose lobe, the processus masculinus (Fig. 10-3B), on the distal end of the peduncle; and at least in one form, *Mesopodopsis*, in addition to the processus masculinus, an accessory unsegmented process is developed in a manner akin to a third flagellum (Fig. 10-3D).

The antennae (Fig. 10-2B) have three segments in the protopod. The scaphocerite is a well-developed scale, except in the distinctive genera *Arachnomysis* and *Chunomysis* in which it is developed as a spine (Fig. 10-1E). The basal three or four segments of the flagellum are peduncular in form.

The labrum (Fig. 10-2C) is helmetlike, with the mandibles fitting into the 'cap.' The mandibles (Fig. 10-2D) are strikingly asymmetrical. The left lacinia mobilis serves to guide the right incisor process between itself and the left incisor process. The right lacinia is typically reduced or absent. Posterior to the laciniae are rows of spines [from which the laciniae are developed (Dahl and Hessler, 1982)]. The molar process can display a variety of structural forms, related to diet, or can even be absent altogether. The three-segment mandibular palp is well developed and frequently bears special distal setae for grooming other mouthparts. This palp is so enlarged in the genus *Petalophthalmus* that it is able to assist in gathering and processing food (Fig. 10-3A). The mandibles of lophogastridans are among the most peculiar such appendages seen in malacostracans, with heavy curved surfaces, cusps and processes developed as massive toothlike structures, all shaped to tightly interlock with features on the opposing member of the pair. The arrangement is such that, though lophogastrids are generally thought to be the most primitive of caridoids, their jaws are among the most specialized (Dahl and Hessler, 1982). The rectangular paragnaths are deeply cleft, with their medial margins generally setose and occasionally the lateral lobes rather elaborate (Fig. 10-2E).

The maxillules (Fig. 10-2F) of all mysidacean forms conform to a

Fig. 10-2. Appendages of *Spelaeomysis longipes*. (A) antennule; (B) antenna; (C) labrum; (D) mandible; (E) paragnath; (F) maxillule; (G) maxilla; (H) first thoracopod (= maxillipede); (I) seventh thoracopod; (J) base of male eighth thoracopod with genital papilla; (K) first pleopod; (L) telson and uropod. (Modified from Pillai and Mariamma, 1964)

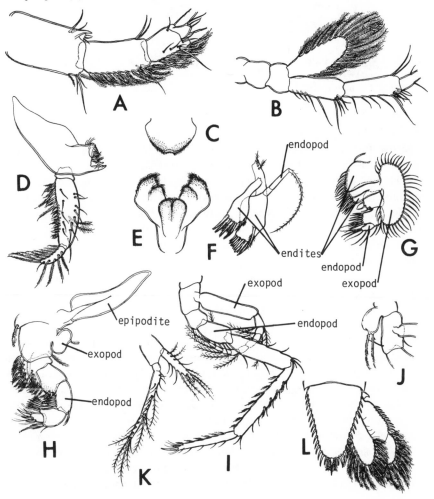

Crustacea

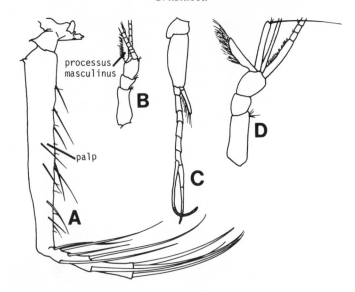

Fig. 10-3. (A) mandible of *Petalophthalmus armiger* (from Mauchline, 1980), (B, C) *Tephromysis louisianae* (from Banner, 1953), (B) antennule; (C) fourth pleopod of male; (D) antennule of *Mesopodopsis slabberi* (From Tattersall and Tattersall, 1951).

general plan. They consist of two well-developed endites, the distal one bearing spinelike setae. *Gnathophausia* (and rarely some mysidans) also have a two-segment endopodal palp directed posteriorly.

The maxillae (Fig. 10-2G) have well-developed setose endites, the distal lobe being deeply cleft. The exopod is developed as a large ovoid flap, while the endopodal palp consists of two large and flattened segments. All forms pretty well conform to this basic plan, though the maxillae of *Petalophthalmus,* in contrast to the specialized mandibular palps of that genus, are only weakly developed.

The first thoracopods are specialized as pediform maxillipedes (Fig. 10-2H). The endopod is shortened over the typical thoracopod length, and sometimes the proximalmost segments possess enditic lobes to match those on the basis. These first maxillipedes also bear a large lamellar epipodite that is carried posteriorly under the carapace. In some taxa, the second thoracopods may also be modified as pediform maxillipedes. However, the Paleozoic suborder Pygocephalomorpha, possess two pairs of maxillipedes that bear large gnathobasic endites and reduced endopodal palps.

The posterior thoracopods (Fig. 10-2I), in their most generalized form, have unmodified endopods, and exopods that consist of a large peduncular segment bearing an annulate flagellum. In the lophogastrid *Eucopia* (Fig. 10-1C), these appendages are subchelate; the three posterior pairs are also rather thin and elongate. Endopodal joints can be fused: sometimes the carpus and propodus beyond the knee or, as in the pygocephalomorphs, the ischium and merus above the knee. Females bear oöstegites on seven (Lophogastrida, Pygocephalomorpha, petalophthalmids, and *Boreomysis*) or on the last two or three (all other Mysidae) thoracopods. The sternites are usually well developed in order to spread the thoracopods apart and accommodate the brood pouch. A single brood chamber is the norm, though in *Heteromysis* the oöstegites bend up at the midline to form separate left and right chambers. In some genera, like *Gastrosaccus* and *Bowmaniella*, the pleura of the first pleomere are enlarged and help form the posterior part of the brood chamber. The lophogastridans also bear a set of dendrobranchiatelike gills on the second through seventh thoracopods; these appear to have probably been originally developed from epipodites, though in some instances they are now borne on the body wall. The eighth thoracopod in males can bear genital papillae medially.

Hessler (1982) has documented a rather unusual array of movements about the coxa–basis joint in mysidacean thoracopods. The abduction–adduction movement occurs around an anteroposterior hinge line (Fig. 10-4A), wherein the posterior condyle projects some distance inward from the

Fig. 10-4. Coxa–basis articulation in *Lophogaster typicus*. (A) View looking ventrad 'through' coxa at coxa–basis articulation; letters 'w' and 'v' mark condylar articulations of the anteroposterior hinge axis around which muscles 1 and 4 abduct while muscles 2, 3, and 5 adduct; (B) sagittal view of coxa–basis joint showing flexible cuticle below 'w'; (C) sagittal view after contraction of muscle 5 as a promotor to pivot the basis around condyle 'v,' with muscles 3 and 4 acting as remotors. (From Hessler, 1982)

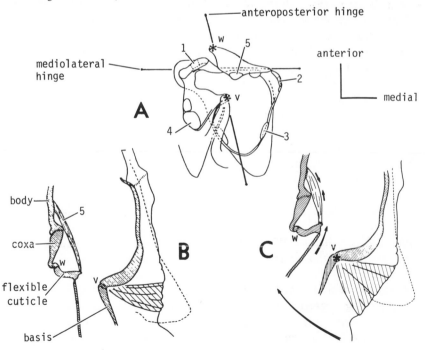

normal posterior face of the protopod. However, a promotion–remotion action is also possible about this same joint and occurs around a mediolateral hinge, oriented at approximate right angles to the anteroposterior axis and located somewhat distal to it. A triangle of flexible cuticle extends below the anterior condyle of the anteroposterior axis to the medioventral hinge axis. Contraction of the promotor muscles (Fig. 10-4B, C) causes the anterior face of the basis to telescope into the coxa while pivoting on the posterior condyle. This double action about a single joint is a unique talent of mysidaceans and is observed in both the lophogastridans and mysidans.

The pleopods are only well developed as a rule on the lophogastridans, while in mysidans (Fig. 10-2K) they are almost always vestigial in females. When present these appendages are frequently formed as annulate, setose rami. The base of the male endopods are usually developed with a lobe, plate, or process generally thought to be respiratory in function. Occasionally, the third or fourth pair of pleopods can be modified for copulatory purposes (Fig. 10-3C).

The telson (Fig. 10-2L) can exhibit a variety of forms that range from rounded, to triangular, to more or less bifid. The uropods (Fig. 10-2L) have a diaeresis on the exopod, and in Mysidae there is a statocyst near the base of the endopod.

The mouth opens into the esophagus, which is lined with spines pointing inward toward the stomach. The anterior or cardiac stomach can project forward into the anterior reaches of the head and can be quite larger than the posterior pyloric stomach. These chambers are variously lined, depending on species, with ridges equipped with spinose setation. Two sets of diverticula are noted in connection with the pyloric stomach (Fig. 10-5). The typically small dorsal diverticulum is generally a single anteriorly projecting outpocketing, though this organ may be paired in some species. This structure secretes the peritrophic membrane around fecal material before the feces proceed down the midgut. The lateral diverticula or digestive caeca consist of five pairs of digitiform glands opening by a common duct from the pyloris–midgut boundary. The largest pair of the caeca can extend as far back as the fourth pleomere. Food circulates freely among all the pouches of this digestive gland and is both digested and absorbed there. Fecal pellets, once formed, for the most part pass very quickly through the mid- and hindguts to be voided at the anus.

The mysidacean circulatory systems seem to be distinguishable into distinct lophogastridan and mysidan patterns (Fig. 10-6). In both groups (Belman and Childress, 1976; Mauchline, 1980) the heart is located in the thorax. In *Gnathophausia* it extends through most of the thorax, while in mysidans it is limited to the more posterior thoracic segments. Blood leaves the heart in both groups through cephalic and abdominal arteries, 'hepatic' or anterolateral arteries, and sternal or descending arteries. These last are single in mysidans and paired in *Gnathophausia*, and in either case join a single ventral aorta. There are segmental arteries that

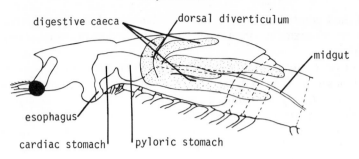

Fig. 10-5. Diagrammatic view of anterior digestive system of a mysid. (From Mauchline, 1980)

arise from the abdominal aorta in the pleon, and in *Gnathophausia* segmental arteries from the heart supply in part the digestive caeca in the thorax. Both groups also have a frontal heart, or cor frontale, in the head that beats in synchrony with the main heart. Blood collects in sinuses and drains into the pericardium in the thorax. One important detour occurs in the anterior region where, in the mysidans, blood first flows into respiratory sinuses in the anterior part of the carapace before returning to the pericardium. Mauchline (1980) points out that much of our knowledge of mysidacean circulatory systems is not consistent between authorities, so further comparative work is needed to adequately understand this system.

Differences in the excretory system are also noted between the two main groups of living mysidaceans. Mysidans have well-developed antennal glands. Lophogastridans, however, have both antennal and maxillary

Fig. 10-6. Circulatory systems of mysidaceans. (A) generalized mysidan; (B) the lophogastridan *Gnathophausia ingens*. (Modified from Mauchline, 1980)

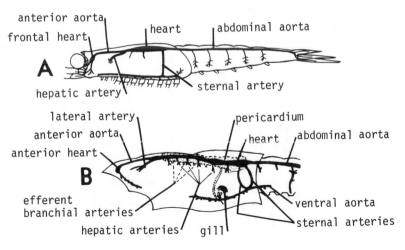

glands (first pointed out by Cannon and Manton, 1927a). Excretion is also carried out by segmental nephrocytes located near the proximal ends of the thoracopods.

The ovaries are located in the thorax, though when gravid they extend into the anterior portions of the abdomen. Developing eggs are held in the dorsal parts of the gonad until sufficient yolk is deposited in the cell. Fertilization then occurs in the brood pouch after the eggs are shed.

The male system is quite elaborate (Fig. 10-7). The testes are developed as paired cords and are located in the ventral portion of the thorax. The developing spermatocytes cause the formation of serial spermatic sacs along the length of the cord, and these in turn open dorsally into a series of spermatic pouches where the spermatids mature to spermatozoa. The pouches are connected dorsally by narrow ducts into the tubular seminal vesicles, and these empty posteriorly into the ejaculatory duct and the genital papillae. The papillae on the eighth thoracopod are usually small conical structures, but in some genera they are so long as to extend forward as far as the cephalon. The sperm are elongate and flagellate.

The brain is complete with proto-, deuto-, and tritocerebra. The thoracic ganglia are generally fused into a single elongate mass, though individual positions of ganglia can usually be discerned. The abdominal ganglia, however, are separate and distinct. Manton (1928a, b) discussed the important issue of the remnants of the seventh abdominal segment. Distinct sixth and seventh abdominal ganglia arise in the course of development of *Hemimysis* and *Lophogaster*. The uropods are associated with the sixth ganglion and are drawn posteriad as development progresses. Eventually the sixth and seventh ganglia fuse, and the uropods become terminal

Fig. 10-7. Male reproductive system as seen in *Archaeomysis* and *Neomysis*. (A) lateral diagram; (B) dorsal diagram. (Modified from Mauchline, 1980)

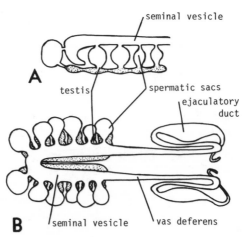

on the abdomen. The groove evident on the last pleomere in front of the uropods of the lophogastrids is a remnant of this fusion.

A peculiar feature of the sensory anatomy in mysidaceans is the pore system to be found on the dorsal cuticle. These pores are associated with gland openings as well as chemo- and tactoreceptors. Mauchline (1980) provides a review of these features. The pores appear to form definite patterns specific to genera and even species. The only thing similar to these structures is the posterodorsal pore seen on the carapace of the Carboniferous belotelsonidaceans.

NATURAL HISTORY   Mysidaceans occupy a number of different habitats but are most commonly observed in epibenthic or inbenthic situations. They are ubiquitous in the waters of the world (fresh, brackish, and marine) and are known to occur at depths down to 7210 m. Although there are nektonic forms, known mysidaceans for the most part seem to have relinquished dominance of the swimming life-style to the euphausiaceans. One of the more interesting aspects of the freshwater radiation of mysids is the development of a modest array of cave-dwelling forms.

Methods of locomotion vary between the two main mysidacean groups. Bottom dwellers of any of the major groups, of course, are ambulatory on their thoracopods, and all forms use the caridoid escape reaction in emergencies. However, lophogastridans in the normal course of events swim with their pleopods. The thoracic exopods in this group are only used to set up respiratory currents around the gills. Mysidans, on the other hand, have well-developed pleopods only on males, and these are sometimes modified for copulatory purposes. Swimming in these forms is achieved by the rotatory beating of the thoracic exopods in order to create currents around the animal that move it forward.

Modes of feeding, as in locomotion, are distinctly different in the two main groups of mysidaceans. The classic study on *Hemimysis* (Cannon and Manton, 1927b) has now been shown to be incorrect. The whirling of the thoracic exopods was said to create a vortex that supposedly sucked particulate matter up the vortex and directed material into the midventral food groove under the thorax. Particulate matter was supposed to move forward to the mouth region. The setae on the mouthparts supposedly served to filter the particulate matter out of the midventral current before transferring material to the mouth.

Attramadal (1981), however, showed in a study of feeding in *Hemimysis lamornae* and *Praunus flexuosus* that there were no filtration currents, no anterior movement of material in a ventral food groove, and no maxillary vibrations to filter material. The mistaken conclusions of Cannon and Manton (1927b) were arrived at because of the too artificial conditions of their original experiments. Attramadal (personal communication) has noted in the course of study of some 10 species of mysids that they feed in one of two ways. Either they actively seize large particles with their thoracic endopods, which form a 'food basket,' or fine particles are trapped on body and limb surfaces and then in the course of cleaning are transferred to the mouth. Attramadal has found that in *P. flexuosus* the antennules and antennae are groomed directly by the mouthparts, while the pleopods and thoracic exopods are groomed by the thoracic endopods, which then transfer material to the mouthparts.

Lophogastrids have a distinctly different pattern (Manton, 1928a). *Gnathophausia* has a maxillary filter but does not use the thoracic exopods to set up a food-gathering current. The respiratory current created by the thoracopodal exopods is directed up under the carapace to bathe the gills. An exhalant current can be directed forward past the maxillae, which typically serve to seal off the front of the respiratory chamber in order to aid the filtering mechanism. *Lophogaster*, on the other hand, lacks a 'food basin' behind the mouth in which to filter food. The massive mandibles, lack of elaborate setation on the mouthparts, and the deflection of the exhalant respiratory current laterad from under the carapace seem to indicate that this genus feeds off large food matter found in the course of scavenging activity on the bottom.

During reproduction the males do not actively search for females. After shedding a previous brood, the female soon molts and is ready to breed again. At that time she apparently produces a pheromone of some kind that stimulates the antennules or the antennular processus masculinus of nearby males. Actual mating is very quick and takes place at night. The male comes to lie under the female either head-to-tail and venter-to-venter, or somewhat doubled up and grasping the anterior part of the abdomen with his antennae. The sperm are either injected into the brood pouch or shed between the mating individuals and swept by currents produced by the thoracic exopods into the marsupium. The copulating pair soon part, and within half an hour the eggs are extruded into the brood pouch and fertilized there.

Incubation period and frequency of mating depends on species and temperature and can range from few weeks to several months. Once mature the females can keep breeding until they die. A variety of breeding strategies to this end has evolved within the group. *Gnathophausia ingens* is the only mysidacean noted to have an especially long life history, taking seven years to mature and then harboring a single brood for more than a year before dying (Childress and Price, 1978). Other species, such as *Mysis relicta*, have relatively long maturation times, over a year and up to two years, and reproduce once or twice before dying. As pointed out by Mauchline (1980), these types of breeding strategies average less than one generation a year. In the most typical annual cycle breeding occurs once in the spring. The adults die later in the summer while the juveniles overwinter to breed the following spring. A variant of this annual cycle occurs in species with two complete generations per year. Breeding occurs in spring, whereupon the adults die and the juveniles quickly mature by late

summer. This group then breeds in autumn and dies, while the second juveniles overwinter and breed the following spring. Another reproductive
strategy is seen in a life cycle that ranges from somewhat less than and up
to more than a year. Each female of this type, however, may breed anywhere from one to three or more times. Broods may occur at three peak
times, or the breeding may be more or less continuous with no distinct juvenile peaks and valleys throughout the year.

Population ratios of male to female vary extensively, ranging from 1:1
to an extreme in one instance of 1:59 (Wigley and Burns, 1971). Rarely do
males outnumber females, though one such case, 1:09, has been reported
for *Heteromysis elegans* by Brattegard (1974). Males in some species perish
immediately after breeding, thus effectively leaving all food resources to
brooding females and juveniles.

There are almost 800 species of mysidaceans. However, complete biogeographic data are either lacking, as is the case for poorly known species
or have simply never been extracted for analysis from the basic literature.
In the case of the latter, the effectiveness of such an analysis would be
rather suspect since many parts of the world have simply not been adequately sampled. Gordon (1957), Mauchline and Murano (1977), and Mauchline (1980) outline the taxonomic literature that would form the basis for
such an analysis.

Some limited analyses, however, can prove enlightening, for example,
those of the hypogean species (Table 10-1). The number of cave-dwelling
forms is limited: several species of Mysinae and the small families Lepido-
mysidae (*Spelaeomysis*) and Stygiomysidae. One can immediately note
that these species are centered around the Caribbean and Mediterranean
seas with some interesting but not improbable exceptions in the Canary
Islands, Peru, and the westernmost Indian Ocean. These distributions are
best understood when considered in light of the late Mesozoic continent of
Pangaea, when all these regions were contiguous around the central and
western Tethys Ocean. Since hypogean forms have limited distributional
abilities, the present biogeography of cavernicolous mysidans probably
represents a vicariance of a late Mesozoic distribution. This biogeographic
pattern in turn serves to place some time limits on the evolution of this portion of the mysidan radiation as well as to indicate areas wherein further
exploration would most likely bring to light more hypogean forms. Analyses of this kind, if begun now for other mysidaceans, would undoubtedly
reveal patterns and present questions that could suggest areas where future
collection should be concentrated.

DEVELOPMENT   The most comprehensive work on mysidacean ontogeny is
that of Manton (1928b) on *Hemimysis lamornae*, with confirmation and
some additional data added by Needham (1937) on *Neomysis vulgaris* and
Nair (1939) on *Mesopodopsis orientalis*. (Manton reviewed the defects and

Table 10–1. Distribution of hypogean mysidans

Taxon	Geographic region
Mysinae	
Anisomysis vasseuri	Madagascar
Antromysis cenotensis	Yucatan, Mexico
A. cubanica	Cuba
A. juberthiei	Cuba
A. pedkorum	Jamaica
A. reddelli	Oaxaca, Mexico
Hemimysis speluncola	France
Leptomysis burgii	France
Troglomysis vjetrenicensis	Yugoslavia
Heteromysoides speluncola	Canary Islands
Lepidomysidae	
Spelaeomysis bottazii	Italy
S. cardiosomae	Columbia, Peru
S. longipes	India
S. nuniezi	Cuba
S. olivae	Oaxaca, Mexico
S. quinterensis	Tamaulipas, Mexico
S. servatus	Zanzibar
Stygiomysidae	
Stygiomysis holthuisi	Lesser Antilles, Puerto Rico
S. hydrutina	Italy

inaccuracies recorded in early works on the subject.) The broad outlines of mysidacean development can be summarized here.

Early cleavage is irregular and occurs within the yolk at sites scattered about the egg. At about the 32-cell stage protoplasm begins to surface, so that by about the 128-cell stage the yolk is entirely interior to the blastomeres. The thicker cells are found at the animal pole and will eventually form the germinal disc, while thinner cells occur at the vegetal pole.

The germinal disc (Fig. 10-8A) is recognized at the time gastrulation occurs. The blastopore region contains presumptive endoderm, mesoderm, and germinal cells. These cell types, however, migrate independently from scattered individual cells in the blastopore region (Fig. 10-8B). The germinal cells leave the blastopore and migrate anteriad into the head or naupliar region to form a genital rudiment (Fig. 10-8C). Two types of mesoderm cells arise. Eight mesodermal teloblasts leave the blastopore and come to lay just below the ectodermal teloblasts. Subsequent to the location of the mesoteloblasts, another group of mesoderm cells migrate anteriad to flank the genital rudiments; these form the head mesoderm. The formation of endoderm is a prolonged process (Fig. 10-8C & E), with cells detaching from the blastopore and wandering into the yolk until the teloblasts cease dividing with the completion of abdomen formation.

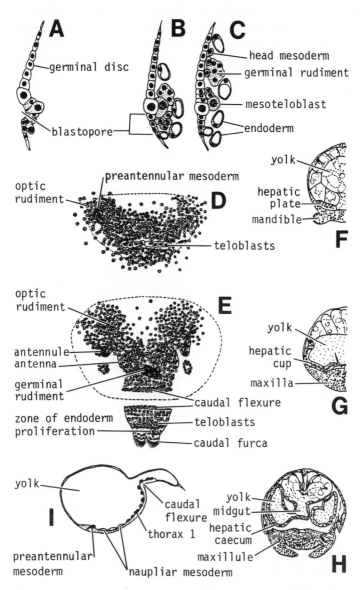

Fig. 10-8. Early development in *Hemimysis lamornae*. (A) initiation of blastopore; (B) gastrulation beginning; (C) gastrulation complete; (D) differentiation of germinal disc; (E) egg-nauplius formation; (F) formation of hepatic plate; (G) formation of hepatic cups; (H) completion of folding to form hepatic caeca; (I) diagrammatic sagittal section of hatchling. (From Manton, 1928b)

119

The ectodermal teloblasts remain on the anterior lip of the blastopore and form a row of cells with a median and seven pairs of flanking teloblasts (Fig. 10-8D). Division of ecto- and mesoteloblasts to form the postnaupliar segments are synchronized. After teloblast activation, a furrow forms in front of the ecto- and mesoteloblasts. This furrow deepens and arcs, allowing the posterior germinal disc to fold forward in order to form the caudal papilla (Fig. 10-8E).

During gastrulation the edges of the germinal disc form the optic rudiments, lateral and somewhat anterior to the blastopore. As the teloblasts begin to proliferate, these optic anlagen move anteriad and mesiad to form a V-shaped thickening in front of the blastopore (Fig. 10-8E). One cell differentiates out of the optic rudiments on each end of the V and sinks inward from the surface to form the preantennal mesoderm (Fig. 10-8D). At this time the anlagen of the naupliar appendages differentiate, first those of the antennules and antennae (Fig. 10-8E), followed thereupon by those of the mandibles. The initiation of these egg nauplius structures in the ectoderm induces the head mesoderm to proliferate and migrate anteriad as a pair of bands. These subsequently bulge up into the appendage anlagen. Also at this time the genital rudiments migrate posteriad out of the naupliar region into the teloblast region (Fig. 10-8I).

The gut develops from several sources. The endoderm only contributes to the midgut epithelium and the anterior midgut middorsal diverticulum that extends into the head. The ectoderm forms the stomodeum, just after hatching, and the proctodeum, just after the last teloblast division. In addition to the muscles of the mandible, the head mesoderm of that segment also gives rise to the hepatic diverticula. The mandibular mesoderm separates into two masses, one forms the jaw muscles and the other produces a plate joined medially behind the stomodeum (Fig. 10-8F). The lateral portions of this plate grow posteriad along the yolk sac, and these sheets eventually hollow out and become cuplike (Fig. 10-8G). The adjacent yolk cells then break down, liberating their contents into the cups. These cups then roll inward to form tubes that unite to the anterior part of the midgut just posterior of the stomodeum (Fig. 10-8H). These tubes are the primary hepatic diverticula, which continue to subdivide in the course of development to form the adult digestive caeca.

All development described to this point takes place for the most part in the egg state, that is, within the vitelline membrane. The embryo hatches into stage 2 when the yolk suddenly absorbs water and swells. Actual hatching occurs when the caudal papilla reflexes and straightens the body, causing the vitelline membrane to rupture (Fig. 10-8I). The caudal furrow at this stage has migrated posterior to the level of the fifth and sixth thoracic segments, when only the first two abdominal segments are clearly demarcated. The hatchling is covered by the naupliar cuticle with the naupliar appendages and caudal furca visible on the surface. Development throughout stage 2 occurs beneath this cuticle, which is lost at the first molt, which initiates stage 3.

Two particularly interesting phenomena occur after hatching during stage 2. First, the preantennal mesodermal masses grow posteriad to the stomodeum (Fig. 10-8J) and spread over the lateral and dorsal aspects of that organ. Hollow tubes develop within the mesoderm, termed the preantennal coelomic sacs, which eventually disappear. At this time an ectodermal intucking forms a median yolk septum or membrane (Fig. 10-9A, B), which divides the cephalic yolk mass in two. This intucking occurs just as a dorsal ectodermal thickening, or dorsal organ, forms and extends down into the yolk (Fig. 10-8A, B). When the resulting septum reaches the

Fig. 10-9. Late development in *Hemimysis lamornae*. (A) initiation of dorsal organ with yolk septum; (B) dorsal growth of preantennulary mesoderm; (C) formation of anterior organs; (D) last division of ectoteloblasts with sixth and seventh ganglion anlagen; (E) posterior growth of vents of sixth pleomere and telson; (F) fusion of sixth and seventh ganglia. (From Manton, 1928b)

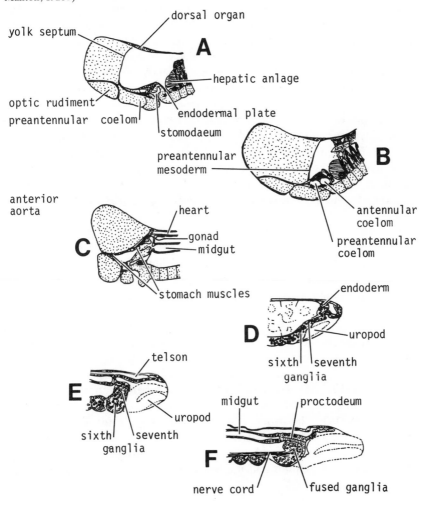

preantennal mesoderm, it induces the mesoderm to grow dorsad toward the dorsal organ (Fig. 10-9B). The dorsal organ grows posteriad toward the heart, whose anterior end is then in the maxillary segment, pulling strands of the preantennal mesoderm along with it. These mesodermal strands then form the walls and floor of the anterior aorta, while the rest of the preantennal mesoderm forms some stomach muscles (Fig. 10-9C).

The second interesting phenomenon that occurs after hatching is to be found in the tail region. The last divisions of the teloblasts result in two cell rows. In the mesoteloblasts these are considered to represent the sixth and seventh abdominal somites. The development of the seventh somite lags far behind that of the sixth, and coelomic sacs appear only briefly after the formation of the proctodeum. The appearance of this latter organ also serves to separate the seventh pleomere mesoderm from the telson mesoderm. While these events are going on, the posterior ectoderm buds a small seventh abdominal ganglionic anlage behind that of the sixth. The uropod anlagen are also associated with the sixth somite (Fig. 10-9D). The venter of the sixth pleomere grows posteriad, which relocates the uropods and sixth ganglia to a position posterior and ventral to the seventh ganglionic anlagen (Fig. 10-9E). Eventually the only exterior indications of a seventh pleomere (Fig. 10-9F) are a groove on the cuticle of the sixth pleomere (referred to above) and an interruption of fibers in the longitudinal muscles of the adult sixth abdominal segment.

Stage 2 is marked externally by development of the trunk appendages and eyes. The termination of this stage is marked by the first molt. Stage 3, with the appearance of stalked eyes and rudimentary appendages, marks a completion of brood pouch development. The hatchlings are all closely packed in the brood pouch and face posteriorly. Occasionally, movements of the brooding female dislodges a hatchling. However, this is easily swept up in locomotory currents and, instead of eating the wayward individual, the female transfers it back into the anterior part of the brood pouch. This 'adoption' also extends to any stray hatchling that may be encountered, not just her own, such that a female may be found brooding eggs and/or hatchlings at several different stages. Stage 3 is ended with the second molt, which occurs at or just after the expulsion of hatchlings from the brood pouch.

FOSSIL RECORD   For a group with a strong preference for benthic habitats, we should not be too surprised at the extent of the fossil record for the mysidaceans. It is especially good for Paleozoic forms.

The best record occurs for the Permo-Carboniferous Pygocephalomorpha (Fig. 10-10A, B). These distinctive reptant, benthic forms were among the most important components of late Paleozoic crustacean communities (Schram, 1981a). The group is characterized by a subtriangular field of sternites in the thorax [paralleled by the form later developed independently in the palinuran decapods of the Meso- and Cenozoic (Schram, 1974)], fusion of ischium and merus in the thoracic endopods, elaborate

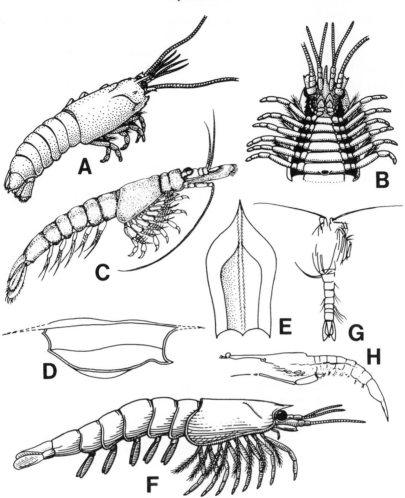

Fig. 10-10. Fossil mysidaceans. (A, B) *Mamayocaris jaskoskii*; (C) *Schimperella beneckii*; (D) *Dollocaris ingens*; (E) *Kilianicaris lerichei*; (F) *Peachocaris strongi*; (G) *Elder ungulatus*; (H) *Francocaris grimmi*. (A, B, F from Schram, 1974; C, D, E, G, H from Hessler, 1969)

development of the tailfan including at least one set of lobelike caudal furca (Fig. 10-10A), and a prominent sexual dimorphism. It was formerly thought that females bore a large seminal receptacle opening on the eighth sternite. Such a structure, however, would be rather unusual in a crustacean. It seems much more likely that this structure (Fig. 10-10B) is a male genital cone, akin to that seen in the tanaidaceans. Four families are currently recognized in this group; however, the characters used in this regard (spination of the margin and carinae on the carapace, and degree of abdominal flexion) probably have not produced an entirely natural classification. Future revision will undoubtedly be necessary as knowledge of the pygocephalomorphs continues to improve.

Fossil Lophogastrida should also be recognized from known material, since these are benthic animals. *Schimperella beneckei* Bill, 1914 from the Triassic of Alsace is similar in its anatomy to that seen in the Lophogastridae (Fig. 10-10C). Hessler (1969) felt that the poor development of abdominal pleura and the presence of processus masculinus on the antennules allied *Schimperella* with the mysidans. However, the well-developed thoracopods and pleopods, lack of uropodal statocysts, incipient (though admittedly weak form of the abdominal pleura), and carapace development are really more like conditions seen in such lophogastridans as *Chalaraspidium* or *Eucopia* (see Fig. 10-1C). An alternative, though less likely, assignment for *Schimperella*, is within the Petalophthalmidae. The presence of a processus masculinus on these fossils might warrant at some future date a separate family status within the Lophogastrida for *Schimperella*.

Dollocaris ingens (Fig. 10-10D) and *Kilianicaris lerichei* (Fig. 10-10E), described by van Straelen (1923) from the upper Jurassic of France, are nothing but poorly preserved carapaces. They have vague structural resemblances to the *Gnathophausia* form, but the higher taxonomic affinities of these fossils must remain uncertain.

One other fossil form that should most probably be allied with the lophogastridans is *Peachocaris strongi* from the Pennsylvanian of Illinois (Fig. 10-10F). Once assigned to the anthracophausiid eocaridaceans, it seems better to ally *Peachocaris* to the other lophogastridans as a separate family. Its well-developed abdominal pleura, pleopods, and thoracopods (this last with a well-developed peduncle at the base of the exopods), and the apparent lack of uropodal statocysts argue for lophogastridan affinities. The apparent lack of any obvious specialization of the first thoracopod as a maxillipede and flaplike pleopods might encourage someone to place *Peachocaris* in its own suborder, but based purely on such primitive characters alone, only a distinct family status is warranted.

Possible mysidan fossils are rare and assignment of these specimens to the suborder uncertain. *Elder ungulatus* Münster, 1839 (Fig. 10-10G) and *Francocaris grimmi* Broili, 1917 (Fig. 10-10H) from the Jurassic Solenhofen Limestone of Germany are too poorly understood to permit an unqualified assignment.

TAXOMONY In light of the arrangement of recent forms outlined by Mauchline (1980) and the remarks above on fossils, the following classification of mysidaceans to family level is offered. The suggestion of Schram (1984), that each of the three major groups represents a separate order, is followed.

In the internal anatomy of the maxillipede region, mysidans are clearly different from lophogastridans. In *Neomysis americana* (Fig. 10-11A) the maxillipede or first thorocopod is not associated with the cephalic limbs, and indeed is separated from the maxillae by a well-developed skeletal bar

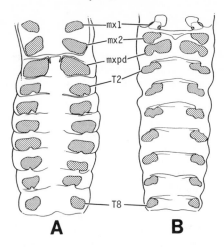

Fig. 10-11. Internal views of the posterior cephalic and thoracic sternites. (A) *Neomysis americana*. Note skeletal bar that separates the thorax from the cephalon. (B) *Gnathophausia ingens*. Note association of the maxillipede with the cephalic region rather than the thorax. Openings into limbs: mx1—maxillules, mx2—maxillae, mxpd—maxillipedes, T2—second thoracopod, T8—eighth thoracopod.

that extends across the anterior part of the thorax. The openings into the thorocopods are associated with weakly developed bars that arise on the posterior rims of the leg openings and extend toward the midline. However, in *Gnathophausia ingens* (Fig. 10-11B) the maxillipedes are more clearly associated with the maxillae than with the second thoracopods. Indeed, no distinct cuticular bars separate the maxillae and maxillipede openings, though a distinct skeletal bar and pair of apodemes separate the maxillules from the maxillae. The openings into the thoracopods are associated with modestly developed bars that arise on the anterior rims of the leg openings and extend across the midline.

Order Lophogastrida Boas, 1883 Pennsylvanian–Recent
    Family Peachocarididae, nov. Pennsylvanian
    Family Lophogastridae Sars, 1857 ?Jurassic–Recent
    Family Eucopiidae Dana, 1852 ?Triassic-Recent
Order Pygocephalomorpha Beurlen, 1930 Carboniferous–Permian
    Family Tealliocarididae Brooks, 1962 Lower Carboniferous
    Family Pygocephalidae Brooks, 1962 Carboniferous–Permian
    Family Notocarididae Brooks, 1962 Permian
    Family Jerometichnoriidae Schram, 1978 Permian
Order Mysida Boas, 1883 (?Jurassic) Recent
    Family Petalophthalmidae Willemoës–Suhm, 1875
    Family Mysidae Latreille, 1803
    Family Lepidomysidae Fage, 1924
    Family Stygiomysidae Caroli, 1937

REFERENCES

Attramadal, Y. G. 1981. On a non-existent ventral filtration current in *Hemimysis lamornae* and *Praunus flexuosus*. *Sarsia* **66**:283–86.

Banner, A. H. 1953. On a new genus and species of mysid from southern Louisiana. *Tulane Stud. Zool.* **1**:1–8.

Belman, B. W., and J. J. Childress. 1976. Circulatory adaptations to the oxygen minimum layer in the bathypelagic mysid *Gnathophausia ingens*. *Biol. Bull.* **150**:15–37.

Bill, P. C. 1914. Über Crustaceen aus dem Voltziensandstein des Elsass. *Mitt. Geol. Landes. v. Elsass-Loth.* **8**:289–338, pls. 10–16.

Boas, J. E. V. 1883. Studien über die Verwandtschaftsbeziehungen der Malakostraken. *Morph. Jahrb.* **8**:485–579.

Brattegard, T. 1974. Additional Mysidacea from shallow water on the Caribbean coast of Colombia. *Sarsia* **57**:47–86.

Broili, F. 1917. Eine neue Crustaceen Form aus dem lithographischen Schiefer des oberen Jura von Franken. *Cbl. Min. Geol. Paläontol., Jahrb. 1917*:426–29.

Calman, W. T. 1904. On the classification of Malacostraca. *Ann. Mag. Nat. Hist.* (7)**13**:144–58.

Cannon, H. G., and S. M. Manton. 1927a. Notes on the segmental excretory organs of Crustacea. *J. Linn. Soc., Lond. Zool.* **36**:439–56.

Cannon, H. G., and S. M. Manton. 1927b. On the feeding mechanism of a mysid crustacean, *Hemimysis lamornae*. *Trans. Roy. Soc. Edinb.* **55**:219–53, 4 pls.

Childress, J. J., and M. H. Price. 1978. Growth rate of the bathypelagic crustacean *Gnathophausia ingens*. 1. Dimensional growth and population structure. *Mar. Biol.* **50**:47–62.

Dahl, E., and R. R. Hessler. 1982. The crustacean lacinia mobilis: a reconsideration of its origin, function, and phylogenetic implications. *Zool. J. Linn. Soc.* **74**:133–46.

Gordon, J. 1957. A bibliography of the order Mysidacea. *Bull. Am. Mus. Nat. Hist.* **112**:283–393.

Hansen, H. J. 1893. Zur Morphologie der Gleidmassen und Mundteile bei Crustaceen und Insecten. *Zool. Anz.* **16**:193–98, 201–12.

Hessler, R. R. 1969. Peracarida. In *Treatise on Invertebrate Paleontology*, Part R, *Arthropoda* **4**(1) (R. C. Moore, ed.), pp. R360–93. Geol. Soc. Am. and Univ. Kansas Press, Lawrence.

Hessler, R. R. 1982. The structural morphology of walking mechanisms in eumalacostracan crustaceans. *Phil. Trans. Roy. Soc. Lond.* (B)**296**:245–98.

Manton, S. M. 1928a. On some points on the anatomy of the lophogastrid Crustacea. *Trans. Roy. Soc. Edinb.* **56**:103–19.

Manton, S. M. 1928b. On the embryology of a mysid crustacean, *Hemimysis lamornae*. *Phil. Trans. Roy. Soc. Lond.* (B)**216**:363–463.

Mauchline, J. 1980. The biology of mysids. *Adv. Mar. Biol.* **18**:1–369.

Mauchline, J., and M. Murano. 1977. World list of the Mysidacea. *J. Tokyo Univ. Fish.* **64**:39–88.

Münster, G. 1839. Über die fossilen langschwänzigen Krebse in Kalkschiefern von Bayern. *Beitr. Petrefaktenk.* **2**:78–81.

Nair, K. B. 1939. The reproduction, oögenesis, and development of *Mesopodopsis orientalis*. *Proc. Ind. Acad. Sci.* **9**:175–223.

Needham, A. E. 1937. Some points on the development of *Neomysis vulgaris*. *Quart. J. Micr. Sci.* **79**:559–88.

Pillai, N. K., and Mariamma, T. 1964. On a new lepidomysid from India. *Crustaceana* **7**:113–24.

Schram, F. R. 1974. Mazon Creek caridoid Crustacea. *Fieldiana: Geol.* **30**:9–65.

Schram, F. R. 1981a. Late Paleozoic crustacean communities. *J. Paleo.* **55**:126–37.

Schram, F. R. 1981b. On the classification of Eumalacostraca. *J. Crust. Biol.* **1**:1–10.

Schram, F. R. 1984. Relationships within eumalacostracan Crustacea. *Trans. San Diego Soc. Nat. Hist.* **20**:301–12.

Tattersall, W. M., and O. W. Tattersall. 1951. *The British Mysidacea.* Ray Society, London.

van Straelen, V. 1923. Les mysidacés du Calloviande la Boulte-sur-Rhône. *Bull. Soc. Geol. Fr.* (4)**23**:432–39.

Watling, L. 1981. An alternative phylogeny of peracarid crustaceans. *J. Crust. Biol.* **1**:201–10.

Watling, L. 1982. Peracaridan disunity and its bearing on eumalacostracan phylogeny with a redefinition of eumalacostracan superorders. *Crust. Issues* **1**:213–28.

Wigley, R. L., and B. R. Burns. 1971. Distribution and biology of mysids from the Atlantic coast of the United States in the NMFS Woods' Hole Collection. *Fish. Bull. NOAA* **69**:717–46.

11

MICTACEA

DEFINITION   Head shield with lateral folds enveloping the sides of the cephalon; lobed eyes may or may not be present; first thoracomere fused to head; antennule biramous; antenna with reduced scale; mandible with well-developed incisor and molar processes, lacinia mobilis, and palp; maxillules and maxillae as setose lobes, no palps; maxillipede without epipods, with distally directed setose endite; pereiopods generally with well-developed, distally setose exopods, females bearing oöstegites, eighth thoracopods of males with penes; pleopods reduced; uropodal rami with two to five segments.

HISTORY   Though privately known for many years from two specimens from the deep-sea west Atlantic, the group was not described until a plentiful source of another species was found in anchialine caves in Bermuda. A series of papers was published together describing the two known species, erecting a new order to accommodate them (Bowman and Iliffe, 1985; Sanders et al., 1985; Bowman et al., 1985). The group has the distinction of being the only one whose existence was predicted on purely theoretical grounds (Schram, 1981, 1984) before it was recognized and described as such.

MORPHOLOGY   The animals are small, slender, and subcylindrical in outline (Fig. 11-1). One species, *Mictocaris halope*, has distinct eye lobes but seems to lack optic elements and also possesses a rostrum. The head is covered by a well-developed head shield that extends laterally to enfold the side of the cephalon.

Fig. 11-1. Body plan of mictaceans. *Mictocaris halope*. (Modified from Bowman and Iliffe, 1985)

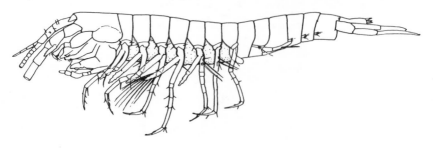

The antennules (Fig. 11-2A) have a large, three-segmented peduncle and two well-developed flagella.

The antennae (Fig. 11-2B) have a small scale on the second segment, and the three basal segments of the long flagellum are pedunclelike.

The labrum is slightly concave anteriorly. The paragnaths are oval setose lobes in *Mictocaris* (Fig. 11-2C) and produced into long slender distal processes in *Hirsutia*. The mandibles (Fig. 11-2D) are slightly asymmetrical. The incisor process is three- or four-cusped, and the left mandible bears a three-cusped lacinia mobilis. These are followed by a spine row in front of a large molar process. The large palp bears batteries of comblike setae on the second and third segments.

The maxillules (Fig. 11-2E) are composed of two quite distinct, distomedially directed endites. The inner lobe is quite setose, the outer is terminally armed with spines and apical teeth.

The maxillae (Fig. 11-2F) are developed with dense arrays of complex setae on their margins of their endites. The inner lobe is single and distinctly separated from the deeply cleft outer lobe.

The maxillipedes (Fig. 11-2G) are large. The basis carries a distally directed lobate endite that is medially and distally equipped with rows of setae. The palp is five-segmented with the merus and carpus distinctly wider than the other segments. In *Mictocaris*, the female may carry oöstegites on this limb.

The second through eighth thoracopods (Fig. 11-2H) are typically natatory limbs with simple, poorly setose endopods associated with well-developed and terminally setose exopods. The second and eighth thoracopods of *Hirsutia* are reported to be uniramous. The females have simple, nonsetose oöstegites on thoracopods 2 through 6, forming a typical brood pouch (*Mictocaris*), or setose oöstegites on thoracopods 3 through 7 directed posteriorly (*Hirsutia*). The males of *Mictocaris* bear a peneal process on the base of the eighth thoracopod.

The pleopods (Fig. 11-2I) are typically reduced to single-segment, biramous flaps, except for the second pair of pleopods in the male that are composed of two segments.

The tailfan (Fig. 11-2J) is quite distinctive. The uropodal rami have two (*Mictocaris*) or up to at least five (*Hirsutia*) segments. The short, narrow, lobate telson has a basal anal opening guarded by a pair of flaps.

Nothing is known at present concerning the internal anatomy of mictaceans.

NATURAL HISTORY   The deep-sea species, *Hirsutia bathyalis* (Sanders et al., 1985) is known only from preserved material collected at a depth of 1000 to 1022 m.

More abundant material exists for *Mictocaris halope*, collected from several anchialine caves in Bermuda (Bowman and Iliffe, 1985). They prefer areas some distance from the sea and shun daylight penetration. The

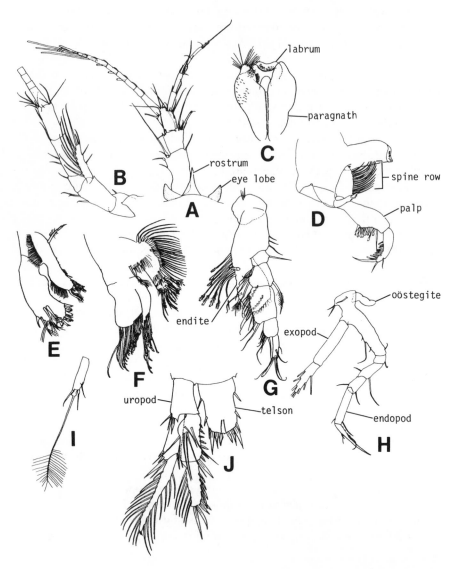

Fig. 11-2. Appendages of *Mictocaris halope*. (A) antennules; (B) antennae; (C) labrum and paragnaths; (D) mandible; (E) maxillule; (F) maxilla; (G) maxillipede; (H) second thoracopod; (I) first pleopod; (J) telson and uropod. (From Bowman and Iliffe, 1985)

animals can be observed in the caves swimming in great numbers (Laurence Abele, personal communication), but in the laboratory *Mictocaris* were observed to spend most of their time on the sides and bottoms of the aquarium or collecting containers they were kept in. A minimal escape reaction was recorded by Bowman and Iliffe (1985), wherein the animals flex the body in order to change the direction of travel before swimming away.

Feeding has not been observed, and *Mictocaris* rejected tubifex worms when offered them in the laboratory (Bowman and Iliffe, 1985). However, from consideration of the anatomy, there is no reason to suppose that the animals would be carnivores. Indeed, their mouthpart morphology and apparent preference for bottoms bespeaks a detritus feeder, and to this end Bowman and Iliffe (1985) point out similarities of the mouthparts of *Mictocaris* to those of the thermosbaenacean *Monodella* (Fryer, 1964). Sanders et al. (1985) noted some features of the pereiopods that suggest *Hirsutia* might be a facultative carnivore.

DEVELOPMENT Minimal information currently is available concerning development (Bowman et al., 1985). The embryos flex dorsally in the course of ontogeny. When development is complete, the juveniles hatch as mancas lacking the last thoracomere.

TAXONOMY Only two species are currently known, but each is sufficiently distinct to warrant being placed in a separate family.

Family Hirsutiidae Sanders et al., 1985
Family Mictocarididae Bowman and Iliffe, 1985

REFERENCES

Bowman, T. E., S. P. Garner, R. R. Hessler, T. M. Iliffe, and H. L. Sanders. 1985. Mictacea, a new order of Crustacea Peracarida. *J. Crust. Biol.* **5**:74–78.

Bowman, T. E., and T. M. Iliffe. 1985. *Mictocaris halope*, a new unusual peracaridan crustacean from marine caves in Bermuda. *J. Crust. Biol.* **5**:58–73.

Fryer, G. 1965. Studies on the functional morphology and feeding mechanism of *Monodella argentarii*. *Trans. Roy. Soc. Edinb.* **66**:49–90.

Sanders, H. L., R. R. Hessler, and S. P. Garner. 1985. *Hirsutia bathyalis*, a new unusual deep-sea benthic peracaridan crustacean from the tropical Atlantic. *J. Crust. Biol.* **5**:30–57.

Schram, F. R. 1981. On the classification of Eumalacostraca. *J. Crust. Biol.* **1**:1–10.

Schram, F. R. 1984. Relationships within Eumalacostracan Crustacea. *Trans. San Diego Soc. Nat. Hist.* **20**:301–12.

12

ISOPODA

DEFINITION   Carapace absent; first thoracomere fused to cephalon (on rare occasions the second fused as well); eyes sessile; antennules typically uniramous; antennae uniramous or sometimes with small exopod; maxillipede with protopodal and proximal endopodal joints with flat lobelike endites directed distally, epipodite typically as a large flap protecting mouthfield laterally; posterior thoracopods uniramous, coxae developed as plates with tendency to fuse to the body; pleopods usually biramous, rami specialized for respiration, second and occasionally first abdominal appendage modified in males; telson typically fused with abdominal somites in varying degrees.

HISTORY   Originally named by Latreille, the group was placed among the myriapods because the initially described isopods were terrestrial oniscoideans that were thought to resemble diplopods. With the recognition of aquatic forms isopods came to be assigned to the crustaceans and were allied by Leach with the amphipods in a taxon, the Edriophthalma. Other workers, beginning with Haeckel and culminating in Reibisch, utilized another taxonomic term, Arthrostraca, for these two and various other groups thought to be related. Calman placed isopods within his superorder Peracarida. Siewing felt isopods were separately derived from amphipods, an opinion that has held sway unchallenged until relatively recently.

MORPHOLOGY   The isopods exhibit the greatest diversity in body plan (Fig. 12-1) noted among any of the crustaceans, except possibly the Reptantia. Generally characterized as depressed or dorsoventrally flattened (an extreme case being the serolid flabelliferans, Fig. 12-1B); some groups, like the anthurids (Fig. 12-1I) and microcerberids (Fig. 12-1G), have a cylindrical body that in some cases is almost vermiform. The phreatoicideans (Fig. 12-1A) vary in form from being rounded in cross section to laterally compressed, reminiscent of amphipods. The gnathiids (Fig. 12-1E, F), though somewhat dorsoventrally flattened, exhibit extreme sexual dimorphism and specialization of body parts. Some of the epicarideans (Fig. 12-1J) are so modified for parasitism that they are little more than reproductive bags or sacks. Nor is the name 'isopod' particularly appropriate, since as a group they are no more isopodous than amphipods with some forms, for example, certain asellotes and valviferans, exhibiting extreme 'amphipody.' A taxon name based on the branchial pleopods would have

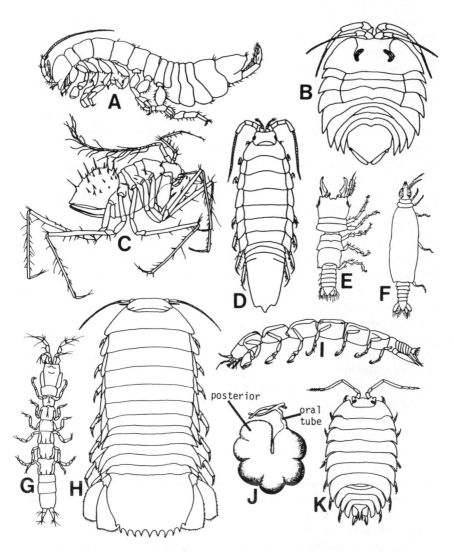

Fig. 12-1. Isopod types. (A) *Neophreatoicus*, Phreatoicidea (from Nicholls, 1942–43); (B) *Serolis*, Flabellifera; (C) *Abyssaranea*, Asellota (from Wilson and Hessler, 1974); (D) *Idotea*, Valvifera; (E) male *Gnathia* and (F) female *Gnathia*, Gnathiidea (E, F, after Monod, 1926); (G) *Microcerberus*, (Microcerberidae) Asellota (from Kaestner, 1970); (H) *Bathynomus*, Flabellifera: (I) *Calathura*, Anthuridea; (J) *Danalia*, Epicaridea (from Kaestner, 1970); (K) *Oniscus*, Oniscoidea (B, D, H, I, K from Hessler, 1969).

been more descriptive, since the nature of the abdominal appendages is one of the few unique external characters that unites members of the group. Another apparent character is biphasic molting (George, 1972), where the posterior part of the body emerges first, followed at some interval by the anterior part of the body.

The cephalon is relatively short, even when it is remembered that the first, and even occasionally the second, thoracomeres are fused to the head. The dorsal and/or anterior aspect of the head is typically marked by the sessile compound eyes. These organs (see, e.g., Nilsson, 1978) are typically well developed, though in cave, deep-sea, and parasitic forms they may be greatly reduced or absent, and trichoniscid oniscoideans substitute clusters of simple ocelli for the compound eyes.

The antennules, with very few exceptions, are uniramous (Fig. 12-2A). The peduncle typically has three segments but may have two or four segments. The flagellum is short. In epicarideans, oniscoideans, and valviferans they may be very reduced.

The antennae (Fig. 12-2B) are typically uniramous. The peduncle is composed of four, five, or six segments. Some asellotes have a small exopodal scale on the third segment, and some anthurideans have a short but many-segmented 'accessory' flagellum mounted on the fourth segment. In some isopods, the antennae are pediform, with few but well-developed joints that come to resemble those of legs, while in some epicarideans the antennae are virtually vestigial. Some authors have devoted much attention to homologies of antennal segments, for example, Wägele (1983) who claims details of setation on segments is important in this regard.

The mandibles (Fig. 12-2E) typically have three-segment palps, though these are absent on oniscoideans and valviferans. The incisor process is typically multidentate. The highly variable lacinia mobilis is developed on both the right and left mandibles in amphisopodids and cirolanids but is usually absent on the right mandible in other forms. The cirolanids are unique in having the molar process movably articulated to the main body of the mandible. Nicholls (1942) reports seeing a slender rodlike structure adjacent to the palp on some New Zealand phreatoicids which he speculates as possibly representing a vestigial exopod. The labrum (Fig. 12-2C) and paragnaths (Fig. 12-2D) are generally rounded lobes, but in epicaridians these form an 'oral cone' that enclose styliform mandibles.

The maxillules (Fig. 12-2F) are, in their most primitive form, thin platelike appendages with two well-developed endites directed ventrally and medially, and very strong setae and spines directed distally. Specializations are numerous, ranging from a platelike to a hooked form, sometimes lacking setae and sometimes with hairlike setae.

The maxillae (Fig. 12-2G) have basically two endite lobes. The proximal or inner one is continuous with the base of the limb and is usually setose along its medial margin. The distal or outer endite is deeply cleft or biramous, directed ventrally, and movably articulated with the base of the

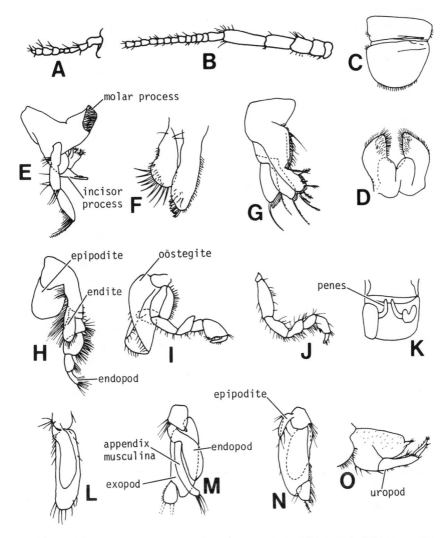

Fig. 12-2. Appendages of *Phreatoicus orarii*. (A) antennule; (B) antenna; (C) labrum; (D) paragnaths; (E) mandible; (F) maxillule; (G) maxilla; (H) maxillipede; (I) second thoracopod of female (= gnathopod); (J) fourth thoracopod; (K) penes at the base of male eighth thoracopods; (L) first pleopod; (M) second male pleopod; (N) third pleopod; (O) lateral view of uropods and pleotelson. (From Nicholls, 1943)

limb. The maxillules and maxillae are vestigial in the epicarideans and gnathiids.

The first thoracopods (Fig. 12-2H) are developed as maxillipedes. The basic form of this appendage includes a short coxa bearing a lamellar epipodite that laterally protects the mouthparts, a large basis with a well-developed medial endite directed distally and often equipped with retinacula to lock with the opposing member, and a stout endopod of five or fewer joints with one or more of these capable of developing enditelike lobes. The isopod maxillipede exhibits a great variety of form. Some taxa have very robust endopods and reduced endites (Fig. 12-3A), slender endopodal palps alone (Fig. 12-3B), or reduced and specialized palps (Fig. 12-3C, E, F). Other forms greatly elaborate the appendage with enditelike lobes on the distal joints of the palp (Fig. 12-3D). In many groups there is a small fringed lobe extending posteriorly into the female brood pouch (Fig. 12-3G), akin to the oöstegites seen on the second maxillipede of cumaceans. The maxillipedes become reduced or highly modified in parasitic or quasiparasitic forms.

The anatomy of the basal portions of the mouthparts has been a vexing issue since the days of Hansen (see, e.g., Hansen, 1893) who suggested that the remnants of three protopodal joints is to be seen on these appendages, that is, precoxa, coxa, and basis (Fig. 12-3N). However, the actual situation is quite variable and is complicated by the fact that the entire region is closely associated with the floor of the cephalon around the mouth. Basal elements have been simply termed sclerites by many authors (e.g., Bastida and Torti, 1973) without reference to possible joint identity (Fig. 12-3L, M, N, O). In some instances these elements have become vestigial (Fig. 12-3L, M, N), are lost entirely with the endites coming to arise independently from an apodeme (Fig. 12-3M), or are fused to each other (Fig. 12-3O). The individual cuticular plates in this region are typically separated by arthrodial membranes. Similar variations are seen at the bases of all the mouthparts, including maxillules, maxillae, and maxillipedes, and caution must be used to guard against always 'finding' three protopodal elements because 'theory' says they should be there (Fig. 12-3K). For example, it can be confusing to try and 'interpret' the articulation of the maxillipedal epipodite to be a sometimes greater, sometimes lesser coxa (Collinge, 1918) that itself might be either whole or with a distal lamella (Fig. 12-3G). This entire issue of basal jointing needs an extensive comparative study with careful attention being paid to the associated muscles (Fig. 12-3O) and the functional morphology of individual endites. Only a few workers have made observations along these lines, for example, Jackson (1926) whose own work suggests that the labeling of these cuticular plates associated with the mouthparts, without adequate reflection on functional matters, may lead to inadequate or incorrect phylogenetic conclusions.

The second thoracopods (Fig. 12-2I) are occasionally subchelate (gnathopods). In the gnathiids the second thoracomeres are also fused to

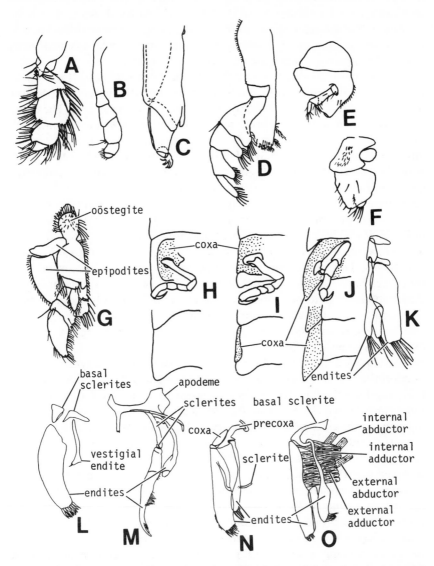

Fig. 12-3. (A, B, C, D, E, F, G) variations of maxillipede form. (A) *Cirolana deminuta*; (B) *Corallana hirsuta*; (C) *Rocinela juvenalis*; (D) *Cymodoce acuta*; (E) *Serolis murrayi*; (F) *Notanthura barnardi*; (G) *Idotea ostrumovi* (from Kussakin, 1979). (H, I, J) illustrating progressive degrees of fusion of coxa to body wall. (H) *Idotea hectica*; (I) *Idotea ochotensis*; (J) *Ciridotea sabini* (from Calman, 1909). (K, L) generalized serolid limbs. (K) maxilla; (L) maxillule (from Bastida and Torti, 1973). (M) maxillule of *Cyathura carinata* (from Wägele, 1979). (N) maxillule of *Idotea pelagica*; (O) maxillule of *Ligia oceanica* (from Jackson, 1926).

the head, and the accompanying thoracopods are specialized as another set of maxillipedes (= second maxillipedes = pyllopods), developed as large flaplike valves covering the nonfunctional mouthfield in the adult males, while in female this limb is a short palplike stenopod to which is proximally attached a large plate that appears to be an oöstegite.

The typical second through eighth thoracopods (Fig. 12-2J) are uniramous, generally ambulatory limbs (= pereiopods). The coxae of these limbs have a tendency to develop as dorsally directed plates. These are movably articulated to the body only in many phreatoicideans and asellotes, while all other groups show varying degrees of fusion of the coxae into the body wall (Fig. 12-3H, I, J). In those instances when the suture lines disappear, it is virtually impossible to distinguish such fused coxae from the body wall proper. The condition of the coxal fusion may vary throughout the life of an individual. Wilson (1980) points out that the coxa of the second thoracopod is normally fused in the asellote *Stenetrium*, but in the molt to a brooding condition in the female the coxa is freed. Though most coxal plates are dorsal, Sheppard (1939) noted ventral coxal plates in valviferans. The anteriormost pereiopods may form subchelae, while most other legs have the dactylus as a stout simple claw. The ischium of isopod thoracopods tends to be more or less elongate but certainly is not particularly short as is the case among amphipods. Thoracopods can on rare occasions be developed as setose paddles used in swimming, as seen for example in some asellotes.

Oöstegites may be found on all thoracopods but more typically occur only on the second through sixth or second through fifth limbs. They are thin broad overlapping plates, which appear to arise from the sternites on those limbs in which the coxae fuse to the body. Numerous variations on the typical oöstegite brood pouch occur. Some forms, for example, some sphaeromatids, carry the eggs inside the body in paired pocketlike invaginations of the cuticle that communicate with the empty marsupium by little slits (see e.g. Harrison, 1984). Other genera have a single internal median brood pocket extending into the abdomen from the border of the sixth and seventh thoracomeres. One family, the Dajidae, do not have oöstegites but rather form a brood pouch from ventral extensions of the body wall. Gnathiids brood eggs and develop young in the body cavity itself. In the flabelliferan genus *Excirolana* the enlarged oviducts serve as uteri, while the oöstegites are reduced to small genital opercula over the gonopore (Klapow, 1970).

The coxa–basis joint is monocondylic on all pereiopods with the condyle projecting into the lumen of the leg from the lateral, or more usually, the posterolateral margin of the coxa. All other leg joints are dicondylic. One or more median sternal processes are noted in the phreatoicideans and anthurideans. Between the maxillipedes, as in cumaceans, these processes can be developed into rather prominent structures.

The pleopods (Fig. 12-2L) are typically biramous, thin-walled, setose appendages employed for swimming and/or respiratory purposes. In aquatic forms, the marginal setae facilitate this dual function, though in some cases (e.g., some gnathiids and cymothoids) the setae are absent. Some epicarideans lose the pleopods or reduce them to the form of dendrobranchiate gills. A feature peculiar to some of the phreatoicideans and almost all flabelliferans is the presence of an epipoditelike accessory lamella (Fig. 12-2N) on at least the third through fifth pleopods. These can be quite elaborate and serve a respiratory function. In some instances (e.g., serolids and sphaeromatids) swimming functions are restricted to the anterior pleopods, while the posterior pairs are respiratory. Sphaeromatids frequently possess transverse plications on the rami of the posterior respiratory limbs to increase the available respiratory surface, and oniscoideans tend to develop ramifying tubules or pseudotracheae within the limbs that exit to the outside by small openings. Many of the flabelliferans develop accessory lamellae or digitiform structures on the pleopods to assist in respiration (e.g., *Bathynomus*, and some cymothoids, aegiids, and cirolanids). Because of the delicate and important nature of the pleopods they are frequently protected by some kind of cover: for example, in anthurideans the first pleopods are enlarged as flaps that cover the more posterior pairs; female paraselloideans fuse the first pleopods to form a single plate, and male paraselloideans also involve the second pleopods in the formation of this plate as well; in asellids and stenetriids the third pleopods form the operculum; in serolids it is the fourth; and valviferans utilize the uropods for protection of the pleopods.

Most isopods modify the second pleopod (Fig. 12-2M), and rarely the first as well, as a copulatory organ or gonopod. The endopod of the second pleopod typically bears a long, often grooved, rodlike organ, the appendix masculina. This presumably receives the sperm and transfers it to the female. In the valviferans and oniscoideans the male first pleopod may have a similar modification to that noted in the second. The asellotes, however, have a unique arrangement in which the first pleopod forms a rather complex apparatus. Epicarideans and gnathiids lack any special modifications of the penis.

The uropods exhibit wide form and location. In the primitive phreatoicideans and asellotes, as well as the oniscoideans (Fig. 12-2O) they are styliform and sometimes ambulatory (phreatoicideans) in function. In anthurideans, flabelliferans, and gnathiids the uropods are wide and flat, forming a tailfan complex with the pleotelson. In valviferans they form opposing flaps to cover the pleopods. In epicarideans they are frequently lost. The isopod anus is terminal or subterminal. Statocysts are often present in the base of the pleotelson of anthurideans.

The digestive system in isopods has only fore- and hindgut portions in the gut proper. These ectodermal derivatives meet typically at the level of the second pereiomere, and a 'midgut' endodermal contribution to the digestive system is confined to the caeca (Holdich, 1974). The digestive caeca are developed as one to three pairs of elongate lobes extending back into the thorax. The structure of the foregut has figured prominently in the phylogenetic speculations of Siewing (beginning in 1951 and in subsequent papers). In point of fact the structure of this region is quite variable (Kunze, 1981 and personal communication) and responsive to functional needs of diet. Ide (1892) described a pyloric funnel and a dorsal anterior diverticulum for *Idotea* identical to that seen in amphipods in the sense of Siewing, and Naylor (1955) provides greater details of this structure in *Idotea* (Fig. 12-4) and felt its form fairly exemplary for omnivorous isopod types. Lateral ampullae armed with bristlelike setae serve to crush food. Dorsal, lateral, and ventral lamellae subdivide the middle portion of the foregut, serve to channel food ventrally and posteriorly, and help prevent regurgitation. The walls of the ventral chamber are grooved to help direct food into the caeca. A somewhat different system is seen in *Eurydice*, a carnivorous form (Jones, 1968) where the esophagus leads into a subdivided dorsal chamber. Two sets of setose lateral folds delineate a central chamber through which food moves from the dorsal into the ventral chamber and thence to the caeca. The flow is regulated by movements of the lateral folds and a movable ventral pyloric ridge. In *Eurydice* a distensible anterior portion of the hindgut is used as a 'crop' or food storage organ. Yet another arrangement of gut structures is seen in asellotes (Schoenichen, 1898) where a series of distinct chambers align themselves anterior to posterior. The first compartment is a chamber whose lobes and processes are armed with setae and spines. This passes medially into a large open section that posteriorly continues dorsally to connect with the hindgut, while the ventral portion opens to the caeca. (Passage between these two portions is regulated by a movable ventral ridge.)

The distinctive feature of the isopod circulatory system is the posterior location of the heart. This would seem to be related to the use of the pleopods as respiratory organs in isopods. However, blood reaches those limbs indirectly through special blood sinuses. The heart lies in the posterior thorax and anterior abdomen and is the center of an elaborate arterial system (Fig. 12-5A). The heart gives rise to an anterior aorta, lateral thoracic arteries (of which those of thoracomeres two through five typically arise from a common artery from the heart), and a pair of abdominal arteries. There is no posterior aorta. The anterior aorta forms a circumesophageal ring, from which arises a subneural artery that proceeds along the length of the body. In many forms, however, this subneural artery is absent, its function taken over by rami from the lateral thoracic and abdominal vessels. Blood eventually collects in a series of sinuses beneath the gut and/or along the ventral body wall and flows posteriorly to the pleopods (Fig. 12-5B). The above scheme is based on generalizations from a fairly across-the-board review of various forms (Delage, 1881). From the pleopods (Fig. 12-5C) efferent sinuses drain blood to the pericardium, and one or two pairs of ostia allow blood to return into the heart.

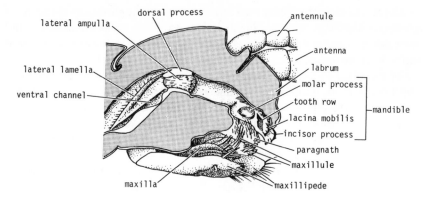

Fig. 12-4. Median section through cephalon of *Idotea emarginata* with mouthparts and foregut typical for isopods. (From Naylor, 1955)

Fig. 12-5. Typical isopod circulatory scheme. (A) arterial system; (B) venous sinuses; (C) circulation into and out of pleopods. (From Delage, 1881)

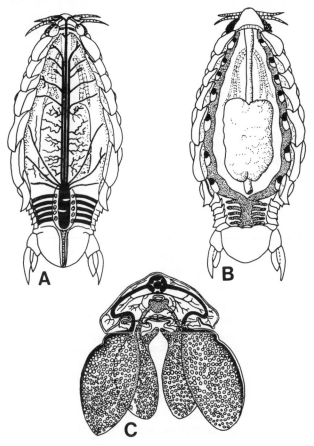

Excretion is achieved by maxillary glands and nephrocytes throughout the body. Vestigial antennal glands have also been noted in asellotes and oniscoideans. In some oniscoideans the maxillary glands are also reduced and may be vestigial.

The central nervous system displays considerable fusion of elements. The maxillary through maxillipedal ganglia are typically fused with the subesophageal ganglion. In addition, the abdominal ganglia display a tendency toward fusion in the anterior part of the pleon. Zimmer (1927) remarks that in *Sphaeroma* and some other genera a seventh abdominal ganglion is found in the telson.

The gonads develop from a series of primordia common to both sexes. These consist of a longitudinal cord in the thorax flanked by a series of segmental mesodermal cords in the third through sixth thoracomeres. In the adult female, the tubes become the ovaries while the cords develop into oviducts and sensory ligaments. In the adult male, the tubes become seminal vesicles, while the anterior cords become testes and the posterior cords become suspensory ligaments and the vasa deferentia. Thus different parts of the basic system develop into sexual organs of either sex and in this way may facilitate the sex changes noted in a few species (see below). The male system opens into a pair of median papillae (penes) or, in gnathiids and oniscoideans, by a single median pene. These penes presumably merely deposit sperm onto the gonopods. Uncertainty as to function exists, however, since actual mating in isopods has seldom been subject to detailed observation.

NATURAL HISTORY   Such a large and diverse group as the isopods have been the subject of a large corpus of observations concerning their biology. They are ubiquitous, easily collected, easy to work with in field or laboratory, and so have lent themselves to a variety of these efforts. It would be difficult to completely review the available literature even if a whole volume were devoted to isopods alone, and only some items of particular phylogenetic interest will be covered here.

Locomotion is regionalized. The uniramous thoracopods are almost exclusively ambulatory. Swimming is achieved by the pleopods or, if there is a further division of labor within the abdomen, by those pleopods that serve a natatory role.

Methods of eating in isopods are different from those noted in any other peracarid group except amphipods; both groups share a great deal in common. Isopods are basically omnivores, with some types specializing in carnivory–scavenging or parasitism, and terrestrial oniscoideans adapted to herbivory. Naylor (1955) recorded the basic mode of function in *Idotea* applicable to all isopods in broad outline. The mandibles bite with a transverse movement, that is, lateromedially. The mandibles alternate in their beat with the maxillules, the maxillules moving to the midline as the mandibles abduct laterally. Occasionally, the maxillules adduct sufficiently to project between the lobes of the paragnaths into the mandibular field. The combined action of these limbs is to scrape and bite food off a surface (plant or carcass) and push it into the esophagus. The removal of debris off these anteriormost mouthparts and the cleaning of their setae is achieved by the maxillae and lateral movements of the endopod of the maxillipedes, assisted to a lesser extent by the second thoracopods.

Green (1957) and Jones (1968) have noted a pattern similar to the above in more carnivorous forms, such as *Chridotea*, *Mesidotea*, and *Eurydice*. In these taxa the anterior three pairs of pereiopods are also involved in manipulating food for the mandibles and maxillules to operate upon and at times may actively assist in maceration by tearing into the food source with the dactylar claws on those pereiopods.

The biting and scraping action of the mouthparts when combined with a muscular sucking type of proventriculus (sometimes termed a cephalogaster) allows easy aptation to a parasitic life-style. In this regard, the dilatable anterior part of the hindgut noted in gorged carnivores like *Eurydice* (Jones, 1968) can be seen as an exaptation to the parasitic habit of long-term storage of a meal followed by slow digestion.

The ubiquitous isopods certainly must be considered among the more successful radiations of eumalacostracans. Some of this success must be attributed to the prolific reproductive habits of the group. For example, Dexter (1977) has found that juveniles of *Excirolana braziliensis* were found at all tide levels (while the adults were present only at the higher tide levels) and that the young composed at least 70% of any *E. braziliensis* population at all times. Though only four to 17 individuals were produced in a brood, depending on the size of the females, reproductive females were present year round though gravid females were never very abundant. Brusca (personal communication) relates that the number of eggs per brood in cymothoids ranges from 200 to 1600.

Johnson (1976) found that in *Cirolana harfordi* reproductive success can be enhanced by the cryptic habits of brooding females. The pregnant mothers tend to remain under their harboring rocks and to eat little if at all during the three to four months they are brooding. Johnson also found that females can breed more than once, probably no more than twice in nature, though three or more times have been recorded in the laboratory.

Mancas are immediately ready to feed and swim as soon as they leave the marsupium. This precocious habit increases reproductive success by minimizing demands on the mother and thus speeding her potential return to the breeding population. However, the greatest mortality in the life cycle occurs among the juveniles, who because of their size must feed often, and among the weakened postbrood females, who may have fasted for months. Both are prime targets for predators.

Some isopods resort to other strategies to increase reproductive success. Burbanck and Burbanck (1974, 1979) report a pattern of protogyny in the anthurid genus *Cyathura* similar to that seen in tanaidaceans. Perhaps not coincidentally both groups are burrowing and tube-dwelling forms. In *Cyathura* there exist primary males, which develop directly from juveniles and secondary males, which metamorphose from postbrood females. Asellotes also exhibit protogyny. Cymothoid flabelliferans and cryptoniscid epicarideans are protandrous hermaphrodites. Several oniscoideans (*Trichoniscus*, *Armadillidium*, and *Cylisticus*) are reported to have parthenogenetic races (e.g., Paris, 1963; Paris and Pitelka, 1962; and Vandel, 1938).

Sexual dimorphism in isopods is for the most part rather subtle. Exceptions to this of course are the distinctly dimorphic gnathiids, and some epicarideans in which males are dwarfs living on the females. In bopyrids this dimorphism is carried one step further where sex is environmentally determined—the first juveniles to settle on a host become females, subsequent settlers become males.

Copulation results in internal fertilization. The male typically mounts the body of the female oriented parallel to her longitudinal axis, flexes his abdomen around her side, and alternately inseminates either side of his mate. Aquatic forms mate between molts of the female, the posterior part of the body sheds first to allow copulation, then the anterior region molts to produce a brood pouch. Within a few hours egg laying occurs. In oniscoideans, the second partial molt can take weeks, months, or even a year or more after mating. In *Thermosphaeroma thermophilum* the male is able to identify a potentially sexually receptive female long before the molt for copulation. Such a male captures the precopulatory female, carries her under his body for up to several days until the copulatory molt occurs, and mates with her before setting her free (Shuster, 1981).

The diversity and success of the Isopoda are manifested in the great diversity of habitats they occupy, be it all manner of marine, brackish, fresh, and truly terrestrial environments. The terrestrial adaptations of some oniscoideans is so complete that most will actually drown if forced under water more than a few hours. Oniscoideans are the only crustaceans found in arid habitats. Especially interesting are the variety of parasitic strategies employed by isopods. The aegid flabelliferans are transitory ectoparasites on marine fish, specialized as blood- and fluid-sucking carnivores. Cymothoids are more-or-less permanent ectoparasites on marine and freshwater fish. Mancas of gnathiids are also blood suckers, while the burrowing adults are incapable of feeding at all. Epicarideans exhibit various degrees of intimacy with their hosts. Bopyrids and dajids are ectoparasites on marine crustaceans and attach to harbored sites on the abdomen or under the carapace in gill chambers. Entoniscids, while technically always remaining 'outside' the host cuticle, cause a pocket to form in which the degenerated parasite that projects into the host crab is protected. Liriopsids mimic rhizocephalans and some of the endoparasitic copepods in that they form a saclike stage inside the host, losing all appearances of a crustacean. Some, like *Danalia* or *Liriopsis*, are hyperparasites, actually parasit-izing the rhizocephalan parasites of other crustaceans.

An important aspect of the radiation of the isopods has been the broad adaptation of the eight suborders to particular habitat types (Table 12-1). The isopods of terrestrial, freshwater, and deep-sea habitats are concen-trated among, respectively, oniscoideans, phreatoicideans, and asellotes. The diverse shallow water habitats, however, have induced specializations of several groups, namely, anthurideans, epicarideans, flabelliferans, gnathiids, and valviferans, though each of these last groups also has a few representatives in the deep sea.

Analysis of biogeographic distributions among isopods has depended on the preferences and outlooks of particular groups of workers. Most workers have preferred to study either ecology or distributional analysis but not a combination of the two. One exception to this has been Brusca and Wallerstein (1979) and Wallerstein and Brusca (1982) who noted strong latitudinal zonation in northeastern Pacific idoteid valviferans. Cer-tain genera were characteristic of particular provinces along the west coast of North America: *Saduria* and *Synidotea* in the Arctic Province, *Edotea* in the Oregonian and Californian provinces, *Colidotea* and *Erichsonella* in the Californian–Cortez provinces, and *Cleantis* and *Eusymmerus* in the tro-pics. The genus *Idotea* had a ubiquitous distribution along the coast, although individual species were restricted to one or a few coastal prov-inces. Experimental studies seemed to indicate that there is a strong tem-perature dependence of some of these taxa. They concluded that fish predation was the significant control on the southern ranges of idoteids south of the Californian Province, while northern range limits were con-trolled by temperature.

Table 12–1. Broad habitat preferences of major isopod groups. (Modified from Hessler et al., 1979). Mild and strong preferences denoted with single and double signs respectively.

Taxon	Terrestrial	Freshwater	Marine Shallow water	Marine Deep sea
Oniscoidea	++			
Phreatoicidea	+	++		
Asellota		+	+	++
Anthuridea			++	+
Epicaridea			++	+
Flabellifera			++	+
Gnathiidea			++	+
Microcerberidea		+	++	
Valvifera			++	+

Workers on the deep-sea isopods have exhibited special interest in seeking to understand the biogeography of asellotes. These studies have been directed toward resolving the issue of the origin of the deep-sea iso-pod fauna. Asellote diversity increases with depth while that of other iso-pods decreases with depth. Kussakin (1973) believed the asellotes evolved into the deep sea from shallow Antarctic waters. Birshtein (1973) thought that the asellotes were a very old group and individual families had evolved into the deep sea from shallow waters all over the world. Hessler and Thistle (1975) and Hessler et al. (1979) argued for a deep-sea origin of the asellote families characteristic of the deep sea with subsequent secondary invasion of Antarctic shallows.

A strong latitudinal component is noted in deep-sea forms. Birshtein (1973) points out that the north Pacific and north Atlantic share 26 genera and four species, while the north Pacific shares only 17 genera and one spe-cies with the southern hemisphere. This would seem to imply a great age for deep-sea asellotes since evolution into the deep sea must have occurred at a time when the northern oceanic basins had easier access to each other. George (1979) observed different thermal tolerances that might bear on this issue. He notes the Antarctic began cooling in the Cretaceous, reach-ing a low point in the Miocene, while the Arctic began cooling only in the last few million years. He found extreme stenothermy in the abyssal amphipod *Eurytheres gryllus* and the Antarctic isopod *Glyptonotus antarc-ticus*, while the Arctic isopod *Saduria entomon* is eurythermic. George felt these differences betray the operation of different physiological strategies. The young Arctic forms retain thermal versatility, while the Antarctic fauna, which may have reached faunal equilibrium 30 million years ago, is more sensitive to narrow temperature requirements.

Birshtein (1973) and Hessler et al. (1979) point out that many deep-sea isopod genera have wide geographic distributions (Fig. 12-6). In asellotes, however, the distributions of particular species seems more often highly endemic. For example, the distributions of the nannoniscids *Nannonis-coides* and *Hebefustis* (Fig. 12-7) were analyzed by Siebenaller and Hessler (1977). Both these genera are for the most part confined to the Atlantic, though in both genera a single species is known from the northwest Pacific (*N. excavatifrons* and *H. robustus*). The rather restricted species distribu-tions of these genera to within the Atlantic basin might suggest an evolu-tion that could be post-Cretaceous, with apparently limited opportunity to disperse out of the Atlantic. However, not all deep-sea isopod species exhi-bit such endemism. While some deep-sea anthurid species (Kensley, 1982) are localized in their distributions, others are very widely dispersed. For example, *Malacanthura truncata* has been collected from the northeast coast of Brazil, west Africa, the Bay of Biscay, and the Davis Strait, and *Oscanthura gracilis* can be found south of Long Island, in northeast Brazil, west Africa, and off the Rio de la Plata.

It must be pointed out, however, that analyses of biogeography are

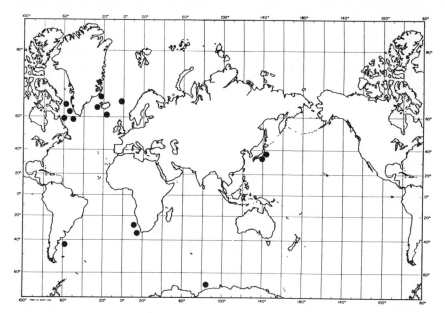

Fig. 12-6. Distribution of a typical ubiquitous asellote genus, *Nannoniscus*. (From Birshtein, 1973)

Fig. 12-7. Distribution of species of *Nannoniscoides* ● and *Hebefustis* ▼. (From Siebenaller and Hessler, 1977)

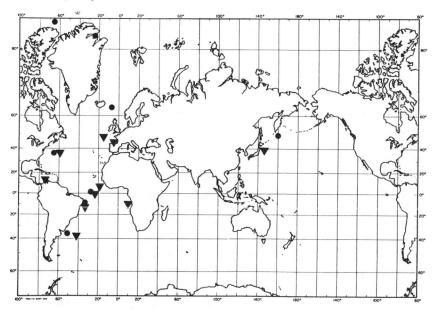

147

only as good as the taxonomic works upon which they are based. As an example, the Atlantic deep-sea faunas have received intensive study, the Pacific less so. Ongoing efforts in collecting the Pacific deep-sea fauna will undoubtedly modify our understanding of genera and species distributions. Meanwhile, current patterns of distribution we think we see might suggest where taxonomic study should be focused. The wide distributions of the anthurideans mentioned above might be those of single wideranging species but also could be the ranges of whole species complexes.

DEVELOPMENT   Embryological development in isopods has received a great deal of attention, and many of the major groups have been examined at least in part. Among the most important studies in this regard are McMurrich (1895) on asellotes and oniscoideans, Strömberg (1965) on the valviferan *Idotea*, Nusbaum (1898), S. G. Nair (1956), and Strömberg (1967) on flabelliferans, Dohrn (1870) on the gnathiid *Gnathia oxyuraea*, and Strömberg (1971) on bopyrid epicarideans. (The best historical review of all work on isopod development is afforded in Strömberg, 1965, and his analysis of *Idotea* ontogeny presents a general model for the most part applicable to isopods as a whole.)

The egg is of a centrolecithal type, and by the eight- to 16-cell stages the nuclei migrate to the surface concentrating on what comes to be the ventral side of the embryo. However, distinct cell boundaries cannot be seen until the 32-cell stage. The initial differentiation of the germinal disc (Fig. 12-8A, B) gives rise to vitellophages that sink out of the cell layer and begin to absorb yolk. An active proliferation and ingrowth of cells results in a mesendodermal plug (Fig. 12-8C) that delineates the blastopore region.

The mesoderm then proliferates anteriad under the ectoderm to form a V-shaped head-mesoderm, and this differentiates into three pairs of masses along the arms of the V, which becomes the mesoderm of the naupliar appendages. While this head-mesoderm is forming, vitellophages differentiate from the mesendodermal plug below the blastopore and migrate into the yolk. Subsequently, vitellophages also differentiate from the area posterior to the ectoteloblasts and migrate into the yolk as well.

As the head-mesoderm is forming, eight mesoteloblasts differentiate in a slightly semicircular line along the anterior margin of the blastopore region (Fig. 12-8E). There then follows a differentiation of a line of ectoteloblasts (Fig. 12-8D), though whether the latter induces the former is not clear. The ectoteloblasts divide twice for every single division of the mesoteloblasts; the divisions of both proceed quite quickly, however, and rapidly give rise to the 16 postnaupliar somites of the body (Fig. 12-8F). The differentiation of the ectodermal portion of the naupliar appendages lags behind the differentiation of their mesoderm from the V-shaped head bands but manifest themselves with the recognition of the metanaupliar limb anlagen from the ectoteloblasts.

After the formation of the head bands and the teloblasts, the posterior

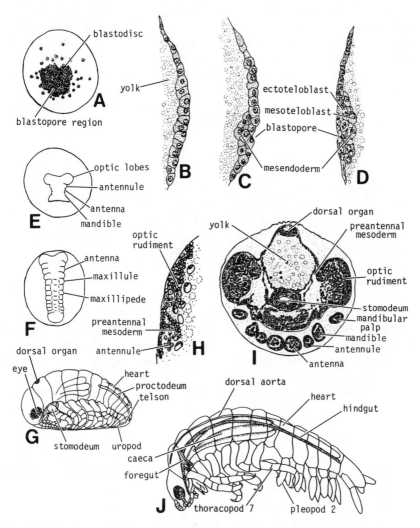

Fig. 12-8. Development in *Irona*. (A) organization of the blastodisc; (B) longitudinal section diagram through A; (C) section at time of mesendoderm formation; (D) section diagram just after ectoteloblast formation; (E) diagram of embryo surface with incipient anlage of nauplian appendages; (F) after rapid division of teloblasts; (G) lateral view of prehatchling with internal organs beginning to form; (H) initiation of preantennal mesoderm; (I) cross section of advanced stage of head formation; (J) manca stage. (From S. G. Nair, 1956)

149

portion of the V that is associated with the mesendodermal mass forms the anlage in the maxillary segment of the hepatic caeca (Fig. 12-8G). S. G. Nair (1956) observed this anlage as a single mass, though Strömberg (1965, 1967) was never able to confirm this from his material. Both K. B. Nair (1939) and S. G. Nair (1956) felt, like Manton (1928) in *Hemimysis*, that these caeca originated from mesoderm, though other authorities (e.g., McMurrich, 1895, and Goodrich, 1939) felt the caeca in isopods were endodermal. Strömberg takes no stand and merely refers to these as mesendodermal. An interesting consequence of the Nairs' position would be that, since the vitellophages and yolk sac disappear by the time maturity is achieved, the adult isopod would have no tissues derived from endoderm.

When the teloblasts commence division, a pair of invaginations from the blastoderm between the optic lobes and antennules form the preantennular mesoderm (Fig. 12-8H). The connection of these inpocketings with the ectoderm is severed and a coelomic pocket is left within the mesodermal masses that come to lie adjacent to the stomodeum. Strands extend ventrally from these masses into the developing labrum to eventually form the labral muscles. Other strands then grow dorsally around the stomodeum (Fig. 12-8I) and send extensions dorsad to the dorsal organ, thus forming primordia of the stomodeal muscles and the anterior portions of the dorsal aorta.

As the thoracic segments develop, posterior bulges have been noted on the developing appendages (S. G. Nair, 1956) which Nusbaum (1891) felt were exopod rudiments. Heart formation begins in the last abdominal segment after teloblast divisions are completed and proceeds forward to the level of the sixth thoracomere (Fig. 12-8G). Developing in parallel with the heart is an inward growth of the proctodeum (Fig. 12-8B). Heart formation *per se* is restricted to the abdomen and posterior thorax, but the basic process of circulatory differentiation continues anteriad to produce the dorsal aorta. The hindgut lies under the heart and dorsal vessel, and eventually becomes enveloped with tissue from the dorsal mesodermal mass from which the circulatory system had developed. The proctodeum continues forward until it encounters the stomodeum in the pyloric region. There they fuse to form a complete gut. The caeca soon connect, and a functional digestive system is then in place (Fig. 12-8J).

S. G. Nair (1956) specifically noted a lack of any transitory seventh abdominal ganglion in *Irona*, but Strömberg (1965, 1967) did note such a structure in *Idotea* and *Limnoria*.

The development of isopods is essentially complete by the time of hatching, and all appendages are present (thoracic and abdominal), except those of the eighth thoracomere, which are lacking (Fig. 12-8J).

FOSSIL RECORD Though predominantly benthic organisms and thus prime candidates for fossilization, isopods have a spotty fossil record. The oldest isopods currently known are the marine Palaeophreatoicidae, the earliest being *Hesslerella shermani* Schram, 1970, which occurs in the Middle Pennsylvanian of Illinois (Fig. 12-9A); and other related genera are noted in the Permian of Germany, the Soviet Union, and possibly as well in Britain (Schram, 1980). An amphisopodid genus, *Protamphisopus*, is recorded from the Late Permian of Germany and the Triassic of Australia (Nicholls, 1943; and Glaessner and Malzahn, 1962), which seems to represent a time of transition of the phreatoicideans from marine to freshwater. Though the record of the Phreatoicidea is quite good in the Late Paleozoic and Triassic and very informative about the group's evolution, no other phreatoicidean fossils are known from then down to Recent time.

The fossil records of other isopod groups (Hessler, 1969) are not nearly as helpful. The Flabellifera are noted in the Triassic of Germany, Jurassic deposits of Britain and Central Europe, and Cretaceous of Brazil, but these are largely sphaeromatids and serolids of uncertain affinity and of the extinct family Archaeoniscidae. Other fossil flabelliferans follow a pattern seen in other isopod groups such as the Oniscoidea and Valvifera and again in the amphipods, that is, a record abruptly beginning in the middle of the Tertiary and extending to the Recent. In the case of the Valvifera, analysis of their cladistic relationships and biogeography (Brusca, 1984) indicates that their evolution as a group must extend back into earliest Mesozoic times, yet there is only a single fossil known for this group, *Proidotea haugi* (Fig. 12-9B) from the Oligocene of Rumania.

Fig. 12-9. (A) *Hesslerella shermani*, the earliest isopod, from the Carboniferous of North America; (B) *Proidotea haugi*, Oligocene of Rumania; (C) *Urda rostrata*, later Mesozoic of central Europe. (From Schram, 1974; Hessler, 1969)

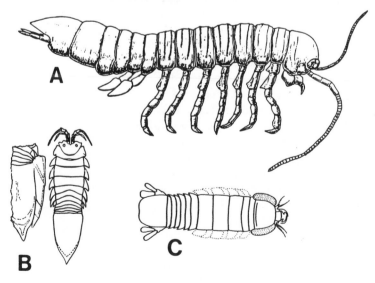

No fossil record is yet known for anthurideans, asellotes, gnathiids, and calabozoideans. Epicarideans are known indirectly from the latter half of the Mesozoic by virtue of fossil decapods exhibiting the characteristic effects of bopyrid infestation, asymmetrically swollen branchial chambers. One Jurassic and Cretaceous genus, *Urda* (Fig. 12-9C), is not assignable to any group with certainty, having been variously assigned to gnathiids, cymothoid flabelliferans, or to an intermediate between them. Except for the phreatoicideans, a lot more material of fossil isopods needs to be discovered before a meaningful contribution can be made to understanding the history of the group.

TAXONOMY   The Isopoda is an extremely diverse group of eumalacostracans. Many groups of isopods are actively being worked on, though few effective across-the-board analyses of characters within or between groups have yet to be performed. The following scheme is modified from that presented by Bowman and Abele (1982).

Infraorder Gnathiidea Leach, 1814 Recent
    Family Gnathiidae Harger, 1880
Infraorder Anthuridea Leach, 1814 Recent
    Family Anthuridae Leach, 1814
    Family Hyssuridae Wägele, 1981
    Family Paranthuridae Menzies and Glynn, 1968
Infraorder Flabellifera Sars, 1882 Triassic–Recent
    Family Aegidae Leach, 1815 Recent
    Family Anuropodidae Stebbing, 1893 Recent
    Family Archaeoniscidae Haack, 1918 Jurassic
    Family Bathynataliidae Kensley, 1978 Recent
    Family Bathynomidae Wood–Mason and Alcock, 1891 Recent
    Family Cirolanidae Dana, 1853 ?Miocene–Recent
    Family Corallanidae Hansen, 1890 Recent
    Family Cymothoidae Leach, 1818 Recent
    Family Keuphyliidae Bruce, 1980 Recent
    Family Limnoriidae Dana, 1853 Recent
    Family Phoratopodidae Hale, 1925 Recent
    Family Plakarthriidae Richardson, 1904 Recent
    Family Serolidae Dana, 1853 ?Triassic–Recent
    Family Sphaeromatidae Milne Edwards, 1840 Triassic–Recent
    Family Tridentellidae Bruce, 1984 Recent
Infraorder Asellota Latreille, 1803 Recent
    Superfamily Aselloidea Rafinesque, 1815
      Family Asellidae Rafinesque, 1815
      Family Atlantasellidae Sket, 1980
      Family Microcerberidae Karaman, 1933
      Family Stenasellidae Dudich, 1924

Superfamily Stenetrioidea Hansen, 1905
  Family Stenetriidae Hansen, 1905
Superfamily Janiroidea Sars, 1899
  Family Abyssianiridae Menzies, 1956
  Family Acanthaspidiidae Menzies, 1962
  Family Dendrotiidae Vanhöffen, 1914
  Family Desmosomatidae Sars, 1899
  Family Echinothambematidae Menzies, 1956
  Family Eurycopidae Hansen, 1916
  Family Haplomunnidae Wilson, 1976
  Family Haploniscidae Hansen, 1916
  Family Ilyarchnidae Hansen, 1916
  Family Ischnomesidae Hansen, 1916
  Family Joeropsididae Nordenstam, 1933
  Family Janirellidae Menzies, 1956
  Family Janiridae Sars, 1899
  Family Macrostylidae Hansen, 1916
  Family Mesosignidae Schultz, 1969
  Family Microparasellidae Karaman, 1933
  Family Mictosomatidae Wolff, 1965
  Family Munnidae Sars, 1899
  Family Munnopsidae Sars, 1869
  Family Nannoniscidae Hansen, 1916
  Family Paramunnidae Vanhöffen, 1914
  Family Pleurocopidae Fresi and Schiecke, 1972
  Family Thambematidae Stebbing, 1913
Superfamily Gnathostenetroidoidea Kussakin, 1967
  Family Gnathostenetroidoidae Kussakin, 1967
Infraorder Calabozoidea Van Lieshout, 1983 Recent
  Family Calabozoidae Van Lieshout, 1983
Infraorder Valvifera Sars, 1882 Oligocene–Recent
  Family Amesopodidae Stebbing, 1905 Recent
  Family Arcturidae Sars, 1899 Recent
  Family Holognathidae Thomson, 1904 Recent
  Family Idoteidae Milne Edwards, 1840 Oligocene–Recent
  Family Pseudidotheidae Ohlin, 1901 Recent
  Family Xenarcturidae Sheppard, 1957 Recent
Infraorder Phreatoicidea Stebbing, 1893 Pennsylvanian–Recent
  Family Amphisopodidae Nicholls, 1943 Triassic–Recent
  Family Nichollsiidae Tiwari, 1958 Recent
  Family Palaeophreatoicidae Birshtein, 1962 Pennsylvanian– Permian
  Family Phreatoicidae Chilton, 1891 Recent
Infraorder Epicaridea Latreille, 1831 Recent
  Family Bopyridae Rafinesque, 1815

Family Dajidae Sars, 1882
Family Entoniscidae F. Müller, 1871
Family Liriopsidae Bonnier, 1900
Infraorder Oniscidea Latreille, 1803 Eocene–Recent
Infraorder Tylomorpha Vandel, 1943 Recent
Family Tylidae Milne Edwards, 1840
Infraorder Ligiamorpha Vandel, 1943 Recent
Section Diplocheta Vandel, 1957
Family Ligiidae Brandt, 1883
Family Mesoniscidae Verhoeff, 1908
Section Synocheta Legrand, 1946 Eocene–Recent
Superfamily Trichoniscoidea Sars, 1899 Eocene–Recent
Family Buddelundiellidae Verhoeff, 1930 Recent
Family Trichoniscidae Sars, 1899 Recent
Superfamily Styloniscoidea Vandel, 1952 Recent
Family Schoebliidae Verhoeff, 1938
Family Styloniscidae Vandel, 1952
Family Titaniidae Verhoeff, 1938
Family Tunanoniscidae Borutskii, 1969
Section Crinocheta Legrand, 1946 Eocene–Recent
Superfamily Oniscoidea Dana, 1852 Eocene–Recent
Family Bathytropidae Vandel, 1973 Recent
Family Berytoniscidae Vandel, 1973 Recent
Family Detonidae Budde–Lund, 1906 Recent
Family Halophilosciidae Vandel, 1973 Recent
Family Olibrinidae Vandel, 1973 Recent
Family Oniscidae Brandt, 1851 Eocene–Recent
Family Philosciidae Vandel, 1952 Recent
Family Platyarthridae Vandel, 1946 Recent
Family Pudeoniscidae Lemos de Castro, 1973 Recent
Family Rhyscotidae Arcangeli, 1947 Recent
Family Scyphacidae Dana, 1853 Recent
Family Speleoniscidae Vandel, 1948 Recent
Family Sphaeroniscidae Vandel, 1964 Recent
Family Stenoniscidae Budde-Lund, 1904 Recent
Family Tendosphaeridae Verhoeff, 1930 Recent
Superfamily Armadilloidea Verhoeff, 1917 Eocene–Recent
Family Actaeciidae Vandel, 1952 Recent
Family Armadillidae Verhoeff, 1917 Recent
Family Armadillidiidae Brandt, 1833 Miocene–Recent
Family Atlantidiidae Arcangeli, 1954 Recent
Family Balloniscidae Vandel, 1963 Recent
Family Cylisticidae Verhoeff, 1949 Recent
Family Eubelidae Budde–Lund, 1904 Recent
Family Periscyphicidae Ferrara, 1973 Recent

Family Porcellionidae Verhoeff, 1918 Eocene–Recent
Family Trachelipidae Strouhal, 1953 Recent
Infraorder uncertain
Family Urdidae Kunth, 1870 Jurassic–Cretaceous

REFERENCES

Bastida, R., and M. R. Torti. 1973. Los isopodos Serolidae de la Argentina. Clave para su reconocimiento. *Physis* (B) **32**(84):19–46.
Birshtein, Ya. A. 1973. *Deep water isopods of the north-western part of the Pacific Ocean.* Indian Nat. Sci. Co. Centre, New Dehli.
Bowman, T. E., and L. G. Abele. 1982. Classification of the Recent Crustacea. In *Biology of the Crustacea*, Vol. I (L. G. Abele, ed.), pp. 1–27. Academic Press, New York.
Brusca, R. C. 1984. Phylogeny, evolution, and biogeography of the marine isopod subfamily Idoteinae. *Trans. San Diego Soc. Nat. Hist.* **20**:99–134.
Brusca, R. C., and B. R. Wallerstein. 1979. Zoogeographic patterns of idoteid isopods in the northeast Pacific, with a review of the shallow water zoogeography of the area. *Bull. Biol. Soc. Wash.* **3**:67–105.
Burbanck, M. P., and W. D. Burbanck. 1974. Sex reversal of female *Cyathura polita*. *Crustaceana* **26**:110–12.
Burbanck, W. D., and M. P. Burbanck. 1979. *Cyathura*. In *Population Ecology of Estuarine Invertebrates*, pp. 293–323. Academic Press, New York.
Calman, W. T. 1909. Crustacea. In *A Treatise on Zoology*, Vol. 7. Adam and Charles Black, London.
Collinge, W. E. 1918. On the oral appendages of certain species of marine Isopoda. *J. Linn. Soc. Zool. Lond.* **34**:65–93.
Delage, Y. 1881. Contribution à l'étude de l'appareil circulatoire des Crustacés edriophthalmes marine. *Arch. Zool. Exper.* **9**:1–173, pls. 1–12.
Dexter, D. M. 1977. Natural history of the Pan-American beach isopod *Excirolana braziliensis*. *J. Zool.* **183**:103–9.
Dohrn, A. 1870. Die embryonale Entwicklung des *Asellus aquaticus*. *Zeit. wiss. Zool.* **17**:221–78, pls. 14, 15.
George, R. Y. 1972. Biphasic molting in isopod Crustacea and the finding of an unusual mode of molting in the Antarctic genus *Glyptonotus*. *J. Nat. Hist.* **6**:651–56.
George, R. Y. 1979. Behavioral and metabolic adaptations of polar and deep-sea crustaceans: a hypothesis concerning physiological basis for evolution of cold adapted crustaceans. *Bull. Biol. Soc. Wash.* **3**:283–96.
Glaesser, M. F., and E. Malzahn. 1962. Neue Crustaceen aus dem niederrheinischen Zechstein. *Fortschr. Geol. Rheinld. u. Westf.* **6**:245–64.
Goodrich, A. L. 1939. The origin and fate of the entoderm elements in the embryogeny of *Porcellia laevis* and *Armadillidium nasatum*. *J. Morph.* **64**:401–29.
Green, J. 1957. The feeding mechanism of *Mesidotea entomon.*, *Proc. Zool. Soc. Lond.* **129**:245–54.
Hansen, H. J. 1893. A contribution to the morphology of the limbs and mouthparts of crustaceans and insects. *Ann. Mag. Nat. Hist.* (6)**12**:417–34.
Harrison, H. K. 1984. The morphology of the sphaeromatid brood pouch. *Zool. J. Linn. Soc. Lond.* **82**:363–407.
Hessler, R. R. 1969. Peracarida. In *Treatise on Invertebrate Paleontology*, Part R, *Arthropoda* 4 (R. C. Moore, ed.), pp. R360–93. Geol. Soc. Am. and Univ. Kansas Press, Lawrence.

Hessler, R. R., and D. Thistle. 1975. On the place of origin of deep-sea isopods. *Mar. Biol.* **32**:155–65.

Hessler, R. R., G. D. Wilson, and D. Thistle. 1979. The deep-sea isopods: a biogeographic and phylogenetic overview. *Sarsia* **64**:67–75.

Holdich, D. M. 1974. The midgut/hindgut controversy in isopods. *Crustaceana* **24**:211–14.

Ide, M. 1892. Le tube digestiv des Edriophthalmes; étude anatomique et histologique. *Cellule* **8**:97–204, pls. 1–7.

Jackson, H. G. 1926. The morphology of the isopod head. 1. The head of *Ligia oceanica*. *Proc. Zool. Soc. Lond.* **1926**:885–911.

Johnson, W. S. 1976. Biology and population dynamics of the intertidal isopod *Cirolana harfordi*. *Mar. Biol.* **36**:343–50.

Jones, D. A. 1968. The functional morphology of the digestive system in the carnivorous intertidal isopod *Eurydice*. *J. Zool.* **156**:363–76.

Jones, D. A. 1976. The systematics and ecology of some isopods of the genus *Cirolana* from the Indian Ocean region. *J. Zool.* **178**:209–22.

Kaestner, A. 1970. *Invertebrate Zoology*, Vol. 3, *Crustacea*. Interscience, New York.

Kensley, B. 1982. Deep-water Atlantic Anthuridea. *Smith Contr. Zool.* **346**:1–60.

Klapow, L. A. 1970. Ovoviviparity in the genus *Excirolana*. *J. Zool., Lond.* **162**:359–69.

Kunze, J. C. 1981. The foregut of malacostracan Crustacea: functional morphology and evolutionary trends. *Amer. Zool.* **21**:968.

Kussakin, O. G. 1973. Peculiarities of the geographical and vertical distribution of marine isopods and the problem of deep-sea fauna origin. *Mar. Biol.* **23**:19–34.

Kussakin, O. G. 1979. *Morskie i solonovatovodnie ravnonogie rakoobraznie (Isopoda) kholodnikh i umerennikh vod cevernogo polyshariya; podotryad Flabellifera.* Akademiya Nauk SSSR, Leningrad.

Manton, S. M. 1928. On the embryology of the mysid crustacean *Hemimysis lamornae*. *Phil. Trans. Roy. Soc. Lond.* (B) **216**:363–463.

McMurrich, J. P. 1895. Embryology of the isopod Crustacea. *J. Morph.* **11**:63–154.

Monod, T. 1926. Les Gnathiidae. *Mem. Soc. Sci. Nat. Maroc.* **13**:1–667.

Nair, K. B. 1939. The reproduction, oögenesis and development of *Mesopodopsis orientalis*. *Proc. Indian Acad. Sci.* (B) **9**:175–223.

Nair, S. G. 1956. On the embryology of the isopod *Irona*. *J. Embryol. Exp. Morph.* **4**:1–33.

Naylor, E. 1955. The diet and feeding mechanism of *Idotea*. *J. Mar. Biol. Assc. U.K.* **34**:347–55.

Nicholls, G. E. 1942–43. The Phreatoicidea. *Pap. Proc. Roy. Soc. Tasmania* **1942**:1–145, **1943**:1–157.

Nilsson, H. L. 1978. The fine structure of the compound eyes of shallow-water asellotes, *Jaera albifrons* and *Asellus aquaticus*. *Acta Zool.* **59**:69–84.

Nusbaum, J. 1891. Beiträge zur Embryologie der Isopoden. *Biol. Zbl.* **11**:42–49.

Nusbaum, J. 1898. Zur Entwicklungsgeschichte des Mesoderms bei den parasitischen Isopoden. *Biol. Zbl.* **18**:557–69.

Paris, O. H. 1963. Ecology of *Armadillidium vulgare*. *Ecol. Monogr.* **33**:1–22.

Paris, O. H., and F. A. Pitelka. 1962. Population characteristics of *Armadillidium vulgare*. *Ecology* **43**:229–49.

Schoenichen, W. 1898. Der Darmkanal der Onisciden und Aselliden. *Zeit. wiss. Zool.* **65**:143–78, pl. 6.

Schram, F. R. 1970. Isopod from the Pennsylvania of Illinois. *Science* **169**:854–55.

Schram, F. R. 1974. Paleozoic Peracarida of North America. *Fieldiana: Geol.* **33**:95–124.

Schram, F. R. 1980. Miscellaneous Late Paleozoic Malacostraca of the Soviet Union. *J. Paleo.* **54**:542–47.

Sheppard, E. M. 1939. The coxal joint and its outgrowth in certain isopod Crustacea. *Ann. Mag. Nat. Hist.* (11)**3**:161–74.

Shuster, S. M. 1981. Sexual selection in the Socorro isopod, *Thermosphaeroma thermophilum*. *Animal Behavior* **28**:698–707.

Siebenaller, J. F., and R. R. Hessler. 1977. The Nannoniscidae: *Hebefustis* and *Nannoniscoides*. *Trans. San Diego Soc. Nat. Hist.* **19**:17–44.

Siewing, R. 1951. Besteht eine engere Verwandschaft zwischen Isopoden und Amphipoden. *Zool. Anz.* **147**:166–80.

Strömberg, J. O. 1965. On the embryology of the isopod *Idotea*. *Ark. Zool.* **17**:421–67.

Strömberg, J. O. 1967. Segmentation and organogenesis in *Limnoria lignorum*. *Ark. Zool.* **20**:91–139.

Strömberg, J. O. 1971. Contribution to the embryology of bopyrid isopods with special reference to *Bopyroides*, *Hemiarthrus*, and *Pseudione*. *Sarsia* **47**:1–46.

Vandel, A. 1938. Le déterminisone du sexe et de la monogénie chez *Trichoniscus provisorius*. *Bull. Biol. France Belgique* **72**:147–86.

Wägele, J. W. 1983. On the homology of antennal articles in Isopoda. *Crustaceana* **45**:29–37.

Wallerstein, B. R., and R. C. Brusca. 1982. Fish predation: a preliminary study of its role in the zoogeography and evolution of shallow water idoteid isopods. *J. Biogeogr.* **9**:135–50.

Wilson, G. D. 1980. Superfamilies of the Asellota and the systematic position of *Stenetrium weddellense*. *Crustaceana* **38**:219–21.

Wilson, G. D., and R. R. Hessler. 1974. Some unusual Paraselloidea from the deep benthos of the Atlantic. *Crustaceana* **27**:47–67.

Zimmer, C. 1927. Isopoda. In W. Kükenthal and T. Krumbach, *Handbuch der Zoologie* **3**:697–766. de Gruyter, Berlin.

13

AMPHIPODA

DEFINITION   Carapace absent; eyes sessile; first thoracomere fused to cephalon; antennules typically biramous and well developed; antennae without scales, typically a five-segment peduncle; mouthparts generally arranged in a compact buccal mass; maxillipedes without epipodites, at least partially fused coxae; thoracopods uniramous, second and third typically as subchelate gnathopods, coxae typically expanded into ventrolateral plates, at least some thoracopods with inner branchial epipodites, oöstegite brood pouch; anterior pleomeres usually with well-developed pleura, posterior pleomeres associated as urosome, rami of first three pleopods annulate, last two pleopods uropodiform; telson typically free, often bilobate.

HISTORY   Latreille erected the order Amphipoda in 1816 for what we today consider gammarideans. However, even at this early date the whale louse *Cyamus* (Fig. 13-1C) was recognized as distinctive and placed within the isopods. This distinctiveness was eventually formalized with the inclusion of the caprellids (Fig. 13-1B) and the cyamids in the Laemodipoda, a taxon originally treated as equal in status with the Isopoda and Amphipoda. Milne Edwards further distinguished the hyperiids (Fig. 13-1I, J) from the gammarids (Fig. 13-1D, E, F). However, it was Dana who in 1852 erected the three 'traditional' suborders of the amphipods: caprellideans, gammarideans, and hyperiideans; these were later joined by the peculiar ingolfiellideans (Fig. 13-1A) that Hansen erected in 1903. The Amphipoda have never lacked for monographers, and a few of the most recent surveys have been produced for ingolfiellideans (Stock, 1976, 1981), hyperiideans (Bowman and Grüner, 1973), gammarideans (e.g., Barnard, 1969; Bousfield, 1982b; Lincoln, 1979), caprellids (McCain, 1968; Laubitz, 1970, 1976), and cyamids (Leung, 1967).

   At one time, Leach united amphipods with isopods into the taxon Edriopthalma. Calmãn (1909) gave these two groups separate status in his classification of the peracarids, and this separation (reinforced as it was by Siewing's consideration of gut and developmental features) has remained unchallenged until recently by Schram (1981, 1984) who felt the shared derived features of amphipods and isopods justify a return to the use of Edriophthalma.

MORPHOLOGY   Though recent taxonomic work on amphipods seems to have resulted in the erection of familial and subfamilial taxa at an increas-

158

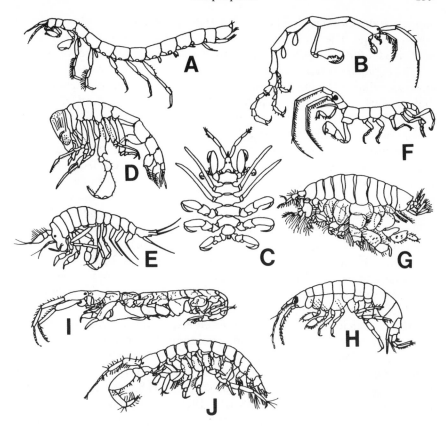

Fig. 13-1. Amphipod body types. (A) *Ingolfiella leleupi* (from Dahl, 1977); (B) *Caprella linearis* (from Dahl, 1977); (C) *Cyamus boopis* (from Hessler, 1969); (D) *Phronima* sp. (from Bowman and Grüner, 1973); (E) *Arachaeoscina* sp. (from Bowman and Grüner, 1973); (F) *Dyopedos* sp. (from Lincoln, 1979); (G) *Pseudohaustorius carolinensis* (from Bousfield, 1973); (H) *Gammarus* sp. (from Lincoln, 1979); (I) *Cerapus* sp. (from Lowry, 1982); (J) *Corophium insidiosum* (from Bousfield, 1973).

ing rate, currently nearly 6000 species (Bousfield, 1982b), the higher taxonomy of the group has remained essentially unaltered since the turn of the twentieth century. Though the possession of three pairs of uropods would seem to distinctly characterize the Amphipoda (Fig. 13-1) (most caprellideans lack these) and the mouthparts form a compact buccal mass, the group is otherwise rather noted for its retention of several primitive features. These include complete cleavage in the eggs, an egg-nauplius stage, development occurring entirely within the egg membranes and the hatchling emerging with a complete set of appendages, biramous antennules, antennal glands, a tendency to not always elaborate the coxal plates, prevalence of dicondylic thoracopodal coxa–basis joints, retention of gills on the thoracopods (this may be a secondary reacquisition), a complete midgut, and a thoracic circulatory system as opposed to the abdominal system in isopods. The probability is high that the possession of most of these primitive features is a primary rather than a secondary reacquisition.

The cephalon is relatively short. The sessile compound eyes can be variously developed (Hallberg et al., 1980) and range from a complete lack of these organs, as in subterranean forms, to the tremendous globular eyes of many (Fig. 13-1D), but not all (Fig. 13-1E), hyperiideans. However, of special note is the fact that the eyes in amphipods lack facets. A rostrum may be present, but more typically, is reduced or absent altogether. The ingolfiellideans are said to have 'eye lobes,' but since these lack any dioptric elements and any nerve supply, the function of these lobes is presently unknown.

The antennules (Fig. 13-2A) have a peduncle of three segments that bears the flagella. The outer branch is typically well developed, and the inner branch (accessory flagellum) is usually somewhat smaller and may sometimes be absent. The primary flagellum usually bears aesthetascs (especially in males). In some gammarideans both antennules and antennae bear cuplike sensory structures called calceoli, variously on peduncular and flagellar segments mostly in the males.

The antennae (Fig. 13-2A) are uniramous, lacking a scale. The peduncle has five segments with a conical opening for the antennal gland on the second joint. Calceoli can also be present on these appendages. In both the antennules and antennae, variations in size, setation, and development of aesthetascs and calceoli are usually sexually dimorphic and are proving of a great taxonomic value (e.g., Bousfield, 1978; Lincoln and Hurley, 1981). In some hyperiideans, for example, mimonectids and proscirids, the antennae in females are reduced to just the tubercles of the antennal gland openings. In some instances the antennae can be pediform (Fig. 13-1I, J).

The labrum may be a simple rounded lobe, or it may be apically notched, the resulting lobules sometimes being asymmetrically developed. The paragnaths (Fig. 13-2B) can also vary from a primitive condition of simple narrow lobes to a derived condition (Fig. 13-3A) wherein they are quite broad with elaboration of accessory 'inner lobes.' The mandibles (Fig. 13-2C) show a great variety of form. A full complement of structures would include a well-developed incisor process, a lacinia mobilis differing on the left and right sides, a spine row between incisor and molar process, a molar process that may be elaborately ridged or toothed, and a three-segment setose palp. In actual practice all these structures may be variously elaborated, reduced, or completely absent.

The maxillules (Fig. 13-2D) can be quite elaborate. The medially directed lobes of the protopod are medially setose and/or terminally spinose. The distal palp may range variously from being very large and robust with one or two joints (Fig. 13-3B) to being absent or vestigial. This limb is noteworthy in that in amphipods it is typically larger and more elaborate than the adjacent maxillae.

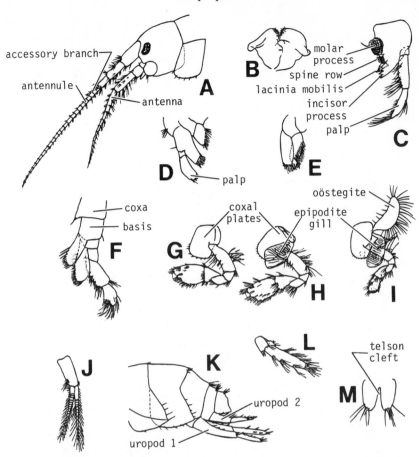

Fig. 13-2. Appendages of *Anisogammarus similimanus*. (A) antennule and antenna; (B) paragnath; (C) mandible; (D) maxillule; (E) maxilla; (F) maxillipede; (G) male second thoracopod (= first gnathopod); (H) male third thoracopod (= second gnathopod); (I) female third thoracopod (= second gnathopod); (J) first pleopod; (K) urosome with first and second uropods; (L) third uropod; (M) telson. (From Bousfield, 1961)

The maxillae (Fig. 13-2E) of amphipods are reduced over the condition seen in other eumalacostracans and are more akin in form to the maxillules seen in other groups. Two setose lobes arise from the base. The setation may be rather elaborate or the lobes themselves may be rather reduced, extremely so in the cyamids where the maxillae generally exist as a single pair of partially fused lobes. This distinctively reduced form of the maxillae seems to be the single feature that unites all amphipods.

The maxillipedes or first thoracopods (Fig. 13-2F) typically have fused coxae, though in ingolfiellideans this fusion is incomplete (Fig. 13-3D). The basis and ischium have well-developed, distally directed endites. In a few families the basal lobes are themselves fused. The endopod palp is

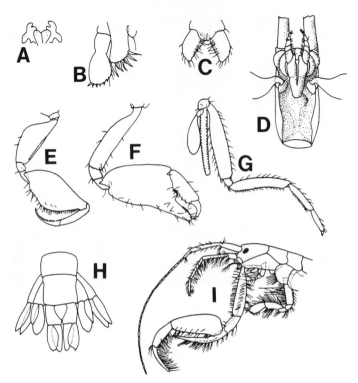

Fig. 13-3. Variant appendage types for amphipods. (A) an elaborate paragnath; (B) maxillule with robust palp (from A.B. Bousfield, 1978); (C) *Lanceola* sp. hyperiidean maxillipede lacking palp (from Bowman and Grüner, 1973); D, E, F, G) *Metaingolfiella mirabilis* (from Ruffo, 1969); (D) maxillipede with incomplete fusion of coxae; (E) carpo-subchelate first gnathopod; (F) carpo-chelate second gnathopod; (G) fifth thoracopod with linear oöstegites; (H) *Pronoe capita*, with lamellate uropod series (from Bowman and Grüner, 1973); (I) *Aorides spinosus*, with chelo-subchelate first gnathopods (from Conlon and Bousfield, 1982).

generally well developed, though in hyperiideans the palp is absent (Fig. 13-3C) and the endites exhibit various degrees of fusion. In cyamids, the entire pair of maxillipedes is frequently reduced to a single unpaired plate, though in young *Cyamus nodosus* the maxillipedes are fully developed.

The more posterior uniramous thoracopods or pereiopods display a great variety of forms and cannot be easily characterized—not even as 'amphipodous,' since they seem no more so taken, as a group, than those seen among certain isopods. The coxae of these limbs tend to be closely associated with the body wall and characteristically are developed as plates. This latter feature, however, is not well expressed in the ingolfiellideans or caprellideans (Fig. 13-1A, B, C). The second and third thoracopods (Fig. 13-2G, H, I) are frequently developed as subchelate or chelate gnathopods and occasionally these can take on rather bizarre forms (Fig.

13-3I). In the ingolfiellideans, these subchelae are formed by opposing the dactylus *and* propodus against the carpus (Fig. 13-3E, F). Gnathopods can assist in food collection but tend to be larger in males, which use them to grasp the females, sometimes for prolonged periods, prior to and during copulation. The posterior thoracic limbs are ambulatory, though in some hyperideans and a few gammarideans these may be chelate or subchelate and raptorial in function (Fig. 13-1P). The fourth and fifth thoracopods are vestigial or highly modified in caprellideans (Fig. 13-1B, C). The pereiopods also bear the oöstegites and coxal gills.

The oöstegites are generally broad and marginally setose (Fig. 13-2I) but may be narrow or 'linear' with few or reduced setae (Fig. 13-3G). Hyperiideans lack marginal setae, and in the genus *Rhabdosoma* the oöstegites are reduced and their function replaced by the gills. The gills are inwardly directed epipodite plates; these may be simple or elaborate with convolutions, ridges, or dendritic processes. The actual formulas for oöstegites and gills on the thoracopods vary from group to group and within groups and are summarized in Table 13-1 for the maximal possible number encountered. Sternal 'gills' are noted in several major groups of amphipods, especially in freshwater forms. These appear to be more osmoregulatory in function, however, than respiratory (Bousfield, 1982a).

Finally, the coxa–basis articulation, so distinctive in other peracarid types (Hessler, 1982), is rather different in the amphipods. The second

Table 13–1. Maximum number of gills and oöstegites and their location in the various subgroups of amphipods. Individual taxa within each of these groups may exhibit variations to the patterns here with fewer numbers of gills and oöstegites than indicated. The '?' indicates valve flap over the genital aperture that may be vestigial oöstegite.

		t_1	t_2	t_3	t_4	t_5	t_6	t_7	t_8
Ingolfiellidea	gills				+	+	+		
	oöstegites			+	+	+			
Gammaridea	gills			+	+	+	+	+	+
	oöstegites			+	+	+	+		
Hyperiidea	gills			+	+	+	+	+	
	oöstegites			+	+	+	+		
Caprellidea	gills			(+)	+	+			
	oöstegites				+	+	?		

through fifth thoracopods have simple dicondylic articulations at that joint. The sixth and seventh limbs, while dicondylic, have a peculiar pulleylike arrangement of rotator muscles (in addition to the to-be-expected promotor/remotor set) that allow some versatility of motion about that joint. Only the eighth thoracopod has the versatile monocondylic joint that is so characteristic of the limbs of other peracarid types, and this may be related to the basis being inside the coxa in this limb.

The pleopods are of two basic types, corresponding to the subregions of the abdomen. The three anterior pleomeres frequently have well-developed pleura, and the pleopods (Fig. 13-2J) are annulate, setose, and biramous. They are utilized for swimming in nektonic forms, and their constant beating in all amphipods facilitates respiratory currents around the thoracic gills. The three posterior pleomeres form the urosome (Fig. 13-2K), and the last three abdominal appendages are uropods. These typically are styliform (Fig. 13-2L), though in nektonic types, for example, hyperiideans (Fig. 13-3H), the uropods can be lamellate. The telson (Fig. 13-2M) is free when present, and its great variation in form has been extensively employed for taxonomic purposes.

In the ingolfiellideans the last pair of uropods is vestigial. In most caprellideans, except the caprogammarids, the entire abdomen is vestigial. These reductions and losses remove the 'traditional' amphipod common character, the three uropods, from universal application in identifying potential members of the group.

The digestive system in amphipods exhibits great variety among the various groups. The most complex arrangement of the foregut (Martin, 1964; Schmitz and Scherrey, 1983; Icely and Nott, 1984) is seen in the gammarideans and caprellideans (Fig. 13-4A, B). The entrance to the cardiac stomach from the esophagus is flanked by spinose and setose ridges called ampullae, which serve to triturate food and prevent regurgitation. Dorsal and ventral lateral folds delineate dorsal and ventral channels in the anterior cardiac chamber. The pyloric chamber is also divided into a dorsal channel and dorsal and ventral chambers by lateral folds. The ventral chamber, or gland filter, or pyloric filter, is separated into right and left halves by a midventral ridge. Digestate from the cardiac stomach passes through the ventral gland filter to the digestive caeca that open into it posteriorly. The dorsal pyloric chamber is continuous anteriorly with the cardiac region, and passes solid indigestibles into the midgut over the caecal openings through a pyloric funnel composed of a series of valves and ridges.

The caeca are numerous. There are generally two lateral pairs of caeca extending the length of the body. However, in some gammarideans and most hyperiideans there is only a single pair, and in caprellideans the ventral pair is vestigial. There is also an anteriorly directed middorsal caecum projecting over the foregut from the midgut. This caecum is single in gammarideans (though Schmitz and Scherrey, 1983, record three caeca in *Hya-*

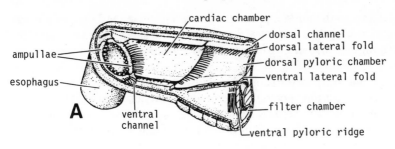

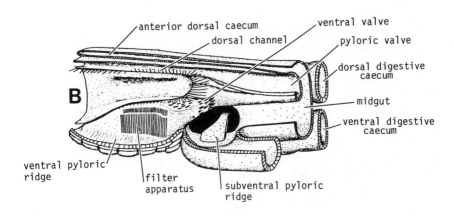

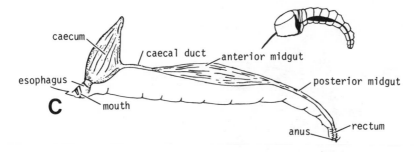

Fig. 13-4. Gut anatomy of amphipods. (A, B) anterior and posterior foregut in *Marinogammarus obtusatus* (from Martin, 1964); (C) gut of the hyperiid *Cystisoma* (from Brusca, 1981).

lella), but the diverticula are paired in caprellideans and some hyperiideans. At the posterior end of the midgut of gammarideans is another, generally paired, set of caeca whose function is unknown.

The foregut of hyperiideans (Fig. 13-4C) presents a different structural plan (Brusca, 1981; Icely and Nott, 1984). The only foregut region (Brusca, personal communication) is apparently the esophagus, lined with sclerotized plates. This leads into a 'caecum,' a deeply furrowed chamber that extends dorsally as a blind pouch and that is emptied posteriorly by a 'caecal' duct into the midgut proper. The caecum, duct, and midgut are not lined with cuticle. The midgut is distended to the sides anteriorly but narrows to a tubular form along its posterior length. A short lined rectum terminates at the anus.

The amphipod proctodeum is generally short. However, in caprellideans, perhaps not unexpectedly, the hindgut extends anteriad to the level of the seventh thoracomere. Schmitz and Scherrey (1983) record in *Hyallela* an elaborate musculature that distends the hindgut and flushes the rectum with water during evacuation. Little is recorded about the guts of ingolfiellideans, though Siewing (1963) and Dahl (1977) observed that the foregut is greatly elongate.

The circulatory pattern (Delage, 1881) consists of a thoracic heart that pumps blood into anterior and posterior aortae. There may or may not be other anterior vessels. The blood, both anterior and posterior, empties eventually into a ventral sinus from which the epipodal gills are supplied. The efferent gill drainage then returns blood into the spacious pericardium to begin the cycle anew (Fig. 13-5).

Excretion is achieved by antennal glands, which are of normal size in marine forms, especially large in fresh and brackish water forms, and very small or vestigial in terrestrial types. There are also segmental nephrocytes or 'coxal glands' adjacent to the pericardial sinus. In addition, the posterior caeca of the gut have been long said to be excretory in nature, since they are sometimes noted to contain calcareous masses. However, the issue

Fig. 13-5. Generalized amphipod circulatory system in lateral and thoracic cross-sectional views. (Derived from Delage, 1881)

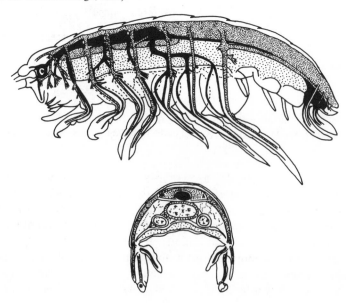

does not seem to have ever been definitively resolved, and Graf and Michaut (1980) exclude on ultrastructural grounds consideration of these caeca as Malphighian tubule homologs. The sternal gills may also be osmoregulatory.

The reproductive systems in amphipods are for the most part rather simple. The cylindrical testes are located in the fourth through seventh thoracomeres, though Brusca (1981) reports very small gonads in the hyperiid *Cystisoma* restricted to only near the caecal duct. The vasa deferentia have glandular walls and open to the outside on penis papillae on the eighth thoracic sternum. The ovaries are also cylindrical organs connected at their midlength by oviducts to gonopores on the sixth thoracomere. Brusca (1981) recorded on the eighth thoracomere the presence of male type thoracopods on female *Cystisoma*, and Calman (1909) noted the occurrence of eggs in testes of *Orchestia*. There is generally no modification of appendages to facilitate copulation in amphipods, though talitroids, bogidielloids, and ingolfiellideans modify their anterior pleopods presumably to facilitate sperm transfer. Oöstegites are progressively developed in individual females through several molts. An internal brood sac in *Cystisoma* is found in the third thoracomere, while the oöstegites of the fourth and fifth segments are used to sweep the eggs into the brood pouch. The small flaplike coverings of the genital openings of the caprellid females are thought to possibly be vestigial oöstegites.

The central nervous system of amphipods exhibits extensive fusion of elements. The ganglia of the mouthparts and maxillipedes are fused. The pereional and three anterior pleonal ganglia are free, but in gammarideans the urosomal ganglia are fused and located in the fourth pleomere, while in hyperiideans the third pleonal and urosomal ganglia are fused into a single unit. The brain of hyperiideans is distinctive in having prominent winglike optic lobes. In addition to the various sense organs already noted, there are statocysts in the dorsal cephalon of gammarideans and hyperiideans.

NATURAL HISTORY   Locomotion occurs in a great many ways among amphipods. Swimming is achieved mainly by the first three abdominal appendages and by flexion of the tailfan in pelagic forms. Ambulatory motion can occur in many different ways. Some benthic amphipods slide along on their sides; the abdomen alternately flexes and extends while the sixth through eighth thoracopods of the prone side push forward. Most benthic amphipods walk like other eumalacostracans with the venter down and thoracopods pushing along the substrate. *Cerapus* (Fig. 13-1I), *Corophium* (Fig. 13-1J), and other tube dwellers can also pull themselves along by using only the flexing action of their proximal antennal joints after the flagellar spines have 'grabbed' the substrate. A saltational locomotion is achieved among many amphipods by using the styliform uropods to push off against the substrate as the abdomen is quickly extended. Caprellids move about like inchworms, grabbing the substrate with their gnathopods, flexing the body dorsally in an arc, grasping the substrate with the posterior thoraco-pods while the gnathopods are released, and then extending the body flat again. Burrowing can be done either by an active scooping action of the pereiopods, or in some instances (e.g., haustoriids) by the scouring effect of the current under the body set up by the pleopods and thrusting of the uropods.

Amphipod feeding is also quite varied but is basically built around the lateral adducting motion of the triturating mandibles, an action that amphi-pods share with isopods (Watling, 1983). Carnivory and/or scavenging is among the most common feeding modes. Food is captured and held by the pereiopods, with the gnathopods serving to tear and shred fragments from the prey that are then passed forward to the mouthparts. The action of the incisor processes and laciniae of the mandibles and of the gastric ampullae serve to further fragment and prepare food for digestive action. The preda-ceous caprellids are not actively stalking carnivores or scavengers as are the gammarideans and hyperiideans. Rather, they passively wait for prey to come by, attached as they are by their posterior pereiopods to some sub-strate, whereupon they grab and rend the prey with their gnathopods. The rapacious hyperiid *Parathemisto gracilipes* is pragmatically either a carni-vore, scavenger, or cannibal as opportunity presents but has been reported (Kane, 1963) to actively eviscerate fish by attaching to the prey's belly and chewing away with mouthparts at the still swimming victim's stomach.

A variety of modes of filter feeding occurs among amphipods, facili-tated by the distinctive current pattern (Fig. 13-6) around the amphipod body (Dahl, 1977). Some caprellids use the antennae as filters. There are special 'swimming setae' on the antennae, which when removed have only slight effect on swimming (a decrease of 9%) but have greater impact on filter ability (Caine, 1977, 1979). This antennal feeding in caprellids clearly

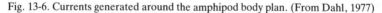

Fig. 13-6. Currents generated around the amphipod body plan. (From Dahl, 1977)

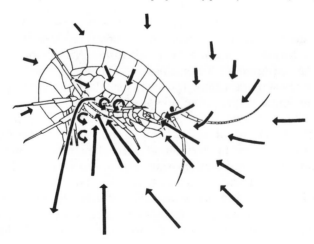

appears linked with grooming behavior, since the antennae are cleaned by the gnathopods and maxillipedes. This led Caine (1976) to suggest filter feeding in caprellids might have evolved from the grooming behavior of more predatory forms. Only limited work has been done on grooming in amphipods, but the variety of setal types evident on gnathopods is related in many instances to some kind of grooming function (Holmquist, 1982).

Other forms of amphipod filter feeding are outlined by Dennel (1933), Hunt (1925), and Kaestner (1970). Gammarideans can use setose anterior pereiopods to extract a filtrate from the respiratory current produced by the pleopods beneath the thorax. The maxillipedes clean these pereiopodal setae and pass the food particles forward to the mouthparts. Another more phyletically interesting mode has been noted in *Ampelisca* and *Haustorius* (Kaestner, 1970), which utilizes a maxillary filter. In this method the maxillae act as the pump in a manner reminiscent to that seen in lophogastrids and brachycaridans. However, unlike these other groups, the filtration arises from a rotary, scooping movement of the maxillae sweeping material out of the general body currents in toward the mouthfield from just lateral to it. The additional lateromediad action of the maxillipedes and maxillules serves to sweep food particles off the maxillae and push them between the lobes of the paragnaths. The rotary action of the maxillae in this mode of filtering is functionally unique (Dennell, 1933) and, like the other modes of filter feeding in amphipods, is independently developed several times within the group.

Other amphipod methods of feeding are as grazers, who scrape nutritive films off sand grains (e.g., some haustoriids and perhaps ingolfiellideans); and as detritus feeders, who sweep in sediment with antennae and then process the material with gnathopods and mouthparts (e.g., *Corophium*). Interestingly, true parasitism in amphipods is quite rare and restricted to ectoparasitism. Dahl (1960) recorded such for *Hyperia galba* on jellyfish, cyamids apparently feed on skin and mucous secretions of whales, a smattering of various gammarideans are known to infest the surfaces of fish, and others (in commensal fashion) inhabit the channels of sponges and tunicates (Kaestner, 1970). However, unlike isopods, amphipods as a group are noticeably depauperate of parasitic types.

Reproductive behaviors in amphipods are relatively simple and straightforward (e.g., Bousfield, 1973). In pelagic mating groups males are well equipped with sensory structures like eyes, aesthetascs, and calceoli, apparently to assist them in finding suitable females. Typically, in benthic groups there can be a precopulatory mounting period when the male places himself dorsal or lateral to the female, grasps her coxal plates or thoracic pleura with his gnathopods (Borowsky, 1984), and thus remains attached to await her molting. In many fossorial forms, this grasping period is virtually aborted and mating occurs in the water column with no prenuptial activities. In many burrowing forms, the males are often smaller than females, with copulation apparently occurring in situ in the sediment.

Actual copulation has been observed in *Gammarus* (e.g., Kinne, 1954). In molting (Kane, 1963; Sexton, 1924) the cuticle splits behind the cephalon due to extreme flexure of the body, followed by a splitting at the side along the coxal plates. The final castoff is then achieved with spasms of flexion. The male then grasps the female, venter to venter, and rapidly fans underneath the posterior margin of the brood pouch with his pleopods while extruding his sperm. The eggs are simultaneously, or soon after, laid and fertilized as they leave the gonoducts. Development commences immediately, and its speed is directly proportional to temperature. The life cycle appears to be short (Blevgad, 1922; Dennell, 1933) and typically involves a single reproductive event. Within these confines, Croker (1967) noted several ways sympatric species of amphipods are able to increase their reproductive efficiency, such as by increasing size of broods, numbers of females, and survivability of hatchlings, and by varying the time of maturation.

Steele (1967), in *Stegocephalus inflatus*, and Lowry (personal communication), in several lysianassoids, recorded protandry. In the latter group, both primary and secondary males were noted as well as females with penal processes. Several species seem to undergo this phenomenon: *Acontiostoma marionis*, *A. tuberculata*, *Scolopostoma prionoplax*, and *Stomatocontion pungapunga*. The phenomenon may be even more widespread than suspected. Thomas and Barnard (1983) believed that the males of *Anamixis hanseni* underwent an extreme metamorphosis in one molt from a condition that is so different it had been given an entirely different name, *Leucothoides pottsi*, while the female of the species apparently exhibited only the *Leucothoides*-morph. This is probably another case of protandry. The changes involved in the metamorphosis are so extreme, however, that the genera involved had been placed in separate families. Likewise, Lowry (personal communication) has found that the secondary male of *Ocosingo borlus* was so different that it had been placed in a separate genus as *Fresnillo fimbriatus*. Not only is the recognition of this protandric polymorphism important for eventually coming to understand amphipod reproductive strategies, it has already proven to be of staggering consequence for amphipod systematics!

Amphipods live in virtually all permanent aquatic media, but have done less well than isopods in pure terrestrial situations. The single exception to this is the southern hemisphere talitrids (Bousfield, 1958, 1982c, 1984; Hurley, 1968), but these have succeeded on land only by remaining in cryptic, moist, frost-free, leaf-litter habitats. Amphipods generally manifest a talent for remaining inconspicuous. They are frequently a major part of a community's biomass (see, e.g., Croker, 1967), and typically adopt cryptic habits either by burying themselves in sediment, hiding in algae or plant debris, camouflaging themselves in body form, hitching rides on hosts, or building protective domiciles. Some of these adaptations are especially intriguing. Crane (1969) observed that the amphipod *Pleustes platypa* mimics the gastropod *Mitrella carinata* by altering its body outline and color as well as adjusting its body posture to match that of the mollusk. Amphithoids generally prefer to make little domiciles of plant materials, but one species, *Pseudamphithoides incurvaria*, makes a portable 'pod' from modified algal fronds (Lewis and Kensley, 1982), which effectively provides it with a bivalved 'carapace.' Several amphipods build tubes, but at least one of them in the genus *Cerapus* comes to resemble (Lowry, 1982) members of another tube-building group, the tanaids (Fig. 13-1I).

Unlike isopods, the biogeographic distribution of amphipods has received considerable attention in the literature, so much so that to effectively summarize it is difficult in so short a space here. However, some particularly pregnant concepts should be mentioned.

In large part, the interest in amphipod biogeography arises from a need to provide some time frame for the evolution of the group given the paucity of its fossil record (Bousfield, 1981, 1983). These analyses would seem to indicate a possible Late Paleozoic origin for the group, though the earliest known fossils are only Eocene. Certain faunas of amphipods seem very ancient indeed, for example, the subterranean amphipods of Texas. This assemblage (Holsinger and Longley, 1980) contains a few crangonyctids (with a present-day northern hemisphere distribution that implies a pre-Jurassic Laurasian origin), and several hadziids and a bogidiellid (which are circumtropical Gondwanan and Tethyan types, see e.g. Stock, 1984). The entire groundwater fauna of these Texas wells has at least 22 species, which is more than twice as large as the most complex cave faunas known and implies a very old and stable habitat. Holsinger and Longley concluded the fauna of these wells is an ancient one that developed from an invasion of the groundwaters when that part of Texas was last covered by an epeiric sea in the Cretaceous. It thus represents a refugium of at least 70 million years.

The issue of refugia arises again in examining the distributions of the cave genus *Stygobromus*. In this instance, several species are endemic to glaciated areas in North America. Some species (Cordilleran region and Alberta) bear only distant relationships to species in nearby nonglaciated regions. The stygobromids have very limited dispersal capacities; thus it is postulated (Bousfield and Holsinger, 1981; Holsinger, 1980, 1981) that these amphipods have occupied their cave refuges under the great continental glaciers of the Pleistocene for the last 155,000 to 700,000 years. Of course dispersal can and has occurred in the stygobromids, but such instances are rather distinctive in character. In the Colorado Rocky Mountains the stygobromids have fewer species per unit area than those of the geologically older and more stable Appalachian region (Ward and Holsinger, 1981), which would seem to indicate that the Cordilleran forms are relatively more recent arrivals by an actively, albeit slowly, dispersing group.

Assessing the limiting factors in amphipod distribution patterns is not always easy. Keith (1969) studied aspects of certain species of *Caprella* and concluded that since they were quite versatile in terms of their dietary intake, food resources could not play any important role in limiting the distribution of these forms. In some instances temperature seems to play some role for some groups; for example, gammarideans are adapted to cold and are therefore noted for their boreal and cold-temperature preferences (Barnard, 1976).

In many instances, however, sheer chance and opportunity also seems to play an important role in explaining amphipod distributions. Vicariant patterns for some amphipods are rather striking, and consideration of changing paleobiogeography through time has something to offer in understanding the history of at least some groups (Bousfield, 1983, 1984). Stock (1980) considered the genus *Pseudoniphargus*. Though species of this genus range from tidal flats to 1000 m and occupy salinities from 0 to 36 ‰, they seem locked into a western Tethyan track, being known presently only from the western Mediterranean and the adjacent regions of the Atlantic.

The truly terrestrial talitrids also seem to exhibit an ancient vicariant track. Though talitrids are cosmopolitan in supralittoral habitats, the truly terrestrial species are found mainly on Gondwanan land masses (but are apparently missing on mainland South America). This might seem to imply a possible origin and history for terrestrial talitrids linked to events on Mesozoic Gondwanaland. However, Hurley (1968) would discount a purely vicariant explanation for terrestrial talitrids distributions for one that favors multiple origins and dispersal of members of this ecotype, a thesis amplified by Bousfield (1984).

Dispersal of course can play and has played a role in establishing amphipod taxa. For example, Barnard (1960) believed the genus *Paraphoxus* originated in the northern Pacific and subsequently spread across Central America into the North Atlantic before the isthmus arose. Mills (1965), in examining the distribution of northern hemisphere ampeliscids, recognized two centers of endemism: one in northern and western Europe, the other essentially in the waters around North America. Though he felt these concentrations were due to Mesozoic origins and early Tertiary radiations of distinctly separated groups of ampeliscids in the northeastern Pacific and the Tethys, the current overlap of the two faunas in the sub-Arctic and northern and western Europe is due to dispersal and mixing of the two faunas.

Some unique distribution patterns present peculiar problems of their own. For example, certain distinct species swarms are noted in isolated lake basins like Lake Titicaca and Lake Baikal. In Lake Baikal, Bazikalova (1945) recorded nearly 300 species in 40+ genera of amphipods in the lake, representing nearly 25% of the world's freshwater amphipod fauna. Then there is the incredible distribution (Noodt, 1965; Stock, 1977, 1981) of the genus *Ingolfiella* extending from the deep-sea off Greenland up to 2000 m in the Andes (though Karaman, 1959 felt that *Ingolfiella* should be broken into four genera). Although interesting in their own right, amphipod species swarms and ubiquitous distributions are probably of greater import for the study of evolutionary mechanisms than for biogeography *per se*.

Finally, it might be observed that 'the clock is running out' for effective study of amphipod biogeography (Fig. 13-7). In his examination of the wood boring family Cheluridae, Barnard (1959) noted that in a small group otherwise characterized by endemic distributions one species, *Chelura terebrans*, was not only a characteristic element in North Atlantic faunas in both Europe and North America, it was also found in South Africa, Australia, and New Zealand. Since this taxon appears to be restricted to waters in which the winter isotherm is not above 22°C, it was felt that British exploration and colonization in the last few centuries may have had as much to do with the distribution of *C. terebrans* as did nature.

DEVELOPMENT   The most recent comprehensive work on amphipod development is the study of *Gammarus pulex* by Weygoldt (1958). The first divisions are total and produce four micromeres and four macromeres. After this point cell divisions are no longer synchronous, and after the 16-cell stage cleavage is only superficial. As cells are migrating on the surface to form the germinal disc, some cells break away from the blastoderm to form vitellophages (Fig. 13-8B). The first thing differentiated on the completed blastoderm is the dorsal organ (Fig. 13-8A, B, C), which delineates the

Fig. 13-7. Zoogeographic distribution of chelurids in relation to the 22°C winter isotherm. ● *Chelura terebrans*, * *Tropichelura insulae*, ▲ *Nippochelura brevicauda*. (From Barnard, 1959)

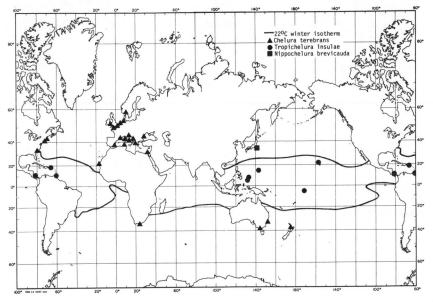

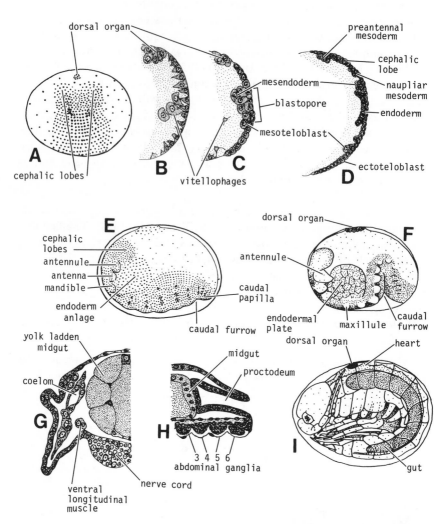

Fig. 13-8. Ontogenetic development in *Gammarus pulex*. (A) early organization of germinal disc; (B) formation of dorsal organ and proliferation of vitellophages; (C) early gastrulation; (D) late gastrulation, parasagittal section; (E) egg-nauplius with initiation of caudal furrow; (F) embryonic metanauplius with gut formation; (G) cross section with mesoderm differentiation; (H) median sagittal section of abdomen terminus; (I) completed development just prior to hatching. (From Weygoldt, 1958)

174

anterior border of the embryo. At that point gastrulation (Fig. 13-8C) then quickly leads to a differentiation of naupliar mesoderm, mesendoderm, and primary germ cells (Fig. 13-8D). The cephalic lobes form anterior to the mesendoderm.

The preantennal mesoderm invaginates as a pair of inpocketings from the anterior borders of each of the cephalic lobes (Fig. 13-8D). This preantennal material subsequently develops in a manner akin to that seen in other eumalacostracans, with the invaginations growing inward to the stomodeum and labrum and then sending elements dorsally to eventually form the anterior portions of the aorta. As the preantennal mesoderm arises, the naupliar mesoderm differentiates and soon induces the development of the naupliar appendage anlagen on the ectoderm. Thus, a distinct egg-nauplius stage comes to exist, while the external differentiation of metanaupliar anlagen lags (Fig. 13-8E).

The teloblasts arise posterior to the blastopore (Fig. 13-8C), a somewhat unusual position for a eumalacostracan. There are eight mesoteloblasts, and subsequent to their formation the ectoteloblasts differentiate, though these latter do not form a clearly recognized row of cells as seen in other crustacean groups. The formation of a caudal furrow occurs only after nine segments have been laid down by the teloblasts.

The mesoderm of the maxillulary segment forms into a compact mass of cells (Fig. 13-8E, F) which form the anlage that eventually gives rise to the digestive caeca and midgut. This mass grows to engulf the yolk sac and then effectively forms the midgut. The caeca arise later as evaginations from the midgut epithelium. The body mesoderm grows dorsad along the side of the body (Fig. 13-8G) and develops a series of coelomic pouches. The lateral wall of the pouch forms the dorsal longitudinal muscles, while the medial parts form the pericardial septum.

There is neither a distinct seventh abdominal somite nor ganglia developed in *Gammarus* (Fig. 13-8H). Weygoldt reported a transitory furrow that appears on the sixth ganglion, which he felt was a remnant of the fusion of a sixth and seventh ganglia. The urosomal ganglia fuse soon after their formation.

The development of *Gammarus* takes place entirely within the egg membrane (Fig. 13-8I) and at hatching results in a completely formed juvenile with all appendages in place. However, in certain hyperiideans (Laval, 1980) the juveniles hatch without a complete eighth thoracopod and thus virtually resemble mancas.

FOSSIL RECORD Though modern amphipods live in all types of environments, many of which would be ideal for forming fossils [and there is no reason to suppose (Bousfield, 1982a) that this ubiquity might have been any different in the past] amphipods have one of the most disappointing fossil records among any of the Eumalacostraca. The earliest amphipod fossils of the genus *Palaeogammarus* (Fig. 13-9A) are from the Baltic

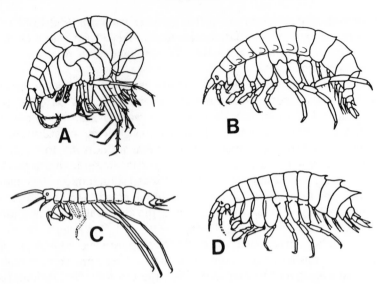

Fig. 13-9. Fossil amphipods. (A) *Palaeogammarus sambiensis*, Upper Eocene, the earliest known amphipod; (B) *Andrussovia sokolovi*, Middle Miocene; (C) *Hellenis saltatorius*, Lower Miocene; (D) *Praegmelina sambiensis*, Upper Eocene. (From Hessler, 1969)

amber of Late Eocene–Early Oligocene age. However, the most extensive fossil amphipod fauna comes from the Miocene of the Caspian region of the USSR and includes the genera *Andrussovia*, *Hellenis*, and *Praegmelina* (Fig. 13-9B, C, D), and a species of '*Gammarus*.' There are other materials from other localities (various species assigned to '*Gammarus*' and '*Melita*') but it is noteworthy that all fossil amphipods found so far are from Europe. There are some 18 species recognized in seven genera (Hurley, 1973; Bousfield, 1982a), and these are classified in five families and four superfamilies.

TAXONOMY   Though quite numerous, amphipods, with a few exceptions, pretty well all conform to a basic plan. Even so, there has been little consensus on how to classify them. An extreme position maintains that all are subsumable under a single taxon, Gammaridea, which then becomes virtually synonymous with the concept Amphipoda. A more reasonable view that presently enjoys fairly wide support is to recognize four major groups. Within these, three groups have only modest controversy within them: Caprellidea, Hyperiidea, and Ingolfiellidea.

The Gammaridea, however, are the subject of great controversy (see, e.g., Bousfield, 1978; Barnard and Karaman, 1980) with no consensus in sight on the arrangement of families and superfamilies (not even within a paper; see, e.g., Barnard and Karaman, 1983). Bowman and Abele (1982) essentially followed Barnard (1969). On the other hand, this book has opted for a taxonomic arrangement similar to Bousfield's (1983), not only to present an alternative to the above but also because the methodology used to arrive at this particular arrangement is repeatable and thus the methodology's results (i.e., the actual classification) are testable.

I would not necessarily agree with the exact methods nor this particular taxonomy of the gammarideans in all details. Bousfield and associates use basically phenetic techniques that utilize numerical averages to group taxa. Such an approach would give at best a 'first estimate' of relationships. While it does try to distinguish advanced from primitive characters, it also produces higher taxonomic groupings that have within their ranks 'exceptions' to the defining characters. Currently, only a few authorities are beginning to look at character matrices in the context of basic body plans using rigid cladistic techniques. It would seem that in the long run only such an analysis will hold forth the promise of producing a classification of amphipods that is at once natural and whose basic assumptions are obvious. For example, preliminary analysis of amphipod taxa using a Wagner 78 program by myself and R. C. Brusca produced a cladogram of relationships among amphipod taxa quite at odds with anything currently in the literature.

The precarious nature of amphipod taxonomy at the specific level is illustrated by the findings of Pinkster (1983) in part of the *Gammarus pulex* group, who described a species that could not be effectively diagnosed. *G. stupendus* could not be keyed using the usual criteria because distinct polymorphism in the taxon appeared to indicate three species, whereas hybrid experiments revealed the three types to be conspecific. In addition *G. stupendus*, because of its extreme variation, could not be distinguished from geographically adjacent species, *G. fossarum* and *G. iberica* but was not able to hybridize with them.

> Infraorder Ingolfiellidea Hansen, 1903 Recent
> Family Ingolfiellidae Hansen, 1903
> Metaingolfiellidae Ruffo, 1969
> Infraorder Caprellidea Leach, 1814 Recent
> Section Caprellida Bousfield, 1979
> Superfamily Phtisicoidea Vassilenko, 1968
> Family Paracercopidae Vassilenko, 1968
> Phtisicidae Vassilenko, 1968
> Dodecadidae Vassilenko, 1968
> Superfamily Caprelloidea White, 1847
> Family Caprogammaridae Kudrjaschov and Vassilenko, 1966
> Aeginellidae Vassilenko, 1968
> Caprellidae White, 1847
> Section Cyamida Bousfield, 1979
> Superfamily Cyamoidea White, 1847
> Family Cyamidae White, 1847
> Infraorder Hyperiidea Milne Edwards, 1830 Recent

Section Physosomata Pirlot, 1929
  Superfamily Lanceoloidea Bovallius, 1887
    Family Lanceolidae Bovallius, 1887
      Chuneolidae Woltereck, 1909
      Microphasmidae Stephenson and Pirlot, 1931
  Superfamily Scinoidea Stebbing, 1888
    Family Archaeoscinidae Stebbing, 1904
      Scinidae Stebbing, 1888
      Mimonectidae Bovallius, 1885
      Proscinidae Pirlot, 1933
Section Physocephalata Bowman and Grüner, 1973
  Superfamily Vibiloidea Dana, 1852
    Family Vibiliidae Dana, 1852
      Cystosomatidae Willemoës–Suhm, 1875
      Paraphronimidae Bovallius, 1887
  Superfamily Phronimoidea Dana, 1853
    Family Hyperiidae Dana, 1852
      Dairellidae Bovallius, 1887
      Phrosinidae Dana, 1853
      Phronimidae Dana, 1853
  Superfamily Lycaeopsoidea Chevreux, 1913
    Family Lycaeopsidae Chevreux, 1913
  Superfamily Platysceloidea Bate, 1862
    Family Pronoidae Claus, 1879
      Anapronoidae Bowman and Grüner, 1973
      Lycaeidae Claus, 1879
      Oxycephalidae Bate, 1861
      Platyscelidae Bate, 1862
      Parascelidae Bovallius, 1887
Infraorder Gammaridea Latreille, 1803 Eocene–Recent
  Superfamily Eusiroidea Stebbing, 1888 Recent
    Family Pontogeneiidae Stebbing, 1906
      Calliopiidae Sars, 1893
      Eusiridae Stebbing, 1888
      Paramphithoidae Stebbing, 1906
      Amathillopsidae Pirlot, 1934
      Bateidae Stebbing, 1906
      Paraleptamphopus family group
  Superfamily Oedicerotoidea Lilljeborg, 1865 Recent
    Family Oedicerotidae Lilljeborg, 1865
      Exoedicerotidae Barnard and Karaman, 1983
      Paracullispiidae Barnard and Karaman, 1983
  Superfamily Leucothoidea Dana, 1852 Recent
    Family Pleustidae Buchholz, 1874
      Amphilochidae Boeck, 1872

Leucothoidae Dana, 1852
Anamixidae Stebbing, 1897
Maxillipiidae Ledoyer, 1973
Colomastigidae Stebbing, 1899
Pagetinidae K. H. Barnard, 1932
Laphystiopsidae Stebbing, 1899
Nihotungidae Barnard, 1972
Cressidae Stebbing, 1899
Stenothoidae Boeck, 1871
Thaumatelsonidae Gurj., 1938
Superfamily Talitroidea Costa, 1857 Recent
Family Hyalidae Bulycheva, 1957
Dogielinotidae Gurjanova, 1954
Hyalellidae Bulycheva, 1957
Najnidae Barnard, 1972
Ceinidae Barnard, 1972
Talitridae Costa, 1857
Eophliantidae Sheard, 1938
Phliantidae Stebbing, 1899
Temnophliantidae Griffiths, 1975
Kuriidae Barnard, 1964
Superfamily Crangonyctoidea Bousfield, 1977 Eocene–Recent
Family Paramelitidae Bousfield, 1977 Recent
Neoniphargidae Bousfield, 1977 Recent
Niphargidae S. Karaman, 1962 Recent
Crangonyctidae Bousfield, 1973 Eocene–Recent
Superfamily Phoxocephaloidea Sars, 1891 Recent
Family Urothoidae Bousfield, 1979
Phoxocephalidae Sars, 1891
Platyischnopidae Barnard and Drummond, 1979
Superfamily Lysianassoidea Dana, 1849 Recent
Family Lysianassidae Dana, 1849
Uristidae Hurley, 1963
Superfamily Synopioidea Dana, 1853 Recent
Family Synopiidae Dana, 1853
Family Argissidae Walker, 1904
Superfamily Stegocephaloidea Dana, 1852 Recent
Family Stegocephalidae Dana, 1852
Acanthonotozomatidae Stebbing, 1906
Ochlesidae Stebbing, 1910
Lafystiidae Sars, 1893
Superfamily Pardaliscoidea Boeck, 1871 Recent
Family Pardaliscidae Boeck, 1871
Stilipedidae Holmes, 1908
Hyperiopsidae Bovallius, 1886

Astyridae Pirlot, 1934
Vitjazianidae Birstein and Vinogradov, 1955
Superfamily Liljeborgioidea Stebbing, 1888 Recent
  Family Liljeborgiidae Stebbing, 1888
    Sebidae Walker, 1908
    Salentinellidae Bousfield, 1977
    Paracrangonyctidae Bousfield, 1982
Superfamily Dexaminoidea Leach, 1814 Recent
  Family Atylidae Lilljeborg, 1865
    Anatylidae Bulycheva, 1955
    Lepechinellidae Schell., 1926
    Dexaminidae Leach, 1814
    Prophliantidae Nicholls, 1940
Superfamily Ampeliscoidea Bate, 1861 Recent
  Family Ampeliscidae Bate, 1861
Superfamily Pontoporeioidea Sars, 1882 Recent
  Family Pontoporeiidae Sars, 1882
    Haustoriidae Stebbing, 1906
Superfamily Gammaroidea Leach, 1814 Oligocene–Recent
  Family Acanthogammaridae Garjej., 1901 Oligocene–Recent
    Anisogammaridae Bousfield, 1977 Recent
    Gammaroporeiidae Bousfield, 1979 Recent
    Gammaridae Leach, 1814 Oligocene–Recent
    Pontogammaridae Bousfield, 1977 Miocene–Recent
    Typhlogammaridae Bousfield, 1979 Recent
    Mesogammaridae Bousfield, 1977 Recent
    Macrohectopidae Sowinsky, 1915 Recent
    Behningiella–Zernovia family group? Recent
    Iphiginella–Pachyschesis family group? Recent
Superfamily Melphidippoidea Stebbing, 1899 Recent
  Family Melphidippidae Stebbing, 1899
    Hornellia–Cheirocratus family group
    Megaluropus family group
    Phreatogammaridae Bousfield, 1982
Superfamily Hadzioidea Karaman, 1932 Recent
  Family Hadziidae Karaman, 1932
    Melitidae Bousfield, 1973
    Carangoliopsidae Bousfield, 1977
Superfamily Bogidielloidea Hertzog, 1936 Recent
  Family Bogidiellidae Hertzog, 1936
    Artesiidae Holsinger, 1980
Superfamily Corophioidea Dana, 1849 Pleistocene–Recent
  Family Ampithoidae Stebbing, 1899 Recent
    Biancolinidae Barnard, 1972 Recent
    Isaeidae Dana, 1853 Recent

Ischyroceridae Stebbing, 1899 Recent
Neomegamphopidae Myers, 1981 Recent
Aoridae Stebbing, 1899 Recent
Cheluridae Allman, 1847 Recent
Corophiidae Dana, 1849 Pleistocene–Recent
Podoceridae Stebbing, 1906 Recent

REFERENCES

Barnard, J. L. 1959. Generic partition in the amphipod family Cheluridae, marine wood borers. *Pac. Naturalist* **1**(3):3–12.

Barnard, J. L. 1960. The family Phoxocephalidae in the eastern Pacific ocean, with analysis of other species and notes for the revision of the family. *Allan Hancock Pac. Exped.* **18**(3):1–375.

Barnard, J. L. 1969. The families and genera of marine gammaridean Amphipoda. *U.S. Nat. Mus. Bull.* **271**:1–535.

Barnard, J. L. 1976. Amphipoda from the Indo-Pacific Tropics: a review. *Micronesica* **12**:169–81.

Barnard, J. L., and G. S. Karaman. 1980. Classification of gammarid Amphipoda. *Crustaceana Suppl.* **6**:5–16.

Barnard, J. L., and G. S. Karaman. 1983. Australia as a major evolutionary centre for Amphipoda. *Mem. Aust. Mus.* **18**:45–61.

Bazikalova, A. Ya. 1945. Amfipodi ozera Baikala. *Trudi Baikalskoy Limnologischeskoy Stantzii* **11**:1-445.

Blevgad, H. 1922. On the biology of some Danish gammarids and mysids. *Repts. Dan. Biol. Stn. Copenhagen* **28**:1–103, 4 pls.

Borowsky, B. 1984. The use of male gnathopods during precopulation in some gammaridean amphipods. *Crustaceana* **47**:245–50.

Bousfield, E. L. 1958. Distributional ecology of the terrestrial Talitridae of Canada. *10th Int. Congr. Entomol. Proc.* **1**:883–98.

Bousfield, E. L. 1961. New records of fresh water amphipod crustaceans from Oregon. *Nat. Mus. Canada, Nat. Hist. Papers* **12**:1–7.

Bousfield, E. L. 1973. *Shallow Water Gammaridean Amphipoda of New England.* Comstock, Ithaca.

Bousfield, E. L. 1978. A revised classification and phylogeny of amphipod crustaceans. *Trans. Roy. Soc. Can.* (4)**16**:343–90.

Bousfield, E. L. 1981. Evolution in north Pacific coastal marine amphipod crustaceans. In *Evolution Today, Proc. 2nd Int. Congr. Syst. Evol. Biol.*: 68–89.

Bousfield, E. L. 1982a. Amphipoda: Paleohistory. In *McGraw-Hill Yearbook of Science and Technology 1982–83*:96–100.

Bousfield, E. L. 1982b. Amphipoda: Gammaridea. In *Synopsis and Classification of Living Organisms* **2**:254–85.

Bousfield, E. L. 1982c. Amphipoda Superfamily Talitroidea in the northeastern Pacific region. 1. Family Talitridae. Systematics and distributional ecology. Publ. Biol. Oceanogr., *Nat. Mus. Nat. Sci., Ottawa* **11**:1–73.

Bousfield, E. L. 1983. An updated phyletic classification and paleohistory of the Amphipoda. *Crust. Issues* **1**:257–78.

Bousfield, E. L. 1984. Recent advances in the systematics and biogeography of landhoppers of the IndoPacific region. *Proc. Symp. Pacific Biogeogr.*, Bishop Mus., Honolulu.

Bousfield, E. L., and J. R. Holsinger. 1981. A second new subterranean amphipod crustacean of the genus *Stygobromus* from Alberta, Canada. *Can. J. Zool.* **59**:1827–30.

Bowman, R. E., and L. G. Abele. 1982. Classification of the recent Crustacea. *Biol. Crust.* **1**:1–27.

Bowman, T. E., and H.-E. Gruner. 1973. The families and genera of Hyperiidea. *Smith. Contr. Zool.* **146**:1–64.

Brusca, G. J. 1981. On the anatomy of *Cystisoma. J. Crust. Biol.* 1:358–375.

Caine, E. A. 1976. Cleaning mechanism of caprellid amphipods from North America. *Mar. Behav. Physiol.* **4**:161–69.

Caine, E. A. 1977. Feeding mechanisms and possible resource partitioning of the Caprellidae from Puget Sound. *Mar. Biol.* **42**:331–36.

Caine, E. A. 1979. Functions of swimming setae within caprellid amphipods. *Biol. Bull.* **156**:169–78.

Calman, W. T. 1909. Crustacea. In *Treatise on Zoology* (E. R. Lankester, ed.). Adam and Charles Black, London.

Conlon, K. E., and E. L. Bousfield. 1982. The superfamily Corophioidea in the North Pacific region. I.3. Family Aoridae: systematics and distributional ecology. *Nat. Mus. Nat. Sci., Publ. Biol. Oceanogr.* **10**:77–101.

Crane, J. M. 1969. Mimicry of the gastropod *Mitrella carinata* by the amphipod *Pleustes platypa. Veliger* **12**:200, pl. 36.

Croker, R. A. 1967. Niche diversity in five sympatric species of intertidal amphipods. *Ecol. Monogr.* **37**:173–200.

Dahl, E. 1960. *Hyperia galba*, a true ectoparasite on jellyfish. *Arbok Univ. Bergen* **1959**:1–8.

Dahl, E. 1977. The amphipod functional model and its bearing upon systematics and phylogeny. *Zool. Scripta* **6**:221–28.

Delage, Y. 1881. Contribution à l'étude de l'appareil circulatoire des Crustacés edriophthalmes marine. *Arch. Zool. Exper.* **9**:1–173, pl: 1–12.

Dennell, R. 1933. The habits and feeding mechanism of the amphipod *Haustorius arenarius. Zool. J. Linn. Soc. Lond.* **38**:363–88.

Graf, F., and P. Michaut. 1980. Fine structure of midgut posterior caeca in the crustacean *Orchestia* in intermolt: recognition of two distinct segments. *J. Morph.* **165**:261–84.

Hallberg, E., H. L. Nilsson, and R. Elofsson. 1980. Classification of amphipod compound eyes—the fine structure of ommatidial units. *Zoomorph.* **94**:279–306.

Hansen, H. J. 1903. The Ingolfiellidae, a new type of Amphipoda. *Zool. J. Linn. Soc. Lond.* **28**:117–32, pls. 14, 15.

Hessler, R. R. 1969. Peracarids. In *Treatise on Invertebrate Paleontology*, Part R, *Arthropods 4*, Vol. I (R. C. Moore, ed.), pp. R360–93. Geol. Soc. Am. and Univ. Kansas Press, Lawrence.

Hessler, R. R. 1982. The structural morphology of walking mechanisms in eumalacostracan crustaceans. *Phil. Trans. Roy. Soc. Lond.* (B)**296**:245–98.

Holmquist, J. G. 1982. The functional morphology of gnathopods: importance in grooming, and variation with regard to habitat, in talitroidean amphipods. *J. Crust. Biol.* **2**:159–79.

Holsinger, J. R. 1980. *Stygobromus canadensis*, a new subterranean amphipod crustacean from Canada, with remarks on Wisconsin refugia. *Can. J. Zool.* **58**:290–7.

Holsinger, J. R. 1981. *Stygobromus canadensis*, a troglobitic amphipod crustacean from Castleguard Cave, with remarks on the concept of cave glacial refugia. *Proc. 8th Int. Congr. Speleol.* **1**:93–5.

Holsinger, J. R., and G. Longley. 1980. The subterranean amphipod crustacean fauna of an artesian well in Texas. *Smith. Contr. Zool.* **308**:1–62.

Hunt, O. D. 1925. The food of the bottom fauna of the Plymouth fishing grounds. *J. Mar. Biol. Assc. U.K.* **13**:560–99.

Hurley, D. E. 1968. Transition from water to land in amphipod crustaceans. *Am. Zool.* **8**:327–53.

Hurley, D. E. 1973. An annotated checklist of fossils attributed to the crustacean Amphipoda. *N.Z. Oceanogr. Inst. Records* **1**:211–17.

Icely, J. D., and J. A. Nott. 1984. On the morphology and fine structure of the alimentary canal of *Corophium volutator*. *Phil. Trans. Roy. Soc. Lond.* (B)**306**:49–78.

Kaestner, A. 1970. *Invertebrate Zoology*, Vol. III. Interscience New York.

Kane, J. E. 1963. Observations on the molting and feeding of a hyperiid amphipod. *Crustaceana* **6**:129–32.

Karaman, St. L. 1959. Über die Ingolfielliden Jugoslaviens. *Biol. Glasnik* **12**:63–80.

Keith, D. E. 1969. Aspects of feeding in *Caprella californica* and *Caprella equilibra*. *Crustaceana* **16**:119–24.

Kinne, O. 1954. Die Bedeutung der Kopulation für Eiablage und Häutungsfrequenz bei *Gammarus*. *Biol. Zentralbl.* **73**:190–200.

Laubitz, D. R. 1970. Studies on the Caprellidae of the American North Pacific. *Nat. Mus. Nat. Sci., Publ. Biol. Oceanogr.* **1**:1–89.

Laubitz, D. R. 1976. On the taxonomic status of the family Caprogammaridae. *Crustaceana* **31**:143–150.

Laval, P. 1980. Hyperiid amphipods as crustacean parasitoids associated with gelatinous zooplankton. *Oceanogr. Mar. Biol. Ann. Rev.* **18**:11–56.

Leung, Yuk-Maan. 1967. An illustrated key to the species of whale-lice, ectoparasites of Cetacea, with a guide to the literature. *Crustaceana* **12**:279–91.

Lewis, S. M., and B. Kensley. 1982. Notes on the ecology and behavior of *Pseudamphithoides incurvaria*. *J. Nat. Hist.* **16**:267–74.

Lincoln, R. J. 1979. *British Marine Amphipoda: Gammaridea*. British Museum (Nat. Hist.), London.

Lincoln, R. J., and D. E. Hurley. 1981. The calceolus, a sensory structure of gammaridean amphipods. *Bull. B.M.(N.H.)* **40**:103–16.

Lowry, J. K. 1982. Department of Marine Invertebrates (crustaceans and coelenterates). *Aust. Mus. Ann. Rept. 1980–81*:29–31.

Martin, A. L. 1964. The alimentary canal of *Marinogammarus*. *Proc. Zool. Soc. Lond.* **143**:525–44.

McCain, J. C. 1968. The Caprellidae of the western North Atlantic: U.S. *Nat. Mus. Bull.* **278**:1–147.

Mills, E. L. 1965. The zoogeography of north Atlantic and north Pacific ampeliscid amphipod crustaceans. *Syst. Zool.* **14**:119–30.

Noodt, W. 1965. Interstitielle Amphipoden der konvergenten Gattungen *Ingolfiella* and *Pseudingolfiella* aus Südamerika. *Crustaceana* **9**:17–30.

Pinkster, S. 1983. The value of morphological characters in the taxonomy of *Gammarus*. *Beaufortia* **33**:15–28.

Ruffo, S. 1969. Descrizione di *Metaingolfiella mirabilis* delle acque sottersanee del salento nell'Italia meridionale. *Museo Civico Storia Naturale Verona Mem.* **16**:236–60.

Schmitz, E. H., and P. M. Scherrey. 1983. Digestive anatomy of *Hyella azteca*. *J. Morph.* **175**:91–100.

Schram, F. R. 1981. On the classification of Eumalacostraca. *J. Crust. Biol.* **1**:1–10.

Schram, F. R. 1984. Relationships within eumalacostracan Crustacea. *Trans. San Diego Soc. Nat. Hist.* **20**:301–12.

Sexton, E. W. 1924. The molting and growth stages of *Gammarus*, with description of the normals and intersexes of *G. chevreauxi*. *J. Mar. Biol. Assc. U.K.* **13**:340–401.

Siewing, R. 1963. Zur morphologie der aberranten Amphipodengruppe Ingolfiellidae und zur Bedeutung extremer Kleinformen für die Phylogenie. *Zool. Anz.* **171**:76–91.

Steele, D. H. 1967. The life cycle of the marine amphipod *Stegocephalus inflatus* in the northwest Atlantic. *Can. J. Zool.* **45**:623–28.

Stock, J. H. 1976. A new member of the crustacean suborder Ingolfiellidea from Bonaire, with a review of the entire suborder. *Stud. Fauna Curaçao* **50**:56–75.

Stock, J. H. 1977. The zoogeography of the crustacean suborder Ingolfiellidea with descriptions of new West Indian taxa. *Stud. Fauna Curaçao* **55**:121–46.

Stock, J. H. 1980. Regression model evolution as exemplified by the genus *Pseudoniphargus*. *Bijd. Dierkunde* **50**:105–44.

Stock, J. H. 1981. The taxonomy and zoogeography of the family Bogidiellidae with emphasis on the West Indian taxa. *Bijdr. Dierk.* **51**:345–74.

Stock, J. H. 1984. First record of Bogidiellidae from the Pacific: *Bogidiella capricornica* new species from the Great Barrier Reef. *Bull. Mar. Sci.* **34**:380–85.

Thomas, J. D., and J. L. Barnard. 1983. Transformation of the *Leucothoides* morph to the *Anamixis* morph. *J. Crust. Biol.* **3**:154–57.

Ward, J. V., and J. R. Holsinger. 1981. Distribution and habitat diversity of subterranean amphipods in the Rocky Mountains of Colorado. *Int. J. Speleol.* **11**:63–70.

Watling, L. 1983. Peracaridan disunity and its bearing on eumalacostracan phylogeny with a redefinition of eumalacostracan superorders. *Crust. Issues* **1**:213–28.

Weygoldt, P. 1958. Die Embryonalentwicklung des Amphipoden *Gammarus pulex pulex*. *Zool. Jahrb. Abt. Anat.* **77**:51–110.

14

SPELAEOGRIPHACEA

DEFINITION   Body long and cylindrical; carapace fused to first thoraco-mere, covering the second; ocular lobe without visual components; antennules biramous, well developed; antennae with small scaphocerite; mandible with lacinia mobilis and unjointed palp; maxillipede with large, cuplike epipodite; thoracopods 2 through 4 with flaplike setose exopods, 5 through 7 (sometimes 8) with branchial exopods; pleopods 1 through 4 well developed, 5 vestigial; uropods biramous and flaplike.

HISTORY   The living form, *Spelaeogriphus lepidops* Gordon, 1957, is known only from Bat Cave in Table Mountain, Cape Town, South Africa. It currently is found there in the freshwater stream that flows through the cave. For all its intrinsic phylogenetic import, hardly any work has been done on *S. lepidops*, nor has there been a concerted search for other forms. Gordon (1957, 1960) described the anatomy, and Grindley and Hessler (1971) have examined the respiratory mechanisms.

One fossil form is also known. *Acadiocaris novascotica* (Copeland, 1957) was originally thought to be a syncarid. Brooks (1962) recognized its distinctive character, and Schram (1974) redescribed and identified it as a spelaeographacean.

MORPHOLOGY.   The body of *Spelaeogriphus lepidops* is long and slender and generally circular in cross section (Fig. 14-1). The carapace is short, dorsally extends over the second thoracic segment, laterally overlaps part of the third, and has a faint cervical groove. There is a movable ocular lobe on the cephalon, but no pigment or optic structures are evident.

Fig. 14-1. *Spelaeogriphus lepidops*, lateral oblique view. (From Gordon, 1960)

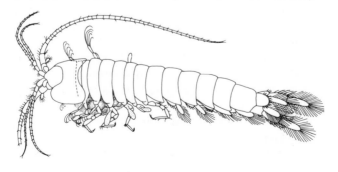

The antennule is composed of a three-segment peduncle. The second segment of the male is modified on its distal medial surface, bearing a field of conical papillae (Fig. 14-2B). The flagella are moderate in length, with one branch somewhat more delicate than the other.

The antenna has a peduncle of four segments, two short protopodal segments, and two longer distal elements. The short, oval, scalelike scaphocerite arises from the protopod. The flagellum is almost as long as the entire body of the animal.

The mandible lies behind a large labrum (Fig. 14-2C) that is decorated with a terminal cluster of fine setae. The mandible (Fig. 14-2D) has a single-segment palp; a lobate incisor area bearing a cuspid incisor process and lacinia mobilis distally, and a marginal row of brush setae; and a well-developed ridged molar process. There are prominent finely setose paragnaths (Fig. 14-2E).

The maxillule (Fig. 14-2F) has a small slender palp, a prominent spinose outer lobe, and a more slender inner lobe with three setal clusters on its distal tip.

The maxilla (Fig. 14-2G) is well developed, with three lobes bearing a complex array of setae. The third lobe is broadly bifurcate and bears pinnate, chelate, and serrate setal types. The second lobe carries simple and pencillate setae.

The maxillipede (Fig. 14-2H) has an enlarged medial plate developed from the basis. This attaches to its counterpart on the opposite side by means of small retinacula. The palp is large, somewhat sinuous in form, and complexly setose. There is a large cuplike epipod extending back under the carapace.

The first six pereiopods consist of moderately setose endopods and variously developed exopods. Those of thoracomere two through four (Fig. 14-2I) have a moderately long, two-segment setose exopod. Those of thoracomeres five through seven (Fig. 14-2J) have a lobate, branchial exopod. The last pereiopod typically does not bear an exopodite, but on males this limb does have a tubular penal process arising from the coxa. Öostegites are borne on the first five pereiopods in the female.

The first four pleopods are similar and natatory and consist of broad protopods with stout setae on the lateral margin and two grappling retinacula on the distal inner margin. The rami are both paddlelike and highly setose (Fig. 14-2K). The fifth pleopod is reduced and partially concealed by the fifth abdominal pleuron (Fig. 14-2L).

The uropods are broad and flaplike (Fig. 14-2M). The protopod distally bears spines. The exopod has two segments, with the proximal segment armed with spines medially and apicolaterally, and the distal segment oval and setose. The endopod is also oval and rather setose.

The telson is broadly rounded and armed with apical spines. The anus is midventral on the telson.

Nothing is known concerning the internal morphology of *S. lepidops*.

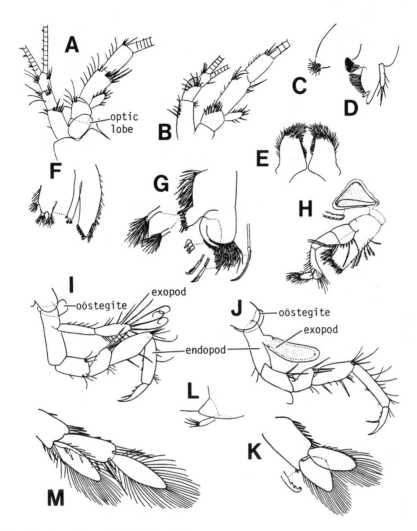

Fig. 14-2. Appendages of *S. lepidops*. (A) antennule, antenna, and optic lobe of female; (B) antennule and antenna of male; (C) labrum, with distal setal tuft; (D) mandible; (E) paragnaths; (F) maxillule; (G) maxilla; (H) maxillipede, with cuplike epipodite; (I) thoracopod 4; (J) thoracopod 5; (K) pleopod 4; (L) vestigial pleopod 5 under edge of pleomere pleuron; (M) uropod. (From Gordon, 1957)

NATURAL HISTORY   Gordon (1957) reported that females could contain up to 10 to 12 eggs in the brood pouch. However, no juvenile forms have ever been described.

Information on behavior is also rare. Grindley and Hessler (1971) examined the respiratory system. The maxillipede epipodite and the posterior thoracic exopods are branchial. A constant flow of water to effect gas exchange is achieved by rapid and sustained beating of the anterior thoracic exopods. The resultant current interestingly flows from anterior under the carapace to posterior over the appendages. (Usually, the flow across gills enclosed by a carapace in malacostracans is posterior to anterior.)

Les Watling (personal communication) reports *S. lepidops* rather randomly searches the pool floor for food (apparently processing detritus) and does not seem to groom itself. It appears that the low pH of the stream minimizes fouling.

FOSSIL RECORD   *Acadiocaris novascotica*, from the lower Carboniferous (Mississippian) of Canada, though obviously a spelaeogriphacean (it is similar to the South African species in regard to antennular and antennal peduncles, and carapace arrangement), possesses some distinct primitive features relative to the South African form (Fig. 14-3). There is no optic notch in the carapace, the thoracic endopods are robust and well developed, and all the pleopods are apparently natatory and functional.

TAXONOMY   Gordon (1957) immediately recognized the separate ordinal status of the spelaeogriphaceans. However, she, Siewing (1960), and Grindley and Hessler (1971) all remarked on certain morphologic peculiarities shared with tanaidaceans, namely, the nature of the carapace and antennal protopod, the cuplike maxillipedal epipodites, and a general similarity of paragnaths. Schram (1981) suggested in his taxonomy for the eumalacos-

Fig. 14-3. *Acadiocaris novascotica*, reconstruction. (From Schram, 1974)

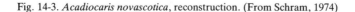
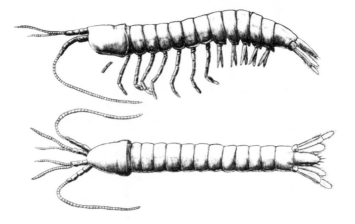

tracans that these two groups, as well as the cumaceans, be united into a taxon Hemicaridea. The two spelaeogriphacean species are placed (Schram, 1974) into separate families: the Spelaeogriphidae Gordon, 1957, and the Acadiocarididae Schram, 1974.

REFERENCES

Brooks, H. K. 1962. The Paleozoic Eumalacostraca of North America. *Bull. Am. Paleo.* **44**:163–338.

Copeland, M. J. 1957. The Carboniferous genera *Palaeocaris* and *Euproops* in the Canadian maritime provinces. *J. Paleo.* **31**:595–99.

Gordon, I. 1957. On *Spelaeogriphus*, a new cavernicolous crustacean from South Africa. *Bull. B.M.(N.H.) Zool.* **5**:31–47.

Gordon, I. 1960. On a *Stygiomysis* from the West Indies, with a note on *Spelaeogriphus*. *Bull. B.M.(N.H.) Zool.* **6**:285–323.

Grindley, J. P., and R. R. Hessler. 1971. The respiratory mechanism of *Spelaeogriphus* and its phylogenetic significance. *Crustaceana* **20**:141–44.

Schram, F. R. 1974. Paleozoic Peracarida of North America. *Fieldiana: Geol.* **33**:95–124.

Schram, F. R. 1981. On the classification of Eumalacostraca. *J. Crust. Biol.* **1**:1–10.

Siewing, R. 1960. Neuere Ergebnisse der Verwandtschaftsforschung bei den Crustaceen. *Wiss. Zeit. Univ. Rostock (Math.–Naturwiss.)* **9**:343–58.

15

TANAIDACEA

DEFINITION Carapace covers only and fused to first two thoracomeres, forming a lateral branchial cavity; antennular and antennal flagella typically reduced in living forms; eyes on small lobes when present; first thoracopods (= maxillipedes) with large segmented branchial epipodite; second thoracopods as large chelipedes; anterior thoracopods (excepting maxillipedes) sometimes with only weak exopods in adult, one or four sets of oöstegites on the sixth or third through sixth thoracopods; abdomen reduced in length over that of thorax, with pleotelson in living forms.

HISTORY The first tanaidacean was described in 1808 by Montagu, but early opinions as to affinities of these animals was divided between either allying them with amphipods or isopods. Some authorities (e.g., Dana) treated them as a separate taxon, the Anisopoda, intermediate between both groups. Sars ranked them as a tribe of isopods, the Chelifera, a placement that persisted among many authors until relatively recently, even in the face of contrary evidence. It was Claus (1887) who recognized and documented the separate position of the group, though the actual name Tanaidacea comes from Hansen. Lang (1956) divided the group into the Dikonophora and the Monokonophora, but Sieg (1980, 1983, 1984) has striven to achieve a more natural classification, which is used here.

MORPHOLOGY The body form of tanaidaceans is subcylindrical to somewhat flattened (Fig. 15-1). The carapace covers and is fused to the first two thoracomeres; the six posterior thoracomeres remain free. The sides of the carapace project down and around the sides of the body to form a branchial chamber. The abdomen is characteristically reduced in size and length relative to the thorax, except in highly specialized forms like *Filitanais* where they are subequal; whereas the extinct anthracocaridomorphs have six free pleomeres, the living tanaidaceans all have a pleotelson with normally five free pleomeres. The eyes, when present, are on lobes. Lauterbach (1970) reported remnants of eye muscles in *Tanais cavolinii*. Anderson et al. (1978) recorded a very generalized arrangement of structures within the ommatidium and noted various similarities with other peracarids in regard to two reduced corneagen cells and with peracarids and eucarids in regard to eight retinula cells.

The antennules (Fig. 15-2A) are prominent organs. The peduncle consists of four segments; the proximalmost joint is typically the largest. The

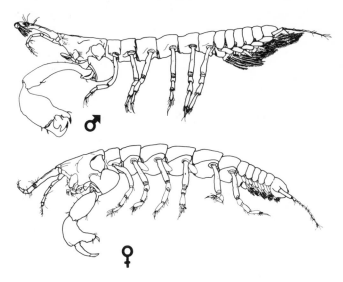

Fig. 15-1. Sexual dimorphism in *Neotanais micromorpher*. (From Gardiner, 1975)

anthracocaridomorphs and apseudomorphs have two well-developed flagella, but the tanaidomorphs and neotanaidomorphs have only a single,
greatly shortened flagellum. The flagella can bear aesthetascs on the distal
joints.

The antennae (Fig. 15-2B) have a two-segmented protopod, and the
proximalmost segments of the flagellum are very peduncular. The distal
segments of the antennal flagellum are reduced in size and number, though
larger in males than in females. In apseudomorphs a small exopodal scale
may be present on the second segment of the peduncle, but this structure is
frequently absent on most neotanaidomorphs and tanaidomorphs.

The mandibles (Fig. 15-2C) are of a rolling type, obliquely hinged by
one or two condyles. A three-segment palp is generally only developed in
apseudomorphs. The incisor process, with lacinia mobilis adjacent to a row
of lifting spines, is rather widely separated from a columnlike molar process. The paragnaths exhibit a wide variety of form and range from a single
simple and incompletely divided lobe to a complex structure (as in apseudids and neotanaids) with several lobes on either side with terminal articles
(Fig. 15-2D).

The maxillules (Fig. 15-2E) display a dichotomy of form. In primitive
types, like apseudids, two well-developed endites are directed medially,
while a distinct two-segment palp is developed laterally. In tanaidomorphs,
the proximal endite is absent and occasionally the palp is reduced or
absent.

The maxillae (Fig. 15-2F) also vary in form. The primitive condition has
a large setose coxal endite associated with a somewhat smaller and deeply

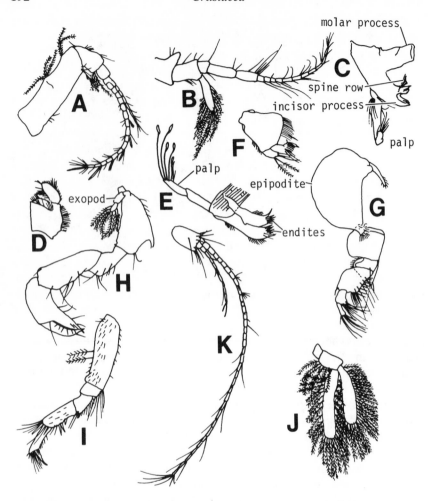

Fig. 15-2. Appendages of *Apseudes hermaphroditicus*. (A) antennule; (B) antenna; (C) mandible; (D) paragnath; (E) maxillule; (F) maxilla; (G) maxillipede; (H) chelipede; (I) sixth thoracopod; (J) pleopod; (K) uropod. (From Lang, 1953)

cleft basal endite. In advanced manifestations of this limb it is reduced to a simple small lobe. It is important to note that in the nonfeeding copulatory males the mouthparts are vestigial.

The first thoracopods (Fig. 15-2G) are developed as maxillipedes. These may be separated at their coxae or exhibit progressive degrees of fusion of their protopodal joints. The most distinctive feature of this limb is the elaborate epipodite, typically with a proximal stalk and a distal membranous lobe. This epipodite projects posteriorly into the branchial chamber under the carapace and has a respiratory function.

The second thoracopod (Fig. 15-2H) is typically termed a chelipede. The terminus is usually formed as a chela, though occasionally as a sub-chela (especially in males).

The third through eighth thoracopods (= pereiopods of some authors) are divided into groups; the first three face anteriorly and the last three face posteriorly. These appendages have well-developed endopods whose distal segments are armed with spines or other types of elaborate setae (Fig. 15-2I). Occasionally spinning glands are present, opening on the dactylus of some limbs, which assist in tube building and spin 'lifelines,' not unlike some arachnids, that are helpful in locomotion. The chelipedes and first pereiopods in some types (e.g., kalliapseudids) bear small exopods, present on all thoracopods in the manca of *Kalliapseudes*. In females, oöstegites occur on the third to sixth thoracopods, but in the tanaidomorphs the oöstegites are on the sixth thoracopod [a small set of oöstegites has also been reported on the chelipedes of *Tanais* (see Kaestner, 1970, p. 401)]. Hessler (1982) noted two types of coxa–basis articulations. The third through fifth thoracopods have a dicondylic joint with pivot condyles projecting deeply into the leg lumen, while the sixth through eighth thoracopods have a monocondylic joint with the pivot projecting into the lumen from the lateral edge of the coxa.

Pleopods (Fig. 15-2J) are variously one- or two-branched, simple or setose, present or absent (especially in females). The terminal uropods can be long and flagelliform (Fig. 15-2K), moderate and biramous, or short and uniramous. The telson is fused as a pleotelson with the sixth segment. Sometimes more segments than just the sixth are fused; for example, in *Curtipleon* the abdomen and telson form one solid unit. However, in the extinct anthracocaridomorphs the telson was free.

The digestive system has proven to be quite variable among species so far investigated, which reflects wide-ranging adaptations to the dietary preferences of each species. The stomach may be elaborately developed or greatly reduced, and there may be either one or two pairs of digestive diverticula.

The circulatory system possesses some minor peculiarities that merely reflect the inherent distortions of the tanaidacean body plan. The heart is the largest structure in the system extending through most of the pereional segments. The cephalic or anterior aorta gives rise to a series of respiratory arteries into the carapace region and a vascular ring around the stomach and esophagus. Posteriorly a pair of abdominal aortae supply the pleon.

Excretion is achieved by maxillary glands, though functional antennal glands appear in the course of development in the larval forms (Claus, 1887; Lauterbach, 1970).

The reproductive organs are fairly standard. Ovaries extend through the pereional segments and open to the sides of the sixth thoracic sternites. The testes are short organs and open directly into a median seminal vesicle. The male gonopores are always paired; however, they can open either into a single or paired median cones on the eighth thoracic sternite. Secondary sexual polymorphism (Fig. 15-1) is very pronounced within tanaids, sometimes extreme. Males have larger eyes, more antennular aesthetascs, well-developed pleopods, enlarged chelipedes, and reduced mouthparts. Females reduce or lose the pleopods. The polymorphism is complicated because individuals can molt through a series of copulatory and noncopulatory stages, each different in its morphology (see discussion below on mating).

The nervous system typically has the segmental ganglia clearly separated, though there is a tendency to fuse the anteriormost postoral and pleonal ganglia. Claus (1887) originally observed that there is a rudimentary seventh ganglion in embryos of *Apseudes* (also noted by Scholl, 1963, in *Heterotanais*); and Sieg in Hessler (1983, p. 160) points out that, unlike what Manton noted in *Hemimysis lamornae*, the nerves to the uropodal anlagen in the tanaids arise from this seventh ganglion and not the sixth.

NATURAL HISTORY   Tanaidaceans have undergone extensive modification of their basic structural plan toward their main life-style as tube-building or tunnel-dwelling animals (Johnson, 1982). The cross section of the body, development of bristles and spines on the body and appendages, V-shape bend of the thoracopods, and variations in pleopod development are all responses to various aspects of a tube- and tunnel-dwelling habit.

Tanaidaceans spend most of their lives in their tubes or tunnels, so consequently movement outside of their burrows is not very effective. Limited swimming is achieved by the pleopods (for those that have them), and ambulatory motion is performed by the pereiopods and/or the chelipedes dragging the body along the bottom. For those species with spinning glands, the secretion of a 'lifeline' facilitates surface movement. Even so, the out-of-tube locomotory abilities are so poor and defensive behaviors almost nonexistent that mortality of surface-roaming tanaids is rather high. Johnson and Attremadal (1982) report that copulatory males of *Tanais cavolinii* are virtually at the mercy of their chief predator, *Idotea pelagica*. The chelipedes serve no defensive function but are used only for food getting and processing or in males for grasping the female's chelipedes during copulation.

Feeding in tanaidaceans takes two forms (Dennell, 1937). Most feeding appears to be raptory. Large detrital chunks or small organisms like nematodes and diatoms are either gathered in by the chelipedes or swept up on the maxillary filter current. The chelipedes and maxillipedes then shred the food items and pass the processed material forward to the mouthparts. The copulatory males of tanaidomorphs have atrophied mouthparts and do not feed.

Some tanaidaceans are able to carry on some limited degree of filter feeding. The current that brings particulate matter into the maxillary area is produced from two sources. The primary mode of current generation is related to the posteriad or ventrad motion of the maxillipedes and maxillae. This effectively drops the 'floor' of the filter chamber and sucks in water and detritus into the space between the maxillules and the maxillae. A secondary filter current is linked to the generation of the respiratory current. Respiration takes place in the space under the carapace. The movement of the maxillipedal epipodite creates a suction that pulls water in under the carapace from wherever it can get in. The prime flow (Fig. 15-3) of water is through the respiratory opening located on the posterior of the carapace. The beating of the exopods on the chelipedes and third thoracopods facilitates this movement. However, an additional flow into the respiratory chamber occurs from an anterior direction, past the mouthparts, and over an accessory flap on the maxillipedal epipodite. The exhalant respiratory current is directed out of the branchial chamber ventrally; the accessory epipodite flap prevents the backwash of water into the feeding chamber. In both of these instances of filtration, it is the setation on the maxillae that traps the detrital particles while the maxillipedes and maxillules transfer food forward to the mandibles. Filter feeding is not the prime source of tanaidacean nutrition, and indeed it is impossible in those forms with atrophied maxillae and that lack proximal maxillulary endites.

It is in their reproductive habits that tanaidaceans exhibit a number of unique and interesting adaptations. Due to the vulnerability of the itinerant rutting copulatory males it appears that, at least in some species, the mancas develop into copulatory females first, and only after they endure at least one brood cycle as a female do they undergo a series of molts into a copulatory male phase (Johnson, 1982). The metamorphosis is at least in part governed by the number of males in a population (Highsmith, 1983), which in turn is affected by the intensity of predation (Mendoza, 1982).

A wide range of strategies has evolved, however, to maximize reproductive potential (Lang, 1957; Sieg, 1983a), and these result in sexual polymorphs of various degrees and kinds (Fig. 15-4). There may be either very little or somewhat marked sexual dimorphism in some tanaidaceans coupled with separate and nonconvertible male and female sexes, each sex developing from different juvenile types (Fig. 15-4A). Such patterns would be viewed as primitive. In other groups, males and females are so distinct

Fig. 15–3. Schematic diagram of water currents around *Tanais cavolinii*. (After Johnson and Attramadal, 1982)

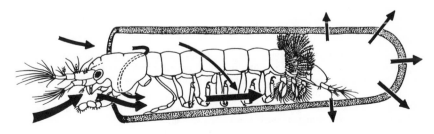

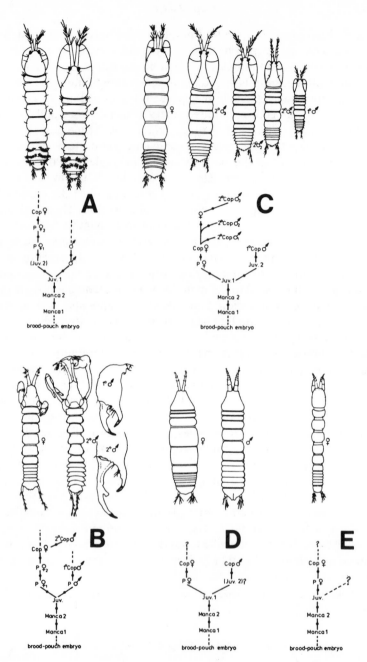

Fig. 15-4. Sexual strategies of tanaidaceans. (A) similar males and females develop from distinct juveniles (the most primitive condition); (B) very distinct males and females, with a post-reproductive female metamorphosing to a secondary male; (C) a type in which any one of an ongoing series of successive females may molt into a male, each with a distinctive form; (D) very distinctive males and females developing from separate juveniles; (E) parthenogenetic conditions with complete loss of males. (From Sieg, 1983a)

196

as to at first glance appear to be in different species (Fig. 15-4D). Even more advanced conditions are seen in which the copulatory female can molt into a male (Fig. 15-4B) or, in more complex life cycles, can molt into any number of ongoing copulatory females, any of which can in turn, after reproducing, molt into males (Fig. 15-4C). The ultimate condition is one in which males are dispensed with altogether, the females reproducing by parthenogenesis (Fig. 15-4E).

A phenomenon unique among crustaceans was recorded by Johnson (1982) in *Tanais cavolinii*. Two sets of eggs are produced. One set is laid in the ovisacs of the sixth thoracopods and allowed to develop; the others are held within the female in a diapause. If the first set of developing eggs is aborted for some reason, the second set of eggs allows the female to quickly remate and start a reproductive cycle anew. If the first set of young reach the manca II stage, the reserve eggs are aborted and their yolk deposited into the ovisac. The mancas then gorge themselves on this second yolk reserve, whereupon they immediately lyse the ovisac (in a manner not yet understood) and escape the mother into the burrow. The mother harbors these manca III forms in her dwelling tube, but they complete their development independent of her by subsisting only on the second reserve of yolk that bloats their guts. Once they molt into the first juvenile or intermediate stage, the young burrow through the mother's tube and take up a completely independent existence in a tube of their own. Such a 'lactation' phenomenon has not been noted in any other tanaidaceans as yet.

Mating (Bückle–Ramirez, 1965; Johnson, 1982) follows a basic pattern in all forms studied so far. The copulatory male roams on the bottom seeking a copulatory female. He then tears his way into her tube or tunnel and makes advances to her. Initially repulsed, he persists until the female is ready, whereupon he grasps her chelipedes with his own while they are in a venter-to-venter position. Fertilization takes place either into the brood pouch or ovisac (whichever is the case). For those species with a rather open brood pouch, those eggs or hatchlings that get dislodged from the marsupium are either allowed to develop on their own on the floor of the burrow or are promptly eaten by the mother but are not replaced by the female into the brood pouch as is, for example, the case in mysids.

Tanaidaceans live in a variety of habitats from freshwater, to intertidal, to deep-sea conditions, from polar to tropical realms around the world. In almost all instances, however, they are inbenthic forms quietly living in their little tubes or tunnels. Gardiner (1975) thought perhaps some neotanaids at least lived exposed on the bottom, ambling about in search of food and mates. Johnson (1982), however, feels that this suggestion is not justified. Peculiar forms, like *Pagurapseudes*, which mimics hermit crabs in that it lives in empty gastropod shells, are rare deviations from the normal inbenthic habits.

Since the generic and specific taxonomy of the group is still basically in revision, little biogeographic analysis has been performed. Sieg (1983a) points out that since tanaidaceans lack free-swimming larvae to faciliate dispersal, 'extension of range' depends largely on how far the juvenile moves when it burrows out of the maternal tube or tunnel before building its own. Some species, like *T. cavolinii*, which can live in tubes that float in algal mats, can have virtually ubiquitous distributions.

This limited dispersal ability may be reflected in the distribution of tanaidaceans noted in the fossil record. Like other Paleozoic eumalacostracans, the anthracocaridomorphs seem endemic to the tropical island continent Laurentia (Schram, 1977). Even after the Late Paleozoic breakup of Pangaea, they have not yet been found in Mesozoic rocks in areas other than those that were in the tropics of the eastern Tethys. Further discoveries of fossil tanaidaceans and analysis of the distributions of living forms may confirm if their dispersal out from their Paleozoic center of origin was indeed a slow one.

DEVELOPMENT   The most complete studies concerning tanaidacean early embryology are those of Claus (1887), who made observations on adult and larval *Apsuedes latreilli*, Scholl (1963), who studied the development of *Heterotanais oerstedi*, and Dohle (1972), who worked on *Leptochelia*. Initial segmentation (Fig. 15-5A) is incomplete, and it is not until the eight-cell stage that the nuclei and their surrounding cytoplasm rise through the yolk to the surface. The micromeres at the vegetal pole (Fig. 15-5B, C) form the germinal disc. The changes of gastrulation occur in a quick sequence of events, and distinct cell and tissue types are not readily distinguishable. Endoderm formation occurs in two distinct phases. First wandering vitellophages move into the yolk from the blastopore region subsequently followed by cells forming a mesendodermal mass (Fig. 15-5D). As in isopods, however, the endodermal components in the embryo only form a yolk sac; the final adult gut is entirely ectodermal in origin. Primary germ cells also migrate inward from the germinal disc (Fig. 15-5D, F).

As gastrulation ensues, a half ring of ectoteloblasts (Fig. 15-5E) forms on the anterior and lateral borders of the zone of immigration. Just caudad of these cells, and associated with the mesendodermal mass, is a row of eight mesoteloblasts. Soon after their formation during the gastrulation process the teloblasts begin to bud off (Fig. 15-5F) the postmandibular segments. No caudal furrow or flexure is formed.

The initiation of somite formation is carried on simultaneously with the craniad migration of the naupliar mesoderm (Fig. 15-5F). Thus there is no distinct egg-nauplius. The naupliar mesoderm produces not only the antennular, antennal, and mandibular mesodermal components but also the labral mesoderm, stomodeal muscles, and anterior aorta as well since there does not appear to be a distinct preantennular mesoderm formed in *Heterotanais*. The naupliar appendage anlagen develop as those of the postmandibular region arise (Fig. 15-5I, J).

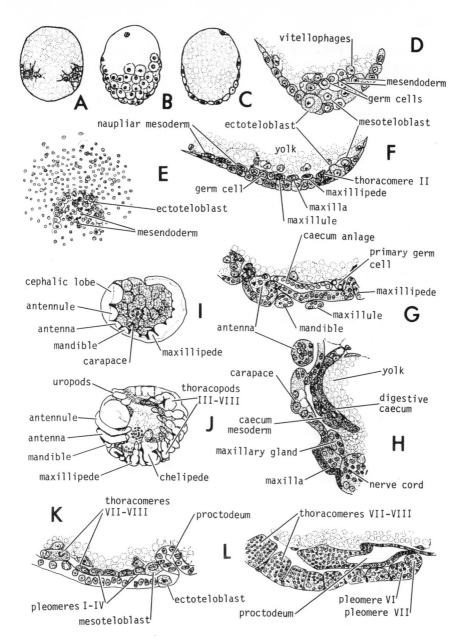

Fig. 15-5. *Heterotanais* development. (A) early cleavage; (B, C) external view and sagittal section, respectively, at completion of germinal disc formation; (D) initiation of gastrulation; (E) surface view of blastopore region showing teloblast ring; (F) early 'metanauplius' phase, note lag in naupliar somite formation; (G) cephalic differentiation underway, anlage of digestive caecum present (section at equivalent of I); (H) well-advanced formation of caecum (section at stage between that seen in (I, J); (I) lateral view of embryo after naupliar somites finally differentiate; (J) advanced embryo, note lack of caudal furrow and flexure; (K) early stage in abdominal formation with proctodeum initiation; (L) advanced abdomen with seven ganglial anlagen. (From Scholl, 1963)

The primary germ cells initially lie within the mesendodermal plate in the maxillulary segment (Fig. 15-5F). Subsequently, they migrate (Fig. 15-5G) to the posterior edge of the maxillipedal mesoderm and eventually come to lie under the pericardium in the third thoracomere as genital anlagen.

The somatic mesoderm divides very early into distinct dorsal and ventral masses. The ventral mass produces the ventral longitudinal body and appendage muscles. The dorsal mass eventually differentiates into heart, pericardium, dorsal longitudinal muscles, and blood cells. The gut diverticula begin as plates or anlagen (Fig. 15-5G) between the mandibular mesoderm and the lateral organs formed by the ectoderm, which round out as cuplike structures as they grow mediad, and come to envelope the yolk mass. The gut proper is formed by stomodeal and proctodeal ingrowths that meet in the area of the developing digestive caeca (Fig. 15-5K).

One final point of interest is noted in the formation of the abdominal ganglia. Seven distinct ganglia are formed (Fig. 15-5L). Scholl (1963) specifically mentions that these ganglia form independently of any legs, which apparently come to be affiliated with the six anterior abdominal ganglia; but Sieg (1983, p. 160) recalls an observation made by Claus (1887) that the innervation of developing uropods actually comes from the seventh ganglion.

FOSSIL RECORD   Of all the peracarid groups, the tanaidaceans have, if not the best, at least the most interesting fossil record (Schram, 1974, 1979; Schram et al., 1986). The earliest tanaidacean is *Anthracocaris scotica* (Fig. 15-6A) from the Lower Carboniferous of Scotland. This and several subsequent fossil forms have six free abdominal somites, that is, no pleotelson. These include *Cryptocaris hootchi* (Fig. 15-6B) from the Middle Pennsylvanian (Upper Carboniferous) of Illinois; *Ophthalampseudes rhenanus*, from the Upper Permian of Germany. The earliest fossil tanaidacean with a pleotelson is *Jurapseudes friedericianus*, from the Middle Jurassic of Germany, which can be compared to living apseudomorphs. Other apseudomorph fossils from the Triassic of Hungary and the Jurassic of Bulgaria and Germany can be assigned to *J. acutirostris* and *Carlclausus emersoni*. *Cretitanais giganteus* comes from the Lower Cretaceous of Germany. It would appear, however, that perhaps through much of their early history tanaidaceans did not possess a pleotelson, a most peculiar quirk to be explained.

Fig. 15-6. Carboniferous anthracocaridomorph, *Anthracocaris scotica*. (From Schram, 1979)

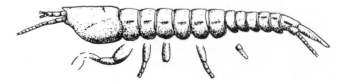

The variations noted in anatomy among these primitive forms indicates a rather diverse radiation, probably comparable in its own way to that seen in the modern groups.

TAXONOMY The classification of tanaidaceans has undergone a series of major adjustments in the course of its history. Lang (1956) strove to bring some order to a disparate array of families and divided the group into the Monokonophora and Dikonophora based on whether there is a single or paired genital cones on the male eighth thoracic sternite. This taxonomic arrangement served a useful purpose at the time, but Sieg (1980) pointed out the polyphyletic nature of Lang's Dikonophora (genital cones are also seen in the pygocephalomorph mysidaceans) and proposed the alternative scheme of classification used here, modified in light of consideration of the fossil forms (Schram et al., 1986).

Infraorder Anthracocaridomorpha Sieg, 1980 Carboniferous–Lower Permian
    Family Anthracocarididae Schram, 1979 Carboniferous–Permian
    Family Cryptocarididae Sieg, 1980 Pennsylvanian
Infraorder Apseudomorpha Sieg, 1980 Jurassic–Recent
    Superfamily Jurapseudoidea Schram et al., 1986 Jurassic–Lower Cretaceous
        Family Jurapseudidae Schram et al., 1986
    Superfamily Apseudoidea Schram et al., 1986 Recent
        Family Apseudellidae Gutu, 1972
        Family Apseudidae Leach, 1814
        Family Cirratodactylidae Gardiner, 1972
        Family Gigantapseudidae Kudinova–Pasternak, 1978
        Family Kalliapseudidae Lang, 1956
        Family Whiteleggiidae Gutu, 1972
        Family Metapseudidae Lang, 1970
        Family Pagurapseudidae Lang, 1970
        Family Sphyrapidae Gutu, 1980
        Family Tanapseudidae Bacescu, 1978
Infraorder Neotanaidomorpha Sieg, 1980 Recent
        Family Neotanaidae Lang, 1956
Infraorder Tanaidomorpha Sieg, 1980 Cretaceous–Recent
    Superfamily Cretitanaoidea Schram et al., 1986 Lower Cretaceous
        Family Cretitanaidae Schram et al., 1986
    Superfamily Tanaoidea Dana, 1849 Recent
        Family Tanaidae Dana, 1849
    Superfamily Paratanaoidea Lang, 1949 Recent
        Family Agathotanaidae Lang, 1971
        Family Anarthruridae Lang, 1971
        Family Leptognathiidae Sieg, 1973

Family Nototanaidae Sieg, 1973
Family Paratanaidae Lang, 1949
Family Pseudotanaidae Sieg, 1973
Family Pseudozeuxidae Sieg, 1982

Even this classification can only be considered tentative in nature, especially in regard to families. Sieg (1982, 1983, 1984) points out certain areas in which future work (especially within the Apseudomorpha) should resolve present problems with polyphyletic taxa.

REFERENCES

Anderson, A., E. Hallber, and S. B. Johnson. 1978. The fine structure of the compound eye of *Tanais cavolinii*. *Acta. Zool. Stockh.* **59**:49–55.
Bückle-Ramirez, L. F. 1965. Untersuchungen über die Biologie von *Heterotanais oerstedi*. *Zeit. Morph. Ökol. Tiere* **55**:714–82.
Claus, C. 1887. Über *Apseudes latreilli* und die Tanaiden II. *Arb. Zool. Inst. Univ. Wien* **7**:139–220.
Dennell, R. 1937. On the feeding mechanism of *Apseudes talpa* and the evolution of the peracaridan feeding mechanism. *Trans. Roy. Soc. Edinb.* **59**:57–78.
Dohle, W. 1972. Über die Bildung und Differenzierung des postnauplien Keimstreifs von *Leptochelia* sp. *Zool. Jahb. Anat.* **89**:503–66.
Gardiner, L. F. 1975. The systematics, postmarsupial development, and ecology of the deep-sea family Neotanaidae. *Smith. Contr. Zool.* **170**:1–265.
Hessler, R. R. 1982. On the structural morphology of walking mechanisms in eumalacostracan crustaceans. *Phil. Trans. Roy. Soc. Lond.* (B)**296**:245–98.
Hessler, R. R. 1983. A defense of the caridoid facies: wherein the early evolution of the Eumalacostraca is discussed. *Crust. Issues* **1**:145–164.
Highsmith, R. C. 1983. Sex reversal and fighting behavior: coevolved phenomena in a tanaid crustacean. *Ecology* **64**:719–26.
Johnson, S. B. 1982. Functional models, life history, and evolution of tube-dwelling Tanaidacea. Doctoral Dissertation, Univ. Lund.
Johnson, S. B., and Y. G. Attramadal. 1982. A functional-morphological model of *Tanais cavolinii* adapted to a tubiculous life strategy. *Sarsia* **67**:29–42.
Kaestner, A. 1970. *Invertebrate Zoology*, Vol. 3. *Crustacea*. Interscience, New York.
Lang, K. 1953. *Apseudes hermaphroditicus*, a hermaphroditic tanaid from the Antarctic. *Ark. Zool.* (2)**4**:341–50, 4 pls.
Lang, K. 1956. Neotanaidae, with some remarks on the phylogeny of the Tanaidacea. *Ark. Zool.* (2)**9**:469–75.
Lang, K. 1957. The postmarsupial development of the Tanaidacea. *Ark. Zool.* (2)**4**:409–22, 4 pls.
Lauterbach, K-E. 1970. Der Cephalothorax von *Tanais cavolinii*, ein Beitrag zur vergleichenden Anatomie und Phylogenie der Tanaidacea. *Zool. Jahrb. Anat.* **87**:94–204.
Mendoza, J. A. 1982. Some aspects of the autecology of *Leptochelia dubia*. *Crustaceana* **43**:225–40.
Scholl, G. 1963. Embryologische Untersuchungen an Tanaidacean (*Heterotanais oerstedi*). *Zool. Jahrb. Anat.* **80**:500–54.
Schram, F. R. 1974. Late Paleozoic Peracarida of North America. *Fieldiana: Geol.* **33**:95–124.

Schram, F. R. 1977. Paleozoogeography of Late Paleozoic and Triassic Malacostraca. *Syst. Zool.* **26**:367–79.

Schram, F. R. 1979. British Carboniferous Malacostraca. *Fieldiana: Geol.* **40**:1–129.

Schram, F. R., J. Sieg, and E. Malzahn. 1986. Fossil Tanaidacea. *Trans. San Diego Soc. Nat. Hist.* 21.

Sieg, J. 1980. Sind die Dikonophora eine polyphyletische Gruppe? *Zool. Anz.* **205**:401–16.

Sieg, J. 1982. Über ein 'connecting link' in der Phylogenie der Tanaidomorpha. *Crustaceana* **43**:65–77.

Sieg, J. 1983. Evolution of Tanaidacea. *Crust. Issues* **1**:229–56.

Sieg, J. 1984. Neuere Erkenntnisse zum natürlichen System der Tanaidacea. *Zoologica* 136: 1–132.

16

CUMACEA

DEFINITION   Carapace short, fused to at least first three thoracomeres, can fuse with up to six, laterally enclosing a branchial cavity, with lateral lappets that extend anteriad and mediad to form a pseudorostrum; eyes generally fused, located on an anterior occasionally bell-shaped lobe; mandibles without palps; anterior three thoracopods as maxillipedes, the first with elaborately lobed branchial epipod and exopod extending forward under pseudorostrum as siphon, the second with fused coxae from which arise elongate endopods; posterior thoracopods often biramous; pleopods generally absent on females and sometimes reduced or absent on males; telson may be either free or fused with the sixth pleomere.

HISTORY   Although there are currently about 1000 species of cumaceans, the literature on them is not extensively developed beyond the alpha-taxonomy stage. The first cumacean recorded in the literature, a *Diastylis*, was noted in 1779 by Lepechin, named then *Oniscus scorpioides*. H. Milne Edwards erected the genus *Cuma* (from which the name Cumacea arises), but he erroneously maintained for some years, in the face of contrary evidence, that 'cumaceans' were merely larvae. The preeminent taxonomic work in the group remains that of Sars, particularly his monograph on the Cumacea of Norway (1900); subsequent taxonomic work has built upon this, as for example in the monographic treatments of Stebbing (1913), Lomakina (1958, 1973), or Jones (1963, 1976). However, the general biological understanding of the group has not significantly advanced over that outlined by Sars (1900) or Oezle (1931). Stebbing (1913) presented an outline of the history of the cumaceans' rather peripatetic taxonomic assignments through time: Fabricius thought they were amphipods; Lepechin placed them in the isopods; Pennant, Goodsir, and van Beneden allied them to the macrurans while H. Milne Edwards, Nicolet, Dana, and Agassiz thought them larval macrurans; Say, Demerest, Latrielle, and Kossman assigned them to the schizopods; DeKay, Bate, and Danielssen claimed they were stomatopods; Huxley and Boas opted for mysidacean affinities; while Latrielle at an earlier time and Stimpson went so far as to place them among the branchiopods. Most recently Schram and Watling have allied cumaceans to tanaids, thermosbaenaceans, and spelaeogriphaceans in the Brachycarida. Certainly, the Cumacea have been among the most difficult of groups to place.

MORPHOLOGY   The cumaceans possess one of the most distinctive anatomical plans among the Eumalacostraca (Fig. 16-1). The carapace and anterior thorax are rather inflated, while the posterior thorax and abdomen are quite slender. The carapace is fused to at least the first three thoracomeres, though in individual cases it can be joined to the fourth, fifth, and sixth segments as well. Anteriorly the carapace has a pair of lateral lappets or lobes that typically extend anteriad and mediad to form a pseudorostrum. In the genera *Eudorella* and *Eudorellopis* the lappets are extended upward, and no pseudorostrum is formed. The carapace is developed laterally to enclose a branchial chamber. The optical components of the eyes are typically reduced or absent. The eyes themselves, when present, are generally fused on a single ocular lobe that arises from the frontal lobe of the cephalon and lies between the pseudorostral lappets. Only in the developing embryos and in the genus *Nannastacus* are the eyes separate. The eyes of males are typically more fully developed than those of the females.

The antennules (Fig. 16-2A) have a three-segment peduncle. The flagella are short and unequally developed. The outer or main flagellum has, at most, six articles, and can bear aesthetascs on the distal joints. The inner or accessory flagellum is often uniarticulate, never exceeds four joints, and may be absent.

The antennae (Fig. 16-2B) are reduced in females but are usually well developed in males (Fig. 16-2C). They have a five-segment peduncle and a long flagellum. The male antennae are bent backward to extend along the body, with the outer edges of the fourth and fifth segments highly setose. In some forms, the antennae of the male are modified for clasping the female.

The epistome and broad labrum are closely allied to the bases of the antennules and antennae. The mandibles (Fig. 16-2D) lack any palps. The incisor process is toothed on its distal inner edge. Just behind this tip on the left mandible is a well-developed lacinia mobilis, which is sometimes present on the right mandible in a rudimentary state. The laciniae are located

Fig. 16-1. Sexual dimorphism in *Bodotria ulchella*. (A) female; (B) male. (From Jones, 1976)

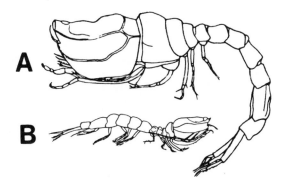

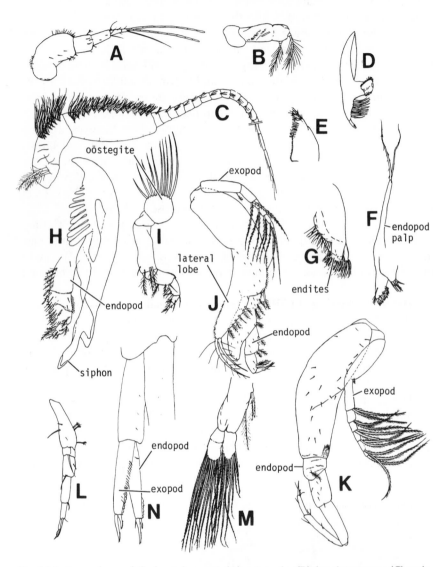

Fig. 16-2. Appendages of *Cyclaspis bacescui*. (A) antennule; (B) female antenna; (C) male antenna; (D) mandible; (E) half of paragnath; (F) maxillule; (G) maxilla; (H) first maxillipede; (I) second maxillipede of female; (J) third maxillipede; (K) fourth thoracopod (= first pereiopod); (L) eighth thoracopod; (M) pleopod of male; (N) uropod. (From Omholt and Heard, 1982)

at the distal ends of well-developed spine rows. The molar process is generally columnar but can be occasionally long and styliform. The paragnaths (Fig. 16-2E) are simple lobes separated by a deep medial cleft. The cleft is bound by short fine setae proximally and can have more elaborate setae and spines at the distal tips of the lobes.

The maxillules (Fig. 16-2F) have a flattened protopod with two medially directed setose endites. The exopod is absent, but the endopod is developed as a posteriorly directed palp with one or two distal spines or setae. This palp extends back into the branchial chamber; its vibrations may facilitate the movement of respiratory currents over the gills.

The maxillae (Fig. 16-2G) also have broadly flattened protopods, the medial margins of which are quite setose. Two endite lobes are located distally while laterally there is a thin plate variously termed a flabellum or exopod. (The exact number of protopodal segments is ambiguous in both the maxillules and maxillae with different opinions having been expressed by various authors. The descriptions presented here have avoided judgment on these matters.)

The first thoracopods are distinctly modified as the first maxillipedes (Fig. 16-2H). The short coxa bears an elaborate posteriorly directed epipodite, which is developed with lammellate fingerlike lobules arranged either in a row or in an open spiral. The exopod is apparently directed forward to extend under the pseudorostrum as a siphon. The siphons serve to direct water out of the branchial chamber. The endopod is a five-segment palp located distally on the basis. The basis itself typically bears a slight lobelike endite and carries two or three medial retinacula or coupling hooks so that the opposing bases lock together. A median cephalic process arises at the level of the maxillipedes and divides the maxillary filter chamber into right and left halves.

The second maxillipedes (Fig. 16-2I) have their coxae fused at the midline. The greatly elongate basis and endopod extend forward under the mouthfield. The fused coxae in females bear posteriorly directed setose flaps, considered rudimentary oöstegites, and these setae project back into the brood chamber.

The third maxillipedes (Fig. 16-2J) somewhat resemble the pediform pereiopods. However, the basis is greatly elongated and frequently marked with a lateral lobe, and occasionally the ischium may also be so equipped. The endopod is shortened over that seen in the pereiopods. The exopod is typically always present, except in some diastylids.

The last five pairs of thoracopods are similar in form (Fig. 16-2K) but vary in orientation. These pereiopods have small coxae and elongated bases. The endopod typically has five well-developed segments, though the fifth thoracopods (=second pereiopods) frequently fuse the ischium and basis. Exopods occur only on the first and usually a few of the following pereiopods in the female, while males generally have exopods on all of the pereiopods except occasionally the last (Fig. 16-2L). The first two pereiopods extend forward under the mouthfield, and the first of these has been reported helpful in conveying food to the mouth. The rest of the pereiopods, especially the last three pairs of posteriorly directed limbs, are used in digging. Oöstegites are on the third through sixth thoracopods.

Hessler (1982) reports that the coxa is fused with the body on the anterior thoracopods but is articulated with the body on thoracopods 6 through 8. The coxa and basis articulate by a single condyle that extends into the lumen of the limb from the posterior or posterolateral rim of the coxa (Fig. 16-3).

The pleopods (Fig. 16-2M) consist of short coxae, elongated bases, and two short rami (the exopods with one segment, the endopod with two). Females lack developed pleopods except in *Archaeocuma*, while males may vary from having a complete set of five pleopods to as few as a single pair or none at all.

The telson may be either free or fused with the sixth pleomere as a pleotelson. The uropods (Fig. 16-2N) are styliform. The protopod bears two rami, the exopod has two segments, and the endopod anywhere from one to three segments.

Oezle (1931) offers the best comprehensive review of cumacean internal anatomy. The short esophagus enters a thick-walled stomach, the anterior part of which is chitinous and armed with long spines. Into the posterior portion of the stomach empty anywhere from one to four pairs of short digestive caeca, the walls of which are inflated with masses of secre-

Fig. 16-3. Monocondylic articulation of coxa and basis in *Diastylis rathkei*. (From Hessler, 1982)

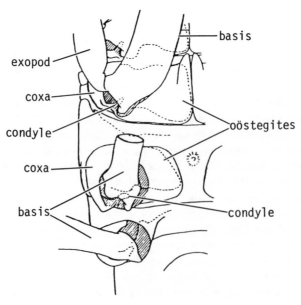

tory cells. The midgut is very short, while the hindgut extends throughout most of the body and sometimes may be coiled in the thoracic region.

The heart is situated in the third through seventh thoracomeres and is notable in being reinforced with circular bands of muscles. There is a single pair of ostia in the posterior region of the heart. The arterial system (Fig. 16-4) extends out from the heart and is marked by the cephalic heart, or cor frontale, in the anterior aorta. The venous sinuses, particularly those of the epipodal branchiae and those associated with the epithelia of the carapace, empty oxygen-enriched blood into the pericardium and thence through the ostia into the heart.

Excretion is achieved by maxillary glands, which have been compared in structure to those seen in tanaidaceans (Oezle, 1931).

The gonads are tubelike organs that flank the gut and lay adjacent to the pericardium. The oviducts open on the inner surface of the coxae on thoracopod 6, and the vasa deferentia open on the sternum of thoracomere 8. The sperm are rather distinctive and have a long tail and a cluster of cilia attached to the base of the head.

The central nervous system consists of a brain, with distinct proto-, deuto-, and tritocerebral portions, and a ventral nerve cord with 16 distinct pairs of ganglia (one for each of the maxillae and the trunk segments, a most primitive arrangement).

NATURAL HISTORY Feeding in cumaceans can take on a multiplicity of modes. Some species (Dixon, 1944) have been recorded as picking up sand grains with the first pereiopods and passing them to the maxillipedes. The maxillipedes then hold the particles in the mouthfield while the maxillae and mandibles scrape off any organic films on the grains. *Diastylis rathkei* sweeps organic-rich muds into the mouthfield with its maxillipedes and then swallows the accumulated mass. The genus *Campylaspis* has a styliform molar process on the mandible and, combined with the fact that foraminiferan and crustacean parts have been found in the stomach of these species, suggests a more rapacious habit.

The process of filter feeding has been studied in great detail by Dennell (1934) for *D. bradyi*. The vibration of the maxillules and maxillae provide the filtration pump. The ventrad and posteriad movement of these limbs sucks water in from the ventral and anterior direction. This suction is facilitated by similar movement of the maxillipedes. A flow of water from a lateral direction is prevented by the maxillary exopod flaps and 'mat' setae on the maxillipedes. As the maxillules and maxillae return anteriad, water is flushed out of the filter chamber laterally into the excurrent respiratory flow, food particles being trapped on the maxillulary setae. The maxillipedal endites act to sweep food from the maxillules between the paragnaths into the mouth.

Feeding is collateral to, but independent of, the respiratory cycle (Fig. 16-5). The vibration of the maxillipedal epipodite (about 40 beats per

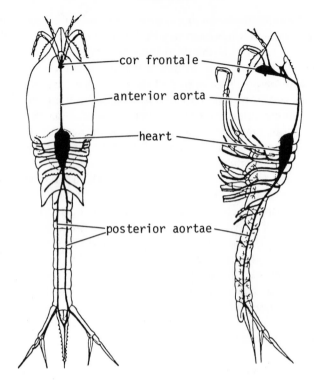

Fig. 16-4. Circulatory system of *Diastylis rathkei*. (From Oezle, 1931)

minute) facilitates the movement of water needed for respiration (Dennell, 1934). On the forward stroke the epipodite is a transverse shelf extended across the branchial chamber medially from the body wall and laterally to the carapace. As the epipodite moves dorsad, water above the limb is compressed in the upper chamber and pushed forward out along the siphon channel formed by the maxillipedal exopods beneath the pseudorostrum. At the same time water is sucked into that part of the branchial chamber below the epipodite from under the ventral margins of the carapace. On the recovery stroke, the epipodite twists allowing the water to mix from the inflated lower and deflated upper chambers. The epipodite then moves ventrad and into a position where it can twist back across the branchial

Fig. 16-5. Current flow around a cumacean body. (From Dennell, 1934)

chamber to start the cycle all over again. This excurrent respiratory flow helps, but it is not completely necessary, to facilitate the filter flow.

Locomotion is typically achieved by swimming with the pereiopods and, if present, the pleopods. The shallow water cumaceans remain buried in the substrate during the day and apparently restrict their wanderings to the night. What locomotory patterns that might exist among deep-sea forms, the most abundant type of cumacean, is not known. Most of the differences in anatomy evident between males and females indicate the former are more mobile and actively seek the somewhat more sedentary females. Males typically have better developed eyes, long antennal flagella, exopods on almost all the pereiopods, and often strong development of the pleopods.

However, knowledge of actual courtship and mating among cumaceans is incomplete. It is typically stated in the literature that copulation occurs during nocturnal swarming, although actual mating has only been witnessed in laboratory *Mancocuma stellifera* (Gnewuch and Crocker, 1973). There are no field data to evaluate the implication inferred above that male mobility may be an important factor in cumacean mating behavior.

Overviews of reproductive strategies among cumaceans are only just beginning to be arrived at (Corey, 1981). Unlike tanaidaceans, who have exploited sexual versatility of single individuals, cumaceans regulate the fecundity of the female. Several variants have been noted. Arctic forms (*Diastylis goodsiri*) possess long life spans of apparently many years and have large females who produce but a few large eggs after delayed maturity. Temperate littoral forms exhibit several different strategies. Some species, like *Pseudocuma longicornis*, have low fecundity and small eggs but breed more or less continuously throughout the year. Other species, like *Diastylis polita* or *Lamprops quadriplicata* breed only once but produce large numbers of small eggs. This latter pattern is also noted in species, like *Diastylis sculpta* and *Iphinoe trispinosa*, that breed twice a year as two separate generations, with relatively high fecundity and small eggs.

Deeper water forms seem to favor strategies with lower fecundity. Some species, like *Hemilamprops rosea* and others, will raise two broods from one generation of females. Other species will breed only once per year in the winter with few eggs but will produce large eggs (e.g., *Leucon nasica*), while others will breed once but produce only a few small eggs (e.g., *Iphinoe serrata*). These last would seem to be at something of a potential competitive disadvantage. Deep water spring–summer breeders utilize a 'shallow-water' strategy with large numbers of small eggs (e.g., *Diastylis lucifera*). Just what these various strategies mean in terms of overall life history aptations is not yet clear, and further insight might be gained if some correlation between mating and brooding behavior can be made.

Cumaceans are almost entirely marine in their habits, though there are a few brackish and freshwater forms; and, while they are found from littoral to abyssal depth, their greatest diversity seems to be in the deep sea (e.g., Jones, 1973; Jones and Sanders, 1972; Reyss, 1973, 1978). The consensus is that cumaceans are probably as important as the isopods in the ecology of the deep sea in terms of both the numbers of individuals and diversity of species. As would be suspected, however, little is known of the biology of these deep-sea forms.

Cumaceans are almost entirely benthic animals. They utilize their posterior thoracopods to burrow into the bottom sediments. Because of problems in filtering the incurrent respiratory waters and the manner in which specific solutions to this problem are achieved, the distribution of species seem to be determined by the particle size of the substrate (Dixon, 1944). Once buried, cumaceans then either remain concealed and filter food items out of the currents that they generate, or they venture out onto the surface to seek food as the need arises. These generally cryptic habits serve a protective function, and in this they resemble tanaidaceans. In view of their habitat preferences, cleaning of the body is most important and is achieved by extreme flexion of the abdomen and brushing the body with the uropod (Dixon, 1944). Though almost always burrowers, one species, *Iphinoe fagei*, has been reported taken in great numbers in plankton tows (Hart and Currie, 1960). Watling (personal communication) notes this is not uncommon in temperate or tropical waters if the tow is taken between an hour after dusk and an hour before dawn.

The biogeographic distributions of cumaceans are only just coming to be appreciated, but it would appear that little significant progress in this area is likely to be achieved until such time as the alpha-taxonomy of the group levels off and stabilizes.

DEVELOPMENT   Research in this field has been somewhat limited and concentrated on describing gross development of form (Dohrn, 1870; Sars, 1900). Very little that is useful has been done on the ontogeny of germ layers and organs (Butchinskiy, 1893, 1895; and Grschebin, 1910).

The cleavage seems to be superficial with the nucleated cells coming to surround the inner glandular yolk. A ventral germinal disc is formed. Like tanaidaceans the definition of somites and leg anlagen of the naupliar and postnaupliar segments seems to occur at about the same time (Fig. 16-6A). Thus the development of the head is delayed while that of the teleoblasts is accelerated such that the embryo becomes dorsally flexed, that is, no caudal furrow or papilla forms.

Some workers (e.g., Jones, 1963) record that hatching occurs at a nauplius stage, but no mention of such an early event is made in any of the classic studies of cumacean development (Dohrn, 1870; and Sars, 1900). Hatching occurs at a relatively advanced metanaupliar stage (Fig. 16-6B). The hatchling undergoes three molts in the brood pouch before a manca stage is reached. The first molt (Fig. 16-6C) produces an individual with clearly delineated appendages in the series back to those of the fourth pereiopods. The ventrolateral carapace folds are prominent, and little or no

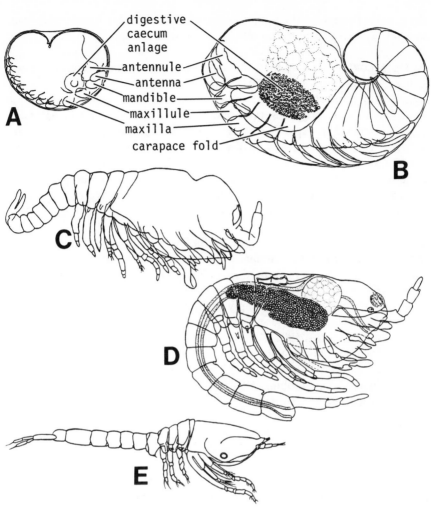

Fig. 16-6. Development of *Cumopsis goodsiri*. (A) early stage within the egg; note nearly equal development of naupliar and metanaupliar appendage anlagen; (B) stage just after hatching; (C) after first larval molt; (D) after second larval molt; (E) manca stage, after third larval molt. (From Dohrn, 1870)

distinct segmentation is visible dorsally from the maxillipedes forward. The rostral lappets are barely formed and the digestive caeca are present but have not completely devoured the yolk. The second molt (Fig. 16-6D) adds no more appendages, but perfection of the cumacean form is evident. The rostral lappets are marked, the component outline of the first maxillipedes is established, and the digestive system is near completion. The final molt (Fig. 16-6E) produces a manca form that lacks only the last pereiopod and any pleopods, though the anlagen of these appendages are visible below the cuticle. Except for the unique cumacean features, the brood patterns of this development seem to resemble those noted for the tanaidaceans.

FOSSIL RECORD   Though they are benthic animals that prefer the fine-grained substrates ideal for fossilization, cumaceans have a poor fossil record. Bachmeyer (1960) described a 'cumaceoid' form, *Palaeocuma hessi*, which is poorly preserved and of which little is known. Malzahn (1972) described two species of a well-preserved Permian genus *Ophthalm-diastylis* from Germany, characterized by a pair of well-developed lobed eyes; similar forms are also to be found in the Pennsylvanian of North America (Schram, unpublished). Besides the presence of lobed eyes, these Paleozoic forms lack a pseudorostrum and, thus, effectively represent a very primitive stage in cumacean development.

TAXONOMY   Although at one time (Stebbing, 1913) it was suggested that there were 26 families of cumaceans, the consensus of modern cumacean workers has settled on eight families of living forms. However, no systematic analysis of characters throughout the group has been performed, so the sequence of families does not imply any phylogenetic arrangement.

> Bodotriidae Scott, 1901
> Ceratocumatidae Calman, 1905
> Diastylidae Bate, 1856
> Gynodiastylidae Stebbing, 1912
> Lampropidae Sars, 1878
> Leuconidae Sars, 1878
> Nannastacidae Bates, 1966
> Pseudocumatidae Sars, 1878

A ninth family has been described, Archaeocumatidae Bacescu, 1972, but Jones (1976) feels that this seems to be assignable to lampropids.

REFERENCES

Bachmeyer, F. 1960. Eine fossile Cumaceenart aus dem Callovien von La Voulet-sur-Rhône. *Ber. Schweiz. Pal. Ges., Ecolog. geol. Helv.* **53**:422–26.
Butchinskiy, P. 1893. Zur Embryologie der Cumaceen. *Zool. Anz.* **16**:386–87.
Butchinskiy, P. 1895. Nablyudyeniya nad embrionalnim razvitiyem Malacostraea. *Zapiski Novorossiickago Obshchyesgva Yestyestvoispitatyelyei* **19**:1–216.
Corey, S. 1981. Comparative fecundity and reproductive strategies in seventeen species of Cumacea. *Mar. Biol.* **62**:65–72.
Dennell, R. 1934. The feeding mechanism of the cumacean crustacean *Diastylis bradyi*. *Trans. Roy. Soc. Edinb.* **58**:125–42.
Dixon, A. Y. 1944. Notes on certain aspects of the biology of *Cumopsis goodsiri* and some other cumaceans in relation to their environment. *J. Mar. Biol. Assoc. U.K.* **26**:61–71.

Dohrn, A. 1870. Untersuchungen über Bau und Entwicklung der Arthropoden. *Jena Zeit. Naturwiss.* **1870**:54–81, 2 pls.

Gnewuch, W. T., and R. A. Crocker. 1973. Macroinfauna of northern New England sand. I. The biology of *Manocuma stellifera. Can. J. Zool.* **51**:1011–20.

Grschebin, S. 1910. Zur Embryologie von *Pseudocuma pectinata. Zool. Anz.* **35**:808–13.

Hart, T. J., and R. I. Curie. 1960. The Benguela Current. *Disc. Repts.* **31**:127–298.

Hessler, R. R. 1982. The structural morphology of walking mechanisms in eumalacostracan crustaceans. *Phil. Trans. Roy. Soc. Lond.* (B)**296**:245–98.

Jones, N. S. 1963. The marine fauna of New Zealand: crustaceans of the order Cumacea. *N.Z. Dept. Sci. Ind. Res. Bull.* **152**:1–82.

Jones, N. S. 1973. Some new Cumacea from the deep Atlantic. *Crustaceana* **25**:297–319.

Jones, N. S. 1976. *British Cumaceans.* Academic Press, London.

Jones, N. S., and H. L. Sanders. 1972. Distribution of Cumacea in the deep Atlantic. *Deep-sea Res.* **19**:737–45.

Lomakina, N. B. 1958. Kumovye raki morei SSSR. *Opredelitel po Faune SSR* **66**:1-301.

Lomakina, N. B. 1973. Cumacea of the Arctic region. *Stud. Mar. Fauna* **6**:97–140.

Malzahn, E. 1972. Cumaceenfunde aus dem neiderrheinischen Zechstein. *Geol. Jahrb.* **90**:441–62.

Oelze, A. 1931. Beitrage zur Anatomie von *Diastylis rathkei. Zool. Jahrb. Abt. Anat. Ontog.* **54**:235–94, 1 pl.

Omholt, P. E., and R. H. Heard. 1982. *Cyclaspis bacescui,* new species from the eastern Gulf of Mexico. *J. Crust. Biol.* **2**:120–9.

Reyss, D. 1973. Distribution of Cumacea in the deep Mediterranean. *Deep-sea Res.* **20**:1119–23.

Reyss, D. 1978. Cumaces de profondeur de l'Atlantique Nord. Famille des Lampropidae. *Crustaceana* **31**:1–21, 72–84.

Sars, G. O. 1900. *An Account of the Crustacea of Norway.* III. *Cumacea.* Bergen Museum, Christiania.

Stebbing, T. R. R. 1913. Cumacea. *Das Tierreich* **39**:1–210.

17

THERMOSBAENACEA

DEFINITION   Carapace reduced in size (especially in males), fused to first thoracomere, free above all others, as a dorsal brood pouch in females; first thoracopod as maxillipede, thoracopods 2 to 8 biramous and lacking epipodites; pleopods markedly reduced or absent; uropods and telson as a tailfan.

HISTORY   Since thermosbaenaceans have only been recognized in this century (Monod, 1924), most of our knowledge of the group has arisen in the post-World War II period (e.g., see Stock, 1976). Originally placed within the peracarids, Siewing (1958) felt that their lack of oöstegites and peculiar mode of brooding justified a separate position for thermosbaenaceans and erected the superorder Pancarida to accommodate them. This latter suggestion has not met with universal acceptance, and the position that they are an order of the Peracarida finds about equal use in the literature (e.g., see Hessler, 1969, p. R366). However, the growing recognition of laciniae mobiles in various stages of development in many different eumalacostracan groups and nectiopodan remipedes has minimized the usefulness of these structures as taxonomic indicators and thus weakened the effectiveness of the one traditional unifying character of the peracarids. Other views have been advanced from time to time. Sars (1929) wanted to ally thermosbaenaceans and bathynellaceans into a group, the Anomostraca; and Glaessner (1957) suggested comparison of thermosbaenaceans to stomatopods.

MORPHOLOGY   The genus *Thermosbaena*, because of the reduction of thoracopod complement and formation of a pleotelson, is rather atypical for the group. The widely distributed genus *Monodella* (Fig. 17-1A, B) or the Caribbean genus *Halosbaena* (Fig. 17-1C) are more typical in form.

The carapace is short, being fused to only the first thoracomere, but covering at least the next one or two segments as well. In females, however, the carapace is greatly enlarged dorsally to accommodate brooded eggs. The carapace may not cover the anterior head or procephalon, as in *Monodella* (Fig. 17-1A, B).

The antennules (Fig. 17-2A) have a three-segment peduncle that can be rather setose. There are two short flagella, the larger of which bears long aesthetascs.

The antennae are uniramous (Fig. 17-2B). The peduncular segments

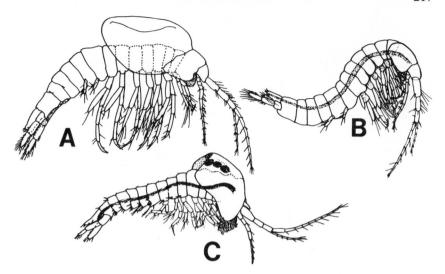

Fig. 17-1. (A, B) *Monodella sanctaecrucis.* (A) female; note dorsal development of carapace as a brood chamber; (B) male; (C) *Halosbaena acanthura.* (From Stock, 1976)

grade into the flagellum, though Stock (1976) considers the proximal five joints to be peduncular.

The labrum is relatively elongate, with terminal spines or setae. The mandible (Fig. 17-2C) has a well-developed three-segment palp, the distal joints of which are armed with large pectinate spines. The body of the mandible is divided into incisor and molar processes between which are a lacinia mobilis and a row of spines. The lacinia appears to be reduced in *Halosbaena* to a row of minute setae. The paragnath is a deeply cleft lobe.

The maxillules (Fig. 17-2D) are composed of two-segment protopods, which bear two large setose endites and a small endopodal palp.

The maxillae (Fig. 17-2E) are complex appendages. The protopod is a large lobe with typically three complexly setose endites directed medially and ventrally. The endopodal palp is short, and the exopod is a small lobe, sometimes closely associated with the endopod base.

The first thoracopod is developed as a maxilliped (Fig. 17-2F). The protopod has well-developed setose endites and an epipodite. The endopod and exopod can be well developed and exhibit some sexual dimorphism, but in *Halosbaena* these are very reduced.

The second thoracopod is uniramous (Fig. 17-2G). The third through eighth thoracopods (Fig. 17-2H) are biramous and without epipodites. The exopod may be one- or two-segmented. The protopod of the eighth thoracopod in the males bears a well-developed penis. *Thermosbaena* possesses only thoracopods 2 through 6. The first two pleopods (Fig. 17-2I, J, K) are single-segment structures with terminal spines. Pleopods 3 through 5 are absent.

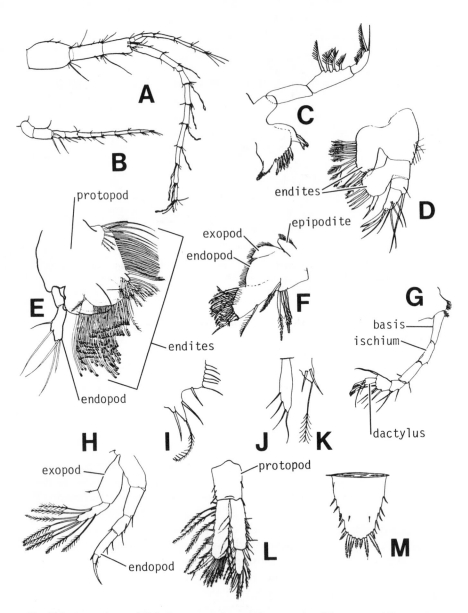

Fig. 17-2. Appendages of *Halosbaena acanthura*. (A) antennules; (B) antenna; (C) mandible; (D) maxillule; (E) maxilla; (F) maxillipede; (G) first pereiopod; (H) second pereiopod; (I) first pleopod attached to part of pleuron; (J) second pleopod; (K) first pleopod; (L) uropod; (M) telson. (From Stock, 1976)

The telson (Fig. 17-2M) is an oval lobe with rather spinose margins. In *Thermosbaena*, the telson is fused with the last pleomere. The uropods (Fig. 17-2L) are well developed and bear complex setae along their margins. The exopod is two-segmented, and the endopod is a single short segment.

The internal anatomy is known in only modest detail (Siewing, 1958). There is a cardiac and pyloric stomach, and near the latter a pair of long digestive caecae arise that extend back to the posterior part of the abdomen. The midgut itself is quite large and is apparently always loaded with food. The anus appears to be terminal (Barker, 1962; Por, 1961).

The carapace is thin and filled with blood and thus acts as a respiratory organ. Blood also circulates in and out of the thoracic exopods and these act as respiratory organs as well.

Because of the importance of the carapace in respiration, the heart is short and occupies only the first thoracomere. It has an ostium that directly drains the carapace sinus. The arterial system appears to be rather limited, with a short posterior aorta and a short plexus of vessels to supply the cephalic organs.

Barker (1962) provides a detailed analysis of the reproductive system in *Thermosbaena*. The testes are markedly anterior and occupy a position in the posterior cephalon and first thoracomere. The vasa deferentia lead posteriorly from the testes to the sixth pleomere, and then turn anteriad to the eighth thoracomere. The vasa are dilated terminally to form seminal vesicles, the actual sizes of which are determined by sperm content. The sperm are reported to be 'filiform.' The immature ovary extends from the first to seventh thoracomeres and, when mature, it extends into the first pleomere. However, during copulation the ovary swells to the level of the fourth pleomere, and the subterminus of the oviduct in the seventh thoracomere acts as a seminal receptacle.

The brain is relatively advanced, with proto-, deuto-, and tritocerebral components, all occupying a supraesophageal position. The nerve cord contains discrete ganglia for every postmandibular somite.

NATURAL HISTORY Fryer (1964) provides an excellent analysis of general life habits in *Monodella argentarii*. Thermosbaenaceans are basically benthic animals scurrying around on the substrate, with little bursts of activity between periods of nonmovement. They can swim and do so frequently, orienting themselves upside down. However, when they swim they use both the thoracic endopods as well as exopods to glide along. The maxillipedal epipodite and the pereiopodal exopods beat when the animal is at rest, and the degree of movement is related to O_2 level in the water. The pereiopods are used only for locomotion and respiration.

Feeding is entirely maxillary; the animals dine on detritus. The gathering of food is irregular, and food moves constantly and very slowly through the gut. The robust terminal setae on the maxillules act as scrapers to accumulate detritus. The maxillae serve to sweep the food forward to the oral region, where a rolling action of the molar process of the mandibles force the food through the mouth into the esophagus. The maxillipedes act to protect and clean the maxillae and to enclose the side of the mouth field to prevent food from squeezing laterally out of the mouth region.

The maxillipedes exhibit sexual dimorphism (except in *Thermosbaena*). The endopod in the male is a five-segment clasper, whereas in the female it is reduced.

Actual copulation and egg laying have never been observed. The internal reproductive anatomy discussed above indicates that fertilization is internal. It has been suggested that the female is on her back when eggs are actually laid. This would allow them to fall back into the carapace brood pouch above the thorax as they leave the gonopore.

Thermosbaenaceans seem to prefer thermal and/or fresh to haline groundwaters or, in rare instances, to cave pools. Brunn (1939) observed in the case of *Thermosbaena mirabilis* that the thermal pools used as public hot baths, in which the animals are found, are periodically 'sterilized' with a poison. Since the shrimp readily repopulate after these cleaning procedures, it is assumed the prime habitat source must be deep in the underground thermal system.

The biogeographic distribution of thermosbaenaceans is circum-Mediterranean and Caribbean (Fig. 17-3). The pattern is all the more interesting in that there are several trans-Atlantic links of closely related taxa, for example, *Monodella texana* from several Texas localities (Stock and Langley, 1981) and *M. sanctaecrucis* in the West Indies, and several *Monodella* species in the Mediterranean; or *Halosbaena acanthura* from Curaçao being most closely related to *Limnosbaena finki* of Yugoslavia (Stock, 1976). These tracks would seem to indicate the operation of plate tectonics in the breakup of the distribution of a very old group. Maguire (1965) felt that the original thermosbaenacean distribution was Tethyan and that the breakup of Pangaea in post-Jurassic time has effectively separated formerly contiguous faunas. Stock (1976, 1982), however, though emphasizing the old nature of the Thermosbaenacea, does not feel we can entirely discount relatively recent movements of members of the group. Curaçao is late Miocene–early Pliocene in age and was almost completely submerged in Pleistocene time during transgressive periods of high sea level. In this regard, *Halosbaena acanthura* is found in haline waters in the Pleistocene limestones and coral debris of the island (Stock, 1976), and the genus is also known from marine habitats (Stock, 1982).

The distribution of thermosbaenaceans is congruent with that of cyclopoids, the mysid *Stygiomysis*, microparasellid and stenaselline isopods, and hadziid amphipods. The latter also occur in the 'eastern Tethys' in the Indo-Pacific. Though not entirely mutually exclusive, hadziids and thermosbaenaceans are only rarely found together. This may be related to the predatory proclivities of the amphipods (Stock, 1982).

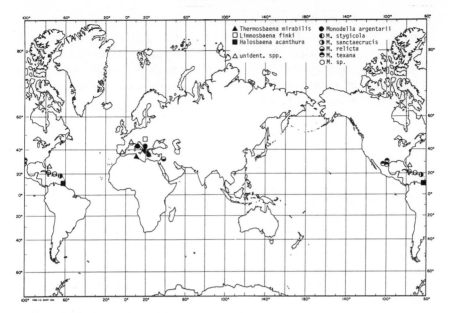

Fig. 17-3. Biogeographic distribution of thermosbaenaceans.

DEVELOPMENT Barker (1962) and Stella (1955, 1959) have offered, respectively, observations on development in *Thermosbaena mirabilis* and *Monodella argenterii* though a great deal of detailed work yet remains to be done. Early stages of cleavage occur in the ovary, and the earliest developmental stage ever seen in the brood pouch is a 12-cell form. The egg is yolky, and although the early cleavages are total, the cytoplasm soon becomes superficial. Barker (1962) was only able to make some general observations as to external development but made frequent comparisons to similarities in the development of *Hemimysis*.

Thermosbaenacean ontogeny has been divided by Stella into marsupial and postmarsupial phases. The earliest marsupial phase exhibits poorly developed body segmentation but has a clear development of limb buds. In the late marsupial stages the embryos are capable of slight movements, have five pairs of pereiopods but no pleopods. At this point the young are shed from the brood pouch. In *Thermosbaena* development beyond this point simply consists of increase in size and perfection of form. The monodellids, however, essentially hatch as mancas and pass through three postmarsupial stages. The first stage has a five-segment external flagellum on the antennule, five pairs of pereiopods and anlagen of the sixth, pleopods, and the male maxillipede developed as a 'clasper.' The second postmarsupial stage finds all the yolk absorbed from the gut, six pairs of pereiopods, anlagen of the seventh, and the abdomen and tailfan completely formed. The third postmarsupial stage has a 10-segment flagellum on the antennule, and seven pairs of pereiopods—essentially a juvenile.

TAXONOMY Taramelli (1954) divided the thermosbaenaceans into two families. The Thermosbaenidae (with a single paedomorphic species *T. mirabilis*) have five pairs of pereiopods, a pleotelson, and no sexual dimorphism in the maxillipede. The Monodellidae (with presently seven species in three genera) possess seven pairs of pereiopods, six separate abdominal segments, and dimorphism in the maxillipedes.

REFERENCES

Barker, D. 1962. A study of *Thermosbaena mirabilis* and its reproduction. *Quart. J. Micro. Sci.* **103**:261–86.
Brunn, A. F. 1939. Observations on *Thermosbaena mirabilis* from the hot springs of El Hamma, Tunisia. *Vidensk. Medd. Naturh. Foren. Kjøb.* **103**:493–501.
Fryer, G. 1964. Studies on the functional morphology and feeding mechanism of *Monodella argentarii*. *Trans. Roy. Soc. Edinb.* **64**:49–90.
Glaessner, M. F. 1957. Evolutionary trends in Crustacea. *Evolution* **11**:178–84.
Hessler, R. R. 1969. Peracarida. In *Treatise on Invertebrate Paleontology*, Part R, *Arthropoda* **4**(1) (R. C. Moore, ed.), pp. R360–93. Geol. Soc. Am. and Univ. Kansas, Lawrence.
Maguire, B. 1965. *Monodella texana*, an extension of the range of the crustacean order Thermosbaenacea to the western hemisphere. *Crustaceana* **9**:149–54.
Monod, T. 1924. Sur un type nouveau de Malacostracé: *Thermosbaena mirabilis*. *Bull. Soc. Zool. Fr.* **49**:58–68.
Por, F. D. 1962. Un nouveau Thermosbaenacé, *Monodella relicta* dans la depression de la Mer Morte. *Crustaceana.* **3**:304–310.
Sars, G. O. 1929. Description of remarkable cave crustacean. J.Fed.Malay. States. Mus. **14**:339–59.
Siewing, R. 1958. Anatomie und Histologie von *Thermosbaena mirabilis* ein Beitrag zur Phylogenie der Rieke Pancarida. *Akad. Wiss u. Liter. Abh. Math.-Naturw. Kl.* **1957**:197–270.
Stella, E. 1955. Behavior and development of *Monodella argentarii*, a thermosbaenacean from an Italian cave. *Verk. Int. Ver. Limnol.* **12**:464–66.
Stella, E. 1959. Ulteriori osservazioni sulla riproduzione e lo sviluppo di *Mondella argentarii*. *Rio. Biol. Perugia.* **51**:121–44.
Stock, J. H. 1976. A new genus and two new species of the crustacean order Thermosbaenacea from the West Indies. *Bijd. Dierk.* **46**:47–70.
Stock, J. H. 1982. The influence of hadziid Amphipoda on the occurrence and distribution of Thermosbaenacea and cyclopod Copepoda in the West Indies. *Pol. Arch. Hydrobiol.* **29**:275–82.
Stock, J. H., and G. Langley. 1981. The generic status and distribution of *Monodella texana*, the only known North American thermosbaenacean. *Proc. Biol. Soc. Wash.* **94**:569–78.
Taramelli, E. 1954. La posizione sistematica die Thermosbenacei quale risulta dall studio anatomico di *Monodella argentarii*. *Monit. Zool. Ital.* **62**:9–27.

18

EUPHAUSIACEA

DEFINITION   Carapace covers entire thorax, fused to underlying thoraco-meres; thoracopods in adults unspecialized as maxillipedes, biramous with setose flaplike exopods, tuffy epipodite gills not covered by carapace, the last one or two pairs frequently reduced; no special brooding structures, eggs generally shed free, young hatch as nauplii in the first of a long series of larval stages.

HISTORY   Euphausiacea were affiliated through most of the nineteenth century with Mysidacea in a taxon Schizopoda. Boas (1883) was the first to separate these groups, aligning euphausiaceans with the decapods. Dana (1852) erected separate genera for the euphausiid late larval stages, and it was the inimitable Claus (1884) who first recognized the larval nature of these forms. Though the group is limited in size, presently some 83 species, their great commercial importance in regard to the whaling and fishing industries has generally dictated that periodic exhaustive surveys be updated concerning their biology (e.g., Sars, 1885; Mauchline and Fisher, 1969; Lomakina, 1978). Potential development of krill fisheries for direct human consumption will continue to dictate strong interest in this ubiqui-tous group.

MORPHOLOGY   The carapace is fused to the thoracic segments and covers the entire thorax (Fig. 18-1). Laterally, however, the carapace does not extend beyond the leg bases, allowing the coxa and its epipodite gills (fre-quently referred to as podobranchs) to be exposed. The rostrum is typically short, sometimes almost nonexistent. The stalked compound eyes are the most prominent features of the head and in some taxa have distinct dorsal and ventral lobes. The cuticle is thin with little mineral or sclerotin inclu-sions. Lomakina (1978) reports a kinetic area in the cephalon, the eyes and both antennal segments forming a protocephalon that movably articulates with the rest of the head.

The antennules (Fig. 18-2A) are composed of a three-segment pedun-cle and two flagella. The antennules extend anteriorly, parallel to each other and are linked by interlocking setae on the basal peduncular seg-ment. In males, the base of the flagella may be somewhat inflated and more setose than in females. The antennules can exhibit considerable variation on this basic form and are important in delimiting species groups.

The antennae (Fig. 18-2B) possess single-segment peduncles, which

223

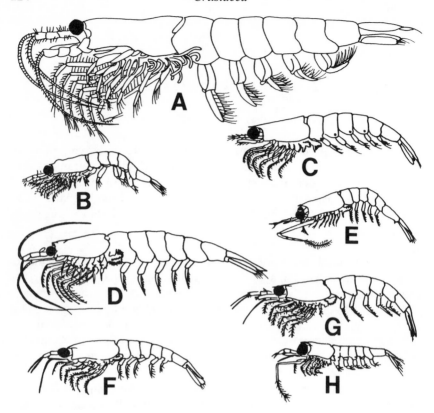

Fig. 18-1. Important species of Euphausiacea. (A) *Euphausia superba*; (B) *E. pacifica*; (C) *E. crystallorophias*; (D) *Meganyctiphanes norvegica*; (E) *Thysanoëssa macrura*; (F) *T. inermis*; (G) *T. raschii*; (H) *T. longipes*. (From Mauchline and Fisher, 1969)

seem to be formed from two segments. The ventral surface of these basal peduncles contain conelike openings for the antennal glands. There is a scalelike scaphocerite with variously developed setose inner margins. The flagellum has a base of three large peduncular segments.

The mandibles (Fig. 18-2C) have distinct but fused molar and incisor processes and a large three-segment palp. A lacinia mobilis is noted on the mandibles of the furcilia larval stage. The labrum is small and triangular (Fig. 18-2D); the paragnaths have a deep, finely setose cleft (Fig. 18-2E).

The maxillules (Fig. 18-2F) are foliaceous with two large setose endites. Sometimes a setose exite or pseudexite lobe arises from the basal endite. The single-segment lobelike endopodal palps (two-segmented in *Bentheuphausia* and in the larvae) are located distally.

The maxillae (Fig. 18-2G) are foliaceous and typically are built around a three-segment base, the second and third segments bearing bilobed setose endites. The distal segment bears a single endopodal, lobelike palp (three-segmented in *Bentheuphausia)* and in some species an exopodal lobe.

The thoracopods (Fig. 18-2H) generally conform to a basic pattern. The coxa laterally carries an epipodite gill; those of the first appendage are simple flaps but those on the following limbs are elaborate branchial structures comparable in development to the podobranchs of the decapodous groups. The exopod is composed of a peduncular base and a one-segment setose ramus (these can be missing in some species and sexes). The

Fig. 18-2. Appendages of euphausiaceans. (A) antennules, *Tessarabranchion oculatus;* (B) proximal elements of antenna, *Bentheuphausia amblyops;* (C) mandible, *B. amblyops;* (D) labrum, *B. amblyops;* (E) paragnaths, *B. amblyops;* (F) maxillule, *Euphausia krohnii;* (G) maxilla, *Thysanoëssa gregaria;* (H) first thoracopod, *E. krohnii;* (I) pleopod, *B. amblyops;* (J) telson and uropods, *B. amblyops;* (K) chela, *Stylocheiron abbreviatum;* (L) second thoracopod, *Nematoscelis megalops;* (M, N, O, P, Q, R) developmental sequence of petasma, *E. superba* (from Dzik and Jazdzewski, 1978); (S, T) ventral and dorsal views of the thelycum region. (Unless otherwise stated, all modified from Lomokina, 1978)

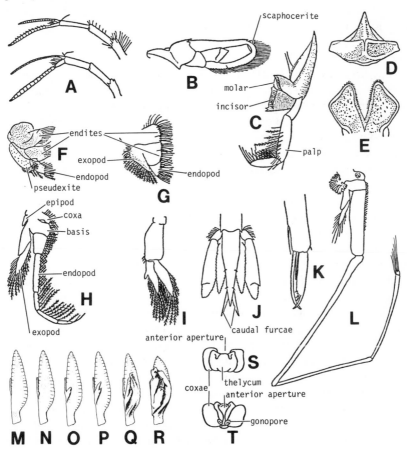

five-segment setose endopod can be variously developed. Though there are no maxillipedes as such in adult euphausiaceans, the anterior thoracopods can be quite specialized. The second thoracopods are very elongate in *Nematoscelis* (Fig. 18-2L) and frequently so in *Thysanoëssa*. The third thoracopods are well developed in *Nematobrachion* and *Stylocheiron* (this latter can also have the first and second thoracopods quite reduced), and *Tessarabrachion* has both the second and third greatly developed. The second and third thoracopods can also carry 'chelae' distally. These 'claws' (Fig. 18-2K) are formed by opposing the dactylus with a spine or group of spines projecting distally from the propodus. With the exception of *Bentheuphausia*, the eighth and sometimes the seventh thoracopods display varying degrees of reduction. The gonopores on the sixth appendage of the female open into a special seminal receptacle or thelycum (Fig. 18-2S, T) developed around the leg bases and sternites.

The pleopods (Fig. 18-2I) are generally biramous flaps. The first and second pleopods in the male are developed into a petasma (Fig. 18-2M, N, O, P, Q, R) wherein the rami are developed as lobes (some with secondary lobes, hooks, and spines) all folded up on each other and rather species-specific in form. The structures evident on the male petasma and the female thelycum are among the most complex copulatory morphologies seen in crustaceans.

The uropods (Fig. 18-2J) are elongate setose ovals, and only the exopods of *Bentheuphausia* have a diaeresis. The telson is a very elongate, subtriangular structure armed distally with a pair of caudal furcae or lobes.

The mouth is connected by a short esophagus to the stomach. This latter is composed of two parts: a large anterior cardiac portion and a smaller posterior pyloric region. In filter feeding forms, the walls of the stomach are highly muscular, the motion of which helps to macerate the food. Enzymes are released into the cardiac stomach and mixed with the food, and primary digestion takes place there. Subsequently food is forced through the setal filter in the pyloric stomach. In carnivorous forms like *Stylocheiron* and *Nematobrachion* the stomach is larger than in filterers, has a poor development of the filtering setae, and there are large chitinous denticles in the cardiac stomach to macerate food. The digestive caeca are compact bodies consisting of many short ramified diverticulae, similar to what is seen in decapods.

The heart is a small flattened organ in the posterior thoracic region. Two pairs of ostia connect it with the pericardial sinus. Several arteries (Fig. 18-3) issue forth from the heart to supply the head, liver, and posterior and ventral regions. Though the arterial system is rather complex, Lomakina (1978) reports that the venous circulation is only poorly understood. Blood appears to collect ultimately in a posteroventral cephalothoracic sinus and then flows to the pericardium (Fig. 18-4). The euphausian system has a number of unique features, namely, a short heart and a system generally with little trace of segmentation in the thoracic region. How-

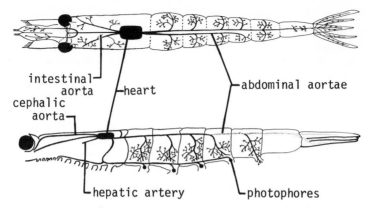

intestinal aorta
cephalic aorta
heart
abdominal aortae
hepatic artery
photophores

Figure 18-3. Arterial system of *Meganyctiphanes norvegica*; note asymmetry in regard to the intestinal and abdominal aortic vessels. (From Mauchline and Fisher, 1969)

ever, it bears resemblances to the general structure and position of the heart in decapods and has the same number of descending and abdominal arteries as found in mysids.

Excretion is achieved with antennal glands in the adults; however, a maxillary gland functions in the larvae.

The gonads in males may be either paired tubes, or, more frequently, a single horseshoe-shaped organ with several lobes along its length. The vasa deferentia can be expanded to form seminal vesicles and open either on the sternite of the last thoracomere or on the coxal base of the eighth thoracopods. The sperm are simple round cells and are generally deposited in a flattened, bottlelike spermatophore. The ovaries extend along the length of the thorax from just below the heart to above the caeca and are

Fig. 18-4. Venous sinuses of *M. norvegica*. (From Mauchline and Fisher, 1969)

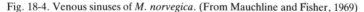

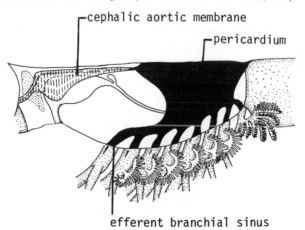

cephalic aortic membrane
pericardium
efferent branchial sinus

connected to each other anteriorly. In gravid females the ovaries swell and extend back into the anterior portions of the abdomen. The oviducts open onto the coxae of the sixth thoracopods, which are in turn generally associated with a thelycum formed by development of a sternal and two lateral coxal plates.

The nervous system has a generalized form. The circumesophageal connectives are long, and Lomakina (1978), in contrast to other reports, records that all the cephalic and thoracic ganglia are separate and distinct, not fused. The ganglia and connectives are generally divided as well.

The eyes in five of the 11 genera are multilobed (Fig. 18-5A); only in *Bentheuphausia* can they be said to be imperfectly developed. The form of the eyes is correlated with the presence or absence of enlarged thoracopods. Genera with normal, unspecialized thoracopods have single-lobed compound eyes (*Bentheuphausia*, *Euphausia*, *Meganyctiphanes*, *Nyctiphanes*, *Pseudeuphausia*, and *Thysanopoda*) while those with elongate and chelate thoracopods have multilobed eyes (*Nematobrachion*, *Nematoscelis*, *Stylocheiron*, *Tessarabrachion*, and *Thysanoëssa*). Greater or lesser development of thoracopods within a genus are marked by corresponding variations in the development of the eyes.

Photophores are located at several places on the body: on the eye stalks, the coxae of the second and seventh thoracopods, and the bases anterior pleopods. The structure of these organs is eyelike (Fig. 18-5B). The cells of the central striped body possess the bioluminescent ability. The light produced by these organs appears to be very important for schooling and possibly serves in sexual identification.

NATURAL HISTORY   The euphausiaceans are pelagic forms, generally spending their entire lives in the water column and, with the exception of a few species, are rarely seen on or near the bottom.

The pelagic habit constrains their modes of feeding, since they are in one way or another essentially omnivorous filterers either of phytoplankton, zooplankton, or detrital material from bottom sediments (Table 18-1). Euphausiaceans were thought to be typical maxillary feeders. In this regard, the adults were thought to differ little from the pattern first manifested in the calyptopis larval stage. Hamner et al. (1983) observed episodic filtration in *Euphausia superba* wherein the thoracopods formed a feeding basket. The setose endopods are thrown out anteriorly to envelop a mass of water and algae in a cage formed by the interlocking endopodal setae. As the legs are drawn under the body, the flaplike exopods flatten against the sides of the basket to effectively seal it. Algae are collected inside the basket as the endopods are flexed and water is expelled through the setal cage. The cycle is repeated at a rate of one to five times per second. When a food bolus of sufficient size is gathered, the action of the mouthparts transfers the bolus to the mandible palps. The palps press the food mass against the body of the jaws as ingestion begins. Animals were

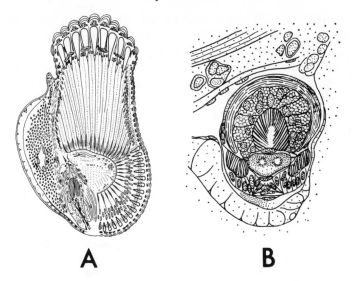

A   B

Fig. 18-5. (A) section of eye complex, *Stylocheiron suhmi;* (B) photophore, *Nematoscelis microps*. (From Mauchline and Fisher, 1969).

also seen to use the endopods as rakes, scraping the underside of ice masses in which algae had been frozen, gathering up bits of thawed material into the food basket.

Forms that feed on bottom sediment are still essentially filterers of suspended material. Individuals can either approach the bottom at a steep angle beating their pleopods to stir material off the bottom into suspension, or they plow into the bottom at a low angle with their antennae and then rise off the surface pulling the accumulated material into suspension. In both cases feeding actually occurs on the suspended sediment.

Table 18–1. Food preferences of euphausiacean genera. (Derived from Mauchline, 1967)

Genera	Detritus	Micro plankton	Macro plankton	Amphipods	Micro crustaceans
Bentheuphausia		+	+		+
Thysanopoda	+	+	+		+
Meganyctiphanes	+	+	+		+
Nyctiphanes	+	+	+		+
Pseudeuphausia	+				
Euphausia		+	+	+	+
Thysanoëssa	+	+	+	+	+
Nematoscelis		+			+
Nematobrachion		+			
Stylocheiron	+				

Some species prefer a more carnivorous diet. Although no active 'hunting' is known among euphausiaceans, macroplankton can be swept into the food basket by the action of the thoracic appendages. The animal is then pierced by the mandibular cusps and the stouter setae on the maxillules, and the soft internal parts are then sucked out before the prey remnant is discarded from the mouth region. The cardiac stomachs of species that prefer carnivorous diets have fewer setae and more lightly constructed stomach teeth for maceration than do the phytoplankton and detritus feeders.

Food passes from the muscular cardiac stomach, where digestion occurs, through the pyloric stomach, into the midgut and hindgut. Both peristaltic and retroperistaltic actions have been noted in the gut. Further digestion occurs in the posterior midgut and absorption occurs in the anterior midgut (Mauchline and Fisher, 1969). Little, if any, digested food is actually absorbed in the caeca, which are mainly secretory organs.

Much attention has been directed toward understanding euphausiacean reproduction and development because of the present and increasing economic importance of the group. Copulation occurs sometime before the females are ready to spawn. During copulation the male transfers the spermatophore to his petasma which then places the spermatophore into the female thelycum. Copulation can take place repeatedly such that several spermatophores can be in the thelycum. They are secured there by a cement produced in glandular walls of the vasa deferentia. Actual egg maturation in the female gonads takes place some weeks after copulation.

At spawning the eggs are usually freely released into the water, that is, in 57 of the 83 species; the other species attach the eggs as a mass to the setae of the posterior thoracopods where they hatch as nauplii. The freely broadcast eggs will frequently sink, and naupliar hatching occurs at a depth other than that in which the adults are found. The nauplii migrate to the surface. However, in the course of a succession of larval stages, the young euphausiaceans then make their way back to the depth characteristic of the adults of that species. Feeding by the larvae does not occur until the relatively advanced calyptopis stage is reached, and up to that time the larvae live off yolk reserves.

Though euphausiaceans are essentially natant forms, their distribution can be quite complex. Living in an open ocean situation, their bathymetric and biogeographic distribution is controlled by a variety of factors related to temperature, pressure, light, food availability, salinity, and currents. A variety of studies has been done on vertical distribution (see, e.g., Mauchline and Fisher, 1969) that reveals rather distinct vertical stratification as well as horizontal regionalization (Fig. 18-6). There thus appears to be a strict partitioning of resources in the sea to minimize competition between species. The vertical migrations evidenced in the course of larval development may be another aspect of this, that is, minimizing competition between adults and their developing young.

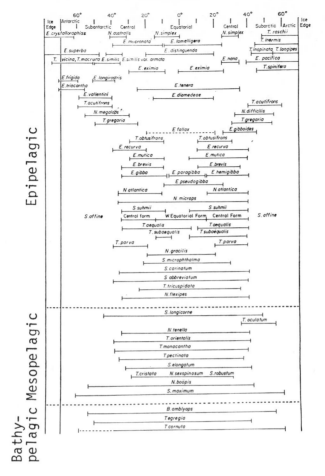

Fig. 18-6. Horizontal and vertical ranges of Pacific Ocean euphausiaceans. (From Mauchline and Fisher, 1969)

Euphausiacean biogeographic distribution presents an interesting study in contrasts. Over 50 of the 83 presently recognized species occur in all oceans, while there are only seven endemic Atlantic species, 17 Pacific endemics, and one unique to the Indian Ocean. However, species can be very restricted, for example, *Meganyctiphanes norvegica*, among the most intensively studied of forms, is reported by Abele (1982) to occur only in the North Atlantic north of the 15°C isotherm at 100 to 1500 m, and *Nyctiphanes couchii* is confined to the west coast of Europe from about 60°N extending south to North Africa and into the northern half of the Mediterranean Sea. Frequently, the spawning areas are more restricted than the total distributional area of a species; for example, *N. couchii* does not spawn north of the Skagerrak and, though it occurs in the southwestern

Baltic, it does not breed there. Distribution of many species clearly mirror temperature and current regimes (Fig. 18-7).

DEVELOPMENT   The most comprehensive review of euphausiacean development is by Taube (1909, 1915). The cleavage is total and in the first three divisions preserves traces of spiral cleavage, whereafter it takes on a secondary radial pattern. The presumptive areas in early ontogeny are not clear. A large blastocoel is formed during cleavage (Fig. 18-8B). By the 32-cell stage the 2 D macromeres (Fig. 18-8C) project into the blastocoel and temporarily cease dividing so that in the 62-cell (Fig. 18-8D) and later stages these presumptive endoderm cells become entirely incorporated into the interior. These cells then differentiate into the endoderm and the germ cells (the latter differentiate at a very early stage). The cells that border the blastopore differentiate into mesenchyme cells, which migrate inward (Fig. 18-8E) to fill the blastocoel and form the mesoderm (Fig. 18-8F).

The eggs are generally small and dense and, as they slowly sink after shedding, they develop quickly to the nauplius stage. These lack complete setation on the antennae necessary for feeding, a nauplius eye, and a fully developed gut. The larval sequence (Fig. 18-9) begins with two naupliar stages after which a metanauplius is passed through prior to a series of calyptopis (protozoea) larvae. The thoracic segments in the calyptopis stage are very short (except in *Bentheuphausia*), and feeding in the calypto-

Fig. 18-7.— Biogeographic distribution of *Euphausia brevis* (shaded) and *E. eximia* ●. (From Mauchline and Fisher, 1969)

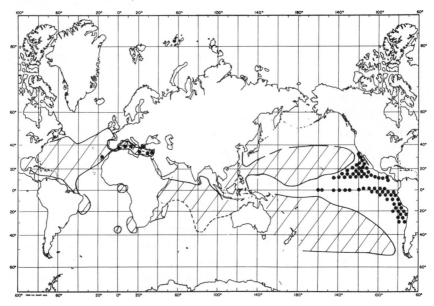

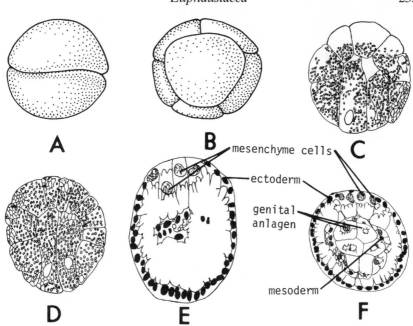

Fig. 18-8. Early development of euphausiaceans. (A) egg; (B) 8-cell stage; (C) 32-cell stage, early invagination; (D) 62-cell stage; (E) early mesoderm formation; (F) late gastrula. (Derived from Taube, 1909, 1915)

pis is maxillary, with the assistance of a maxillipede. Not until the first of six furcilia (zoea) stages are the post-first-thoracopod (maxillipede) appendages added. The development from this point on is directed toward a gradual unfolding of the adult condition.

An interesting aspect of larval development in this group is its individual variability (see. e.g., Mauchline and Fisher, 1969; Makarov, 1974). The major variations are in the furcilia stages, and the differences noted are essentially those arising from the possibility of multiple alternative pathways in the course of molting and the gradual development toward the adult condition. Variations in pattern can occur between simultaneous populations of a species at different localities and between essentially a single population lineage during successive parts of the breeding season.

FOSSIL RECORD   Although a number of authors have alluded to possible Carboniferous euphausiacean fossils (e.g., Dzik and Jazdzewski, 1978; or Schram, 1981), none of these suggestions are especially good ones. The genus *Peachocaris* is more likely a mysidacean, though forms such as *Anthracophausia* or *Essoidia* might be related to Euphausiacea.

TAXONOMY   The Euphausiacea are separable into two families. The monotypic Bentheuphausiidae is the more primitive in that it possesses developed

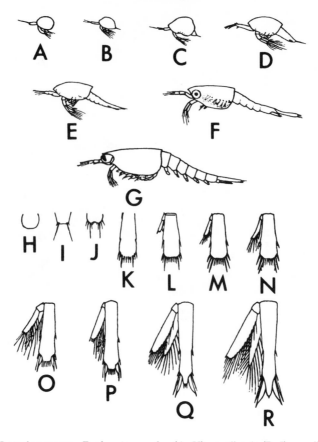

Fig. 18-9. Larval sequence, *Euphausia superba*. (A, H) nauplius 1; (B, I) nauplius 2; (C, J) metanauplius; (D) calyptopis 1; (E, K) calyptopis 2; (F, L) calyptopis 3; (G, M) furcilia 1; (N) furcilia 2; (O) furcilia 3; (P) furcilia 4; (Q) furcilia 5; (R) furcilia 6. (From Dzik and Jazdzewski, 1978)

seventh and eighth thoracopods and lacks a petasma and well-developed eyes. The Euphausiidae have typically reduced the posterior thoracomeres, and possess well-developed petasmae and eyes. Colosi (1917) divided the euphausiids into subfamilies (Thysanopodinae, Euphausiinae, Nematoscelinae, and Stylocheirinae). However, Zimmer and Grüner (1956) merely ranked the genera in an order that they felt approximated their affinities, and subsequent authors have followed this practice, such as Mauchline and Fisher (1969) and Lomakina (1978) (the latter's ranking is somewhat different than the others).

Thysanopoda
Meganyctiphanes
Nyctiphanes

Euphausia
Pseudeuphausia
Nematobrachion
Thysanoëssa
Tessarabrachion
Nematoscelis
Stylocheiron

REFERENCES

Abele, L. G. 1982. Crustacean biogeography. In *Biology of Crustacea*, vol. II, (L. G. Abele, ed.) , pp. 241-304. Academic Press, New York.

Boas, J. E. V. 1883. Studien über die Verwandtschaftsbeziehungen der Malakostraken. *Morph. Jahrb.* **8**:485–579.

Claus, C. 1884. Zur Kenntniss der Kreislaufsorgane der Schizopoden und Dekapoden. *Arb. Zool. Inst. Wien.* **5**:271–318.

Colosi, G. 1917. Crostacei II Euphausiacei. *Raccolte planct. fatta della R. Nave 'Liguria'* **2**:165–205.

Dana, J. D. 1852. Crustacea. *U.S. Expl. Exp.* **13**:1–685.

Dzik, J., and K. Jazdzewski. 1978. The euphausiid species of the Antarctic region. *Pol. Arch. Hydrobiol.* **25**:589–605.

Hamner, W. M., P. P. Hamner, S. W. Strand, and R. W. Gilmer. 1983. Behavior of Antarctic krill, *Euphausia superba*: chemoreception, feeding, schooling, and molting. *Science* **220**:433–5.

Lomakina, N. B. 1978. *Euphausiids of the World Oceans. Studies of the fauna of the USSR.* Acad. Sci. USSR, Leningrad.

Makarov, R. R. 1974. Dominance of larval forms in euphausiid ontogenesis. *Mar. Biol.* **27**:93–99.

Mauchline, J. 1967. Feeding appendages of the Euphausiacea. *J. Zool., Lond.* **153**:1-43.

Mauchline, J., and L. R. Fisher. 1969. The biology of Euphausiids. *Adv. Mar. Biol.* **7**:1–454.

Sars, G. O. 1885. Report on the Schizopoda collected by H.M.S. Challenger during the years 1873-1876. *Challenger Rept. Zool.* **13**:1–228.

Schram, F. R. 1981. On the classification of Eumalacostraca. *J. Crust. Biol.* **1**:1–10.

Taube, E. 1909. Beiträge zur Entwicklungsgeschicte der Euphausiden. I. Die Furchung des Eies bis zur Gastralation. *Zeit. wiss. Zool.* **92**:427–64.

Taube, E. 1915. Beiträge zur Entwicklungsgeschichte der Euphausiden. II. Von der Gastrala bis zum Furciliastadium. *Zeit. wiss. Zool.* **114**:577–656.

Zimmer, C. and H.-E. Gruner. 1956. Euphausiacea. In Dr. H. G. Bronns' *Klassen und Ordnungen des Tierreiches* **5**(1), **6**(3):1–286. Akad. Verlag, Leipzig.

19

AMPHIONIDACEA

DEFINITION   Carapace thin, fused to thorax, greatly developed laterally in females; first thoracopod as maxillipede, thoracopods two through seven essentially uniramous in adults, with exopods reduced to stumps, eighth thoracopod absent in females; first pleopod in female large, extending toward anterior of thorax, forming brood pouch under thorax with sides of carapace.

HISTORY   These animals were initially described under the generic name *Amphion* from forms that eventually proved to be larvae (Milne Edwards, 1832). It was not until Zimmer (1904) that the adults were recognized and described under the name *Amphionides*. However, Koeppel (1902) made the prescient deduction that *Amphionides* held its eggs in a thoracic brood chamber, something not seconded until Williamson (1973). Several species have been described, but all were synonymized into the single species *A. reynaudii* (see, e.g., Williamson, 1973). Originally placed in the carideans, a separate ordinal position was recognized by Williamson (1973).

MORPHOLOGY   These are small to modest-sized pelagic forms (Fig. 19-1). The carapace is very thin, with a fringe of setae along the margins. In males the carapace extends laterally to about the level of the leg bases, but in females the carapace proceeds ventrally to such a degree that it practically completely envelopes the thoracopods. The rostral spine is small, and frequently there is a postrostral spine. The eyes are stalked but weakly developed.

The antennular peduncle (Fig. 19-2A) is composed of three segments, lacks statocysts, and is distally equipped with two thin flagella.

The antennae have rather distinctive peduncles (Fig. 19-2B). The basal segment of the protopod is greatly inflated and laterally armed with brush setae. The scaphocerite is a huge round scale, with marginal brush setae. The two basal segments of the flagellum are peduncular but small.

The mandibles (Fig. 19-2C) are rather reduced and are typically formed as weak molar lobes, with no incisor lobe or palps. In females, the mandibles have been termed vestigial, over which the labrum and paragnaths form opposed simple convex and concave flaps, respectively.

The maxillules are vestigial (Fig. 19-2C). The maxillae (Fig. 19-2D) have a protopod with three enditic lobes without setae, a short single-joint palp, and a large scaphagnathite with marginal brush setae.

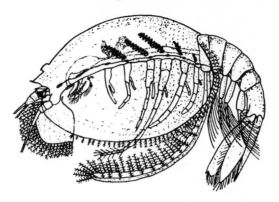

Fig. 19-1. *Amphionides reynaudii*, adult female. (From Heegaard, 1969)

The first thoracopods are developed as maxillipedes (Fig. 19-2E). The coxa bears a podobranch. The setose exopod is about twice as long as the endopod, has the proximal segment markedly inflated laterally, with the four distal segments variously reduced in length and width as one proceeds distally. The endopod is distally setose, with the five segments of the branch successively narrowing distally.

The posterior thoracopods (Figs. 19-2F, G, H, I) are markedly separated from the maxillipedes. The second thoracopod is very reduced, both exo- and endopods are short and indistinctly jointed. The third through seventh thoracopods are essentially similar in form, though different in size. The protopods are long, approaching half the length of the total appendage, with large coxae and very elongate bases. The adult exopods are vestigial, without setae, and apparently functionless. The segmented endopods bear three to five joints, and those of the fifth thoracopods are very long and terminally equipped with brush setae and peglike spines. Thoracopods three through seven bear pleurobranchs. The eighth thoracopod is absent in females, and uniramous in males.

In the female, the uniramous first pleopod (Fig. 19-2J) is very long and marginally armed with setae. The proximal one-fifth of the ramus is turned up along the margins to form a trough when it is extended forward under the thorax. The first pleopod in males is biramous (Fig. 19-2K) but with a reduced endopod. The posterior pleopods (Fig. 19-2L) are biramous, marginally setose, and bear a large appendix interna on each limb. The uropods (Fig. 19-2M) are oval flaps and marginally setose. The telson (Fig. 19-2M) is terminally pointed and not setose, except for a pair of terminal hairs.

The digestive system is moderately well developed in the male. In females the system appears to be vestigial, with no recognizable stomach and reduced glands and diverticula.

The heart (Fig. 19-3A) is large and is found in the posterior portion of

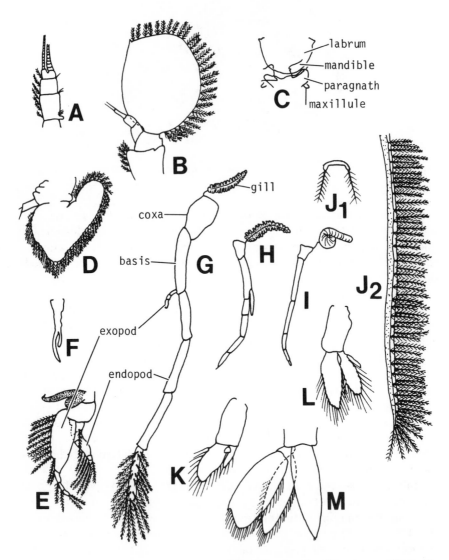

Fig. 19-2. Appendages of *Amphionides reynaudii*. (A) antennule; (B) antenna, (C) mouth region; (D) maxilla; (E) maxillipede; (F) second thoracopod; (G) fifth thoracopod; (H) sixth thoracopod of typical adult form; (I) eighth thoracopod of male; (J_1) crosssection and (J_2) lateral view of female first pleopod; (K) male first pleopod; (L) second pleopod; (M) uropods and telson. (From Heegaard, 1969)

238

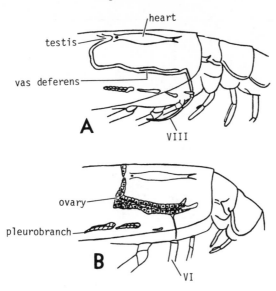

Fig. 19-3. (A) male reproductive system, (B) female reproductive system. (From Williamson, 1973)

the thorax. The testes are located middorsally (Fig. 19-3A) and the vasa deferentia open on the coxae of the eighth thoracopods. The ovary is an elongate structure in the posterior region of the thorax (Fig. 19-3B); the oviducts open on the coxae of the sixth thoracopods. Very little is known of the nervous and excretory systems, or body and appendage musculature.

NATURAL HISTORY *Amphionides* is pelagic. Most records of females are at about 2000 m or more (Williamson, personal communication), though a few have been taken near the surface. The larvae are common in the upper 30 m and at night concentrate in the 0 to 10 m range. Only three males have ever been noted, and Williamson (1973) suggests they are short-lived and not common or, being better swimmers, they may generally escape collecting nets.

The zoeal or amphion larvae and the males have well-developed mouthparts and guts. However, the reduced mouthparts and vestigial guts of the adult female (Fig. 19-1) indicate these do not feed but live off food reserves.

Nothing is known about modes of feeding, nor details of reproduction or locomotion. It is assumed the first pleopods and carapace act as a brood pouch, but females have never been collected with eggs in this structure. The setae of the pleopod and carapace margins appear to lock and serve to form a subthoracic brood chamber open at its anterior end.

A. reynaudii is a ubiquitous species, being found worldwide between 35°N to 35°S (Fig. 19-4).

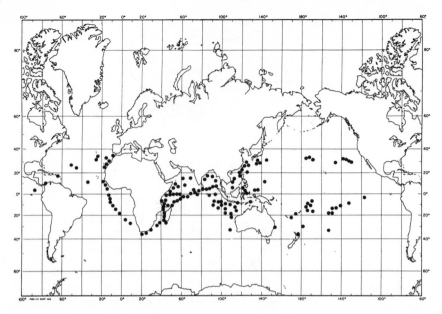

Fig. 19-4. Biogeographic distribution of *Amphionides reynaudii*.

DEVELOPMENT   *Amphionides* passes through a number of larval stages, termed zoeae by Williamson (1973) and mysis stages by Heegaard (1969). Each stage represents a typical gradual transition to the adult condition (Fig. 19-5). The amphion larvae have been compared to the zoeae of carideans (e.g., Gurney, 1936); however, they bear a reduced number of telson processes, have hepatic caeca, totally lack chelae and, thus, are not related to those natant decapods.

There is not total agreement on the number of stages in a complete developmental sequence. Heegaard (1969) recorded 13 stages and affords the most detailed descriptions of larval anatomy, but Williamson demonstrated that Heegaard's stage 12 and 13 are merely the female and male of the final zoeal stage. Gurney (1936, 1942) felt there were nine zoeal stages. Williamson (1973) suggests that the number of larval stages may in fact vary in different regions and in different individuals from the same region,

Fig. 19-5. Amphion larvae. (A) stage 1; (B) stage 5. (From Heegaard, 1969)

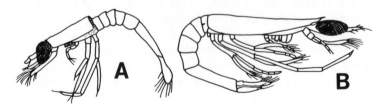

reminiscent to such variations seen in euphausiaceans. Clear sexual dimorphism is noted by Williamson in stage 9 onward in the Heegaard sequence.

TAXONOMY At various times, four different species have been ascribed to *Amphionides*. However, Gurney (1936), Heegaard (1969), and Williamson (1973) all agree that all these are synonyms of a single ubiquitous, albeit quite variable, species, *Amphionides reynaudii*.

REFERENCES

Gurney, R. 1936. Larvae of decapod Crustacea. Part II, Amphionidae. *'Discovery' Repts.* **12**:392–99.

Gurney, R. 1942. *Larvae of Decapod Crustacea.* Ray Society, London.

Heegaard, P. 1969. The larvae of decapod Crustacea. The Amphionidae. *'Dana' Repts.* **77**:1–82.

Koeppel, E. 1902. Beiträge zur Kenntnis der Gattung *Amphion. Arch. Naturgesch.* **68**:262–98.

Milne Edwards, H. 1832. Note sur un nouveau gerre de Crustacés de l'ordre des Stomatopodes. *Ann. Soc. Ent. France* **1**:336–40.

Williamson, D. I. 1973. *Amphionides reynaudii*, representative of a proposed new order of eucaridan Malacostraca. *Crustacean.* **25**:35–50.

Zimmer, C. 1904. *Amphionides valdiviae. Zool. Anz.* **28**:225-228.

20

DENDROBRANCHIATA

DEFINITION   Carapace dorsally fused to thorax, head shield free of anterior cephalon; protocephalon with ocular plate and subdivided epistome articulating with itself, epistomal bars anterior to labrum; thoracopods various, first two or three as maxillipedes, posterior five or six pediform and biramous, fourth through sixth chelate; gills dendrobranchiate, pleurobranchs appear in ontogeny after arthrobranchs; gastric mill, with strongly armed large median tooth, a series of lateral teeth; pleopods without appendix internae; pleura of first pleomere overlap those of second, pleomeres locked together with midlateral hinges hidden between third and fourth pleomeres but exposed on all others; female with thelycum, males with petasma; eggs typically shed free, developing into nauplii.

HISTORY   The taxonomic assignment of dendrobranchiates has been in flux (as has all the decapod eucarids). H. Milne Edwards (1834) treated them as part of the Macrura, and Boas (1880) placed them within his Natantia as the Peneidea. It was Bate (1888) who first coined the term when he wrote of 'macrura dendrobranchiata normalia' (as opposed to the 'aberantia,' that is, mysidaceans and euphausiaceans). Beurlen and Glaessner (1930) allied dendrobranchiates with stenopids and astacideans in the Trichelida. Gordon (1955) discussed the possibility of aligning euphausiaceans, dendrobranchiates, and carideans into a single group Natantia, apparently an opinion shared by Calman and Gurney to some degree. Mauchline and Fisher (1969) discussed the several points of similarities of dendrobranchiates with euphausiaceans (especially sergestoids) but maintained the separate status of both groups. However, Burkenroad (1963, 1981) opted to place these as a distinct and separate suborder of decapods, Dendrobranchiata, and this arrangement has been followed by Glaessner (1969), St. Laurent (1979, who prefers to use the term Penaeidea), and Felgenhauer and Abele (1983).

MORPHOLOGY   The detailed work of Young (1959) provides a survey of general dendrobranchiate morphology. The anterior head, or protocephalon, is free of the head shield and projects forward under the rostrum (Fig. 20-1). The protocephalon encompasses the eyes, both pairs of antennae, and the labrum. This region is movable or adjustable in relation to the fixed posterior cephalon. The compound eyes are well developed on movable stalks. The carapace has a weakly developed cervical groove; and the

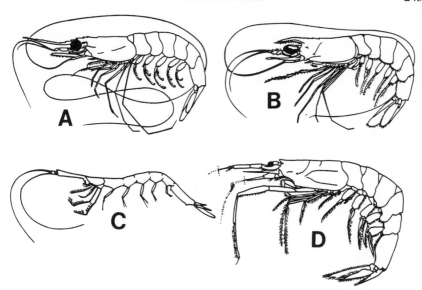

Fig. 20-1. (A) *Tanypeneus caribeus*, a *Xiphopeneus*-like peneid (from Pérez Farfante, 1972); (B) *Hymenopeneus robustus*, a solenocerine aristeid (from Pérez Farfante, 1977); (C) *Lucifer* sp., a sergestoid; (D) *Sergestes geminus*, a sergestid (from Judkins, 1978).

surface can be marked with various spines, a fine hairlike surface orna-ment, and a rostrum (which can be weakly or strongly developed, with denticulae).

The antennules (Fig. 20-2A) have a peduncle of three segments. The basal segment of the antennule is frequently notched to accommodate the eyes and bears the statocyst. The two flagella are modest in size and gener-ally circular in cross section, though in the benthic infaunal genus *Soleno-cera* the flagella are troughlike with a U-shaped cross section. The statocyst may contain either a sand grain or secreted statolith, depending on whether the species are, respectively, benthic or pelagic. Some sexual dimorphism is noted in the antennular flagella (Fig. 20-2B).

The antennae are prominent structures (Fig. 20-2C). The proximalmost segment is a small incomplete ring, while the second segment is large and heavily sclerotized. The basal segments of the flagellum are typically enlarged as a continuation of the peduncle. The flagellum itself is typically very long. The antennal scale or scaphocerite is large, flat, and ovoid.

The labrum (Fig. 20-2D) is modest in size and sculpted to accommodate the molar and incisor lobes of the mandible. The mandibles are distinctive structures (Fig. 20-2F) with well-developed molar and incisor surfaces, but the frequently large, two-segmented palp is distally a broad, flat, setose structure. The paragnaths are two symmetrical, simple, oval lobes (Fig. 20-2E).

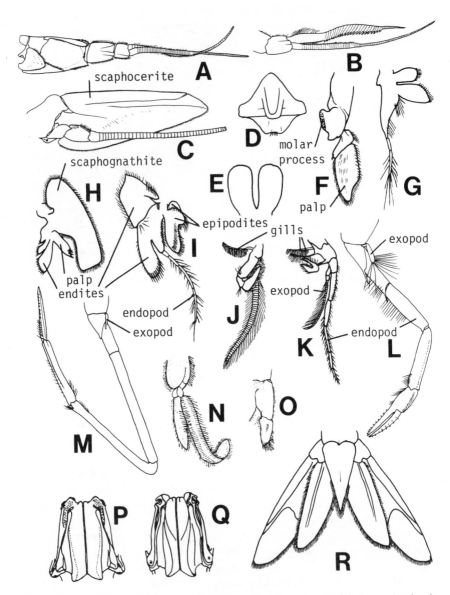

Fig. 20-2. Appendages of *Penaeus setiferus*. (A) antennule of female; (B) antennule of male; (C) antenna; (D) labrum; (E) paragnath; (F) mandible; (G) maxillule; (H) maxilla; (I) first maxillipede; (J) second maxillipede; (K) third maxillipede; (L) first pereiopod; (M) fourth pereiopod; (N) female pleopod; (O) appendix masculina; (P, Q) petasma; (R) telson and uropods.

The maxillules (Fig. 20-2G) have a large basal protopodal segment with two large, setose, endite lobes. The margins of the basal unit are equipped with arrays of simple and brush setae, and the distal endopod is also armed with brush setae.

The maxillae (Fig. 20-2H) have a number of endite lobes on the protopod to assist in food processing. The scaphognathite, or gill bailer, is directed laterally with short dorsal and longer ventral lobes and is marginally setose. This appendage is so extensively modified that to suggest homologies with the various components of other limbs is a questionable exercise.

The anterior maxillipedes, like the maxillae, are extensively modified. The first maxillipede (Fig. 20-2I) has a greatly inflated setose protopodal endite. In addition, two thin large epipodites overlap and extend laterally. The endopod is a thin palp. There may be a small rudimentary (?) arthrobranch near the base of the appendage (the use of the term arthrobranch here is merely a convention according to Burkenroad, 1981).

The second maxillipede (Fig. 20-2J) is somewhat pediform. The four-segment endopod is paddle-shaped. The flagellar exopod is very setose. A large podobranch arises laterally from the basis. The somite of this limb typically bears an arthrobranch and pleurobranch.

The third thoracopod (= maxillipede) (Fig. 20-2K) is typically very pediform and in many instances varies hardly at all from the pereiopods. The coxa carries the gills laterally. The exopod is flagellar and setose; the endopod is a modestly developed stenopod. One could consider this appendage, in those instances where it is little modified, merely the first of six pereiopods.

All the pereiopods (Fig.20- 2L, M) conform to a basic plan. The actual total number of gills per somite varies with family (see Table 20-1). All legs tend to arise along the ventral midline. The exopods are developed to various degrees or may be entirely absent (e.g., in *Sicyonia*). An ischiobasis marks the base of the endopod, and the first three pereiopods are chelate. The sternites of the sixth and frequently the seventh and eighth

Table 20–1. Gill formulas of dendrobranchiates illustrating maximal, intermediate, and impoverished variants. r = rudimentary, 0 = absent, e = epipodite, p = podobranch, a = arthrobranch, pl = pleurobranch. (From Burkenroad, 1981)

Maxillipede			Pereiopod				
1	2	3	1	2	3	4	5
e + p a?	e + p a pl	e + p a pl	e + p a pl	e + p a pl	e + p a pl	e + p a	pl	pl
1 + 0 r	1 + 1 1 1	1 + 1 2 1	1 + 1 2 1	1 + 1 2 1	1 + 1 2 1	1 + 0 2	1	1
1 + 0 0	1 + 1 1 1	0 + 0 2 1	1 + 0 2 1	1 + 0 2 1	0 + 0 2 1	0 + 0 1 + r 0	0
1 + 0 0	1 + 0 0 0	0 + 0 1 0	0 + 0 1 0	0 + 0 1 0	0 + 0 0 0	0 + 0 0	0	0

thoracomeres in the females are developed as a thelycum or seminal receptacle (Fig. 20-3A).

The pleopods (Fig. 20-2N, O) are generally biramous (specialized as uniramous in a few forms like *Sicyonia*). The endopod of the first pleopod in the male (Fig. 20-2P, Q) is developed as a petasma for sperm transfer. This can be quite elaborate and even asymmetrical from side to side (as in *Metapeneopsis*). The pleopodal appendixes are sometimes absent (Fig. 20-2M).

The pleura of the first abdominal segment overlap those of the second. The articular hinge-points on the sides of the abdominal segments are exposed to view on all but between the third and fourth segments, where they are hidden under the cuticle margin. The telson (Fig. 20-2R) is long and triangular, and the margins are sometimes marked by several pairs of movable spines, the most distal of which can be fused to the tip to provide a trifid form. The uropodal rami are long, thin, setose ovals (Fig. 20-2O).

A modest esophagus leads into the cardiac stomach. The gastric mill is centered about a large median tooth typically equipped with curved denticles along its margins and a battery of flanking lateral teeth (Soto, 1980; Felgenhauer and Abele, 1983). The large digestive caeca, or hepatopancreas, are filled with glandular secretory cells. The midgut is lined with columnar epithelium and serves primarily as an area of absorption.

The heart is a large structure in the posterodorsal median area of the thorax. Two pairs of dorsal ostia and one pair of lateral ostia drain the pericardium into the heart. The arterial system is rather elaborate (Fig. 20-4).

Fig. 20-3. *Hymenopeneus robustus*. (A) thelycum; (B) spermatophore attached to thelycum. (From Pérez Farfante, 1977)

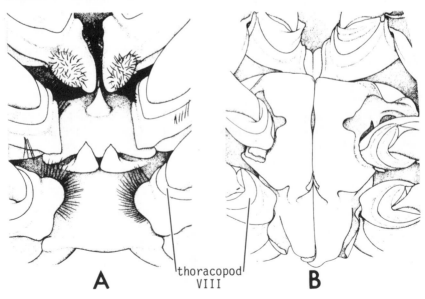

A

thoracopod
VIII

B

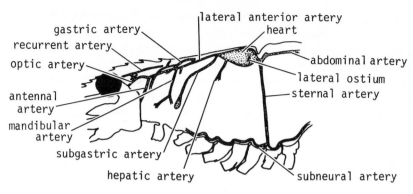

Fig. 20-4. The arterial circulation in *Penaeus setiferus*.

The principal anterior vessels are the paired lateral anterior arteries, which give rise to vessels supplying the sub- and supragastric regions, extensive branchiostegal capillaries into the carapace, the mouthparts, antennae, eyes, and brain. The hepatic arteries arise from the anteroventral portion of the heart. A single posterior dorsal aorta proceeds into the abdomen, which just posterior to the heart gives rise to a ventrally descending sternal artery, which in turn forms subneural arteries along the nerve cord. All the posteriorly arising vessels are clearly segmental in nature. The venous circulation is apparently open, the venous sinuses draining toward the pericardium.

Besides respiration by the thoracic gills, the rich blood supply of the branchiostegite would seem to indicate that the inner surface of the carapace serves to augment gas exchange.

Excretion is by antennal glands. These structures are rather large and diffuse, with a dorsal portion positioned above the supraesophageal ganglion and a ventral portion extending up into the peduncle of the antenna.

The testes are paired, multilobed, and lie dorsal to the digestive caeca. The vas deferens is marked by a terminal bulb near the gonopore that secretes the spermatophore (Fig. 20-3B). The sperm itself (Nath, 1942) is spherical and filled with a single large nucleus. The ovary has an anterior region that lies adjacent to the stomach and caeca, and a thin dorsal extension that parallels the midgut into the abdomen. The female gonopore opens on the third pereiopodal coxa just anterior of the thelycum.

The brain of *Penaeus* presents some primitive features. The tritocerebrum is subesophageal in position, the nerves to the antennae arising from the circumesophageal connectives. However, the ganglia of the mouthparts and anterior maxillipedes are fused into a single mass. The ganglia of the third maxillipedes and posterior appendages are developed as separate masses along a single ventral nerve cord. Thus, while the brain region tends to retain certain primitive features such as a distinct tritocerebrum, the loss of distinct pairs of ganglia and fusion of adjacent segmental elements in the nerve cord are advanced features. A large sixth abdominal ganglion lies posteriorly in the sixth pleomere, near the border with the telson, and four pairs of nerves arise from it. The anterior pair supply muscles, the second and third supply the uropods, and the fourth the telson. This is similar to a condition noted in tanaids and unlike a more typical abdominal arrangement where each ganglion has two pairs of nerves: the anterior into the pleopods and the posterior into the muscles. The terminal ends of the optic tracts are associated with neurohormonal organs, the sinus gland and X-organ. No experimental work has been done on these endocrine structures in dendrobranchiates, though it is assumed they have similar functions to those verified in higher decapods.

NATURAL HISTORY   For a group of such great economic importance some crucial aspects of the biology of the dendrobranchiates has attracted relatively little attention. Almost all the efforts in the literature have been directed toward systematics and the comparative gross anatomy of the thelyca and petasmae. Exacting details concerning the various modes of locomotion, feeding, and reproduction are lacking.

Dendrobranchiates occupy a diversity of habitats ranging from completely nektonic, for example, *Funchalia*, to epibenthic and almost completely benthic forms, for example, *Penaeus* and *Solenocera*. The Aristeinae and Solenocerinae are chiefly oceanic forms, while the Penaeinae and Sicyoninae are mostly littoral types.

Modes of feeding undoubtedly differ based on habitat preferences. For example, Williams (1965) records the bottom sediments preferred by *Penaeus duorarum* and their penchant for preferring to remain buried. Burkenroad (1939) observed *P. duorarum* to feed nocturnally by digging trenches in the bottom with an action of the pereiopods and flush the resultant trough by means of vigorous beating of the pleopods. The animal itself actually feeds on the benthic worms exposed to the trench from the surrounding sediment. On the other hand, the closely related *P. setiferus* appears to be a detritus feeder, though on occasion its gut is packed with the remains of pelecypods (Felgenhauer, personal communication). The more nektonic species have very setose maxillae and anterior maxillipedes and apparently are filter feeders, though detailed studies as to the exact method of filtration are lacking. *Solenocera*, another infaunal type that prefers to remain buried in bottom sediments, has a rather distinctive morphology to facilitate the movement of currents around the body (Burkenroad, 1934). The antennal flagellum has a trough along the dorsal surface. A channel is formed wherein a water current passes down the flagellar trough, through an atrium between the mandibular palps and antennal peduncles, into the gill chamber. Fluid movement is achieved by the beating of setae on the distal segment of the antennal peduncle. However, the resultant current appears to be largely respiratory in function rather than for feeding.

Though great attention has been paid by various workers to the times of egg brooding and to seasonal migrations of dendrobranchiate populations, little attention has been directed toward the study of modes of copulation. If such observations could be made, they would undoubtedly prove of importance in understanding the evolution of the group, since species distinctions are frequently based on structural variants in the thelyca and petasmae among forms that otherwise have few differences in gross body form. For example, intriguing issues need attention in considering peculiar morphologies like that seen in *Metapeneopsis*, with its asymmetrical petasma whose elements wind around each other to form a tight telescoping spiral on one side. Heldt (1938) noted that females with spermatophores can be found in any season, though they áre less numerous in winter and most numerous in summer; egg laying typically occurs at night.

Biogeographic distributions of dendrobranchiates, though available in the literature, have never been extracted and reviewed as a whole. The distribution of oceanic forms, like the aristeines and solenocerines, are frequently cosmopolitan, though not always so, and typically have irregular and patchy distributions within their ranges. Littoral groups tend to be more localized in their distribution. Burkenroad (1936) remarked on the general similarity of North American Atlantic and Pacific littoral peneid faunas to each other. The oceanic forms of the two oceans, however, are rather distinctive. Atlantic ocean forms off North America have a strong Indo-Pacific and southeast Atlantic component. Throughout the Tertiary the middle American seaway was largely one of shallow water that allowed a continuum of littoral species to evolve between the east and west coasts. The open-ocean and deeper-water American forms were effective in different oceans at opposite ends of the world, the Atlantic faunas being connected to the Pacific forms through an early Tethyan and subsequent Indo-Pacific/South Atlantic mediary.

DEVELOPMENT   Little has been done on the early embryology of dendrobranchiates. Brooks (1882) observed in *Lucifer* and Heldt (1938) noted in *P. setiferus* that the cleavage was total and (at least in the latter) the antennular anlagen appeared after the antennal and mandibular. However, a great deal is known about larval development after hatching (see, e.g., Brooks, 1882; Gurney, 1942; Heldt, 1938).

Like the euphausiaceans, the dendrobranchiates typically shed their eggs freely into the water (though there are some exceptions such as *Lucifer*, which broods them a short time on the posterior thoracopods). They also resemble euphausiaceans in that they have similar larval series (Table 20-2) that begin with a nauplius. Different terms are used between groups for the various larval stages (Fig. 20-5). However, this does not detract from the fact that equivalent stages in each group bear general resemblances to each other. Naupliar, metanaupliar, and protozoeal stages have antennal locomotion, zoeal stages have thoracopodal locomotion,

Table 20–2. Comparison of similarity of larval stages between the Euphausiacea and the superfamilies of Dendrobranchiata. (Modified from Gurney, 1942).

Euphausiacea	Penaeoidea	Sergestoidea
Nauplius	nauplius	nauplius
Metanauplius (2)	metanauplius (2–7)	?
Calyptopis (3)	protozoea (3)	'elaphocaris' (3)
Furcilia (3–5)	zoea (2–16)	'acanthosoma' (2)
Cyrtopia	postlarva	mastigopus

and the postlarval stages (Fig. 20-5F) finally arrive at pleopodal locomotion (Gurney, 1942).

Within the dendrobranchiates the actual number of stages within larval phases can vary depending on species. The nauplius (Fig. 20-5A) and the metanaupliar stages do not feed and are passed through relatively quickly, within 24 to 68 hours; and a stage is arrived at where the first two maxillipedes are developed, the telson is bifurcate, and the carapace bud is developed. The protozoeal (elaphocaris) stages (Fig. 20-5A, D) eventually achieve a separation of distinct pleomeres, development of the rostrum,

Fig. 20-5. Dendrobranchiate larval types. (A) nauplius, *Sicyonia*; (B) protozoea 2, *Gennadas*; (C) zoea, *Gennadas*; (D) elaphocaris 3, *Sergestes cornutus*; (E) acanthosoma 2, *S. atlanticus*; (F) mastigopus, *S. robustus*. (From Gurney, 1942)

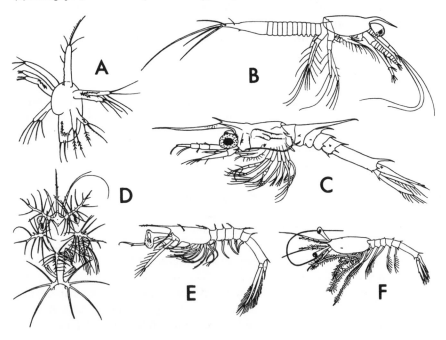

and the appearance of most (if not all) leg rudiments and the uropods. The zoeal (acanthosoma) stages (Fig. 20-5C, E) result in the appearance of all appendages and a tailfan resembling the adult form and the eventual functioning of the pleopods. The postlarval or juvenile condition is achieved when the pereiopodal exopods are reduced or lost and the abdominal appendages begin to function in locomotion; through a gradual series of molts the eventual adult condition is achieved.

Though the pattern of appendage development between the two dendrobranchiate superfamilies, the peneoids and sergestoids, are essentially identical, the gross morphologies of their forms are rather distinctive. The sergestoid larvae tend to be more spinous and are thus sometimes given in the literature the distinctive names elaphocaris (for protozoea) and acanthosoma (for zoea). (These were originally generic names.) In addition, the larval sergestoids are characterized by the appearance of the nonfunctional uropods in the elaphocaris stage before the pleopods, which do not appear until the acanthosoma.

FOSSIL .RECORD   The fossil record of dendrobranchiates is moderately good, mainly because of the presence of a wide array of types in the Jurassic Solenhofen Limestone of Germany and similar *Lagerstatt* deposits in the Upper Cretaceous of Lebanon (see Glaessner, 1969, for a recent summary). Though the preservation of fossils is exceptional in these beds, several other types of dendrobranchiate forms are not that well preserved and have been lumped (Balss, 1922) with catchall generic assignments to *Penaeus* or placed in the fossil genus *Antrimpos*.

All the Mesozoic dendrobranchiate material is in need of reexamination and revision, and until that occurs the relationships to the modern families and subfamilies cannot be clearly established. The genus *Aeger*, however, seems distinct enough to warrant a separate family (Burkenroad, 1963).

The diverse expression of dendrobranchiate types in Mesozoic faunas implies an early radiation of these forms. They undoubtedly took origin back in the Paleozoic, though no fossil dendrobranchiates have yet been recognized from that time period.

TAXONOMY   The family level relationships within the Dendrobranchiata have been in flux with our maturing understanding of the group. Burkenroad (1981) formally recognized superfamilies within the group and clearly defined family, subfamily, and tribal relationships with the superfamilies (Burkenroad, 1983). Pérez Farfante (1977) indicated a different family treatment but never formally defined any of her taxa. The Burkenroad scheme is followed here.

The peneoids are characterized by the possession of pleurobranchs on at least the second and third maxillipedes and generally a limited loss of gills on the pereiopods (many still retain three per somite) with at least 11 sets of gills in total. The sergestoids entirely lack pleurobranchs and have never more than two gills per somite with the total number of gills never more than seven pairs.

Much spleen has been vented over the spelling of the genus *Penaeus* vs. *Peneus*. Burkenroad (1983) was of the opinion that *Peneus* was proper and that the entry of *Penaeus* on the list of official generic names was improperly done. No matter how much this author believes Burkenroad is correct, by virtue of the action of the International Commission of Zoological Nomenclature, *Penaeus is* the official spelling. However, there is no reason why the vernacular forms like 'peneid' should not reflect the simpler spelling. In addition, an attempt some years ago to alter the spelling of all derivative names on the official list; for example, *Parapeneus* to *Parapenaeus* came to naught. And so the use of the simpler spelling for such names should be preferred and is used here, for example, *Metapeneopsis* rather than *Metapenaeopsis*.

Superfamily Penaeoidea Rafinesque, 1815 Triassic–Recent
    Family Aristeidea, Alcock, 1901 Cretaceous–Recent
        Subfamily Solenocerinae Wood–Mason and Alcock, 1891 Recent
        Subfamily Aristeinae Alcock, 1901 Cretaceous–Recent
            Tribe Aristeae Bouvier, 1908 Recent
            Tribe Benthysicymae Bouvier, 1908 Upper Cretaceous–Recent
    Family Penaeidae Bate, 1881 Jurassic–Recent
        Subfamily Penaeinae Burkenroad, 1934 Jurassic–Recent
            Tribe Penaeae Burkenroad, 1936 Jurassic–Recent
            Tribe Parapeneae Burkenroad, 1983 Recent
            Tribe Trachypeneae Burkenroad, 1983 Recent
        Subfamily Sicyoninae Ortmann, 1901 Upper Cretaceous–Recent
    Family Aegeridae Münster, 1839 Triassic–Jurassic
Superfamily Sergestoidea Dana, 1852 Recent
    Family Sergestidae Dana, 1852
        Subfamily Sicyonellinae Burkenroad, 1983
        Subfamily Sergestinae Bate, 1881
    Family Luciferidae Burkenroad, 1983

REFERENCES

Balss, H. 1922. Studien an fossilen Decapoden. *Paleontol. Zeit.* **5**:123–47.
Bate, C. S. 1888. Report on the Crustacea Macrura collected by the H.M.S. Challenger during the years 1873–1876. *Rept. Sci. Results Voyage H.M.S. Challenger. Zool.* **24**:1–942.
Beurlen, K., and M. F. Glaessner. 1930. Systematik der Crustacea Decapoda auf Stammesgeschichtlicher Grundlage. *Zool. Jahrb. Abt. f. Syst.* **60**:49–84.
Boas, J. E. V. 1880. Studier over Decapoderns Slaegtskabsforhold. *Vidensk. Selsk. Kristianaia, Skrifter* **(5)6**:25–210.

Brooks, W. K. 1882, *Lucifer*: a study in morphology. *Phil. Trans. Roy. Soc. Lond.* **173**:57–137, pls. 1–11.

Burkenroad, M. D. 1934. The Penaeidea of Louisiana with a discussion of their world relationships. *Bull. Amer. Mus. Nat. Hist.* **68**:61–143.

Burkenroad, M. D. 1936. The Aristeinae, Solenocerinae and pelagic Peneinae of the Bingham Oceanographic collection. Material for the revision of the oceanic Peneidae. *Bull. Bingham Oceanogr. Coll. New Haven* **5**(2):1–151.

Burkenroad, M. D. 1939. Further observations on Penaeidae of the northern Gulf of Mexico. *Bull. Bingham Oceanogr. Cruise* **6**:1–62.

Burkenroad, M. D. 1963. The evolution of the Eucarida in relation to the fossil record. *Tulane Studies in Geology* **2**(1):3–16.

Burkenroad, M. D. 1981. The higher taxonomy and evolution of Decapoda. *Trans. San Diego Soc. Nat. Hist.* **19**:251–68.

Burkenroad, M. D. 1983. Natural classification of Dendrobranchiata, with a key to recent genera. *Crust. Issues* **1**:279–90.

Felgenhauer, B. E., and L. G. Abele. 1983. Phylogenetic relationships among the shrimp-like decapods. *Crust. Issues* **1**:291–311.

Glaessner, M. F. 1969. Decapoda. In *Treatise on Invertebrate Paleontology*, Part R, *Arthropoda* **4**(2) (R. C. Moore, ed.), pp. R399–533. Geol. Soc. Am. and Univ. Kansas Press, Lawrence.

Gordon, I. 1955. Systematic position of the Euphausiacea. *Nature* **176**:1–934.

Gurney, R. 1942. *Larvae of Decapod Crustacea*. Ray Society, London.

Heldt, J. H. 1938. La reproduction chez les crustaces decapodes de la famille des Peneides. *Ann. Inst. Oceanogr.* **18**:31–206.

Judkins, D. C. 1978. Pelagic shrimps of the *Sergestes edwardsii* species group. *Smith. Contr. Zool.* **256**:1–34.

Mauchline, J., and L. R. Fisher. 1969. The biology of Euphausiids. *Adv. Mar. Biol.* **7**:1–454.

Milne Edwards, H. 1834. Histoire naturelle des Crustaces comprenant l'anatomie, la physiologie, et la classification de ces animaux, Vol. I. Paris.

Nath, V. 1942. The decapod sperm. *Trans. Nat. Inst. Sci. India* **2**:87–135.

Pérez Farfante, I. 1972. *Tanypeneus caribeus*, a new genus and species of the shrimp family Penaeidae from the Caribbean Sea. *Bull. Mar. Sci.* **22**:185–95.

Pérez Farfante, I. 1977. American solenocerid shrimp of the genera *Hymenopeneus*, *Haliporoides*, *Pleoticus*, *Hadropeneus* n. gen., and *Mesopeneus* n. gen. *Fishery Bull.* **75**:261–346.

Soto, L. A. 1980. Estudio morphologico del estomodeo de *Penaeus aztecus*. *An. Centro Cienc. Mar. Limnol. Univ. Natl. Auton. Mexico* **7**:119–50.

St. Laurent, M. de. 1979. Vers une nouvelle classification des Crustaces Decapodes Reptantia. *Bull. Office Nat. Pêches, Tunisie* **3**:15–31.

Williams, A. B. 1965. Marine decapod crustaceans of the Carolinas. *U.S. Fish and Wildl. Serv., Fish. Bull.* **65**:1–298.

Young, J. H. 1959. Morphology of the white shrimp *Penaeus setiferus*. *U.S. Fish and Wildl. Serv., Fish. Bull.* **59**:1–168.

21

PROCARIDIDEA

DEFINITION   Carapace covers entire thorax, fused to underlying thoraco-meres; protocephalon with ocular plate and subdivided epistome articulat-ing with itself; thoracopods biramous, with well-developed epipodites, first two thoracopods modified as maxillipedes, third through eighth pediform without chelae; gills phyllobranchiate; well-developed gastric mill, strong median tooth, large lateral teeth; second abdominal pleura overlap those of first and third.

HISTORY   First described from anchialine pools on Ascension Island in the South Atlantic (Chace and Manning, 1972), the only other known localities for these shrimp are the Hawaiian Islands on Maui and Hawaii (Holthuis, 1973; Schram and Felgenhauer, 1982) and anchialine caves in Bermuda (Manning, personal communication). All these authors placed *Procaris* within the carideans. Although phylogenetically interesting, the procarids have obtained only limited treatment in the literature. Provenzano (1978) made some observations on feeding behavior, and Felgenhauer and Abele (1983), Abele and Felgenhauer (1985) examined various aspects of their ecology, anatomy, especially in the protocephalon and foregut.

MORPHOLOGY   The carapace is extremely thin, and smooth, except for a weak cervical groove. With its rounded dorsal profile and short rostrum, it has a bulbous appearance (Fig. 21-1). The stalked eyes are reduced and

Fig. 21-1. *Procaris ascensionis*. (From Chace and Manning, 1972)

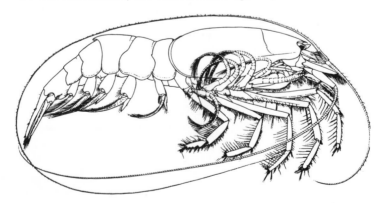

bear carotinoid pigment. It is assumed from the absence of ommatidia that *Procaris* is blind. The cuticle is thin, but in nature the animals have an orange iridescence. The Hawaiian species is highlighted by a blue phosphorescent gastric region.

The antennules (Fig. 21-2A) have short, broad, three-segmented peduncles. The stylocerite is large, pointed, and reaches to the distal end of the second segment. The flagella are long, the lateral one almost twice the length of the entire body.

The antennae (Fig. 21-2B) have short peduncles surmounted by an oval scaphocerite. The flagellum is long, with the distalmost of three peduncular joints longer than the proximal two.

The mandible (Fig. 21-2C) is massive, with a three-segment setose palp. The incisor process is scooplike, and the molar process is rather reduced and armed with small spherical tubules. The large labrum is rectangular, while the deeply cleft paragnaths (Fig. 21-2D) are distally attenuated, symmetrical, and sinuous.

The maxillules (Fig. 21-2E) have well-developed setose endites and a short single-segmented palp.

The maxillae (Fig. 21-2F) have well-developed, deeply cleft and heavily setose endites, and a distal slender palp. The scaphognathite is relatively small, with all margins setose.

The thoracopods resolve themselves into two groups. The first two are specialized as maxillipedes, and the next six as pediform pereiopods. The first maxillipede (Fig. 21-2G) has a broad rounded endite, a stout club-shaped endopod palp, an annulate exopod, and a simple epipodite.

The second maxillipede (Fig. 21-2H) is semipediform. The five-segmented endopod is short and stout, with the blunt rounded dactylus diagonally attached to the propodus. The exopod is large and annulate, while the epipod is a small simple lobe.

The rest of the thoracopods conform to a basic plan. The third thoracopod (= third maxillipede) differs from the other pereiopods (Fig. 21-2I, J) only in its being slightly more setose. All these posterior thoracopods have well-developed five-segmented endopods, large annulate exopods, and large lobate epipods. The fourth through seventh thoracopods are associated with a single phyllobranchiate pleurobranch (Table 21-1).

The pleopods are all basically alike (Fig. 21-2K). The endopods are short and only weakly developed, while the exopods are annulate and well developed. The pleopods lack any appendices internae.

The setose uropods (Fig. 21-2L) are rounded and shorter than the telson. The exopods have a set of lateral spines just anterior to a circular diaeresis. The telson (Fig. 21-2L) is subrectangular, the terminal margin is armed with four pairs of large spines, and the dorsal surface has four spiney processes. The second abdominal pleura overlap those of the adjacent segments.

Little is currently published concerning internal anatomy in *Procaris*

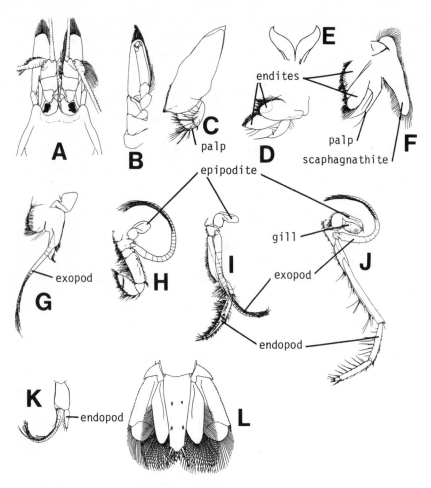

Fig. 21-2. Appendages of *Procaris ascensionis*. (A) anterior cephalon with antennules and eyes; (B) base of antenna; (C) mandible; (D) paragnaths; (E) maxillule; (F) maxilla; (G) first maxillipede; (H) second maxillipede; (I) third thoracopod; (J) fourth thoracopod; (K) first pleopod; (L) uropods and telson. (From Chace and Manning, 1972)

Table 21–1. Gill formula of *Procaris*. (From Holthuis, 1973, as modified by Burkenroad, 1984)

	Maxillipede			Pereiopod				
	1	2	3	1	2	3	4	5
Pleurobranchs	0	0	0	1	1	1	1	0
Arthrobranchs	0	0	0	0	0	0	0	0
Podobranchs	0	0	0	0	0	0	0	0
Epipodite	0	0	0	1	1	1	1	0

except for the foregut (Fig. 21-3). There is a well-developed median tooth equipped with a V-shaped array of accessory denticules along its margins. The lateral teeth are large and bilobed. The floor of the foregut is equipped with numerous denticles guarding the entrance to the pyloric stomach.

Details concerning the reproductive system are just emerging (Felgenhauer, personal communication). The ovaries extend into the abdomen. Eggs are about 0.3 mm in diameter and few in number (25–50 in an oviduct). The spermatophore is produced in the distal end of the vas deferens and appears to be a single-lobed structure rather than bilobed as in carideans.

NATURAL HISTORY   These peculiar shrimp live in rocky anchialine habitats very near the shoreline, which, though they receive freshwater runoff, have subterranean connections to the sea and respond to tidal fluctuations. Their native habitat is in caves and openings deep among the rocks where they exist in great numbers (Abele and Felgenhauer, 1985). These shrimp apparently only rarely wander near the surface. In the surface pools, the creatures scamper over rocks, seeming to constantly test the rock surface with the tips of their pereiopods, all the while not stopping their progress.

Abele and Felgenhauer (1985) reported segregated populations of *Procaris ascensionis* below the surface. Small individuals spend most of their time in crevices and among the alga *Valonia* and other filamentous algae. Large individuals spend a great deal of time swimming about in open water.

Fig. 21-3. Scanning electron micrograph of cardiac stomach of *Procaris ascensionis*. Note the prominent V-shaped median tooth and the bifid lateral teeth. (Courtesy of B. E. Felgenhauer)

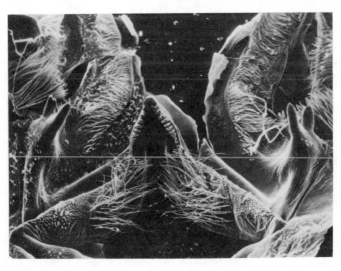

Swimming is achieved by metachronal beating of the thoracic exopods and pleopods (Abele and Felgenhauer, 1985). The animals move equally well right side up or upside down. Thoracic endopods are used in walking across substrates.

Abele and Felgenhauer (1985) report that of 15 individuals examined for gut contents, seven were filled only with diatoms and plant matter. The stomachs of seven other *Procaris* were filled exclusively with crustacean parts belonging to the amphipod *Melita* and an atyid caridean *Typhlatya rogersi*. One individual had both plant and animal remains in its stomach.

Feeding involved *Procaris* probing its habitat with extended thoracic endopods (Provenzano, 1978); mechano- and chemoreceptors on the pereiopods seem to be involved in gathering prey (Abele and Felgenhauer, 1985). Prey are pulled in by the endopods and grasped in a cage formed by the pereiopods on the underside of the thorax, and *Procaris* then moves off the bottom to swim upside down and consume the prey.

Nothing is known of reproductive biology. Sexes are not distinguished externally, nor do there appear to be any brooding adaptations. Abele and Felgenhauer (1985) record a sex ratio based on sectioned specimens of 45 to 1, females to males. It is suspected by almost all workers who have studied *Procaris* that the eggs may be shed free and that the larvae possibly develop planktonically.

DEVELOPMENT   Nothing is known of the ontogeny of *Procaris*, though, should this information become available, it may greatly help in elucidating the phylogenetic position and taxonomic relationships of procarids.

FOSSIL RECORD   No clearly procarid fossils have yet been recognized. Felgenhauer and Abele (1983) point out, however, certain similarities between *Procaris* and *Udorella agassizi* of the Upper Jurassic.

TAXONOMY   Two species have been recognized to date, *Procaris ascensionis* and *P. hawaiiana*. A third species from Bermuda is yet to be described. Originally placed within the Caridea on the basis of the anterior and posterior overlapping of the second abdominal pleura, the distinct cephalic, gastric, and thoracopodal anatomy of *Procaris* justifies its assignment to a separate taxon.

REFERENCES

Abele, L. G., and B. E. Felgenhauer. 1985. Observations on the ecology and feeding behavior of the anchialine shrimp *Procaris ascensionis*. *J. Crust. Biol.* **5**:15–24.
Burkenroad, M. D. 1984. A note on branchial formulae of Decapoda. *J. Crust. Biol.* **4**:277.
Chace, F. A., and R. B. Manning. 1972. Two new caridean shrimps, one representing a new family, from marine pools on Ascension Island. *Smith. Contr. Zool.* **131**:1–18.

Felgenhauer, B. E. and L. G. Abele. 1983. Phylogenetic relationships among shrimp-like decapods. *Crust. Issues* 1:291–311.

Holthuis, L. B. 1973. Caridean shrimps found in land-locked saltwater pools at four Indo-west Pacific localities, with description of one new genus and four new species. *Zool. Verh. Leiden* **128**:1–48.

Provenzano, A. J. 1978. Feeding behavior of the primitive shrimp, *Procaris. Crustaceana* **35**:170–6.

Schram, F. R., and B. E. Felgenhauer. 1982. Collecting *Procaris* in the Hawaiian Islands. *Environment Southwest* **498**:18–21.

22

CARIDEA

DEFINITION Carapace fused to thorax; thoracopods variously biramous or uniramous, first maxillipede usually with lateral setose expansion ('caridean lobe') on base of exopod, third maxillipede variable with three to five segments, pereiopods biramous (especially in primitive families), one or other of first two pereiopods predominantly chelate (sometimes sub- or achelate); gills phyllobranchiate, pleurobranchs appear before arthrobranchs in ontogeny; gastric mill variable or absent; pleopods typically with appendixes internae and masculinae; pleura of second pleomere usually overlap those of first and third, two anterior and two posterior pleomeres typically hinged with middle articulation lacking a hinge; eggs brooded on female pleopods, hatch as zoeae.

HISTORY The lack of clear unambiguous characters by which a taxon Caridea can be defined has been in part responsible for the rather chaotic history of this group. Very early workers simply lumped all long tailed (macrurous forms) together. Dana (1852) coined the term Caridea for the natant, nondendrobranch macrurans. Boas (1880) divided decapods into what he felt were natural groups, Natantia and Reptantia, and coined the term Eukyphotes for 'caridean' forms. Several subsequent German systematists and Burkenroad (1981) used variations of this term (e.g., Eucyphidea, Eucyphida, Eukyphida). Huxley (1878) preferred to divide decapods on the basis of gill type and placed carideans within his Phyllobranchiata. This was a system essentially followed by Bate, who treated carideans as a series of separate tribes. Recent authors have varied in their opinions and treatments. Beurlen and Glaessner (1930) placed carideans within their taxon Heterochelida. Burkenroad (1963) located carideans within his Pleocyemata, and this was followed by Glaessner (1969). However, de St. Laurent (1979) treated carideans coequal with peneids and reptants; and Burkenroad (1981) felt 'eukyphids' were coequal with euzygids (= stenopids), reptants, and dendrobranchs. Felgenhauer and Abele (1983a), who also removed the procarids from the carideans, felt that the taxon Caridea (they preferred the Dana term) was probably a taxonomic catchall and possibly polyphyletic. Thus, there appears to be a need for a taxon Eukyphida but not quite in the sense of Burkenroad (1981). Certainly the extremely heterogenous nature of this group makes it difficult to survey it at all concisely.

MORPHOLOGY   In dealing with such a diverse group as that which carideans represent (Fig. 22-1), it seems best to resolve anatomical observations around a representative primitive form and note some major variations on that plan. To this end we might consider (Holthuis, 1955) in this respect oplophorids or pasiphaeids as representing a possible array of generalized types (Fig. 22-1A).

The carapace bears a rostrum that is typically modestly developed. It is frequently rather large and serrate in some families, but in extremes it can be tiny, as in alpheids and atyids, or a broad deep keel, as in phycetocarids (Fig. 22-1B). A variety of body outlines are exhibited (Fig. 22-1), albeit sharing a basic 'shrimplike' form. The eyes are stalked and well developed (Fincham, 1984).

The antennules (Fig. 22-2A) have three-segmented peduncles. The basal segment typically bears a stylocerite. The flagella are moderately developed, though they can be extremely long and delicate structures.

The antennae (Fig. 22-2B) have a large protopod that bears a well-developed scaphocerite and a flagellum whose basal joints are peduncular in form.

The mandibles (Fig. 22-2C) may or may not have palps. Palps, when present, may range from one to three segments. The jaws may or may not

Fig. 22-1. Body forms seen in carideans. (A) *Leptochela bermudensis* (from Gurney, 1939); (B) *Physetocaris microphthalma*; (C) *Stylodactylus amarynthis*; (D) *Pterocaris typica*. (B,C, D from Holthuis, 1955)

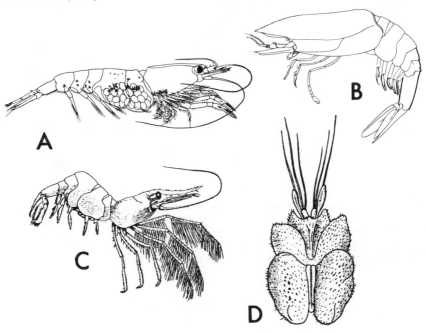

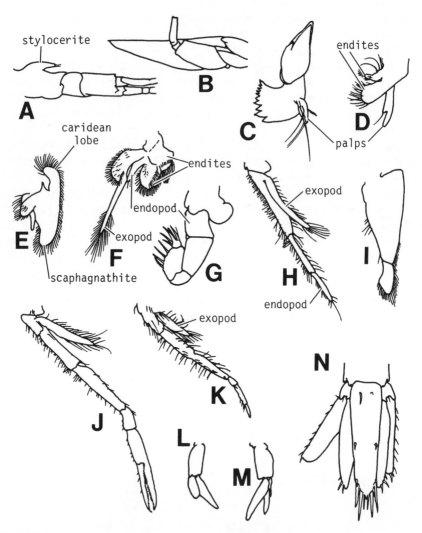

Fig. 22-2. Representative caridean appendages. (A) antennule; (B) antenna; (C) mandible; (D) maxillule; (E) maxilla; (F) first maxillipede; (G) second maxillipede; (H, I) third maxillipedes; (J) first pereiopod; (K) third pereiopod; (L) first pleopod; (M) second pleopod; (N) uropods and telson. (F, *Palaemon elegans* from Höglund, 1943; I, *Pasiphaea semispinosa*, from Holthuis, 1951; all others *Leptochela hawaiiensis*, from Chace, 1976)

have distinct molar and incisor processes (e.g., pasiphaeids have only a well-developed incisor process, while gnathophyllids possess only the molar process). If both processes are present, they may be either distinct or fused. Höglund (1943, pl. 2) illustrates processes on early stages of mandibles that resemble laciniae mobili, and Dahl and Hessler (1982) saw laciniae on the larvae of the pandalid *Dichelopandalus* and the crangonid *Pontophilus*.

The maxillules (Fig. 22-2D) are thin, reduced structures. They typically have one or two spinose endites arising from a small protopod and a small single-segmented palp. The palp may be simple or bear various arrangements of spines, hooks, or setae.

The maxillae (Fig. 22-2E) are dominated by the endites and scaphognathite. The setose endites are typically well developed, though they may be reduced as in the pasiphaeids or processids. The proximal endite may be missing in the adult, but the distal endite is typically deeply cleft. The palp is small and lobelike. The scaphognathite is well developed both anteriorly and posteriorly, the posterior aspect being the principal branch and is either broadly rounded or long and tapering.

The first maxillipedes (Fig. 22-2F) have the thin coxa and basis bearing broad endite lobes, though these may be barely developed in some families. The endopod is a simple lobe. The exopod is flagelliform, though not always annulate, and is typically marked proximally by a setose broadening or lobe ('caridean lobe'). Felgenhauer and Abele (1983a) point out, however, that in some families the lobe is absent, for example, pasiphaeids and some crangonids, in others it is only weakly developed, for example, some hippolytids and some alpheids, and the lobe has also been noted in some reptants, namely pagurids.

The second maxillipedes (Fig. 22-2G) have a five-segment endopod that is strongly curved to bring the propodus and dactylus in opposition to their counterparts across the mouth field. In this arrangement the dactylus comes to lie effectively posterior to the propodus, and the former frequently articulates on the median side of the latter. Stylodactylids have a peculiar double segment articulating side by side distally. The exopod is flagellate but is absent in pasiphaeids. The coxa can carry an epipodite and podobranch.

The third thoracopods (= third maxillipedes) (Fig. 22-2H) are somewhat pediform. The endopod typically has three segments, an ischiomerus, carpus, and propodus, with the dactylus absent or reduced to a terminal spine. The endopod can be modified away from a leglike form, for example, by reducing the distal joints and broadening the ischiomerus to a wide plate (Fig. 22-2I). An exopod is present and is frequently annulate. Epipods and a vestigial podobranch can also be found on this limb.

The posterior thoracopods or pereiopods exhibit wide ranges in form. A suite of families generally considered primitive (oplophoroids, most bresilioids, and pasiphaeids) possess exopods, and these can range from small flaps (atyids) to annulate flagelliform rami. The first two pereiopods (Fig. 22-2J) are frequently subequal and chelate. The first pereiopods are sometimes shortened and carried recurved over the mouth field, may have the chelae greatly reduced or absent (e.g., thalassocarcinids), or bear modified chelae (e.g., psalidopodids as scissors). Likewise the second pereiopods can have chelae reduced (e.g., hippolytids or alpheids), greatly enlarged (e.g., campylonotids), or absent (e.g., psalidopids). One group, the crangonids, do not have chelae at all but rather possess subchelae on the first two pereiopods.

The third through fifth pereiopods are ambulatory and usually achelate (Fig. 22-2K). However, these can be rather reduced in size, especially the last two pairs (e.g., pasiphaeids), may be very long and thin (e.g., nematocarcinids), or in one case can be chelate (the bresiliid *Pseudocheles*). Gills are phyllobranchiate in form, and like euzygids the arthrobranchs develop after the pleurobranch. Table 22-1 presents variant gill formulas within the group.

The pleopods (Fig. 22-2L and M) typically have appendixes internae on limbs 2 through 5. The males have an appendix masculina to facilitate spermatophore transfer on the second pleopod, which can on occasion be reduced.

The pleura of the second pleomere typically overlap those of the first and second. These pleura are variously developed, however (see, e.g., the alpheid *Pterocaris* as an extreme, Fig. 22-1D), and the anterior overlap is absent altogether in some psalidopodids and glyphocrangonids. Hinge points are present between all the pleomere boundaries, except between segments 3 and 4, though all segments are hinged in *Gylphocrangon*. The uropods may or may not have a diaeresis (Fig. 22-2N), and the telson is quite variable in form and ornament.

Little work has been published on internal anatomy of carideans. Felgenhauer and Abele (1983a and personal communication) have offered the most recent series of observations on the foregut. Unlike that seen in other groups (where cardiac structure is rather uniform within groups), the carideans exhibit great variety in this area. The pasiphaeids (Fig. 22-3A) have

Table 22–1. Gill formulas of carideans illustrating maximal, intermediate, and impoverished variants. r = rudimentary, 0 = absent, e = epipodite, p = podobranch, a = arthrobranch, pl = pleurobranch. (From Burkenroad, 1981, 1984)

Maxillipede			Pereiopod				
1	2	3	1	2	3	4	5
e + p a?	e + p a pl	e + p a pl	e + p a pl	e + p a pl	e + p a pl	e + p a pl	pl
1 + 0 0	1 + 1 0 0	1 + r 1 1	1 + 0 1 1	1 + 0 1 1	1 + 0 1 1	1 + 0 1 1	1
1 + 0 0	1 + 0 0 0	1 + 0 0 r	0 + 0 r 1	0 + 0 r 1	0 + 0 r 1	0 + 0 r 1	1
0 + 0 0	0 + 0 0 0	0 + 0 0 0	0 + 0 0 1	0 + 0 0 1	0 + 0 0 1	0 + 0 0 1	r

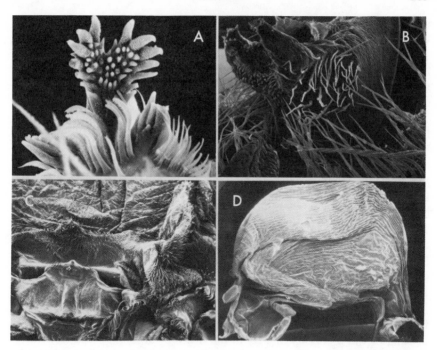

Fig. 22-3. Variations noted in cardiac stomach of carideans. (A) *Leptochela bermudensis* median tooth; (B) *Atya innocous*, bifid median tooth with denticles in saddle; (C) *Palaemonetes kadiakensis*, with reduced median tooth without associated lateral teeth; (D) *Saron marmoratus*, note lack of dentary ossicles. (From Felgenhauer and Abele, 1983a)

a unique arrangement with a tall, distinctively bifid tooth arising from the ventral inferior cardiac ossicle with a large number of scalelike denticles on its distal surface, a pair of massive lateral teeth, and has an elaborate pyloric valve. The atyids (Fig. 22-3B), stylodactylids, and rhyncocinetids have a bifid median tooth on the urocardiac ossicle with numerous denticles lining the saddle between the double tips, flanked by a series of stout lateral teeth. The pyloric stomach of some atyids has an elaborate convoluted membrane filling the lumen. The palaemonids (Fig. 22-3C), hippolytids, and thalassocarids have an absent or a very small median tooth with minute denticles, which is not flanked by lateral teeth. Other carideans (viz., gnathophyllids, alpheids, ogyridids, crangonids, glyphocrangonids, pandalids, processids, oplophorids, psalidopodids, campylonotids, eugonatonotids, bresiliids, and physetocarids) have no gastric mill in the cardiac stomach (Fig. 22-3D). Coombs and Allen (1978) suggest that forms without a gastric mill use swallowed sand grains to facilitate grinding. The caridean midgut can be very long, extending to the last somite of the body.

The excretory antennal gland in carideans has a bladderlike vesicle that sends off a series of lobate extensions into the body cavity. These can unite with their counterparts above the stomach to form a median vesicle.

The ventral nerve cord can exhibit considerable modification. Some forms fuse all the postoral cephalothoracic ganglia to form a single ganglionic mass. Though statocysts are the typical otic organs in decapods, some carideans like *Hippolyte* and *Pandalus* lack statocysts altogether.

The gonads are simple, paired tubular organs lying just dorsal to the foregut.

NATURAL HISTORY   Although sharing a generalized shrimplike body plan, the carideans, nonetheless, exhibit a remarkable degree of specialized variations in behavior and habitat preferences. This is in part the reason for the difficulty in trying to assess relationships within the group and without. There have been relatively few detailed biological studies of caridean types; rather work on the group has been essentially dominated by alpha taxonomic and ontogenetic efforts. Some notable exceptions have been studies of the biology of *Palaemon elegans* (Höglund, 1943; Forster, 1951) and *Atya innocous* (Felgenhauer and Abele, 1983b).

Carideans occupy a variety of habitats: pelagic, benthic, epibenthic, fresh, brackish, and marine situations. Swimming in adults is achieved by beating of the pleopods; the rapid tail-flicking caridoid escape reaction is reserved for emergency situations. Walking on a substrate is achieved by the ambulatory thoracopods. Forms that live on soft muddy bottoms commonly have the posterior pereiopods developed as long, thin, widely extended limbs to support them without sinking into the bottom, for example, *Nematocarcinus*. Other forms, such as crangonids, live buried in sandy bottoms and emerge only when feeding.

Feeding is largely directed toward carnivorous or scavenger habits. Like some euzygids, there are grooming species that make a living cleaning or grooming other life forms. Another series of specialized forms filter particulate matter out of the passing water. Among the most interesting in this regard are the atyids (Fryer, 1960, 1977; Felgenhauer and Abele, 1983b), who have specialized setose chelae that when open spread a filter-fan. These atyids collect diatoms, which become impacted in the convoluted membranes of the pyloric stomach. The animals actually feed on the encapsulating mucopolysaccarides on the outside of these diatoms (Felgenhauer, personal communication). The 'stripped' diatoms themselves are voided live in the feces and presumably allowed to regenerate their nutritious coats, thus eventually providing another meal upon which some other shrimp can 'graze.'

All carideans appear to secrete a chitinous peritrophic membrane in the anterior of the midgut that surrounds the material to be voided as feces (Forster, 1953). These membranes have been noted in other crustacean groups and apparently serve to protect the midgut lining from damage by undigestible matter in the feces.

Most forms have separate sexes. *Lysmata seticaudata* (Charniaux-Cotton, 1958) and species of *Pandalus* (Allen, 1959; Hoffman, 1969, 1972) are protandrous hermaphrodites. The onset of the female phase in *Pandalus* depends on when the male characters are repressed (Allen, 1959). There are primary females in which male characters are never allowed to appear; secondary females in which the male characters appear, but the male never functions as such; and hermaphroditic females, who develop after a functional male phase of the life cycle. Cold-water populations in high latitudes in Norwegian waters undergo these changes slowly, becoming mature males after three years and changing to females after six. Populations of *Pandalus* in the British North Sea mature in 18 months and metamorphose at 27 months, the entire life span being three years; and as many as 30% of one year's brood may be primary females. Some individual shrimp have been noted to possess oöcytes in the anterior part of the gonad and spermatocytes in the posterior portions. This protandrous phenomenon is also noted in the testes of other groups, such as the hippolytid *Thor*. Obviously, carideans are capable of some interesting and sophisticated strategies to maximize reproductive potential, and further studies along these lines would prove helpful in understanding the evolution of the group.

Mating has been observed in a number of forms (e.g., Höglund, 1943; Bauer, 1976; Felgenhauer and Abele, 1982). Mating occurs immediately or soon after the female molts. The male mounts and either crawls under the female on his back or flips her over onto her back. The spermatophore is placed on the endopods of the first pleopods and then brushed off onto the female sternites by the appendixes masculinae of the second pleopods. Further mountings by other males can occur, but the female typically holds her pleopods forward to prevent other spermatophores from being implaced on her last thoracic or sometimes first abdominal sternites. Felgenhauer and Abele (1982) report an elaborate courting precopulatory ritual for atyids, a phenomenon considered rare in carideans (Berg and Sandifer, 1984).

Spawning occurs within a few hours after mating. Höglund (1943) and Fisher and Clark (1983) reported that in *Palaemon* the female remains upright and still while the eggs pass out her gonopores and pass back to settle on the pleopods. These are affixed by an externally applied coating produced by special epithelial cells in the pleopodal protopod (Fisher and Clark, 1983). Ishikawa (1885) observed that *Atyephyra compressa* bends its abdomen forward into a 'fishhook' position, and Meyer (1934) reported *Crangon vulgaris* similarly bends the abdomen while lying on its side. Females in breeding state have extra setae on the posterior thoracic sternites to hold the spermatophore and direct the eggs as they exit the genital opening. The female possesses an extensive complex of setae on the pleopods to act as a brood chamber (Fig. 22-4) formed by the abdominal sternites dorsally, the pleopodal epimeres and basal setae laterally, the basal and endopodal setae ventrally, and the medially arranged fourth and fifth

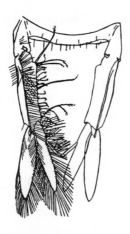

Fig. 22-4. Brood chamber of female *Palaemon elegans*, paired second pleomeres, only left appendage with setae in place. (From Höglund, 1943)

bases posteriorly. Before actual spawning and all through the brooding period, the female grooms the pleopods and egg masses using specialized setae on the thoracic appendages.

Carideans are especially devoted to cleaning and grooming the appendages, body surface, and gill chambers (e.g., Felgenhauer and Schram, 1978; Bauer, 1979) and, if prevented from doing so, become quickly fouled by a variety of organisms, mainly protistans (Felgenhauer and Schram, 1978). This seems to correspond with similar behaviors noted in other natant decapod types (Bauer, 1981), but whether there is a significance to this beyond immediate functional need requires further study.

Few overall studies have been made concerning caridean biogeography, although the basic data for such are available in the primary taxonomic literature. One group that has received some attention in this regard is the hypogean forms, both in continental habitats (Hobbs et al., 1977) and Indo-Pacific islands (Maciolek, 1983). These cave fauna types appear to be old, since their genera are for the most part unique to caves (Holthuis, 1956); at least one species, *Antecaridina lanensis* was suggested to be old enough so that its distribution is the possible result of vicariance through plate tectonics (Monod, 1975). However, island forms can also be dispersed as larvae or juveniles (Smith and Williams, 1981), and Maciolek (1983) argued that the isolated occurrences of many of these hypogean forms are probably only scattered and incomplete records of taxa that are actually widespread in both submerged and emergent isolated rock habitats throughout the Indo-Pacific tropics. Clearly, much useful information is to be garnered from the caridean literature that might not only lend insight into caridean history but also act as a guide to potential lines of productive field work.

DEVELOPMENT Early ontogeny has been examined in a variety of forms: germ layer differentiation in *Crangon vulgaris* (Weldon, 1892), teloblast

formation in *Heptacarpus rectirostris* (Oishi, 1959), and the complete ontogeny in *Palaemon idae* (Aiyer, 1949), *Caridina laevis* (Nair, 1949), and an
exceptionally fine work on *Palaemonetes varians* (Weygoldt, 1961).

The cleavage is superficial but equal (Fig. 22-5A, B). The divisions are
regular until about the 128-cell stage. At that point the yolk is surrounded
by a single-cell layer. Gastrulation begins with the proliferation of vitellophages into the yolk from a blastopore (Fig. 22-5C).

There follows the formation of a U-shaped germinal band (Fig. 22-5F)

Fig. 22-5. Ontogenetic development in carideans. (A) egg; (B) early cleavage; (C) initiation
of gastrulation; (D) proliferation of vitellophages; (E) late gastrulation, with mesendodermal
plate; (F) initiation of egg-nauplius on germinal disc, early teloblast formation; (G) lateral
view of caudal papilla development; (H) early differentiation of preantennal mesoderm and
stomodeum; (I) formation of anterior aorta; (J) teloblast formation complete, (K, L) cross
section through caudal papilla, (K) from proximal region near base of caudal furrow; (L) from
distal section near terminus of papilla. (A, B, C, D, E, G, H, I, K, L from Weygoldt, 1961; F,
J from Oishi, 1959)

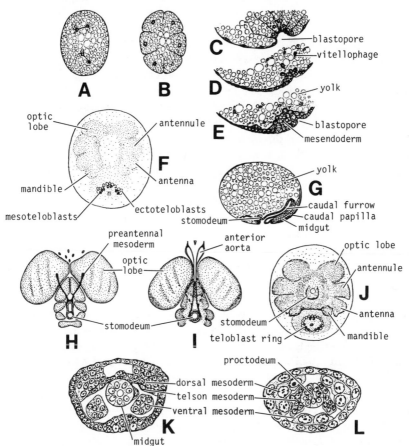

that extends toward the anterior portion of the embryo from the germinal disc. This differentiates into three paired regions: an anteriormost optical lobe, a middle section, and a ventral plate flanking the blastopore. The middle section then sequentially differentiates into the antennular, antennal, and mandibular anlagen. The ventral plate thickens around the blastopore and forms the mesendoderm (Fig. 22-5E).

The 19 ectoteloblasts form in a ring around the blastopore. The eight mesoteloblasts differentiate from the mesendoderm and form in a circle below the ectoteloblasts (Fig. 22-5F). At this point the caudal papilla is completed and flexes under the embryo, before beginning to bud off the metanaupliar segments (Fig. 22-5G). The teloblasts bud off all the postnaupliar segments except for the maxillary segments, which differentiate later from the 'unsegmented' region between the naupliar segments and the teloblasts.

The endoderm is represented by the vitellophages and the unpaired endodermal plate. The former wander about absorbing yolk and forming a yolk sac. The latter forms the midgut within the caudal papilla (Fig. 22-5G). The digestive caeca arise as paired outpocketings from the anterolateral portions of the yolk sac.

The preantennal mesoderm and stomodeum begin to form just as the naupliar anlagen have appeared. The former contribute to part of the wall of the stomodeum (Fig. 22-5H) and the anterior aorta (Fig. 22-5I). The naupliar mesoderm differentiates from the anterior border of the mesendodermal plate, but no coelom is formed in any of these segments. The trunk mesoderm is derived from the mesoteloblasts. These bud off blocks of tissue as the somites differentiate in the course of caudal papilla growth. The presumptive telson mesoderm cells are segregated early, as caudal papilla formation is initiated (Weygoldt, 1961) but do not differentiate until the end of development (Fig. 22-5K, L). This differentiation of presumptive telson mesoderm as teloblast formation is going on is important. The telson ectoderm is at least in part developed from the ectoteloblasts (Oishi, 1959). The derivation of telson tissues in carideans would second the notion (Chapter 1) that a telson is best considered as part of a particular kind of functional unit rather than dismissed as an 'unsegmented' part of the body.

The postontogenetic sequence for carideans is more or less typical for that seen in higher decapods. A series of zoeal stages follow upon hatching and in turn are succeeded by a short postlarval sequence before the definitive adult condition is achieved. The sequence (Fig. 22-5) for *Palaemon elegans* (Höglund, 1943) is typical, and Gurney (1942) and Modin and Cox (1965) reported ontogenetic series among carideans of anything from 4 to 11 stages. Gurney (1942) outlined general characteristics of caridean larval development. 1) They hatch with three pairs of functional maxillipedes and typically two pereiopods. 2) Legs typically appear in succession, though in some families (alpheids, hippolytids, and palaemonids) the appearance of leg 5 can precede that of legs 3 and 4. Exopods are present on all legs. 3) The rostrum is cylindrical or compressed laterally, never flattened horizontally. 4) The scaphocerite is typically jointed in stage 1, and the antennal endopod bears a long apical seta that may be fused to the endopod to form a spine. Other, but smaller, such setae and spines are possible. 5) The palp of the mandible rarely appears before the end of the larval sequence. 6) Maxillules typically lack an exopod, while the endopod usually has only two segments or is reduced. 7) Maxillary endopods are typically not segmented and are occasionally greatly reduced. The distal coxal endite is reduced or absent. 8) The first maxillipede has an enlarged and broadened coxa and basis and a small endopod. 9) The second and third maxillipedes have long endopods and frequently three apical setae in stage 1. 10) The telson has six spines on either side.

However, one must be cautious in making generalizations about caridean development. Boas (1880) and Gurney (1942) document the phenomenon of poecilogony in the group, that is, a single species can have different patterns of development depending on environmental factors. Gurney tends to minimize the actual possibility of this occurring and felt that much of what has been reported in the literature could be explained by differing patterns of development in sibling species. However, experiments like that of Edmondson (1929), where atyids hatched in the laboratory in running water emerge as postlarvae while those hatched in standing water emerge as zoeae, would seem to indicate that such reported differences could have adaptive value if manifested in the wild.

Along these same lines, carideans (along with other decapods) manifest temporary degeneration of appendages in the course of development. For example, *P. elegans* develops nonsetose exopods on the second and third maxillipedes, which only reappear in the final molt to the adult stage; or, as in *Atyaephyra*, the mouthparts may degenerate during the postlarval sequence. No clear explanation for such throwbacks has been put forward.

One final peculiarity of caridean development has great import for consideration of evolution within crustaceans. Bouvier (1925) proposed originally that in the atyid genera *Caridina* and *Atya*, which seem to represent a phylogenetic sequence, progeny of one female *Caridina* can contain individuals assignable taxonomically to the genus *Atya*. For example, Charmay (1920) found in broods of *Caridina richtersi* 1 out of 17 individuals were apparently assignable to *Atya edwardsi*. Though Edmondson (1929) largely explained away some of Bouvier's 'evolutionary mutations' as sexual and structural polymorphs within a single variable species, Gurney (1942) felt that it was still probable that *Atya* has evolved independently several times from *Caridina* in various parts of the world. It is also possible these polymorphs may represent sequential precopulatory and copulatory phases in hermaphroditic species akin to similar polymorphs seen in tanaids or amphipods.

FOSSIL RECORD   Glaessner (1969) presents the most recent review of caridean fossils and reveals what very few forms are recognized. Specimens assigned to *Caridina* occur in the Oligocene of France; material referred to as *Oplophorus* and *Notostomus* has been noted in the Jurassic and Cretaceous deposits of Germany and Lebanon; and various palaemonids have been described from the Oligocene and Miocene of Europe. All this material generally leaves much to be desired in regard to preservation and is generally not good enough to be effectively compared to modern forms. Several Mesozoic carideanlike things (*Blaculla* and *Udora* from the famous Jurassic Solenhofen beds and *Gampsurus* from the German Cretaceous) cannot even be assigned to family.

TAXONOMY   The basic suprafamilial taxonomy of this group now in use was ordered by Holthuis (1955), but there is not overall agreement as to details (Felgenhauer, personal communication). In addition, Holthuis himself pointed out his arrangement was not a natural one but was based largely on similarities in thoracopod form. Though he suggested a more natural system probably could be derived from consideration of mouthparts, gills, sexual features, and comparative larval studies, in all the intervening years no such analysis, preliminary to proposing such a system, has been yet published. There are also potential problems in classifying carideans at lower taxonomic levels. Some genera and species cannot be adequately separated on strictly anatomical grounds, and electrophoretic studies indicate that some common genera, such as *Palaemonetes*, may be paraphyletic or perhaps even polyphyletic (Boulton and Knott,1984). The currently accepted scheme of caridean families is given below.

Superfamily Oplophoroidea Dana, 1852 Jurassic–Recent
    Family Oplophoridae Dana, 1852 Jurassic–Recent
    Family Atyidae de Haan, 1849 Oligocene–Recent
    Family Nematocarcinidae Smith, 1884 Recent
Superfamily Stylodactyloidea Bate, 1888 Recent
    Family Stylodactylidae Bate, 1888
Superfamily Pasiphaeoidea Dana, 1852 Recent
    Family Pasiphaeidae Dana, 1852
Superfamily Bresilioidea Calman, 1896 Recent
    Family Bresiliidae Calman, 1896
    Family Disciadidae Rathbun, 1902
    Family Rhynchocinetidae Ortmann, 1890
Superfamily Palaemonoidea Rafinesque, 1815 Oligocene–Recent
    Family Palaemonidae Rafinesque, 1815 Oligocene–Recent
    Family Campylonotidae Solland, 1913 Recent
    Family Gnathophyllidae Dana, 1852 Recent
Superfamily Psalidopodoidea Wood–Mason and Alcock, 1892 Recent
    Family Psalidopodidae Wood–Mason and Alcock, 1892

Superfamily Alpheoidea Rafinesque, 1815 Recent
  Family Alpheidae Rafinesque, 1815
  Family Ogyridae Hayand Share, 1918
  Family Hippolytidae Dana, 1852
  Family Processidae Bate, 1888
Superfamily Pandaloidea Dana, 1852 Recent
  Family Pandalidae Dana, 1852
  Family Thalassocarididae Bate, 1888
Superfamily Physetocaridoidea Chase, 1840 Recent
  Family Physetocarididae Chase, 1940
Superfamily Crangonoidea H. Milne Edwards, 1837 Recent
  Family Crangonidae H. Milne Edwards, 1837
  Family Glyphocrangonidae Smith, 1884
Superfamily incerta sedis
  Family Udorellidae van Straelen, 1924 Upper Jurassic

REFERENCES

Aiyer, P. 1949. On the embryology of *Palaemon idae*. *Proc. Zool. Soc. Bengal* **2**:101–31.

Allen, J. A. 1959. On the biology of *Pandalus borealis* with references to a population of the Northumberland coast. *J. Mar. Biol. Assoc. U.K.* **38**:189–220.

Bauer, R. T. 1976. Mating behavior and spermatophore transfer in the shrimp *Heptacarpus pictus*. *J. Nat. Hist.* **10**:415–40.

Bauer, R. T. 1979. Antifouling adaptations of marine shrimp: gill cleaning mechanisms and grooming of brooded embryos. *Zool. J. Linn Soc. Lond.* **65**:281–303.

Bauer, R. T. 1981. Grooming behavior and morphology in the decapod Crustacea. *J. Crust. Biol.* **1**:153–73.

Berg, A. V., and Sandifer, P. A. 1984. Mating behavior in the grass shrimp *Palaemonetes pugio*. *J. Crust. Biol.* **4**:417–24.

Beurlen, K., and M. F. Glaessner. 1930. Systematik der Crustacea Decapoda auf stammesgeschichtlicher Grundlage. *Zool. Jahrb.* **60**:49–84.

Boas, J. E. V. 1880. Studier over Decapodernes Slaegtskabsforhold. *Vidensk. Selsk. Kristiania, Skrift.* **(5)6**:25–210.

Boulton, A. J., and B. Knott. 1984. Morphological and electrophoretic studies of Palaemonidae of the Perth region, Western Australia. *Aust. J. Mar. Freshw. Res.* **35**:769–83.

Bouvier, E. L. 1925. Les Macroures Marcheurs. *Mem. Mus. Comp. Zool.* **47**:401–72.

Burkenroad, M. D. 1963. The evolution of Eucarida in relationship to the fossil record. *Tulane Stud. Geol.* **2**:3–16.

Burkenroad, M. D. 1981. The higher taxonomy and evolution of Decapoda. *Trans. San Diego Soc. Nat. Hist.* **19**:251–68.

Burkenroad, M. D. 1984. A note on branchial formulae of Decapoda. *J. Crust. Biol.* **4**:277.

Chace, F. A. 1976. Shrimps of pasiphaeid genus *Leptochela* with descriptions of three new species. *Smith. Contr. Zool.* **222**:1–51.

Charmay, D. 1920. Observations sur les Caridines de l'Ile de Maurice, principalement sur le *Caridina richtersi*. *Bull. Mus. Paris* **1920**:473–6.

Charniaux–Cotton, H. 1958. La glande androgène de quelques Crustacés Décapodes et particulièrment de *Lysmata seticaudata* espèce à hermaphrodisme protérandrique fonctionnel. *C. R. Acad. Sci.* **246**:1665–68.

Coombs, E. G., and J. A. Allen. 1978. The functional morphology of the feeding appendages and gut of *Hippolyte varians*. *Zool. J. Linn. Soc. Lond.* **64**:261–82.

Dahl, E., and R. R. Hessler. 1982. The crustacean lacinia mobilis: a reconsideration of its origin, function, and phylogenetic implication. *Zool. J. Linn. Soc. Lond.* **74**:133–46.

Dana, J. D. 1852. Crustacea. *U.S. Exploring Expedition during the years 1838–1842 under the command of Charles Wilkes, U.S.N.* **13**:1–635.

Edmondson, C. H. 1929. Hawaiian Atyidae. *Bull. Bishop Mus.* **66**:1–36.

Felgenhauer, B. E. and L. G. Abele 1982. Aspects of the mating behavior in the tropical freshwater shrimp *Atya innocous*. *Biotropica* **14**:296–300.

Felgenhauer, B. E., and L. G. Abele. 1983a. Phylogenetic relationships among the shrimp-like decapods. *Crust. Issues* **1**:291–311.

Felgenhauer, B. E., and L. G. Abele. 1983b. Ultrastructure and functional morphology of feeding and associated appendages in the tropical fresh-water shrimp *Atya innocous* with notes on its ecology. *J. Crust. Biol.* **3**:336–63.

Felgenhauer, B. E., and F. R. Schram. 1978. Differential epibiont fouling in relation to grooming behavior in *Palaemonetes kadiakensis*. *Fieldiana: Zool.* **72**:83–100.

Fincham, A. A. 1984. Ontogeny and optics of the eyes of the common prawn *Palaemon serratus*. *Zool. J. Linn. Soc. Lond.* **81**:89–113.

Fisher, W. S., and W. H. Clark. 1983. Eggs of *Palaemon macrodactylus*: 1. Attachment to the pleopods and formation of the outer investment coat. *Biol. Bull.* **164**:189–200.

Forster, G. R. 1951. The biology of the common prawn *Leander serratus*. *J. Mar. Biol. Assoc. U.K.* **30**:333–60.

Forster, G. R. 1953. Peritrophic membranes in the Caridea. *J. Mar. Biol. Assc. U.K.* **32**:315–18.

Fryer, G. 1960. The feeding mechanism of some atyid prawns of the genus *Caridina*. *Trans. Roy. Soc. Edinb.* **64**:217–44.

Fryer, G. 1977. Studies on the functional morphology and ecology of the atyid prawns of Dominica. *Phil. Trans. Roy. Soc. Lond.* (B)**277**:57–129.

Glaessner, M. F. 1969. Decapoda. In *Treatise on Invertebrate Paleontology*, Part R, *Arthropoda* **4**(2) (R. C. Moore, ed.), pp. R399–533. Geol. Soc. Am. and Univ. Kansas Press, Lawrence.

Gurney, R. 1939. A new species of the decapod genus *Leptochela* from Bermuda. *Ann. Mag. Nat. Hist.* (11)**3**:426–33.

Gurney, R. 1942. Larvae of Decapod Crustacea. *Ray Society,* London.

Hobbs, H. H., Jr., H. H. Hobbs III, and M. A. Daniel. 1977. A review of the troglobitic decapod crustaceans of the Americas. *Smith. Contr. Zool.* **244**:1–183.

Hoffman, D. L. 1969. The development of androgenic glands of a protandric shrimp. *Biol. Bull.* **137**:286–96.

Hoffman, D. L. 1972. The development of the ovotestis and copulatory organs in a population of protandric shrimp, *Pandalus platyceros*, from Lopez Sound, Washington. *Biol. Bull.* **142**:251–70.

Höglund, H. 1943. On the biology and larval development of *Leander squilla* forma *typica*. *Svesk. Hydrog. Biol. Kommiss. Skrft.* n.s. **2**(6):3–44.

Holthuis, L. B. 1951. The caridean Crustacea of tropical West Africa. *Atlantide Rept.* **2**:7–187.

Holthuis, L. B. 1955. The Recent genera of the caridean and stenopodidean shrimps with keys for their determination. *Zool. Verh.* **26**:1–157.

Holthuis, L. B. 1956. An enumeration of the Crustacea Decapoda Natantia inhabiting subterranean waters. *Vie Milieu* **7**:43–76.

Huxley, T. H. 1878. On the classification and distribution of the crayfishes. *Proc. Zool. Soc. Lond.* **1878**:752–88.

Ishikawa, C. 1885. The development of *Atyaephyra compressa*. *Quart. J. Micro. Soc.* **25**:391–428.

Maciolek, J. A. 1983. Distribution and biology of Indo-Pacific insular hypogeal shrimp. *Bull. Mar. Sci.* **33**:606–18.

Meyer, P.-F. 1934. Ein Beitrag zur Eiablage der Nordseekrabbe *Crangon vulgaris*. *Zool. Anz.* **106**:145–57.

Modin, J. C., and K. W. Cox. 1965. Post-embryonic development of laboratory-reared ocean shrimp, *Pandalus jordani*. *Crustaceana* **13**:197–219.

Monod, T. 1975. Sur la distribution de quelques crustacés malacostraces d'eau douce ou saumâtre. *Mem. Mus. Natl. d'Hist. Nat. Zool.* (A)**88**:98–105.

Nair, K. B. 1949. The embryology of *Caridina laevis*. *Proc. Ind. Acad. Sci.* **29**:211–88.

Oishi, S. 1959. Studies on the teloblasts in the decapod embryo. 1. Origin of teloblasts in *Heptacarpus rectirostris*. *Embryologia* **4**:283–309.

Smith, M. J., and W. D. Williams. 1981. The occurrence of *Antecaridina lanensis* in the Solomon Islands—intriguing biogeographical problem. *Hydrobiol.* **85**:49–58.

St. Laurent, M. de. 1979. Vers une nouvelle classification des Crustacés Décapodes Reptantia. *Bull. Off. Nat. Pêch. Tunisie* **3**:15–31.

Weldon, W. F. R. 1892. The formation of the germ-layers in *Crangon vulgaris*. *Quart. J. Micro. Sci.* **33**:343–63, pls. 20–22.

Weygoldt, P. 1961. Beitrag zur Kenntnis der Ontogenie der Dekapoden: embryologische Untersuchungen an *Palaemonetes varians*. *Zool. Jahrb. Anat.* **79**:223–70.

23

EUZYGIDA

DEFINITION   Carapace fused to thorax; epistome in two parts, narrow anterior portion between antennae, posterior heavily armed with spines on perimeter of circular portion to which labrum attaches; posterior thoracopods (pereiopods) uniramous, fourth through sixth limbs chelate with sixth typically enlarged; gills trichobranchiate, pleurobranchs appear in ontogeny before arthrobranchs; gastric mill with a single knoblike median tooth arising from hastate plate, well-developed lateral teeth; pleopods without appendixes internae, first pair uniramous; pleuron of first pleomere overlaps the second, only three posterior pleomeres hinged by lock joints; eggs brooded on pleopods, hatched as zoeae (or later).

HISTORY   The first stenopid was described in 1811 as *Palaemon hispidus* Oliver, the common Indo-Pacific species. The Mediterranean form was initially described as *Byzenus scaber* Rafinesque, 1814, but this species was later described as *Stenopus spinosus* Risso, 1826. The name *Stenopus* Latreille, 1819, was first used for *S. hispidus*. Though the name *Byzenus* has priority, over 40 authors down to the present have used the name *Stenopus* while fewer than a half dozen authors have made little more than passing references to the name *Byzenus*. This state of affairs caused Holthuis (1946), as first revisor, to opt for continued use of the name *Stenopus*. Determining the systematic affinities of this group have been nearly as complicated. Milne Edwards, de Hann, and Boas treated them as peneids. Huxley (1878) created a separate family for them within his group Trichobranchiata. Bate (1888) treated them as a separate tribe, Stenopidea allied to palinurans and astacideans within Trichobranchiata Normalia. Most subsequent authors, including Calman have placed them as a distinct tribe within Macrura Natantia. Gurney (1924), however, felt that stenopids belonged among the Reptantia based on their larvae, and in this matter he was followed by de St. Laurent (1979) though on morphological grounds. Burkenroad (1981) separated this group as a separate infraorder Euzygida, though he conceded the possibility of some distant relationship to eukyphidans (= carideans). Felgenhauer and Abele (1982) essentially agreed with Burkenroad but preferred to use the name Stenopodidea.

MORPHOLOGY   The carapace and abdominal segments are typically rather inflated or glabrous (Fig. 23-1A) and often possess densely spinulate surfaces (Fig. 23-1B). The rostrum is moderate to large in form. The eyes are

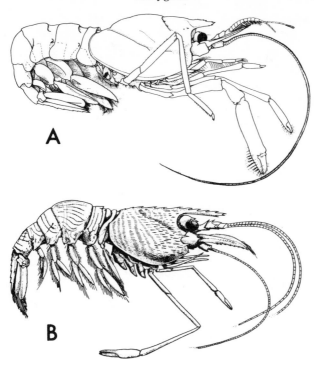

Fig. 23-1. (A) *Spongiocaris semiteres* (from Bruce and Baba, 1973); (B) *Odontozona sculpti-caudata* (from Holthuis, 1947).

stalked and usually well developed (but the corneas can be reduced, as in *Spongiculoides*).

The antennules (Fig. 23-2A) have three peduncular segments. The most proximal segment has developed either a pointed stylocerite or a blunt rounded lobe on the lateral surface. The flagella can be equipped with aesthetascs.

The antennae (Fig. 23-2B) have massive and typically spinose basal peduncular segments. The scaphocerite, especially in *Stenopus* and related genera, is long, slender, and laterally serrate; while other genera have the antennal scale rather broad and rounded. The anterior and median margins are densely setose.

The mandibles (Fig. 23-2C) have the incisor and molar processes fused. The three-segment palp is strongly recurved allowing the terminal joint to lay under the incisor portion of the mandibular gnathobase.

The maxillules (Fig. 23-2D) have two endites. The proximal lobe is setose while the distal one is rather more spinose. The single segment palp is small and slender.

The maxillae (Fig. 23-2E) have deeply cleft endites that bear terminal setae. The slender endopodal palp tapers to its terminus. The scaphognathite

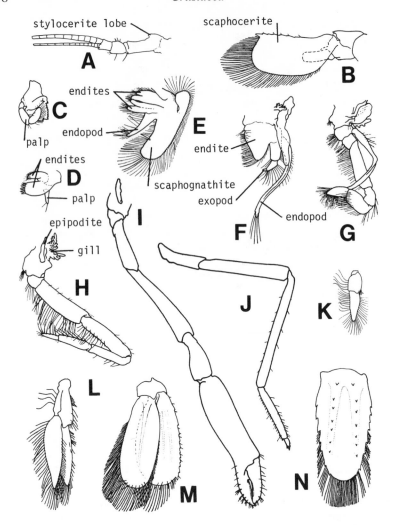

Fig. 23-2. Appendages of *Spongiocaris semiteres*. (A) antennule; (B) antenna; (C) mandible; (D) maxillule; (E) maxilla; (F) first maxillipede; (G) second maxillipede; (H) third thoracopod; (I) third pereiopod; (J) fifth pereiopod; (K) first pleopod; (L) second pleopod; (M) uropod; (N) telson. (From Bruce and Baba, 1973)

is anteriorly broad and bears long plumose setae along all the margins; the posterior lobe is small.

The first two thoracopods are extensively modified as maxillipedes. The first pair (Fig. 23-2F) have broad setose endites on the basis; the coxal endites are reduced. The endopod is a three-segment palp. The exopod is moderately long and whiplike but is typically undivided and without a 'caridean lobe.' The epipodite is moderately developed.

The second maxillipedes (Fig. 23-2G) have the medial margins of the coxa and basis setose. The five-segment endopod is strongly reflexed medially at the carpus–propodus joint. The exopod is long and setose but, like that on the first maxillipede, is not annulate. There is a small, simple epipod and podobranch.

The third thoracopods (Fig. 23-2H) are usually seven-segmented and termed the third maxillipedes. However, they are distinctly pediform and differ from the more posterior thoracopods only in being somewhat more setose medially. An exopod is present on *Stenopus* and related forms, while it is absent on other genera.

The pereiopods are uniramous. The first through third are chelate (Fig. 23-2I), with the third being the largest in a progressive series. The fourth and fifth pereiopods (Fig. 23-2J) are ambulatory, and these are sometimes quite long and slender. The gills per somite are given in Table 23-1.

The first pair of pleopods (Fig. 23-2K) are uniramous, while more posterior pairs (Fig. 23-2L) are biramous. None bear any appendixes internae. In *Stenopus* the third pleopods are the largest in the series, while the fourth and fifth pairs are progressively smaller.

The abdominal pleura all overlap those of the next most posterior segment (Fig. 23-1). The three anterior pleomeres are loosely bound together by intersegmental arthrodial membranes only, but locking joints are present between the fourth, fifth, and sixth segments. The uropods (Fig. 23-2M) lack a diaeresis. The exopod is usually laterally serrate, while the other margins of the rami are usually setose. The telson (Fig. 23-2N) ranges in outline from quadrangular, through rounded, to subtriangular, and bears marginal setae.

Sexual dimorphism is weakly developed in euzygidans. Sexes exhibit only slight differences in the sizes of the first pleopod and joints of the thoracopods, and there is some slight dimorphism in rostral and optic notch development on the carapace.

Little is recorded concerning the internal anatomy of stenopids. Some data are available concerning the anatomy of the gastric mill (Felgenhauer and Abele, 1982). The cardiac stomach is equipped dorsally with a

Table 23–1. Gill formulas of euzygids illustrating maximal, intermediate, and impoverished variants. r = rudimentary, 0 = absent, e = epipodite, p = podobranch, a = arthrobranch, pl = pleurobranch. (From Holthuis, 1946; and Burkenroad, 1981)

Maxillipede			Pereiopod				
1	2	3	1	2	3	4	5
e + p a?	e + p a pl	e + p a pl	e + p a pl	e + p a pl	e + p a pl	e + p a pl	pl
1 + 0 1	1 + 1 1 1	1 + 0 2 1	1 + 0 2 1	1 + 0 2 1	1 + 0 2 1	1 + 0 2 1	1
1 + 0 r	1 + r r 0	1 + 0 1 1	0 + 0 1 1	0 + 0 1 1	0 + 0 1 1	0 + 0 1 1	0
1 + 0 0	1 + r 0 0	1 + 0 1 1	0 + 0 1 1	0 + 0 1 1	0 + 0 1 1	0 + 0 r 1	0

prominent median tooth mounted on a hastate plate and decorated with knoblike denticular processes. This is flanked by a series of peglike lateral teeth (Fig. 23-3).

NATURAL HISTORY   Euzygids prefer substrates and habitats that allow them to be rather cryptic. They frequent coral reefs and rocky bottoms, and several species live entrapped for life within hexactinellid sponges. As a group they apparently prefer shallow waters, since only a few species have been collected below 1000 m.

They appear to feed as scavengers. Members of the genus *Stenopus* are best known for their cleaning of coral reef inhabiting fish. These crustaceans occupy a station to which the fish periodically come to be serviced. Those euzygids that live in glass sponges seem to feed on macroparticles that are pumped through the sponge.

Studies have yet to be made of stenopid locomotion despite the potential phylogenetic importance of such. These creatures are largely reptant in habit but seem to prefer hiding rather than putting themselves in situations where they would be required to retreat in haste. In this regard, the effect of the characteristic loose intersegmental articulations between the anterior pleomeres on the efficiency of their caridoid escape reaction has yet to be assessed.

Euzygids are restricted to tropical and warm temperate waters. Their northernmost locations (Sagami Bay in the Pacific and off Iceland and Ireland in the Atlantic) occur interestingly in areas under some influence from warm water currents arising in tropical latitudes. (The northernmost occurrence, *Spongicoloides profundus* off Iceland, is also the deepest record for this group as well, 1480 m).

There are essentially only three groups within the Euzygida that have wide distributions. These are certain species within the genera *Stenopus*, the genus *Spongiculoides*, and *Microprosthema*. *Stenopus hispidus* is virtually a ubiquitous tropical form (Fig. 23-4) that occurs across the Indian Ocean extending up into the Red Sea and as far south along the African coast as Durban. It ranges through the Indonesian archipelago, across the Pacific to the Hawaiian chain, and is recorded from Formosa and Sagami Bay in the north and Port Jackson in the south. It also is found throughout the east Atlantic and Caribbean tropics. Other species within the genus *Stenopus* are very restricted in their distributions: *S. tenuirostris* being known only from a couple of sites in Indonesian waters, *S. spinosus* from the Mediteranean and northern Red Seas, and *S. scutellatus* from the West Indies and Caribbean. The supposed wide distribution of *Odontozona spongicola* (off India and California) is not verified. The genus *Spongiculoides* is worldwide in distribution, but individual species are rather restricted in occurrence (see, e.g., Goy, 1980; Baba, 1983).

Two species of *Microprosthema* have modestly wide-ranging distributions: *M. semilaeve* is known from localities throughout the West Indies

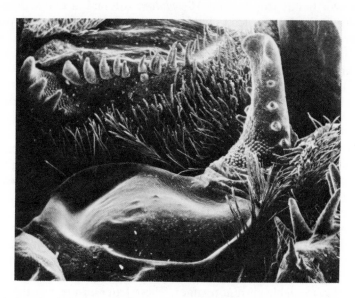

Fig. 23-3. Gastric mill of *Stenopus hispidus*. Note large median tooth with knoblike denticles and peglike lateral teeth. (From Felgenhauer and Abele, 1983)

Fig. 23-4. Some biogeographic distributions of selected stenopodid euzygidans, illustrating some patterns typical for the group.

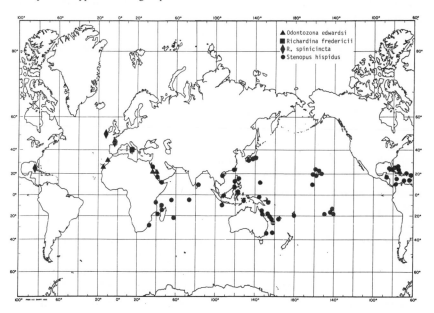

and east Atlantic tropics, and *M. validum* is found in the Red Sea, the western Indian Ocean, across Indonesia, and on the east coast of Australia as far south as Port Jackson. Other species of *Microprosthema* are known only from endemic occurrences in Mauritius and off northwest New Guinea.

All other euzygidans have highly restricted distributions. The stenopodid genera *Richardina* (the Caribbean Dry Tortugas, the southern European Atlantic, and in the Mediterranean off Capri), *Engystenopus* (Bay of Bengal and Indonesia), and *Odontozona* (Morocco through the Red Sea to Indonesian waters) overlap to some extent in their ranges. The spongicolid genera *Spongicola* (Indo-Pacific) and *Spongiocaris* (off Durban and North Island, New Zealand) are endemic to specific regions and have no overlap of ranges.

Several broad generalizations concerning euzygidan biogeography can be drawn from this data. There is a strong Indo-Pacific-Tethyan track (Fig. 23-4) manifested in the group. This is especially evident in the ranges of *Stenopus* and its related genera. Their distributions also seem tightly governed by the presence of suitable temperature ranges and substrate habitats within available current regimes. These latter ensure that their larvae do not disperse far on their own but are highly dependent on where currents take them. The result is that certain areas notably lack euzygids, namely, the south Atlantic and the southeast Pacific. This would seem to indicate that the Euzygida is a very old group, is highly specialized to particular requirements, originated in the tropical Tethys, and has not had any notable success in dispersing out of this range since the vicariance induced by post-Cretaceous continental drift.

Record of post-Cretaceous vicariant events is noted in specific instances. For example, the disjunct distribution (Fig. 23-4) of *Odontozona edwardsi* in Morocco and the Sudan is probably related to the Miocene salinity crisis in the Mediterranean. However, by this same token, the apparent endemism of a species like *Richardina fredricii* to the Mediterranean (Fig. 23-4) would seem to indicate the capacity for relatively recent dispersal and evolution in at least some euzygidans. *R. spinicincta* is found on either side of the Atlantic in the Gulf of Mexico and Northern Europe (Fig. 23-4). Most euzygidans are endemic to specific oceans, ocean basins, or island chains. An example of this are the genera of spongicolids (except *Microprosthema*), which are endemic to particular oceans, or *Richardina*, a north Atlantic-Mediterranean form (Fig. 23-4).

DEVELOPMENT   As with most decapods, a modicum of attention has been paid to the broad patterns manifested in euzygid larval sequences, but few observations have been made on development within the egg. Herrick (in Brooks and Herrick, 1891) noted that the nucleus and immediately surrounding yolk divide during the first cleavage and then migrate to the surface. Subsequent segmentation is superficial, however, though the

ˈsuperficial furrows appear to be quite deep. Invagination, or inmigration, begins at the eighth cell division, and the newly relocated internal cells undergo proliferation. Beyond these general features, however, no details are known.

The general nature of the larval sequence (Fig. 23-5) is very anomuran-like in many respects. This caused Gurney (1924, 1936) to ally stenopids with laomediid thalassinideans, but he later (Gurney, 1942) thought the primitive maxillae noted by Lebour in her 'stenopid A' (a primitive type and possibly a *Richardina*) might preclude anomuran affinities.

Lebour (in Gurney and Lebour, 1941) outlined general diagnostic features of euzygid larval development. 1) They hatch with four pairs of swimming limbs, the first four thoracopods. 2) The abdomen is bent at a right angle at the third pleomere in the later stages. 3) The pereiopodal endopods are nonfunctional, and posterior pairs of thoracopods don't appear until late in development. 4) The rostrum is very long and the supraorbital spines are very prominent, except in the first zoeal stage. 5) The mandibles are very large and fit into notches in the carapace that

Fig. 23-5. Euzygidan larval types. (A, B, C, D) Stenopid C, second through fifth larval stages; (E, F) Stenopid B, ninth and postlarval stages. (From Gurney and Lebour, 1941)

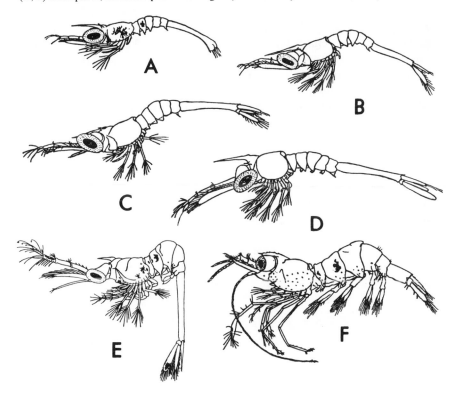

accommodate them. 6) The telson is posteriorly deeply cleft in the first two zoeal stages and has almost a straight posterior margin in later stages. 7) The endopods of pereiopods 1 through 3 originate far from the distal end of the bases. The zoeae attain an incredibly large size before reaching maturity, up to 20 mm.

FOSSIL RECORD *Uncina posidoniae*, a Jurassic form from Germany, is very intriguing (Fig. 23-6). Glaessner (1969) placed this taxon in a separate infraorder between stenopodideans and carideans. Though *Uncina* possessed a large second abdominal pleuron that apparently overlapped both the first and third pleura, it shares several features with the stenopodideans. Its first three pairs of pereiopods were chelate, one of the chelae is very enlarged (Glaessner says '? first'), the pereiopods are uniramous, and the uropods lack a diaeresis. It would appear that *Uncina* bears some sister-group relationship to the living stenopodideans.

TAXONOMY The modern forms are arranged among 25 species in eight genera (Holthuis, 1946, 1955; Bruce and Baba, 1973). These are clearly divisible into two groups, and each should bear familial status: Stenopodidae, characterized by a compressed body, long subtriangular telson terminating in a pair of spines (sometimes with a median spinule between them), uropodal endopods with two dorsal ridges, and the third maxillipede with an exopod (*Stenopus, Engystenopus, Odontozona,* and *Richardina*); and the Spongicolidae, characterized by a body depressed in form (if at all distorted), telson rounded or subquadrangular terminating in three to five subequal spines, uropodal endopod with a single dorsal ridge, and the third maxillipede typically without, or with only a rudimentary, exopod (*Spongicola, Spongicoloides, Spongiocaris,* and *Microprosthema*).

For the reasons given by Burkenroad (1981), and in light of a possible sister-taxon relationship between uncinideans and stenopodideans, the use of the term Euzygida is preferred here.

Family Stenopodidae Huxley, 1879 Recent
Family Spongicolidae nov. Recent

Fig. 23-6. *Uncina posidoniae*, from the Jurassic of Germany. (From Glaessner, 1969)

REFERENCES

Baba, K. 1983. *Spongiculoides hawaiensis*, a new species of shrimp from the Hawaiian Islands. *J. Crust. Biol.* **3**:477–81.

Bate, C. S. 1888. Report on the Crustacea Macrura collected by H.M.S. Challenger during the years 1873–1876. *Challenger Repts. Zool.* **24**:1–942.

Brooks, W. K., and F. H. Herrick. 1981. The embryology and metamorphosis of the Macrura. *Mem. Nat. Acad. Sci.* **5**:321–577.

Bruce, A. J., and K. Baba. 1973. *Spongiocaris*, a new genus of stenopodidean shrimp from New Zealand and South African waters, with a description of two new species. *Crustaceana* **25**:153–70.

Burkenroad, M. D. 1981. The higher taxonomy and evolution of Decapoda. *Trans. San Diego Soc. Nat. Hist.* **19**:251–68.

Felgenhauer, B. E., and L. G. Abele. 1983. Phylogenetic relationships among the shrimp-like decapods. *Crust. Issues* **1**:291–311.

Glaessner, M. F. 1969. Decapoda. In *Treatise on Invertebrate Paleontology*, Part R, *Arthropoda* **4**(2) (R. C. Moore, ed.), pp. R399–533. Geol. Soc. Am. and Univ. Kansas Press, Lawrence.

Goy, J. W. 1980. *Spongicoloides galapagensis*, a new shrimp representing the first record of the genus from the Pacific Ocean. *Proc. Biol. Soc. Wash.* **93**:760–70.

Gurney, R. 1924. Decapod larvae. *Nat. Hist. Rept. Terra Nova Exped. Zool.* **8**:37–202.

Gurney, R. 1936. Larvae of decapod Crustacea. *Discovery Repts.* **12**:377–440.

Gurney, R. 1942. *Larvae of Decapod Crustacea.* Ray Society, London.

Gurney, R., and M. V. Lebour. 1941. On the larvae of certain Crustacea Macrura mainly from Bermuda. *J. Linn. Soc.* **41**:89–181.

Holthuis, L. B. 1947. Biological results of the Snellius Expedition, XIV. The Decapoda Macrura of the Snellius Expedition I. The Stenopodidae, Nephropsidae, Scyllaridae, and Palinuridae. *Temminckia* **7**:1–178.

Holthuis, L. B. 1955. The recent genera of the caridean and stenopodidean shrimps with keys for their determination. *Zool. Verhd.* **26**:1–157.

Huxley, T. H. 1878. On the classification and the distribution of the crayfishes. *Proc. Zool. Soc. Lond.* **1878**:752–88.

St. Laurent. M. de. 1979. Vers une nouvelle classification de Crustacés Décapodes Reptantia. *Bull. Off. Nat. Pêche Tunisie* **3**:15–31.

24

REPTANTIA

DEFINITION   Carapace fused to thorax; epistome as a large plate between antennae and mandibles, extending anteriorly between antennal bases; cuticle (at least the carapace) often well sclerotized and/or strongly mineralized; antennal scale often reduced or absent; cephalothorax with elaborate internal skeleton or endophragmal system; gills tricho- or phyllobranchiate, arthrobranchs and pleurobranchs usually develop simultaneously, pleurobranchs never occur anterior to second pereiopod; first three thoracopods as maxillipedes, first pereiopod typically with large chelae, chelation on second through fifth pereiopods variable; first pleomere shorter than posterior segments, all pleonic segments locked with midlateral hingepoints when pleon well developed, pleura of second pleomere often overlaps those of first; first pair of pleopods uniramous, pleopods with or without appendixes internae and often reduced, eggs brooded on pleopods; larvae hatch as zoeae.

HISTORY   Though the reptant decapods have always been treated as a group of related taxa, attempts at elucidating the relationships among Reptantia have proven it to be a morass. Various names have been given to groups or parts of groups by different authors at different points in time. (An excellent summary is available in Glaessner, 1969 with some additional comments by de St. Laurent, 1979 and Burkenroad, 1981.) For example, the crayfish and clawed lobsters are variously known as Astacidea Latreille, or Macroures Cuirassés Milne Edwards, or Homaridea Boas, or Nephropidea Ortmann. The Anomura of Borradaile are known as the Anomala of de Haan when the Thalassinidea of Latreille are separated from them. Essential differences in classifications arise from differing concepts of what the stem group was like: either some kind of macrurous form (Borradaile, 1907; Bouvier, 1917) or rather some sort of a thalassinoid (Burkenroad, 1963). The extreme diversity of form makes it difficult to study the group as a whole, especially since, frequently in the past, generalizations to the whole were sometimes extended from too narrow a base of specialty. There have been few across-the-board analyses of all groups using a single suite of characters, though rigid attention to character states has apparently begun to yield some elucidation of brachyuran relationships (Guinot, 1977, 1978, 1979; de St. Laurent, 1980a, b). It should be obvious that this approach should be extended to all reptantians in connection with review of all eucarids.

MORPHOLOGY   Great variation exists among reptantians in regard to body form, ranging from 'lobsters,' with a fully developed pleon (Fig. 24-1A), to 'crabs,' with their atrophied abdomens (Fig. 24-1G). The range of variation between these two extremes took substantive form in the taxon Anomura of Borradaile. That taxon contained forms with well-developed abdomens, like the thalassinoids (Fig. 24-1B), and forms with reduced pleons, like the hippoids (Fig. 24-1C), but it also included some forms with highly specialized abdomens, such as the pagurids (Fig. 24-1D), and others, such as the lithodids (Fig. 24-1F) and porcellanids, whose reduced abdomens resembled those of crabs. The brachyurans themselves contain forms, like the raninoids (Fig. 24-1E), which have relatively large abdomens vis-à-vis the typical 'crab' habitus with a distinctly atrophied pleon (Fig. 24-1G).

The carapace in all reptantians is very well developed. It envelops the entire thorax and forms a branchiostegite enclosing a branchial chamber on the side of the cephalothorax. The carapace is typically well sclerotized and mineralized. It is also marked by a series of grooves and channels, and at least the three principal transverse grooves (cervical, postcervical, and branchiocardiac) are thought to be remnants of segment boundaries (Glaessner, 1960; Secretan 1960, 1973). In addition, these are typically in series of one or more longitudinal grooves or linea that represent lines of weak calcification in the carapace to accommodate molting sutures. The epistome on the ventral surface of the cephalon is a large plate extending from between the bases of the antennae to the mandibles. The posterior portion is always heavily sclerotized, but some species can separate the anterior and posterior portions with a membrane (Felgenhauer and Abele, 1983a).

The antennules (Fig. 24-2A) are biramous and composed of a three-segmented peduncle and flagella. These latter can be quite short, for example, in some anomalans and the brachyurans. The advanced brachyuran crabs fold the antennules into grooves in the anterior part of the carapace.

The antennae (Fig. 24-2B) consist of a well-developed peduncle of five segments and a flagellum. The antennal scale, exopod, or scaphocerite is reduced to a spinelike structure or may be totally absent, but it is hardly ever developed as a large flap as in other eucarids. The basal segment of the limb contains the pore of the antennal glands.

The mandibles (Fig. 24-2C) are large well-mineralized structures. Separate incisor and molar processes cannot be easily discernible, and a modest palp of three segments is present. Attached to the base of this limb can be various apodemes and tendons that act as surfaces of insertion for the powerful jaw musculature.

The maxillules (Fig. 24-2D) are thin platelike structures. Two large endites extend mediad from the protopod and have very setose median margins. A laterally directed exite lies above a small lobate endopodal palp.

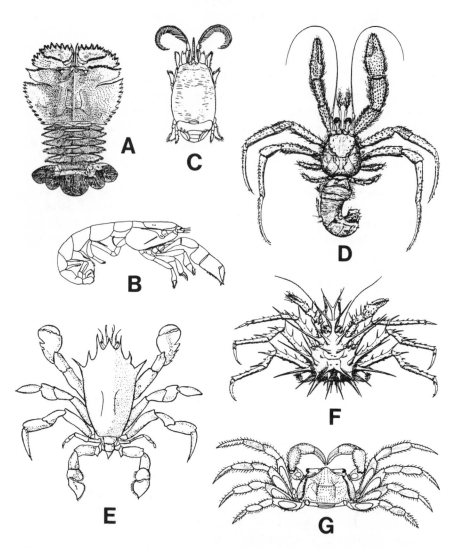

Fig. 24-1. Reptantian body plans. (A) *Ibaccus ciliatus*, a palinuran; (B) *Callianassa* sp., a thalassinidean; (C) *Emerita emerita*, a hippoid anomalan; (D) *Catapagurus doederleini*, a paguroid anomalan; (E) *Lyreidus channeri*, a raninid archaeobrachyuran; (F) *Lithodes turritus*, a paguroid anomalan displaying a convergence to a crab form; (G) *Scopimera globosa*, an eubrachyuran. (From Balss, 1957)

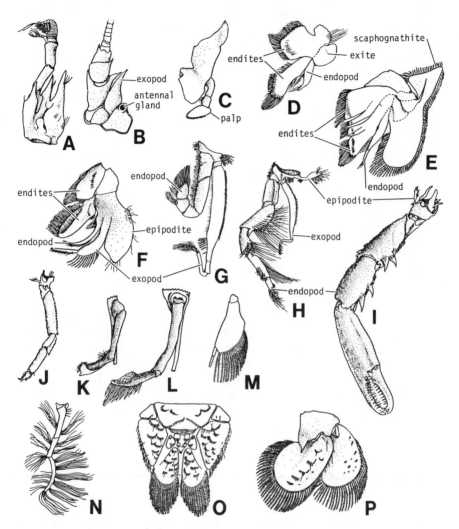

Fig. 24-2. Appendages of *Galathea spinifera*. (A) antennules; (B) antenna; (C) mandible; (D) maxillule; (E) maxilla; (F) first maxillipede; (G) second maxillipede; (H) third maxillipede; (I) first pereiopod; (J) third pereiopod; (K) male first pleopod; (L) male second pleopod; (M) male third pleopod; (N) female third pleopod; (O) telson; (P) uropod. (From Pike, 1947)

The maxillae (Fig. 24-2E) are also thin and platelike limbs. The two protopodal endites are medially directed, have setose medial margins, and are both deeply cleft. The endopodal palp has two segments. A large scaphognathite is developed laterally; its margins are setose, and the dorsal lobe extends back into the branchiostegal chamber. Its vibrations facilitate the flow of a respiratory current over the gills.

The first maxillipedes (Fig. 24-2F) are very similar to the maxillae. The protopod bears two platelike endites with very setose medial margins. The endopodal palp has a single segment. The exopod is a somewhat reduced two-segmented structure and is unadorned, except for marginal setae; however in the pagurids a 'caridean lobe' is noted. The limb also bears a thin lobate epipod.

The second maxillipedes (Fig. 24-2G) are somewhat pediform. The robust endopod has its terminal segments recurved and directed medially and proximally; the medial endopodal margins are very setose. The exopod is flagelliform, with its terminus frequently directed medially.

The third maxillipedes (Fig. 24-2H) are very robust. The endopod bears very setose segments, and the ischium typically can be developed with a serrated median border or crista dentata. The flagelliform exopod is similar to, but somewhat reduced over, that seen in the second maxillipedes.

The pereiopods are uniramous. The first (Fig. 24-2I) are the largest in the series. They are usually chelate, except in the palinurans, which lack pincers. The chelae frequently display asymmetry, for example, in astacideans the chelae are opposed as crushing and tearing claws, and in male fiddler crabs one claw is greatly enlarged and employed as a signaling organ in intraspecific interactions. These limbs are variously used in defense or food procurement, and thus in regard to having a distinct form and function they are more like maxillipedes than ambulatory pereiopods.

The second through fifth periopods (Fig. 24-2J) are typically used as walking legs. In the astacideans the second and third are chelate. In several groups, the fifth limb can be modified as a paddle to facilitate swimming or digging, as in some anomalans and brachyurans. The gonopores are always located on the coxae of the third (female) or fifth (male) pereiopods, except in some of the brachyurans where they may be on the sternites of the respective segments. The joints on the pereiopods have a characteristic arrangement. They alternate on every segment between either acting to promote–remote or abduct–adduct the more distal parts of the limb.

The pleopods are quite variable, though in general it can be said that they are reduced over that seen in other eucarid groups. The first pair can be absent, for example, as in some astacideans and palinurans, or as in some female brachyurans. The other pleopods can also be reduced or absent, as in many anomalans, where for instance in pagurids they are only developed on one side, or in male brachyurans, which completely lack the third through fifth pleopods. Copulatory appendages may or may not be developed. Males, if they develop gonopods, frequently modify the first and second pleopods (Fig. 24-2K, L). However, some anomalans can also develop gonopods on the females as well. Posterior pleopods can also exhibit slight sexual differences (Fig. 24-2M, N).

The telson (Fig. 24-2O), though it may be reduced, is typically broad, never styliform or pointed. The uropods, if present, are biramous with flaplike rami (Fig. 24-2P), and the exopods may have a diaeresis.

One of the most distinctive features of reptantian anatomy is the elaborate development of the gills. These are either trichobranchiate or phyllobranchiate in nature and arise (Table 24-1) either on the side of the body (pleurobranchs), on the arthrodium (arthrobranchs), and/or on the coxa (podobranchs). The arthrobranchs and pleurobranchs generally appear together in development; but in thalassinids the pleurobranchs seem to develop later if they are present at all, while in brachyurans the pleurobranchs appear earlier than the arthrobranchs (if the latter are present) on the second to fourth pereiopods (Burkenroad, 1981).

Another distinctive feature of reptantians is the elaborate development of the endoskeleton, or axial or endophragmal system, of the cephalon and thorax. These structures provide points and surfaces of origin for muscles that operate the mouthparts and the extrinsic muscles of the pereiopods (Schmidt, 1915). Some of this published work, mostly that on the gnathocephalon, is only of a preliminary nature and awaits further study in the several groups of reptantians (Secretan, 1966, 1970, 1977, 1981). The system actually develops from apodemal invaginations of the body wall (Secretan, 1980). This is in contrast to personal observations in the Dendrobranchiata and Caridea where only short apodemal ingrowths from the cuticle around the limb bases arise to accommodate the relevant muscles.

Work on the thoracic skeletomusculature has been more detailed but restricted to studies of animals in only a few groups: some astacideans (Huxley, 1880; Parker, 1889; Schmidt, 1915; Pilgrim and Wiersma, 1963; Raynor, 1965), some palinurans (Parker, 1889; Paterson, 1968), a pagurid (Pilgrim, 1973), and brachyurans (Bourne, 1922; Cochran, 1935). None of these authors have noted the differences that can occur among basic types of thoracic endoskeletons, nor which groups share the basic forms.

The astacideans, thalassinoideans, and glypheids (based on an examination of a prepared specimen of *Neoglyphea inopinata* made available through the courtesy of M. de St. Laurent) are similar (Fig. 24-3A). Muscle compartments of each pereiopod are separated by an interpleurite that extends into the thorax mediad from the body wall and dorsad from the sternite. Midway along these arises an intrapleurite that is directed medioposteriorly. The entire pleurite takes on a Y form, looking at it dorsally, with the muscle compartment thus separated into a lateral and a medial portion. The intrapleurite separates the coxal promotor and remotor

Table 24–1. Gill formulas of reptantians illustrating maximal, intermediate, and impoverished variants. r = rudimentary, 0 = absent, e = epipodite, p = podobranch, a = arthrobranch, pl = pleurobranch. (From Burkenroad, 1981)

	Maxillipede			Pereiopod				
	1 e + p a?	2 e + p a pl	3 e + p a pl	1 e + p a pl	2 e + p a pl	3 e + p a pl	4 e + p a pl	5 pl
Homarida	1 + r r	1 + 1 1 0	1 + 1 2 0	1 + 1 2 0	1 + 1 2 1	1 + 1 2 1	1 + 1 2 1	1
	1 + 0 0	1 + 1 1 0	1 + 1 2 0	1 + 1 2 0	1 + 1 2 r	1 + 1 2 r	1 + 1 2 r	0
	r + 0 0	1 + 0 r 0	1 + 1 1 0	1 + 1 1+r 0	1 + 1 1+r 0	1 + 1 1+r 0	1 + 1 1+r 0	0
Palinura	1 + 0 0	1 + 1 1 0	1 + 1 2 0	1 + 1 2 0	1 + 1 2 1	1 + 1 2 1	1 + 1 2 1	1
	1 + 0 0	1 + 0 0 0	1 + 1 2 0	1 + 1 2 0	1 + 1 2 1	1 + 1 2 1	1 + 1 2 1	1
	1 + 0 0	0 + 0 0 0	r + 0 1 0	1 + 1 2 0	1 + 1 2 1	1 + 1 2 1	1 + 1 2 1	1
Thalassinidea	1 + 0 1	1 + 1 2 0	1 + 1 2 0	1 + 1 2 0	1 + 1 2 1	1 + 1 2 1	1 + 1 2 1	r
	1 + 0 0	1 + r r 0	1 + 1 2 0	1 + 1 2 0	1 + 1 2 0	0 + 0 2 0	0 + 0 2 0	0
	0 + 0 0	1 + 0 0 0	0 + 0 2 0	0 + 0 2 0	0 + 0 2 0	0 + 0 2 0	0 + 0 2 0	0
Anomala	1 + 0 0	r + 0 0 0	1 + 0 2 0	1 + 0 2 0	1 + 0 2 1	1 + 0 2 1	1 + 0 2 1	1
	1 + 0 0	0 + 0 0 0	1 + 0 2 0	1 + 0 2 0	1 + 0 2 1	1 + 0 2 1	0 + 0 2 1	1
	1 + 0 0	0 + 0 0 0	0 + 0 1 0	0 + 0 2 0	0 + 0 2 0	0 + 0 2 0	0 + 0 2 0	1
Brachyura	1 + 0 0	1 + 1 1 0	1 + 1 2 0	1 + 1 2 0	1 + 1 2 1	1 + 1 2 1	1 + 1 2 1	1
	1 + 0 0	1 + 1 0 0	1 + 0 2 0	0 + 0 2 0	0 + 0 0 1	0 + 0 0 1	0 + 0 0 1	0
	1 + 0 0	0 + 0 0 0	r + 0 1 0	0 + 0 2 0	0 + 0 0 0	0 + 0 0 0	0 + 0 0 0	0

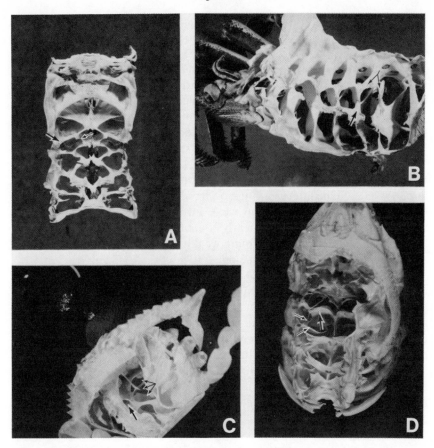

Fig. 24-3. Reptantian thoracic endoskeletons. (A) dorsal view of *Homarus americanus*; (B) lateral view from the midline of *Panulirus interruptus*; (C) dorsal oblique view of *Callinectes bellicosus*; (D) dorsal view of *Blepharipoda occidentalis*. In A, B, and D, large thick arrows denote the interpleurites separating muscle compartments; the small thin arrows designate the intrapleurites. In C, the large thick arrow points to the posterior opening of the remotor muscle compartment of the first pereiopod, which is dorsal and at right angles to the deeper muscle compartments (small thin arrows) of the other pereiopods.

muscles, which take origin in part on the pleurite and in part on the interosternite (an ingrowth dorsad from the median sternite).

The palinurans display a different pattern (Fig. 24-3B). The ventral interpleurite arises from the body wall and the lateral portion of the thoracic sternites. In addition, a dorsal intrapleurite arises from the body wall between the pleurobranchs and has two portions: a medial projection, which bridges the muscle compartment to fuse with the interpleurite, and a dorsal projection, which contacts the medial projection of the next anterior intrapleurite. The effect of this arrangement is to form a system of 'flying buttresses' to the lateral body wall arranged around the laterally directed pereiopods. The center field of the thoracic sternites are free of any apodemal ingrowths.

The pattern seen in palinurans is similar to that seen in galatheid anomalans. However, in these, the interpleurite arises only from the lateral body wall, the sternites being quite unspecialized, and the dorsal intrapleurite apodeme projections are nearly subequal. In addition, extensions from the more anterior pleurites project dorsally and medially to form a bridge across the thorax at the level of the second pereiomere.

The hippoideans and pagurids (Fig. 24-3C) have a different scheme. Prominent interpleurites arise from along the sternites lateral to the midline, while median interosternites occur along the midline and serve to demarcate right from left. The weak intrapleurites arise dorsally from along the body wall and form nearly equal projections anteriorly and posteriorly. The intrapleurites are quite weakly developed in the pagurids but are more prominent in the albuneids.

The brachyurans (Fig. 24-3D) possess a system of more or less completely enclosed muscle compartments. The interpleurites of the second through fifth pereiomeres form a complete wall between chambers and arise from the lateral sternites and the body wall. The ingrowths on the dorsal part of the body wall do not form intrapleurites but are developed to provide a ceiling to the ventral muscle compartments. The muscles of the first pereiopods are arranged differently from those of the second through fifth. The coxal remotors lie in a separate compartment dorsal to the muscle compartments of the second through fifth pereiopods, while the coxal promotors lie in the body chamber somewhat medial and dorsal to the posterior pereiopod muscle compartments.

In all of these cases the insertion of the coxal muscles is facilitated by large apodemes that extend inward into the muscle compartments from around the rim of the coxa. These range from being slightly developed, as in the thalassinoids, to being very large, as for the chelipedes of brachyurans. Clearly all these endoskeletal modifications are related to the principal thrust of the radiation of this group as indicated by their name, that is, reptant locomotion. Further analysis of the skeletomusculature of this system should offer some constructive insights into the adaptive radiation of Reptantia and should assist in clarifying the higher taxonomy of the group.

The structure of the digestive system has received a great deal of attention, especially that of the foregut (Mocquard, 1883; Patwardhan, 1935a, b, c, d; Reddy, 1935; Pike, 1947; Schaefer, 1970; Caine, 1975a, b; Kunze and Anderson, 1979; Ngoc-Ho, 1984). Although differences occur, there is a basic similarity in form. A complex of ossicles in the foregut wall act as insertion points for a musculature that operates the gastric mill (see, e.g., Patwardhan, 1935a, b, c, d; Pike, 1947; or Ngoc-Ho, 1984). Internally, a pattern akin to that seen in the dendrobranchiates and euzygids is evident. A cardiac stomach has a large median tooth and a pair of complex lateral teeth (Fig. 24-4A) opposed to a complex of setose and spinose pads and

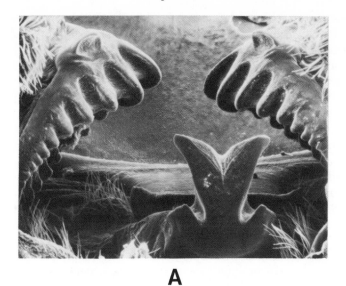

A

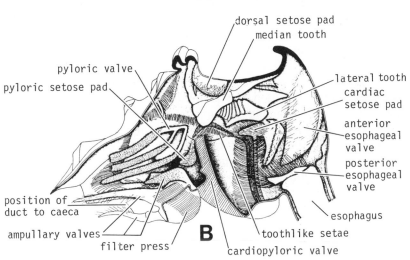

B

Fig. 24-4. Reptantian foregut. (A) SEM of the dorsal median and flanking lateral teeth (courtesy of Dr. B. E. Felgenhauer, Florida State University); (B) diagrammatic view of the interior of the cardiac and pyloric regions of the foregut (from Ngoc-Ho, 1984).

folds (Fig. 24-4B). This serves to thoroughly macerate food, which is only rendered into large chunks by the action of the mandibles and third maxillipedes. The pyloric stomach (Fig. 24-4B) dorsally joins the midgut, and ventrally is developed into a setose filter press or gland filter. This latter serves to further strain and divide food before it enters the digestive caeca. Undigestibles are either regurgitated from the cardiac stomach or pass directly into the midgut for voiding by the anus. The bulk of the actual digestion occurs in the caeca.

The midgut can be reduced to a few millimeters in length, as in brachyurans (Patwardhan, 1935b). The digestive caeca are very well developed. Digestion was originally thought to be extracellular (Vonk, 1960),
but recent work on cell types in the caeca confirms that most of the actual
digestion is intracellular (Barker and Gibson, 1977).

The circulatory system centers around a short heart in the posterior
portion of the cephalothorax (Fig. 24-5), which is enclosed in a spacious
pericardial sinus. Several sets of arteries leave the heart: from the anterior
end an anterior aorta, which can be equipped with a bulbous cor frontale,
lateral cephalic arteries, and hepatic arteries to the gut caeca; from the posterior end a descending artery supplying the legs and gills, and a posterior
aorta. This last is especially interesting in that its walls are muscular, at
least in *Panulirus*. Burnett (1984) felt the muscles now serve to help regulate blood pressure in the entire arterial system, but are a remnant of a
primitive system in which the heart would have extended through the
entire length of the trunk (not unlike the arrangement seen in stomatopods). Indeed, this would agree with the mode of development noted by
Shiino (1950) where the heart and posterior aorta have a common origin.

Excretion is achieved by antennal glands. However, Pellegrino (1984)
has pointed out that the cuticle is quite permeable and can serve to cool the
body by evaporation as well as help eliminate nitrogenous wastes in the
form of ammonia.

The central nervous system (Fig. 24-6) clearly reflects the tendency
toward abdominal reduction. In macrurous forms, the abdominal segments
each contain a set of ganglia. In anomurans, there is a tendency to move
ganglia forward into the thorax, at least that of the first pleomere as in the
thalassinoids, and almost all of them as in some pagurids. In the brachyurans, all the trunk ganglia fuse into a single huge postesophageal ganglion.
Reptantia, as is frequently the case in lower eucarids, have an antennular

Fig. 24-5. Arterial circulatory system of *Panulirus interruptus*. (From Burnett, 1984)

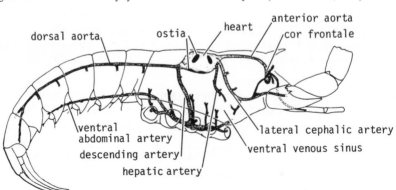

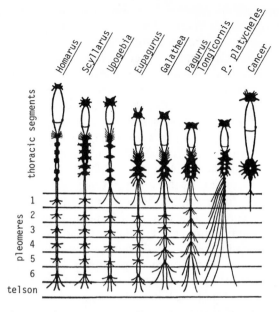

Fig. 24-6. Reptantian central nervous system, illustrating the degeneration and fusion of posterior abdominal elements into a single central thoracic ganglionic center. (Modified from Bouvier, 1889)

statocyst. This may either be open to the outside, using a sand grain as a statolith; or be closed, as in the brachyurans and anomalans, and function without a solid particular statolith.

In the male reproductive system the testes are typically separate and lie for the most part in the thorax. This pattern differs in the pagurids where the testes come to lie totally within the abdomen. In this case they are either fused into a single mass on the left side or are separate but arranged in a line—the right being anterior to the left. The vas deferens is rather complex and resembles in some respects that seen in stomatopods. The efferent duct drains the testes and is continuous with a middle convoluted glandular portion. The terminal vas is specialized as a muscular ejaculatory duct. Calman (1909) records that in all brachyurans, except dromiids, the glandular duct has a series of diverticula distally.

The female reproductive system is much like that of the male in terms of gonad location; pagurids and thalassinids have the organs placed completely in the abdomen. The brachyurans, which utilize internal fertilization, have the terminal part of the oviduct developed as a vagina in order to accommodate the male. Associated with this vagina is a single or double glandular sac, the seminal receptacle. In the dromiids, the receptacles are transient, and only form at the time of copulation.

NATURAL HISTORY   The Reptantia have had a remarkable degree of success, exhibiting great diversity of form and occupying habitats that extend from the deep sea to terrestrial situations, though their greatest success seems to be in relatively shallow water situations.

The most interesting adaptations are those related to terrestrial habitation (see, e.g., Bliss, 1968). The most crucial requirement is to maintain water balance. All the terrestrial reptants return to water for reproduction; however, some forms otherwise will spend most of their lives several kilometers from water, for example, coenobitid anomalans or gecarcinid brachyurans. Reptantians have crucial problems in this regard because of the great permeability of the cuticle (Pellegrino, 1984). Some taxa, such as the coconut crab *Birgus*, actively drink water. Others have specific adaptations to soak up water, for example, the pericardial sacs of some brachyurans (Bliss, 1963), or special tufts of plumose setae on the pereiopodal coxae and cuticular modifications that facilitate capillary uptake of water as in some terrestrial crabs (see, e.g., Felgenhauer and Abele, 1983b). Whatever the adaptations of structure or physiology, these adaptations have been so successful at sustaining reptantians on land that these animals are the only crustaceans that have been able to achieve moderate to large body size in terrestrial situations.

Feeding patterns among reptants are quite diverse. They can be generally characterized as omnivorous, though specific groups exhibit peculiar specializations. Macrurous forms generally are scavenger or low-level carnivores, that is, not actively stalking live prey but not passing it by if randomly encountered. In *Homarus* (Barker and Gibson, 1977) food is procured with the maxillipedes and/or chelipedes and second pereiopod, and presented to the third maxillipede. This maxillipede pushes food toward the mandibles while the other mouthparts move laterad. The mandible grasps and holds the food while the crista dentata of the maxillipede tear the culinary treats into chunks. These are captured by the second maxillipedes and moved by the mouthparts to the mouth. The mandibular palps and paragnaths prevent loss of food while it is being swallowed. Farmer (1974) records a similar morphology of mouthparts for *Nephrops norvegicus*, but his statements on actual feeding are general and mostly suppositions or inferences, and Caine (1975b) records some detailed observations on feeding in *Procambarus* that closely accord with that seen in *Homarus*.

However, other modes of feeding in macrurans are possible. Budd et al. (1978) observed filter feeding in crayfish. In addition to a respiratory current generated by the scaphognathite that flows under the edges of the carapace, over the gills, and exiting anteromedially, a separate filter current exists in juvenile crayfish. This latter is generated by the exopods of the first and second maxillipedes and flows from below and just posterior of the mouth field to anterodorsally beneath the cephalon between the maxillipedes. Setae on the maxillae and first maxillipede trap suspended particles from the water. These particles are in turn removed by the second maxillipede, which acts as a comb, and are stuffed into the mandibular field by the maxillules. This type of filter feeding is obligate in the juveniles, but the adults as well can be facultative filterers.

Hermit crabs can be carnivorous or scavengers, catching food with the chelipedes and rendering it with the mandibles and maxillipedes (Greenwood, 1972; Caine, 1975a; Gerlach et al., 1976; Kunze and Anderson, 1979; Schembri, 1982b). However, hermits also commonly engage in feeding on detritus, shoveling sediment with their chelipedes so that the third maxillipedes can accumulate a bolus, which is then transferred to the mouth by the second maxillipedes (Jackson, 1913; Orton, 1927; Roberts, 1968; in addition to the just-mentioned references). Filtration was noted in *Pagurus* (Gerlach et al., 1976) with *Artemia* nauplii captured by the setose endopods of the third maxillipedes and algae removed by the maxillules and maxillae from the water in a current produced by the exopods of the second and third maxillipedes. This method of feeding was also noted in porcelain crabs (Caine, 1975a; Warner, 1977) and galatheids (Nicol, 1932). In addition, antennal filtration was noted in hermits (Boltt, 1961; Greenwood, 1972) as the setose antennae moved to and fro and were periodically groomed by the maxillipedes.

The occurrence of antennal filtration in hermits is similar to the feeding noted in the sand crab *Emerita analoga* (Efford, 1966). The extremely setose antennae are projected upward into a passing current. Alternately, antennae are cleaned by the maxillipedes.

Brachyurans are extremely versatile in their feeding habits (see, e.g., Warner, 1977). They make very effective scavengers and carnivores and, in regard to the latter, are frequently quite active. Crabs are major predators of mollusks, using their chelipedes to crush or chip away at the shells of their victims. Vermeij (1976, 1977) had originally postulated possible profound effects on molluscan evolution with the advent of true crabs as predators [although supposed associated patterns of biogeographic size variation in crabs in this regard have been rejected (Abele et al., 1981)]. Herbivorous crabs either use the mouthparts to scrape algae (Warner, 1977) or pluck macrophytes with their chelipedes. I have watched spider crabs do the latter in laboratory situations. This involvement of chelipedes in macrophagous activity is important in omnivorous forms and, although many crabs have a crista dentata on the third maxillipede used in shredding food in cooperation with the mandibles, not all do. The leucosiid *Ebalia* lacks a crista dentata, and the mandibles render most of the diet with promotor–remotor slicing actions (Schembri, 1982a). Deposit feeding is a relatively specialized mode of feeding in crabs, but it is similar in form and function to that seen in hermits. Miller (1961) studied this closely for fiddler crabs. Detritus is scooped up by the chelipedes and sorted by the various mouthparts with exhalant water from the gill chamber. A specialized mode of detritus feeding (Patton, 1974) may occur in the coral-inhabiting crab *Trapezia*, which feeds on mucus secreted by the coral and the entrapped particles produced when the coral is poked and irritated by the crab (Knudsen, 1967).

In general, food-getting in Reptantia is a series of elaborations and specializations on a basically cephalic feeding mode that is augmented by the incorporation of the first three to five thoracopods into an enlarged mouth field.

Sexes are separate in Reptantia. Some data seemed to indicate the possibility of protandry in sand crabs (Barnes and Wenner, 1968), but this does not now seem to be the most reasonable interpretation of the data (Wenner and Haley, 1981). Thus sex reversal in reptantians does not seem to occur. The sexual dimorphism is typically marked and can be quite striking, as in sand crabs where the male is a small symbiont near the gonopores of the female. Efford (1967) noted neoteny was a factor in the evolution of male *Emerita*.

Fertilization can be either external or internal. In examples of the former, copulation results in the placement of a sperm mass into an annulus or thelycumlike structure on the female. Actual fertilization occurs much later at the time the eggs are laid. In instances of internal fertilization, as in the brachyurans, copulation generally only occurs when the female is in a soft-shell postmolt phase and thus capable of accepting the insertion of the male gonopods. Precopulatory activities may be almost nonexistent, as in some macrurans. However, some groups, such as fiddler crabs (Crane, 1975; Christy and Salmon, 1984) and hermit crabs (Hazlett, 1966, 1968) can exhibit some elaborate and species-specific precopulatory behaviors. These behaviors have been shown to be linked with specific reproductive strategies (Christy and Salmon, 1984). Eggs are brooded on the female's pleopods. The numbers of eggs can be astounding. Herrick (1896) recorded 97 440 eggs on one female *Homarus americanus*. Though copulation and brooding can take place some distance from water in the case of terrestrial forms, actual hatching and shedding of larvae must occur in water.

One of the most recent comprehensive reviews of the distribution of reptantians as a part of Decapoda as a whole was given by Balss (1957). Since then, distribution patterns observed in specific regions of the world have been analyzed by a number of authors (for a summary, see Abele, 1982, p. 288).

Abele (1982) points out a number of broad patterns seen for all decapods; namely, strong latitudinal gradients with the peak in species numbers occurring in the tropics just north of the equator and declining toward the poles, and within the tropical fauna the richest diversity exists in the Indo-West Pacific region, which possesses both the greatest number of species and the highest percentage of endemics.

Aside from these general patterns, certain specific studies related to decapod distributions have some import to evolutionary theory. There are two competing models of evolution: the Red Queen Hypothesis, which states that evolution is driven by biotic factors (i.e., changes in any species in a community produce evolutionary change in other particular cohabitants of that community) and the Stationary Model, which states that changes in species are driven by abiotic factors (i.e., physical changes in the environment). Stenseth and Maynard Smith (1984) in their review of the issues conclude that evolution must conform to one or the other of these alternatives and that the choice between the two models will have to be made primarily on the basis of paleontological evidence.

However, in a series of papers dealing with modern decapod biogeography, information is available that would seem to question the rigid position of Stenseth and Maynard Smith. It has been traditionally thought that species richness was directly related to habitat heterogeneity (see, e.g., Abele, 1974; Ricklefs, 1979; Heck and Whetstone, 1977). However, Abele and Patton (1976) found that area alone, irrespective of habitat complexity, affects the number of species present in a habitat. This conclusion was extended by Abele and Walters (1979) to question the classic Stability-time Hypothesis of Sanders (1968), in that the number of species in an old and stable environment like the deep sea was shown to be no greater than what would be expected given the area of the deep sea relative to other benthic regions. In addition, constancy of environment does not affect species richness; rather species equilibrium is related to external factors of the environment (Abele, 1976a). Fluctuating environments everywhere are noted for their degree of species richness and general lack of characteristic dominant species (Abele, 1979). Dominant species seem to be those that are specialized to exploit some abundant dimension of their environment (Abele, 1976b).

To generalize, these studies of Abele and his co-workers on decapod distribution appear to imply that it is the physical or abiotic factors of the environment that are most important in determining the size and structure of biologic communities. Such a conclusion would conform to the predictions of the Stationary Model of evolution, and would reinforce the important role that chance must play in evolution. This is not to say that selection is not important, but that its invocation is not justified until the role of chance in the operation of a basically stochastic universe is ruled out.

DEVELOPMENT The number of detailed and effective studies of ontogeny in reptantians is scarce and restricted to analysis of macrurous forms, namely, *Homarus americanus* (Herrick, 1896) and *Panuliris japonicus* (Terao, 1929; Shiino, 1950).

Early segmentation is not visible on the surface, since cleavage furrows do not appear until the eight-cell stage (Fig. 24-7A) and the blastomeres do not develop internal boundaries until about the 16-cell stage (Fig. 24-7B). However, the arrangement of the furrows seems to preserve remnants of the original radial cleavage (Fig. 24-7A). The large blastomeres exude

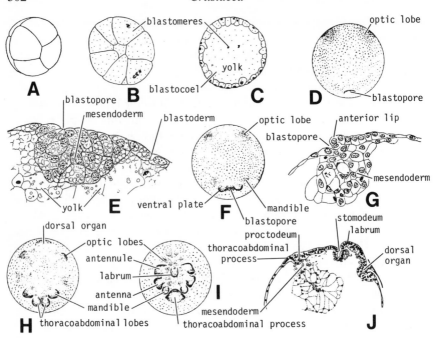

Fig. 24-7. Early ontogeny of *Panulirus japonicus*. (A) spiral cleavage furrows, eight-cell stage; (B) early blastula with yolk-laden blastomeres; (C) late blastula with yolk in blastocoel; (D) initiation of blastopore and optic lobes to delineate germinal disc; (E) early gastrulation; (F) advanced gastrulation with ventral plate forming on blastopore anterior lip; (G) section through blastopore of F; (H) initiation of naupliar anlagen; (I) advanced egg–nauplius stage; (J) schematic section through midline of I. (From Shiino, 1950)

their yolk into the blastocoel and grow smaller as cleavage proceeds (Fig. 24-7C).

Eventually a blastopore is formed (Fig. 24-7D), and the resultant thickening of cells produces a mesendormal plate (Fig. 24-7E). The anterior lip of the pore enlarges (Fig. 24-7F) and grows over the blastopore to form a bilobed thoracoabdominal or ventral plate (Fig. 24-7G). As the blastopore develops, the paired optic lobes begin to thicken, and the three together delineate the limits of the germinal disc. The lobes and pore are eventually connected by a U-shaped thickening of ectoderm, and the U is converted to an O-shaped structure when the interocular transverse band appears (Fig. 24-7D).

The naupliar phase is initiated by the appearance on the ectodermal band of the antennular and mandibular anlagen (Fig. 24-7H), followed later by the antennal. The antennal anlagen are initially biramous, but then form uniramous structures (Fig. 24-7I). All other limb anlagen were said to be uniramous, but Terao (1929) observed the initial antennular anlagen in *P. japonicus* to be weakly horseshoe-shaped. The labrum differentiates, and the stomodeum is initiated at the time the naupliar anlagen are formed (Fig. 24-7J).

The thoracoabdominal process or caudal papilla is formed from the ventral plate. The maxillulary and maxillary segments form on the process, but are soon transferred to the germinal disc where the limb anlagen appear (Fig. 24-8A). The first maxillipede segment and limb are formed on the process before being transferred to the disc (Fig. 24-8C). At the time of hatching, the second maxillipede is already part of the head. There are 18 or 19 ectoteloblasts in a ring around the terminus of the caudal papilla (Oishi, 1960). The full number only gradually differentiate from the ventral plate. Therefore, the dorsal portions of some of the anterior metanaupliar segments are derived from the germinal disc rather than the teloblasts.

The naupliar mesoderm grows forward from the mesendodermal plate beneath the ectodermal bands. The metanaupliar mesoderm arises from eight mesoteloblasts (Fig. 24-8B) arranged within the ectoteloblast ring (Oishi, 1960). The telson mesoderm forms independently from cells that

Fig. 24-8. Late ontogeny of *Panulirus japonicus*. (A) beginning of metanaupliar differentiation, maxillule and maxilla anlagen have differentiated on posterior portion of germinal disc; (B) midsagittal section through the caudal papilla or thoracoabdominal process with teloblasts, proctodeum, and presumptive midgut differentiated; (C) intermediate stage of metanaupliar development; (D) diagrammatic sagittal section through a late metanaupliar stage; (E) cross section of D through head. (From Shiino, 1950)

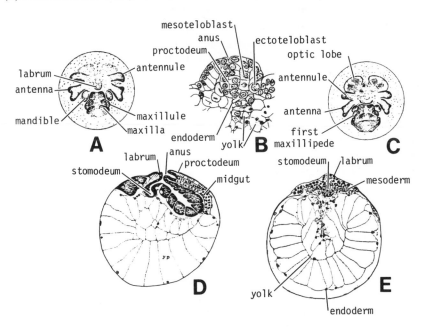

migrate inward from the telson ectoteloblasts. There is no preantennary mesoderm formed at all.

The endoderm forms from the mesendodermal mass and grows to produce a cellular yolk sac that absorbs the free yolk in the blastocoel (Fig. 24-8). Shortly after the blastopore closes and the caudal papilla is differentiated, the proctodeum is initiated on the same spot as the blastopore would have been (Fig. 24-8B). So by the late egg–nauplius stage all the presumptive components of the gut are present.

The brain is formed by the fusion of the distinct optic lobes with the ganglia of the antennules and antenna. Circumesophageal commissures grow out from this to join the subesophageal ganglion. This latter is formed by the fusion of the mandibular, maxillulary, and maxillary ganglia.

The heart and dorsal aorta are initiated together as a single unit from the dorsal somatic mesoderm. No coelomic pockets ever appear in the body except in the formation of the antennal gland. Finally, the gonads arise from mesodermal rudiments in the maxillary segment.

Larval development in reptantians (Gurney, 1942) is remarkably varied and irregular, that is, it seems to be more often than not characterized by deviations in the number or duration of larval stages among related species (see, e.g., Gore, 1985). Patterns range from abbreviated ones, as seen in macrurans, in which mysis forms (Fig. 24-9A, B) with biramous limbs gradually develop into juveniles; to more anamorphic types with sequences of distinct larval forms, namely, zoea (Fig. 24-9D, E) and metazoea, preceding the juvenile. All forms have a juvenile or decapodid stage (Felder et al., 1985), known by a variety of names depending on group such as megalopa, pseudibacus, puerulus (Fig. 24-9C), and glaucothoe.

Because of the great variety and depth of information available about reptantian larvae, much attention has been directed toward using them in phylogenetic analysis (e.g., Williamson, 1976; Rice, 1980). These have led to some interesting insights, especially in regard to brachyuran evolution, but no one has yet attempted to combine larval and adult features into a coherent overview of reptant relationships. Whatever phylogenetic patterns ultimately emerge from a synthesis of larval and adult data, it appears that the variety and irregularity of developmental patterns may have been a major feature in the survival and diverse radiation seen among reptantians (Rabalais and Gore, 1985).

FOSSIL RECORD   The record of reptantians extends from the late Devonian. *Palaeopalaemon newberryi* (Fig. 24-10) from North America is a well-preserved macrurous form, but its exact taxonomic affinities are in doubt. Schram et al. (1978) compared it to both astacideans and palinurans, but Felgenhauer and Abele (1983a) pointed out that the large scaphocerite of *P. newberryi* is more characteristic of natant forms than reptantians. Other Paleozoic fossils are merely carapaces, but hint at the potential diversity yet to be discovered among reptantians of that era: *Imocaris tuberculata*

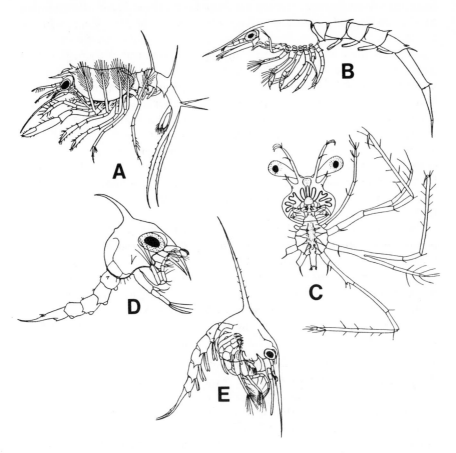

Fig. 24-9. Reptantia larvae. (A) *Nephrops norvegicus*, stage 3 mysis; (B) *Axius stirhynchus*, last mysis; (C) *Scyllarus arctus*, puerulus; (D) *Maia squinado*, first zoea; (E) *Corystes cassivelaunus*, last zoea. (From Gurney, 1942)

Fig. 24-10. Earliest decapod, *Palaeopalaemon newberryi*, late Devonian. (From Schram, et al., 1978)

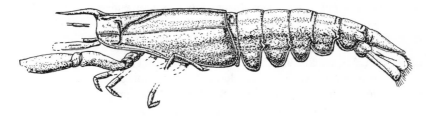

from the Mississippian of America resembles a dromiacean (Schram and Mapes, 1984), and *Protoclytiopsis antiqua* seems to be a glypheid (Schram, 1980).

A fairly good reptant record can only be said to be present from the Triassic onward (Förster, 1967), and these are macrurous forms. Anomurans extend from the Jurassic, but their record is generally a poor one and is better marked by a plethora of chelipedes rather than good body fossils (Glaessner, 1969). The brachyurans also extend from the Jurassic (Glaessner, 1969), but several successive radiations can be detected. The Jurassic crab fauna is largely one of podotreme groups such as dromiaceans. The Cretaceous contains an early divergence of eubrachyuran forms, especially heterotremates. Finally, what is essentially the modern radiation of crabs, especially strong in thoracotremates, was initiated in the Eocene.

A rather cursory examination by me of survivorship curves of malacostracans extending from the Paleozoic to the Recent seemed to indicate that there may have been a definite shift to a high turnover rate of taxa beginning in the Cretaceous. This might indicate that the great success exhibited by crabs may be in some measure due to some intrinsic capacity of the group to develop variations on basic plans very quickly. However, this supposition must be checked when more careful stratigraphic control can be exerted on analyzing malacostracan biostratigraphy.

TAXOMONY   The classification of Reptantia is a morass. A single overview using character analysis on all the groups is badly needed. The system given here is a compromise but tries to incorporate some recent advances in our understanding of the relationships of reptant groups. It is largely based upon the system outlined by Abele (1983) that, at least for the time being, eliminates superfamily groupings. Recent consensus (see, e.g., de St. Laurent, 1979; Burkenroad, 1981; Bowman and Abele, 1982; Abele, 1983) does not recognize 'Anomura' as a distinct taxon; that is, it separates the thalassinideans from the anomalans; and certainly, thoracic endoskeletal anatomy would seem to indicate that the former are perhaps more closely related to the astacideans than to the latter. The arrangement of the brachyurans here incorporates the positions of both Guinot (1978) and de St. Laurent (1980a, b); but the listing here of families within the sections and subsections of brachyurans (especially thoracotremes) is largely alphabetical while awaiting some effective analyses of possible phylogenetic relationships within that group.

Infraorder Astacidea Latreille, 1803 Permian–Recent
    Family Erymidae Van Straelen, 1924 Permian–Cretaceous
    Family Nephropidae Dana, 1852 Jurassic–Recent
    Family Thaumatochelidae Bate, 1888 Recent
    Family Cambaridae Hobbs, 1942 Eocene–Recent
    Family Astacidae Latreille, 1803 Jurassic–Recent
    Family Parastacidae Huxley, 1879

Family Platychelidae Glaessner, 1969 Triassic
Infraorder Thalassinidea Latreille, 1831 Jurassic–Recent
    Family Thalassinidae Latreille, 1831 Recent
    Family Axiidae Huxley, 1879 Jurassic–Recent
    Family Laomediidae Borradaile, 1903 Miocene–Recent
    Family Callianassidae Dana, 1852 Jurassic–Recent
    Family Callianideidae Kossmann, 1880 Recent
    Family Upogebiidae Borradaile, 1903 Jurassic–Recent
    Family Axianassidae Schmitt, 1924 Recent
Infraorder Palinura Latreille, 1803 Triassic–Recent
    Family Cancrinidae Beurlen, 1930 Jurassic
    Family Eryonidae Derlaan, 1841 Jurassic–Cretaceous
    Family Glypheidae Winckler, 1883 Triassic–Recent
    Family Mecochiridae Van Straelen, 1925 Triassic–Cretaceous
    Family Pemphicidae Van Straelen, 1928 Triassic
    Family Palinuridae Latreille, 1803 Jurassic–Recent
    Family Polychelidae Wood–Mason, 1874 Jurassic–Recent
    Family Scyllaridae Latreille, 1825 Cretaceous–Recent
    Family Synaxidae Bate, 1881 Recent
    Family Tetrachelidae Beurlen, 1930 Triassic
Infraorder Anomala Boos, 1880 Jurassic–Recent
    Family Pomatochelidae Miers, 1879 Recent
    Family Diogenidae Ortmann, 1892 Cretaceous–Recent
    Family Coenobitidae Dana, 1851 ?Miocene–Recent
    Family Lomisidae Bouvier, 1895 Recent
    Family Paguridae Latreille, 1803 Jurassic–Recent
    Family Lithodidae Samouelle, 1819 Recent
    Family Parapaguridae Smith, 1882 Recent
    Family Galatheidae Samouelle, 1819 Jurassic–Recent
    Family Aeglidae Dana, 1852 Recent
    Family Chirostylidae Ortmann, 1892 Recent
    Family Porcellanidae Haworth, 1825 Cretaceous–Recent
    Family Albuneidae Stimpson, 1858 Oligocene–Recent
    Family Hippidae Latreille, 1825 Recent
Infraorder Brachyura Latreille, 1803 ?Carboniferous Jurassic–Recent
  Section Dromiacea De Haan, 1833 ?Carboniferous Jurassic–Recent
    Family Eocarcinidae Withers, 1932 Jurassic
    Family Prosopidae von Meyer, 1860 Jurassic–Recent
    Family Dromiidae De Haan, 1833 Paleocene–Recent
    Family Dynomenidae Ortmann, 1892 Jurassic–Recent
    Family Homolodromiidae Alcock, 1899 Recent
  Section Archaeobrachyura Guinot, 1977 Jurassic–Recent
    Family Cymonomidae Bouvier, 1897
    Family Dakoticancridae Rathbun, 1917 Cretaceous
    Family Homolidae De Haan, 1839 Jurassic–Recent
    Family Latreilliidae Stimpson, 1858 Recent

Family Raninidae De Haan, 1839 Cretaceous–Recent
Family Tymolidae Alcock, 1896 Cretaceous–Recent
Section Eubrachyura de St. Laurent, 1980
    Subsection Heterotremata Guinot, 1977 Cretaceous–Recent
        Family Calappidae De Haan, 1833 Cretaceous–Recent
        Family Dorippidae MacLeay, 1838 Cretaceous–Recent
        Family Leucosiidae Samouelle, 1819 Eocene–Recent
        Family Majidae Samouelle, 1819 Eocene–Recent
        Family Mimilambridae Williams, 1979 Recent
        Family Parthenopidae MacLeay, 1838 Eocene–Recent
        Family Atelecyclidae Ortmann, 1893 Eocene–Recent
        Family Cancridae Latreille, 1803 Eocene–Recent
        Family Corystidae Samouelle, 1819 Recent
        Family Pirimelidae Alcock, 1899 Recent
        Family Thiidae Dana, 1852 Recent
        Family Belliidae Dana, 1852 Recent
        Family Bythograeidae Williams, 1980 Recent
        Family Cancineretidae Beurlen, 1930 Cretaceous
        Family Deckeniidae Bott, 1970 Recent
        Family Geryonidae Colosi, 1923 Eocene–Recent
        Family Goneplacidae MacLeay, 1838 Paleocene–Recent
        Family Pinnotheridae De Haan, 1833 Eocene–Recent
        Family Platyxanthidae Guinot, 1977 Recent
        Family Portunidae Rafinesque, 1815 Eocene–Recent
        Family Potamidae Ortmann, 1896 Miocene–Recent
        Family Retroplumidae Gill, 1894 Eocene–Recent
        Family Xanthidae MacLeay, 1838 Cretaceous–Recent
    Subsection Thoracotremata Guinot, 1977 Eocene–Recent
        Family Gecarcinidae MacLeay, 1838 Pliocene–Recent
        Family Gecarcinucidae Rathbun, 1904 Recent
        Family Grapsidae MacLeay, 1838 Eocene–Recent
        Family Hapalocarcinidae Calman, 1900 Recent
        Family Hexapodidae Miers, 1886 Eocene–Recent
        Family Hymenosomatidae MacLeay, 1838 Recent
        Family Isolapotamidae Bott, 1970 Recent
        Family Ocypodidae Rafinesque, 1815 Eocene–Recent
        Family Palicidae Rathbun, 1898 Recent
        Family Parathelphusidae Alcock, 1910 Recent
        Family Potamocarcinidae Ortmann, 1899 Recent
        Family Potamonautidae Bott, 1920 Recent
        Family Pseudothelphusidae Ortmann, 1893 Recent
        Family Sinopotamidae Bott, 1970 Recent
        Family Sundathelphusidae Bott, 1969 Recent
        Family Trichodactylidae Milne Edwards, 1853 Recent.

REFERENCES

Abele, L. G. 1974. Species diversity of decapod crustaceans in marine habitats. *Ecol.* **55**:156–61.

Abele, L. G. 1976a. Comparative species richness in fluctuating and constant environments: Coral-associated decapod crustaceans. *Science* **192**:461–3.

Abele, L. G. 1976b. Comparative species composition and relative abundance of decapod crustaceans in marine habitats of Panama. *Mar. Biol.* **38**:263–78.

Abele, L. G. 1979. The community structure of coral-associated decapod crustaceans in variable environments. In *Ecological Processes in Coastal and Marine Systems* (R. J. Livingston, ed.), pp. 265–87. Plenum, New York.

Abele, L. G. 1982. Biogeography. In *The Biology of Crustacea*, Vol. 1 (L. G. Abele, ed.), pp. 241–304. Academic Press, New York.

Abele, L. G. 1983. Classification of the Decapoda. In *The Biology of Crustacea*, Vol. 5 (L. H. Mantel, ed.), pp. xxi–iii. Academic Press, New York.

Abele, L. G., K. L. Heck, D. S. Simberloff, and G. J. Vermeij. 1981. Biogeography of crab claw size: assumptions and a null hypothesis. *Syst. Zool.* **30**:406–24.

Abele, L. G., and W. K. Patton. 1976. The size of coral heads and the community biology of associated decapod crustaceans. *J. Biogeog.* **3**:34–47.

Abele, L. G., and K. Walters. 1979. The stability-time hypothesis: reevaluation of the data. *Am. Nat.* **114**:559–68.

Balss, H. 1957. Decapoda. In *Bronns Klassen und Ordnungen des Tierreichs* **5**(1)7. Akad. Verlag, Leipzig.

Barker, P. L., and R. Gibson. 1977. Observations and the feeding mechanism, structure of the gill, and digestive physiology of the European lobster *Homarus gammarus*. *J. Exp. Mar. Biol. Ecol.* **26**:297–324.

Barnes, N. B., and A. M. Wenner. 1968. Seasonal variation in the sand crab, *Emerita analoga* in the Santa Barbara area of California. *Limnol. Oceanogr.* **13**:465–75.

Bliss, D. E. 1963. The pericardial sacs of terrestrial Brachyura. In *Phylogeny and Evolution of Crustacea* (H. B. Whittington and W. D. I. Rolfe, eds.), pp. 59–78. Museum Comp. Zool., Cambridge.

Bliss, D. E. 1968. Transition from water to land in decapod crustaceans. *Am. Zool.* **8**:355–92.

Boltt, R. E. 1961. Antennary feeding of the hermit crab *Diogenes brevirostris*. *Nature* **192**:1099–100.

Borradaile, L. A. 1907. On the classification of the decapod crustaceans. *Ann. Mag. Nat. Hist.* (7)**19**:457–86.

Bourne, G. C. 1922. The Raninidae: a study in carcinology. *J. Linn. Soc. Zool.* **35**:25–79, pls. 4–7.

Bouvier, E. L. 1889. Le système nerveaux des Crustacés décapodes et ses rapports avec l'appareil circulatoire. *Ann. Sci. Nat. Zool.* **7**:73–96, 1 pl.

Bouvier, E. L. 1917. Crustacés décapodes (microures marcheurs) provenant des campagnes des yachts Hirondelle et Princess Alice. *Res. Camp. Sci. Monacco* **50**:1–104.

Bowman, T. E., and L. G. Abele. 1982. Classification of the recent Crustacea. In *The Biology of the Crustacea*, Vol. 1, (L. G. Abele, ed.), pp. 1–27. Academic Press, New York.

Budd, T. W., J. C. Lewis, and M. L. Tracey. 1978. The filter-feeding apparatus in crayfish. *Can. J. Zool.* **56**:695–707.

Burkenroad, M. D. 1963. The evolution of the Eucarida in relation to the fossil record. Tulane Stud. Geol. **2**(1):3–16.

Burkenroad, M. D. 1981. The taxonomy and evolution of Decapoda. *Trans. San Diego Soc. Nat. Hist.* **19**:251–68.

Burnett, B. R. 1984. Striated muscle in the wall of the dorsal abdominal aorta of the California spiny lobster *Panuliris interruptus*. *J. Crust. Biol.* **4**:560–66.

Caine, E. A. 1975a. Feeding and masticatory structures of selected Anomura. *J. Exp. Mar. Biol. Ecol.* **18**:277–301.

Caine, E. A. 1975b. Feeding and masticatory structures of six species of the crayfish genus *Procambarus*. *Form. Funct.* **8**:48–66.

Calman, W. T. 1909. *Crustacea*. In *A Treatise on Zoology* (E. R. Lankester, ed.). Adam and Charles Black, London.

Christy, J. H., and M. Salmon. 1984. Ecology and evolution of mating systems of fiddler crabs (genus *Uca*). *Biol. Rev.* **59**:483–509.

Cochran, D. M. 1935. The skeletal musculature of blue crab, *Callinectes sapidus*. *Smith. Misc. Coll.* **92**:1–76.

Crane, J. 1975. *Fiddler Crabs of the World. Ocypodidae: Genus* Uca. Princeton Univ. Press, Princeton.

de St. Laurent, M. 1979. Vers une nouvelle classification des Crustacés Décapodes Reptantia. *Bull. Off. Nat. Pêches Tunisie* **3**:15-31.

de St. Laurent, M. 1980a. Sur la classification et la phylogénie des Crustacés Décapodes Brachyoures. I. Podotremata et Eubrachyura. *C. R. Acad. Sci. Paris* (D)**290**:1265–68.

de St. Laurent, M. 1980b. Sur la classification et la phylogénie des Crustacés Décapodes Brachyoures. II. Heterotremata et Thoracotremata. *C. R. Acad. Sci. Paris* (D)**290**:1317–20.

Efford, I. E. 1966. Feeding in the sand crab *Emerita analoga*. *Crustaceana* **10**:167–82.

Efford, I. E. 1967. Neoteny in sand crabs of the genus *Emerita*. *Crustaceana* **13**: 81–93.

Farmer, A. S. 1974. The functional morphology of the mouthparts and pereiopods of *Nephrops norvegicus*. *J. Nat. Hist.* **8**:121–42.

Felder, D. L., J. W. Martin, and J. W. Goy. 1985. Patterns in early postlarval development of decapods. *Crust. Issues* **2**:163–225.

Felgenhauer, B. E., and L. G. Abele. 1983a. Phylogenetic relationships among shrimp-like decapods. *Crust. Issues* **1**:291–311.

Felgenhauer, B. E., and L. G. Abele. 1983b. Branchial water movement in the grapsid crab *Sesarma reticulatum*. *J. Crust. Biol.* **3**:187–95.

Förster, R. 1967. Die reptanten Dekapoden der Trias. *Neues Jahrb. Geol. Paleo. Abhl.* **128**:136–94.

Gerlach, S. A., D. K. Ekstrøm, and P. B. Eckardt. 1976. Filter feeding in the hermit crab *Pagurus bernhardus*. *Oecologia* **24**:257–64.

Glaessner, M. F. 1960. The fossil decapod Crustacea of New Zealand and the evolution of the order Decapoda. *N.Z. Geol. Surv. Paleontol. Bull.* **31**:1–63, 7 pls.

Glaessner, M. F. 1969. Decapoda. In *Treatise on Invertebrate Paleontology*, Part R, *Arthropoda* 4, Vol. 2 (R. C. Moore, ed.), pp. R399–533. Geol. Soc. Am. and Univ. Kansas Press, Lawrence.

Gore, R. H. 1985. Molt and growth in decapod larvae. *Crust. Issues* **2**:1–65.

Greenwood, J. G. 1972. The mouthparts and feeding behavior of two species of hermit crabs. *J. Nat. Hist.* **6**:325–37.

Guinot, D. 1977. Propositions pour une nouvelle classification des Crustacés Décapodes Brachyoures. *C. R. Acad. Sci. Paris* (D)**285**:1049–52.

Guinot, D. 1978. Principes d'une classification évolutive des Crustacés Décapodes Brachyoures. *Bull. Biol. Fr. Belg.* **112**:211–92.

Guinot, D. 1979. Problémes practique d'une classification cladistique des Crustacés Décapodes Brachyoures. *Bull. Off. Nat Peches Tunisie* **3**:33– 46.

Gurney, R. 1942. *Larvae of Decapod Crustacea*. Ray Society, London.

Hazlett, B. A. 1966. Social behavior of the Paguridae and Diogenidae of Curaçao. *Stud. Fauna Curaçao* **23**: 1–143.

Hazlett, B. A. 1968. Sexual behavior of some European hermit crabs. *Pubbl. Staz. Zool. Napoli* **36**:238–52.

Heck, K., and G. Wetstone. 1977. Habitat heterogeneity and invertebrate species richness and abundance in tropical sea-grass meadows. *J. Biogeo.* **4**:135–42.

Herrick, F. H. 1896. The American lobster: a study of its habits and development. *Bull. U.S. Fish. Comm.* **15**:1–251, 54 pls.

Huxley, T. H. 1880. *The Crayfish.* C. Kegan Paul & Co., London.

Jackson, H. G. 1913. *Eupagurus. Liverpool Mar. Biol. Comm. Mem.* **21**:1–79.

Knudsen, J. W. 1967. *Trapezia* and *Tetraclita* as obligate ectoparasites of pocilloporid and acroporid corals. *Pac. Sci.* **21**:51–7.

Kunze, J., and D. T. Anderson. 1979. Functional morphology of the mouthparts and gastric mill in the hermit crabs *Clibanarius taeniatus, C. virescens, Paguristes squamosus,* and *Dardanus setifer. Aust. J. Freshw. Res.* **30**:683–722.

Miller, D. C. 1961. The feeding mechanism of fiddler crabs, with ecological considerations of feeding adaptations. *Zoologica* **46**:89–100.

Mocquard, M. F. 1883. L'estomac des Crustacés podophthalmaires. *Ann. Sci. Nat. Zool.* (6)**16**:1–311.

Ngoc-Ho, N. 1984. The functional anatomy of the foregut of *Porcellana platycheles* and a comparison with *Galathea squamifera* and *Upogebia deltaura. J. Zool.* **203**:511–35.

Nicol, E. A. T. 1932. The feeding habits of the Galatheidae. *J. Mar. Biol. Assc. U.K.* **18**:87–106.

Oishi, S. 1960. Studies on the teloblasts in the decapod embryo II. Origin of teloblasts in *Pagurus samuelis* and *Hemigrapsus sanguineus. Embrylogia* **5**:270–82.

Orton, J. H. 1927. On the mode of feeding of the hermit crab *Eupagurus bernhardus* and some other Decapoda. *J. Mar. Biol. Assc. U.K.* **14**:909– 21.

Parker, T. J. 1889. The skeleton of the New Zealand crayfishes. *Stud. Biol. N.Z. Students* **4**: 1–25.

Paterson, N. F. 1968. The anatomy of the Cape rock lobster *Jasus lalandii. Ann. S. Afr. Mus.* **51**:1–228.

Patton, W. K. 1970. Community structure amongst the animals inhabiting the coral *Pocillopora damicornis* at Herron Island, Australia. In *Symbiosis in the Sea* (W. B. Vernberg, ed.), pp. 219–43. Univ. South Carolina, Columbia.

Patwardhan, S. S. 1935a. On the structure and mechanism of the gastric mill in Decapoda. 1. The structure of the gastric mill in *Parathelphusa guerini. Proc. Ind. Acad. Sci.* (B)**1**:183–96, 5 pls.

Patwardhan, S. S. 1935b. On the structure and mechanism of the gastric mill in Decapoda. 2. A comparative account of the gastric mill in Brachyura. *Proc. Ind. Acad. Sci.* (B)**1**:359–75.

Patwardhan, S. S. 1935c. On the structure and mechanism of the gastric mill in Decapoda. 3. Structure of the gastric mill in Anomura. *Proc. Ind. Acad. Sci.* (B)**1**:405–13.

Patwardhan, S. S. 1935d. On the structure and mechanism of the gastric mill in Decapoda. 4. The structure of the gastric mill in reptantous Macrura. *Proc. Ind. Acad. Sci.* (B)**1**:414–22.

Pellegrino, C. R. 1984. The role of desiccation pressures and surface area/volume relationships on seasonal zonation and size distribution of four intertidal decapod Crustacea from New Zealand: implications for adaptation to land. *Crustaceana* **47**:251–68.

Pike, R. B. 1947. *Galathea. Liverpool Mar. Biol. Comm. Mem.* **34**:1–138, 20 pls.

Pilgrim, R. L. C. 1973. Axial skeleton and musculature in the thorax of the hermit crab, *Pagurus bernhardus. J. Mar. Biol. Assc. U.K.* **53**:363–96.

Pilgrim, R. L. C., and C. A. G. Wiersma. 1981. Observations on the skeleton and somatic musculature of the abdomen and thorax of *Procambarus clarkii. J. Morph.* **113**:453–87.

Rabalais, N. N., and R. H. Gore. 1985. Abbreviated development in decapods. *Crust. Issues* **2**:67–126.

Raynor, M. D. 1965. A reinvestigation of the segmentation of the crayfish abdomen and thorax, based on a study of the deep flexor muscles and their relation to the skeleton and innervation. 1. The skeleton and intersegmental membranes. *J. Morph.* **116**:389–412.

Reddy, A. R. 1935. The structure, mechanism, and development of the gastric armature in

Stomatopoda with a discussion as to its evolution in the Decapoda. *Proc. Ind. Acad. Sci.* (B)**1**:650–75.

Rice, A. L. 1980. Crab zoeal morphology and its bearing on the classification of the Brachyura. *Trans. Zool. Soc. Lond.* **35**:271–424.

Ricklefs, R. E. 1979. *Ecology*. Chiron Press, New York.

Roberts, M. H. 1968. Functional morphology of mouthparts of the hermit crabs *Pagurus longicarpus* and *P. pollicaris*. *Chesapeake Sci.* **9**:9–20.

Sanders, H. L. 1968. Marine benthic diversity: a comparative study. *Am. Nat.* **102**:253–82.

Schaefer, N. 1970. The functional morphology of the foregut of three species of decapod Crustacea. *Zool. Afr.* **5**:309–26.

Schembri, P. J. 1982a. The functional morphology of the feeding and grooming appendages of *Ebalia tuberosa*. *J. Nat. Hist.* **16**:467–80.

Schembri, P. J. 1982b. Feeding behavior of fifteen species of hermit crabs from the Otago region, southeastern New Zealand. *J. Nat. Hist.* **16**:859–78.

Schmidt, W. 1915. Die Musculatur von *Astacus fluviatilis* ein Beitrag zur Morphologie der Decapoden. *Z. wiss. Zool.* **113**:165–251.

Schram, F. R. 1980. Notes on miscellaneous crustaceans from the Late Paleozoic of the Soviet Union. *J. Paleo.* **54**:542–7.

Schram, F. R., R. M. Feldmann, and M. J. Copeland. 1978. The Late Devonian Palaeopalaemonidae and the earliest decapod crustaceans. *J. Paleo.* **52**:1375–87.

Schram, F. R., and R. H. Mapes. 1984. *Imocaris tuberculata* from the upper Mississippian Imo Formation, Arkansas. *Trans. San Diego Soc. Nat. Hist.* **20**:165–8.

Secretan, S. 1960. Observations relatives au processus d'evolution des sillons chez les Crustacés Décapodes Macroures. *C. R. Séance Acad. Sci.* **251**:1551–3.

Secretan, S. 1966. Transformations squellettiques liées à la céphalisation chez les Crustacés Décapodes. *C. R. Acad. Sci. Paris* (D)**262**:1062–5.

Secretan, S. 1970. Nouvelles observations sur la mandibule des Crustacés, sa composition et son origine. *C. R. Acad. Sci. Paris* (D)**271**:1888–91.

Secretan, S. 1973. A propos des sillons et d'une mandibule apparente sur des spécimens nouveaux de *Pseudoglyphea etalloni*. *Ann. Paléont. Invert.* **59**:187–201, 2 pls.

Secretan, S. 1977. La notion d'épimère chez les Crustaces Decapodes. Èpimère et connexions épimèro—endophragmales chez les Astacoures. *Bull. Soc. Zool. Fr.* **102**:345–74.

Secretan, S. 1980. Le plan de base du 'squelette axial' d'un Crustaće décapode macroure et sa terminologie. *C. R. Acad. Sci. Paris* (D)**291**:877–80.

Secretan, S. 1981. Constitution de l'arceau endocéphalique du squelette axial d'un Crustaće Décapodes macroure. *C. R. Acad. Sci. Paris* (III)**292**:1059–62.

Shiino, S. M. 1950. Studies on the embryonic development of *Panulirus japonicus*. *J. Fac. Fish. Prefect. Univ. Mie, Otanimachi* **1**:1–221 (in Japanese with English summary).

Stenseth, N. C., and J. Maynard Smith. 1984. Coevolution in ecosystems: Red Queen evolution or stasis. *Evol.* **38**:870–80.

Terao, A. 1929. Embryonic development of the spiny lobster *Panulirus japonicus*. *Japan. J. Zool.* **2**:387–449, pls. 11–15.

Vermeij, G. J. 1976. Interoceanic differences in vulnerability of shelled prey to crab predation. *Nature* **260**:135–6.

Vermeij, G. J. 1977. Patterns in crab claw size: the geography of crushing. *Syst. Zool.* **26**:138–51.

Vonk, H. J. 1960. Digestion and metabolism. In *Physiology of Crustacea* (T. H. Waterman, ed.), pp. 291–316. Academic Press, New York.

Warner, G. F. 1977. *The Biology of Crabs*. Van Nostrand Reinhold, New York.

Wenner, A. M., and S. R. Haley, 1981. On the question of sex reversal in male crabs. *J. Crust. Biol.* **1**:506–17.

Williamson, D. I. 1976. Larval characters and the origin of crabs. *Thal. Jugoslav.* **10**:401–14.

25

LEPTOSTRACA

DEFINITION   Carapace bivalved, with adductor muscle, but without definite hinge line; antennular rami unequal, inner branch flagelliform, outer branch as a scale; antenna uniramous; mandibular gnathobase reduced, palp large; thoracopods phyllopodous, rami variously developed; abdomen of seven segments; six pairs of pleopods, anterior four biramous paddles, posterior two reduced and uniramous; anal segment with terminal anus, bearing pair of broad, oval, setose caudal rami.

HISTORY   Members of the group have been recognized as presenting a taxonomic problem since the early 1800s. Initially treated as 'Macrura' because of their well-developed abdomens, H. Milne Edwards in the early half of the nineteenth century placed *Nebalia* among the phyllopods (though he admitted certain general similarities to mysids), and his assignment prevailed for some decades. Metschnikoff in the middle 1800s, while describing the embryology of *Nebalia*, termed the group 'phyllopodiform decapods.' It was not until Claus (1872,1888) that an extensive comparison and assignment of leptostracans as 'podophthalmid' malacostracans took place. This malacostracan placement was based on possession of movably stalked eyes, the nature of the nervous system, well-developed carapace, an eight-segment thorax, and a general similarity of *Paranebalia* thoracopods to those of schizopodous eumalacostracans. On the other hand, Sars (1887) and Packard (1883) rejected Claus' malacostracan assignment, preferring an alignment close to the phyllopodous branchiopods; but when Calman (1909) sided with Claus the issue seemed settled. However, Calman too, like Sars and Packard, was concerned with the distinctive nature of the phyllopodous thoracopods and their similarity to those of branchiopods and devoted considerable attention to this in his 1909 treatise.

MORPHOLOGY   The leptostracan carapace is laterally compressed to form a bivalved shell enclosing the thoracopods and extending back over the anterior part of the abdomen (Fig. 25-1). Though the carapace does not have a median hinge line as such, it is closed by a large adductor muscle at the level of the maxillae. The rostrum is large, movable, and used to control and restrict the anterior-to-posterior flow of fluid into the carapace chamber. The eyes are stalked and mobile, but when visual pigment cells are present they are generally deeply buried in the tissue of the eye. The surface reticulated corneal network characteristic of higher malacostracans

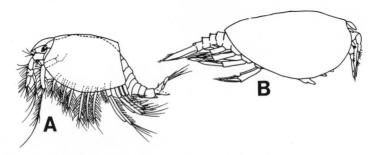

Fig 25-1 (A) *Paranebalia fortunata* (from Wakabara, 1976); (B) *Nebaliopsis typica* (from Cannon, 1960).

is not present. In addition, the eye lobes of *Paranebalia* are anteromedially serrated, and those of *Dahlella* are scimitarlike blades (Hessler, 1984).

The antennules (Fig. 25-2A) are distinctive appendages. The large peduncle has four segments, the most distal of which can bear heavy toothlike spines, as in *Paranebalia*, developed as a serrated blade. The external branch is flagellar. The antennules bear a large number of aesthetascs.

The antennae are uniramous. The peduncle has three or four segments. In *Nebalia* and *Nebaliella* (Fig. 25-3A) the flagellum is long with many small joints, while in *Nebaliopsis* and *Paranebalia* (Fig. 25-2B) the flagellum is short and with few though large joints.

The labrum is not particularly large and the paragnaths are deeply cleft into two distinct lobes. The mandibles (Fig. 25-2C) have rather small molar and incisor regions, but the three-segment palp is very large with the distal one or two segments potentially rather setose.

The maxillules can be variously developed. In *Nebaliopsis* (Fig. 25-3B) they are two-segmented, moderately setose flaps, whereas in *Nebalia* and *Paranebalia* (Fig. 25-2D) they are biramous and bear well-developed endites.

The maxillae are also variously structured. In *Nebaliopsis* (Fig. 25-3C) they are simple broad lobes; in *Nebaliella* (Fig. 25-3D) and *Nebalia* they are somewhat thoracopodlike, while in *Paranebalia* (Fig. 25-2E) the exite is developed as a great palp.

The thoracopods differ in all the genera. *Nebalia* (Fig. 25-3E) has a three- to four-segment endopod, well-developed epi- and exopods, and the complex of setae largely restricted to along the medial edge of the limb. *Paranebalia* (Fig. 25-2F) has a long, thin five-segment endopod, long and narrow epi- and exopods, and complex setae along both the medial and lateral margins of the limb. *Nebaliella* (Fig. 25-3F) has a stout endopod of five segments, a well-developed exopod, no epipod, and simple setae along medial and lateral limb margins. *Nebaliopsis* (Fig. 25-3G) has the endo- and exopods reduced to simple setate lobes and a well-developed epipod.

The four anterior pleopods (Fig. 25-2G) are well developed as swimmerets. The protopod is large, massive, and possesses well-developed muscles. The exopod is a single oval flap. The endopod has a short proximal segment, which is developed medially with appendix internae, and an elongate distal segment. The fifth and sixth pleopods (Fig. 25-2H) are reduced uniramous lobes.

The pleon bears a terminal anus and a pair of broad, setose caudal rami (Fig. 25-2I).

The digestive system can exhibit a great deal of variation in structure.

Fig 25-2. Appendages of *Paranebalia fortunata*. (A) antennule; (B) antenna; (C) mandible; (D) maxillule; (E) maxilla; (F) thoracopod; (G) first pleopod; (H) fifth pleopod; (I) caudal ramus. (From Wakabara, 1976)

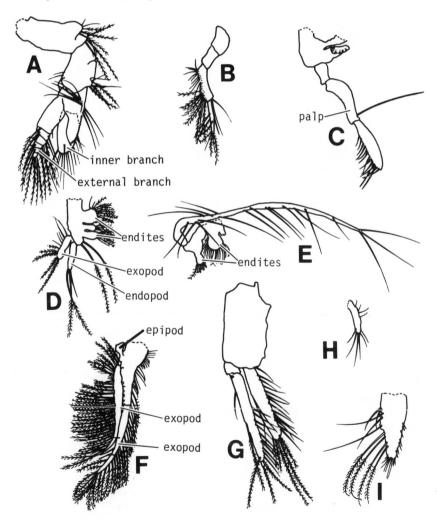

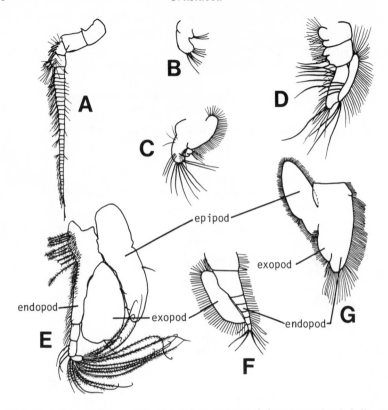

Fig. 25-3. Variant appendage morphology in leptostracans. (A) antenna of *Nebaliella antarctica*; (B, C) respectively, maxillule and maxilla of *Nebaliopsis typica*; (D) maxilla of *Nebaliella antarctica*; (E) thoracopod of *Nebalia pugettensis*; (F) *Nebaliella antarctica*; (G) *Nebaliopsis typica*. (All except E from Thiele, 1927; E from Clark, 1932)

The posteriorly directed mouth opens into the foregut in *Nebalia* and *Nebaliella* (Fig. 25-4B, C), which has moderately well-developed cardiac and pyloric regions. Caeca arise in the anterior region of the midgut with a pair of short anteriorly directed and a pair of long posteriorly directed diverticula. At the extreme posterior end of the midgut a medial rectal caecum extends dorsally into the pleon, being quite large in *Nebalia*. *Nebaliopsis* (Fig. 25-4A) has a distinctly different arrangement. There is little differentiation of the foregut, and the huge ventral midgut caecum fills the great bulk of the thorax and anterior abdomen. The midgut is very slender, while the hindgut is relatively large.

The circulatory system has a heart extending from the posterior cephalon to the level of the fourth pleomere (Fig. 25-5). There are seven ostia, one in the head and then a set in each of the first six thoracomeres. The first two and the last thoracic ostia are paired and lateral, while the ostia of the third through fifth thoracomeres are single and dorsal (a rather unusual

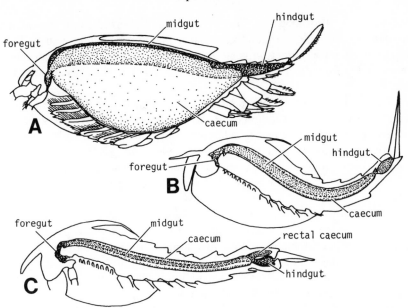

Fig. 25-4. Digestive systems of leptostracans. (A) *Nebaliopsis*; (B) *Nebalia*; (C) *Nebaliella*. (From Cannon, 1960)

arrangement). The anterior, posterior, and 12 paired lateral arteries supply the body organs and muscles. The turgor of the phyllopodous thoracopods, so important in sustaining the feeding mechanism, is maintained by open communication of the appendages with the thoracic hemocoel.

Respiration, such as its needs exist, is achieved over the surface of the thoracic appendages and from a blood sinus extending into the carapace.

Excretion is achieved apparently by small antennal and maxillary glands, as well as eight pairs of glandular tissue associated with the hemal

Fig. 25-5. Internal organs of *Nebalia geoffroyi*. (From Cannon, 1960)

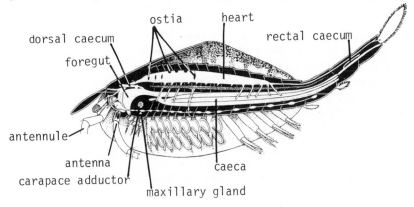

system of the thoracopods. Claus (1888) first reported these tissues responded to carmine dye much like the antennal and maxillary glands. Their apparent excretory function is similar to structures seen in some mysidaceans and bathynellaceans, though no work has been done to actually verify this.

The reproductive system is composed of simple thin tubelike gonads, which extend most of the length of the body, empty near the bases of the sixth and eighth thoracomeres in females and males, respectively. The only apparent secondary sexual characters in the appendages involve the antennules, which generally have a longer flagellum and smaller scale in males than females. Eggs are brooded between the carapace and the thoracopods.

The nervous system is a mosaic of features. The anterior portion of the system is composed of a single cord and completely fused ganglia, as is found in higher malacostracans. In addition, the cephalic and thoracic ganglia are almost nearly fused longitudinally, though distinct ganglionic masses are still discernible. Interganglionic connectives reminiscent of branchiopods are noted in the abdomen.

Hessler (1964) illustrates and discusses the body musculature of *Nebalia pugettensis* (Fig. 25-6). The massive dorsal longitudinal muscles of the thorax and abdomen, and the ventral longitudinal muscles of the abdomen are the largest muscle systems in the body. However, of some interest is the box truss arrangement formed in the thorax by vertical and anteriorly descending dorsoventral trunk muscles, similar in form to such systems seen in the thoraxes of cephalocarids and branchiopods. *Nebalia*, though, appears to lack a posteriorly descending dorso-ventral muscle seen in the latter groups.

NATURAL HISTORY   Many leptostracans, with the exception of the pelagic *Nebaliopsis*, seem to prefer mud bottoms that are low in oxygen content. *Dahlella* lives right down hydrothermal vents. When placed in laboratory tanks or watch glasses with mud, they will swim right to the bottom and burrow in (Cannon, 1927). Sometimes the trunk limbs move after the animal has settled down. However, frequently, nothing of the body is moved at all, and even the heartbeat slows down, which are functional and physiological adaptations to living in low-oxygen environments. However, the group can be found in a variety of habitats ranging from the intertidal to the abyssal deeps.

Swimming is achieved by the four anterior pleopods and caudal rami (Wägele, 1983), the phyllopodous thoracopods being strictly used for feeding. Cannon (1927) performed a detailed analysis of feeding in *Nebalia bipes* (Fig. 25-7). The enveloping carapace forms a perithoracic filtering chamber, with the movable rostrum serving to direct and regulate the flow of detritus-laden water into the chamber. The anterior pleopods and the antennules cooperate to move the animal with forward jerking motions

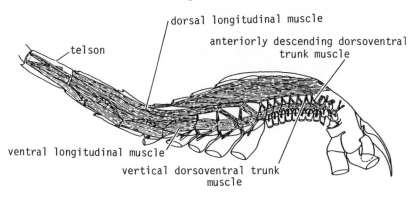

Fig. 25-6. Muscular system of *Nebalia pugattensis*. (From Hessler, 1964)

within the bottom ooze. The antennules also serve to create a low-pressure area near the rostrum, which allows material to flow into this area prior to being sucked into the thoracic chamber by the action of the thoracopods.

The epi- and exopods overlap laterally (Fig. 25-4), preventing the flow of fluid anywhere but up the midventral line adjacent to the endopods. The endopods themselves are closer to their paired mate posteriorly than laterally, thus forming a V-shaped basket. The limbs do not beat in a strict metachronal pattern; the anterior amplitudes are greater while the recovery beats of the posterior limbs are slower than those of the anterior.

Water flows into the carapace chamber anteromedially under the rostrum. The thoracic endopods have four rows of setae (Fig. 25-8). The first and third rows interlock like the elements on a bird feather. These form an interconnected network that helps to prevent particles from escaping (Fig. 25-9). The brush setae on the second row actually trap the particles and pass them up to the food groove where they are moved anteriorly to the mouth. The setae of the second row tend to sweep clogged particles on the

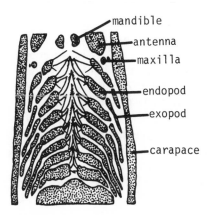

Fig. 25-7. Ventral view of the thoracic feeding chamber. (Modified from Cannon, 1927)

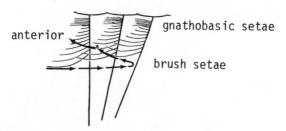

anterior

gnathobasic setae

brush setae

Fig. 25-8. Lateral view of food flow around thoracopods. (Modified from Cannon, 1927)

net formed by the first and third rows, so that they can be trapped on the brush setae of more posterior appendages. The maxillules and maxillae manipulate the food particles between the paragnaths into the mouth. Actual mastication of any large particles is achieved by the maxillary endites rather than the mandibles. These latter are too weakly developed to really chew. The palps of the mandibles are used to hold food in place before it is pushed into the mouth. The palps also maintain food in the maxillary area if mastication is taking place.

There are two significant features about this feeding system: (1) the current producers and the filter are the trunk endopods and (2) the main feeding current passes from anterior to posterior. The latter stands in contrast to the system seen in some eumalacostracans where a maxillary filter traps particles. The leptostracan system is distinctly opposed to the posterior-to-anterior feeding flow of feeding currents apparently seen in branchiopods and brachypod cephalocarids. Wägele (1983) observed that *N. marerubri* is able to feed on carrion, using its antennae to tear chunks of flesh.

Though secondary sexual characters are recognized in the antennules of leptostracans, exact modes of mating are not known. Eggs are generally brooded under the carapace of the female. Most unusual is the early hatching of embryos and their continued brooding under the carapace in a helpless 'embryonized' condition.

The biogeographic distributions of leptostracans indicate widespread genera. However, the taxonomy of the group is so badly in need of revision, little of meaning can be deduced from current knowledge of their distribution. *Paranebalia longipes* occurs in waters 1 to 9 meters in depth around Bermuda, the Virgin Islands, and in Japan and southeast Asia, while *P. fortunata* is apparently restricted to New Zealand at about 600 meters. *Nebaliopsis typica* occurs widely down to 3500 meters in the southern Indo-Pacific and the southern Atlantic. *Nebaliella*, occurring from 9 to 2085 meters has a disjunct distribution, *N. caboti* occurring from Newfoundland to New Jersey, and *N. antarctica* and *N. extreme* from New Zealand and Antarctica. The genus *Nebalia* has six species at present that effectively make the genus ubiquitous in its distribution from the Arctic to various reaches of the southern hemisphere.

DEVELOPMENT   Manton (1934) remains the standard reference on leptostracan embryology. Besides offering a detailed presentation of data and a review and critique of all earlier and more incomplete efforts, she also compared *Nebalia* development to that seen in other crustaceans, predominately concentrating her attention on comparisons with eumalacostracans.

The egg is heavily laden with yolk, and the cytoplasm gradually makes its way to the surface before segmentation begins. A blastodisc is formed at a single point, which then grows outward as a single-layered blastoderm to

Fig. 25-9. Germinal area of *Nebalia bipes* embryos, surface features on the right and deeper layers on the left. (A) gastrulation initiated; (B) symmetry established; (C) beginning of egg-nauplius stage; (D) teloblast formation almost complete; (E) caudal papilla formed. (Modified from Manton, 1934)

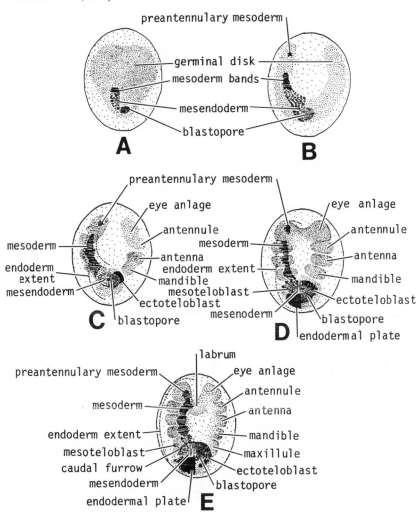

completely enclose the yolk mass (Fig. 25-10A). Eventually a blastopore area thickens, and the thickening extends anteriad to form a germinal disc (Fig. 25-9A). The disc further thickens in its lateral and posterior aspects and begins to thin in its anterior aspect; the resultant V-shaped primordium establishes the planes of symmetry of the embryo (Fig. 25-9B).

The external aspects of development can be easily characterized. The optic lobes and naupliar appendage anlagen thicken (Fig. 25-9C), with the optic lobes moving medially to convert the V-shape of the naupliar embryo into an O-shape (Fig. 25-9D). The area where the optic lobes begin to

Fig. 25-10. Sections of developing embryos of *Nebalia bipes*. (A) sagittal section of germinal disc; (B) sagittal section of gastrulation initiation; (C) parasagittal section with the beginning of teloblast formation; (D) sagittal section after initiation of caudal papilla; (E) transverse section of D. (Modified from Manton, 1934)

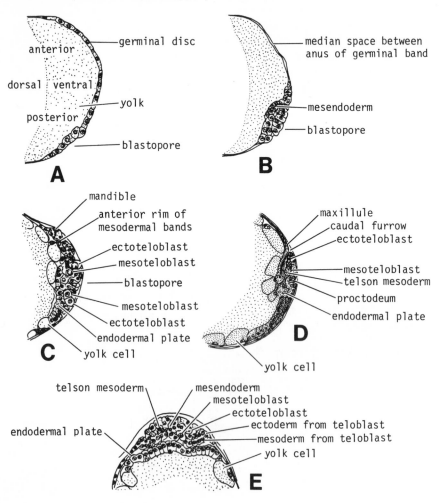

approach each other thicken and grow posteriad to form the labral rudiment (Fig. 25-9E). After forming the maxillulary and maxillary segments, the teloblasts are separated from the head region by a transverse furrow that induces that part of the germinal disc posterior to the furrow to grow forward as the caudal papilla (Fig. 25-9E). The limb buds appear in sequence from anterior to posterior up to the point where the third abdominal limb anlagen appear. The carapace extends to the level of the third thoracomere whereupon the embryo hatches to continue its development within the carapace chamber of the mother.

Internal development commences once the germinal disc is laid down. The blastopore area anteriorly differentiates mesoderm and posteriorly differentiates endodermal tissues (Fig. 25-10B). These posterior endoderm cells are modified to become vitellophages as they grow outward and completely envelope the yolk mass in a second layer below the initial blastoderm (Fig. 25-10C). The endoderm cells at the blastopore, however, merely thicken to form the epithelial cells of the endodermal plate. The plate in turn slowly grows outward by converting vitellophagic yolk cells on its margin to epithelial types. By the time of hatching, the yolk mass is completely enveloped by gut epithelium.

Four distinct mesodermal areas are recognized. The naupliar mesoderm grows forward under the V-shaped ectodermal mass (Fig. 25-9B, C) and contributes to the naupliar anlagen. The preantennulary mesoderm delaminates from below the differentiating optic lobes (Fig. 25-9C, D) and grows medially. The posterior portion of this acronal mesoderm grows past the antennulary mesoderm posteriorly and contributes to the labrum and the stomodeum (Fig. 25-9E). The anterior end of the preantennulary mesoderm grows forward to the ectodermal dorsal organ and, together with that, induces the formation of the cephalic caeca and the anterior aorta. The ecto- and mesoteloblasts begin to form anterior of the blastopore and soon grow to form a ring around the edges of the caudal papilla (Figs. 25-10D, E). The ectoteloblasts are composed of an anterior median and nine lateral pairs of cells, and the mesoteloblasts arise as four pairs of cells below these. The teloblasts form all the walls of the caudal papilla, and after the seventh abdominal somite is delimited they disappear. The fourth mesodermal area (Fig. 25-10E) is that of the anal segment and caudal rami, which arises from cells laying between the teloblasts and the caudal papilla ectoderm.

After hatching (Fig. 25-11), the embryo continues to pass through two molt stages. The primary development prior to attaining a free juvenile stage involves increasing appendage maturation and gut formation. In this latter, the yolk comes to be concentrated in the anterior part of the midgut. The cephalic caeca grow forward as paired diverticula. These structures, unique to leptostracans (as far as malacostracans are concerned), flank the stomodeum. By the time the first embryonic cuticle is shed, the posterior caeca arise as paired simple diverticula from the anterior midgut just

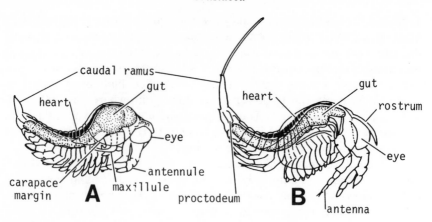

Fig. 25-11. Advanced hatchling stage of *Nebalia bipes*. (From Manton, 1934)

behind the anterior cephalic caeca. Through the free swimming stages these continue to grow and, as the adult condition is approached, they each divide longitudinally into three lobes. At the time of the second embryonic molt a single middorsal caecum arises at the posterior end of the midgut, which eventually comes to grow back over the proctodeum. The proctodeum itself grows in from the blastopore. The anus initially opens dorsally near the tip of the caudal papilla, though it eventually moves to its characteristic terminal position.

FOSSIL RECORD   Only one fossil leptostracan (Fig. 25-12) has been recognized in the literature, the Permian *Rhabdouraea bentzi* (Malzahn 1962). It is distinctly different from any other leptostracan with heavy pleopods and long rodlike caudal rami and has a separate family status (Schram and Malzahn, 1984). However, the fossil is only that of an abdomen and makes comparison to other leptostracans difficult.

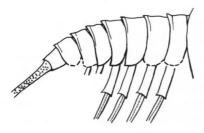

Fig. 25-12. The Permian leptostracan, *Rhabdouraea bentzi*. (From Schram and Malzahn, 1984)

TAXONOMY   Currently five genera with 13 species are recognized among living forms. Dahl (personal communication) relates the group is badly in need of revision, with new genera and species to be erected. A total of three families is currently recognized.

Family Neballiidae Baird, 1850 Recent
Family Nebaliopsidae Hessler, 1984 Recent
Family Rhabdouraeidae Schram and Malzahn, 1984 Permian

REFERENCES

Calman, W. T. 1909. Crustacea. In *A Treatise on Zoology*, Vol. 9 (E. R. Lancaster, ed.). A. & C. Black, London.

Cannon, H. G. 1927. On the feeding mechanism of *Nebalia bipes*. *Trans. Roy. Soc. Edinb.* **55**:355–69.

Cannon, H. G. 1960. Leptostraca. In H. G. Bronn, *Klassen und Ordnungen des Tierreichs* (2nd edition) **5**(1) **4**(1):1–81. Akad. Verlag, Leipzig.

Clark, A. E. 1932. *Nebaliella caboti* n. sp. with observations on other Nebaliacea. *Trans. Roy. Soc. Can.* (3)**26**:217–35.

Claus, C. 1872. Über den Bau und die systemische Stellung von *Nebalia*. *Zeit. wiss. Zool.* **22**:323–30.

Claus, C. 1888. Über den organismus der Nabaliden und die systematische Stellung der Leptostraken. *Arb. zool. Inst. Wien* **8**:1–148.

Hessler, R. R. 1964. The Cephalocarida comparative skeletomusculature. *Mem. Conn. Acad. Arts & Sci.* **16**:1–97.

Hessler, R. R. 1984. *Dahlella caldariensis*, new genus, new species: a leptostracan from deep-sea hydrothermal vents. *J. Crust. Biol.* **4**:655–64.

Malzahn, E. 1962. Beschreibung der Arten, Teil 1. In M. F. Glaessner and E. Malzahn, *Neue Crustaceen aus dem niederrheinischen Zechstein*. Fortschr. Geol. Rheinld. u. Westf. **6**:245–64.

Manton, S. M. 1934. On the embryology of the crustacean *Nebalia bipes*. *Phil. Trans. Roy. Soc. Lond.* (B)**223**:163-238.

Packard, A. S. 1883. A monograph of the phyllopod Crustacea of North America, with remarks on the order Phyllocarida. *12th Ann. Rept. U.S. Geol. Geogr. Surv. of the Territories* **1878**:295–457.

Sars, G. O. 1887. Report on the Phyllocarida collected by H.M.S. Challenger during the years 1873–76. *Challenger Sci. Repts. Zool.* **14**:1–38.

Schram, F. R., and E. Malzahn. 1984. The fossil leptostracan *Rhabdouraea bentzi*. *Trans. San Diego Soc. Nat. Hist.* **20**:95–98.

Thiele, J. 1927. Leptostraken. In *Handbuch der Zoologie* (W. Kukenthal and T. Krumbein, eds.), **3**:567–92. Der Gruyter, Berlin.

Wägele. J. W. 1983. *Nebalia mererubri* aus dem Roten Meer. *J. Nat. Hist.* **17**:127–38.

Wakabara, Y. 1976. *Paranebalia fortunata* n. sp. from New Zealand. *J. Roy. Soc. N.Z.* **6**:297–300.

26

HYMENOSTRACA, ARCHAEOSTRACA, HOPLOSTRACA, AND CANADASPIDIDA

The extinct orders generally assigned to the Phyllocarida exhibit a great range of morphologic variation. Little is known about many of these groups, and at present the *single* feature, and none other, that apparently unites all of these as Phyllocarida is a seven-segment abdomen; not a hinged bivalved carapace (hoplostracans and hymenostracans lack this); not a rostral plate (canadaspidids lack this). So little is known concerning appendages of these orders, despite the generally good preservation of the body, that one is tempted to suspect that they all might share thin phyllopodous appendages similar to those of the living Leptostraca. However, comparison of the limbs of *Canadaspis perfecta*, from the Cambrian of British Columbia, to those of any leptostracan clearly reveals that such generalizations about appendages involve considerable risk. The implication of all this is that although it is convenient to group all these orders together in the Phyllocarida (e.g., Briggs, 1978), some of these groups may have nothing to do with each other.

HYMENOSTRACA

DEFINITION   Carapace without hinge line (apparently bivalved); last abdominal somite subequal to all others; abdomen of seven segments(?); uropods biramous and bladelike; telson narrow and deeply cleft.

MORPHOLOGY   So little is known about *Hymenocaris* (Fig. 26-1). The carapace was ovate and appeared to have a posterior thickened rim. The carapace appeared to cover the thoracic segments only. Nothing is known about appendages. All the abdominal somites were subequal in length, though the posterior three segments appeared to be not quite as deep as the anterior. The telson was narrow and deeply cleft. The caudal rami (uropods according to Bowman, 1971) were narrow, bifid blades. The appearance of the tailfan then is as an array of six 'spines.' The tailfan interpretation favored here is different from the traditional one of multira-

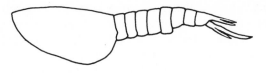

Fig. 26-1. *Hymenocaris vermicauda.*

mous furcae (e.g., Rolfe, 1969), but is influenced by the observations of Bowman (1971).

NATURAL HISTORY   Nothing is known about what hymenostracans may have been doing.

They range in age from Middle Cambrian (possibly lower as well) of Great Britain, Australia, and New Zealand to the Lower Ordovician of Wales.

ARCHAEOSTRACA

DEFINITION   Medium to large in size, carapace bivalved, with a hinge line; antennules apparently biramous; antenna biramous with two flagellar rami; mandibles with massive gnathal lobe; seventh pleomere generally longer than any others; telson as a prolonged posteriorly directed spike; caudal rami uniramous and bladed.

MORPHOLOGY   Only one species, the middle Devonian *Nahecaris stuertzi*, is known at all completely (Fig. 26-2) because of the unusual preservational conditions that prevail in the Hunsruck Slate where it is found.

The carapace formed a large laterally enveloping shield, completely enclosing the thorax, thoracopods, and anterior pleomeres. The rostrum was developed as an articulating plate. The margin of the carapace was frequently developed as a thickened rim reflected to form a doublure. The eyes were stalked.

The little that is known about appendages is summarized by Rolfe (1969). The most complete information comes from *Nahecaris*. The

Fig. 26-2. *Nahecaris stuertzi.* (From Rolfe, 1969)

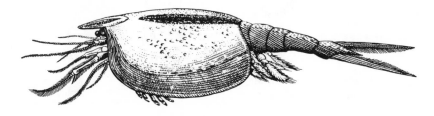

antennules had large peduncles and two long setose flagella (Fig. 26-2). The setose antennae had a flagellate exopod, and an endopod with two large proximal segments and a terminal flagellum (Fig. 26-2). The mandibles had massive molar lobes and typically prominent incisor processes (Fig. 26-3A). What little can be discerned of the thoracopods indicates they apparently did not have stenopods. Only the first five abdominal segments carry pleopods, which were rather nondescript, thin, broad paddles mounted on a two-segment protopod.

The telson was developed as a long spike. What is frequently termed the telson head can have a large anal apparatus that was at least occasionally floored with a ventral platform (Fig. 26-3B). A pair of prominent, broad, uniramous caudal rami flanked the telson (referred to by some authors as furcae).

NATURAL HISTORY   Archaeostracans are almost without exception found in marine rocks, except for a few taxonomic problematica, which we are not even sure are phyllocarids. However, while a marine habit is shared by all, the range of morphologic forms indicate a great radiation into a variety of habitats. Heavily armored forms, like the echinocaridines (Fig. 26-4A) and aristozoids, are assumed to have been benthic reef-dwelling forms (Chlupáč, 1960; Krestovnikov, 1961). Smaller types with well-developed caudal rami and wide geographic distributions, like *Caryocaris* (Fig. 26-4B, C) were probably planktonic. Forms with moderately thin cuticles, such as *Ceratiocaris* (Fig. 26-4D) or *Dithyrocaris*, occur in faunas, which indicate probable benthic or neritic habitats in the calm waters of coastal or lagoonal settings.

Fig. 26-3. (A) medial view of massive mandible of *Ceratiocaris monroei*, palp reconstructed; (B, C) ventral and lateral views of *Shugurocaris cornwallisensis* telson base and seventh pleomere. (From Rolfe, 1969)

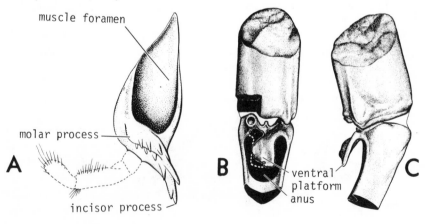

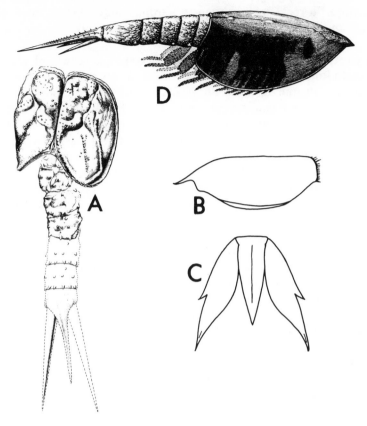

Fig. 26-4. (A) *Echinocaris punctata*; (B) carapace of *Caryocaris maccoyi*; (C) tail of *Caryocaris monodon*; (D) *Ceratiocaris papilio*. (From Rolfe, 1969)

The generally large and massive mandibles found on many archaeostracans do not necessarily indicate carnivorous habits. However, these, combined with gut fillings and a frequently inferred benthic and possible burrowing habit, might indicate a certain degree of scavenging in a diet involving the processing of a large amount of detritus (Rolfe, 1969).

Sexual dimorphism is known in the living leptostracan phyllocarids. However, the possibility of such in archaeostracans has generally been ignored by workers, even though several 'species' typically occur together in fossil faunas and the characters used to delineate species are frequently rather subtle carapace length–height ratios. Only Glaessner (1931) seems to have ever attributed variations within a species, *Austriocaris carinata*, to dimorphism, and this species is not a phyllocarid (Briggs and Rolfe, 1983).

Distribution within the group appears to be worldwide. Archaeostracans can be locally abundant fossils when encountered; however, they are generally too rare to be the basis of practical biogeographic deductions.

They occur in rocks of Cambrian to Permian age, with the most diverse array found in the Devonian (Rolfe and Edwards, 1979).

TAXONOMY   Rolfe (1969) offers the most complete review of genera to date, with some updating herein. Two suborders are recognized. The Ceratiocarina, with four families, are characterized by a carapace without a median dorsal plate and simple rostral plate. The Rhinocarina, with two families, have a median dorsal plate posterior to the rostral plate frequently marked with a longitudinal ridge. Rolfe (1969) recognized 22 genera, with an additional 22 genera of uncertain affinities within the archaeostracans, and another 22 genera of completely uncertain position but that are frequently allied with phyllocarid types. Clearly, this is a group whose understanding can only improve with time. A familial taxonomy is offered below.

Suborder Ceratiocarina Clarke, 1900 Ordovician–Permian
    Family Ceratiocarididae Salter, 1860 Lower Ordovician–?Permian
    Family Echinocarididae Clarke, 1900 Lower Devonian–Lower Mississippian
    Family Pephricarididae van Straelen, 1933 Upper Devonian
    Family Aristozoidae Gürich, 1929 ?Middle Ordovician–Middle Devonian
Suborder Rhinocarina Clarke, 1900 Silurian–Permian
    Family Rhinocarididae Hall and Clarke, 1888 Silurian–Permian
    Family Ohiocarididae Rolfe, 1962 Upper Devonian

HOPLOSTRACA

DEFINITION   Carapace short, without a hinge line; no rostrum; abdominal somites become progressively longer distally in the sequence; telson developed as a long spike; caudal rami tiny.

MORPHOLOGY   The distinctive carapace was subtriangular to axlike in lateral outline and was divisible into two regions: an anteriorly extended 'cephalic hood' and a posterior 'thoracic' region. The carapace did not extend over the three posterior thoracomeres (Fig. 26-5A).
    Cephalic and thoracic appendages are poorly known. Schram (1973) and Schram and Horner (1978) recorded stalked compound eyes in *Kellibrooksia macrogaster* and *Sairocaris centurion* as well as what was termed a peculiar 'antennal lappet,' which appeared to rise under the cephalic hood. However, Rolfe (1981) discovered this lappet was only the medial joint of a raptorial limb whose elements form small subchelae with each other (Fig. 26-5B). This subchelate appendage was presumably either a highly specialized antenna, mouthpart, or a thoracopod whose proximalmost joints have

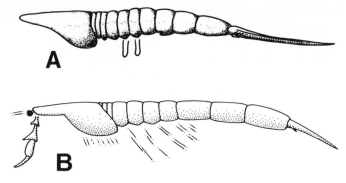

Fig. 26-5. (A) *Sairocaris elongata* (from Schram, 1979); (B) *Kellibrooksia macrogaster*.

not yet been observed. The mandibles were large and well-sclerotized. Faint traces of posterior thoracopods are known in *K. macrogaster*, but details are lacking (which may indicate the appendages were rather thin and foliaceous).

The abdomen was thin and very elongate. Relative length of pleomeres differs between species. However, the abdominal segments generally increase in length going from anterior to posterior in the series. Traces of apparently long, thin pleopods were noted in *K. macrogaster*, and a few broad pleopod flaps were recorded for *S. elongata* (Schram, 1979).

The telson was a long stylet. The base is bulbous and contains the anus. The bulk of the unit, however, was developed as a long, thin spike, whose base was flanked by two tiny caudal rami.

Nothing is known of internal anatomy.

NATURAL HISTORY   The body form of the hoplostracans is the most distinctive recognized among phyllocarids. Schram (1973) drew parallels to stomatopod hoplocaridan anatomy, but felt that the similarity was merely an analogy and did not imply direct relationship. The great raptorial limb is similar in some respects to the small first maxillipede in stomatopods, which acts in that group largely as a grooming and sensory appendage. Rolfe (1981) felt this limb in the sairocarids could have folded back into the anteroventral excavation of the sairocarid carapace, which is similar to what happens to the first maxillipede in stomatopods. Rolfe also observed the distal end of the protopod of this raptorial limb was so broad it implied that the limb might have been used as a smashing appendage in a rapacious food-getting behavior. However, fossorial forms frequently have highly modified antennae that are virtually pediform, and the extreme anterior location of this limb in *Kellibrooksia* might logically argue for this limb's being interpreted as such an antenna.

Hoplostracans are found in rocks derived from fine-grained sediments (shales or shaley limestones). Their overall morphology, that is, long, thin, well-sclerotized forms, implies benthic habits. The sairocarid long tail probably facilitated the animals' burying themselves in the bottom and lying in wait for passing prey.

These forms are known from only a few localities in the Carboniferous of Europe and North America characterized by unusually well-preserved fossils in a very diverse fauna. Little can be said about their biogeography until more is known of their distribution.

DEVELOPMENT   Nothing is known about any possible juvenile stages.

TAXONOMY   At present, two genera with three species are recognized. *Sairocaris* from the lower middle part of the Carboniferous of Scotland and Montana and *Kellibrooksia* from the upper Carboniferous of Illinois. All are placed in a single family Sairocarididae.

CANADASPIDIDA

DEFINITION   Carapace with hinge line; rostral plate absent; first and second antenna uniramous; pleopods lacking.

MORPHOLOGY   *Canadaspis perfecta* is the best known species of the group (Fig. 26-6A). The carapace valves were suboval, tapered somewhat anteriorly and with a straight dorsal hinge line. The eyes were stalked and were borne on an anterior process of the cephalon extending beyond the front of the carapace. In the case of the canadaspids the eyes were separated by a median cephalic spine.

The antennules were a pair of short, narrow, uniramous, unsegmented projections. These antennules arose just between and above the eyestalks.

The antennae were a pair of stout, uniramous, annulate appendages. They were rather setose and extended forward just below and lateral to the eyes.

A very large labrum marks an epistomal region. The mandible presents some peculiar aspects. It seems divisible into anterior incisor and posterior molar regions. Anteriorly, there were two lobes bearing fine spines, while posteriorly a single lobe bears fine, stout, parallel spines.

The 10 appendages posterior to the mandibles were biramous and essentially similar. The first two have been interpreted as maxillules and maxillae and the next eight as thoracopods. There was a rounded endite lobe bearing a dense array of spines, a stout telopod of 14 joints, which on the maxillules and maxillae have medial enditic spines on the proximal segments, and the distal segments each with a pair of very elongate spines. The thoracopods differed from this basic plan in lacking the proximal array of dense spines. All biramous appendages had an additional cluster of spines on the posterior surface of the proximalmost segment. A single oval

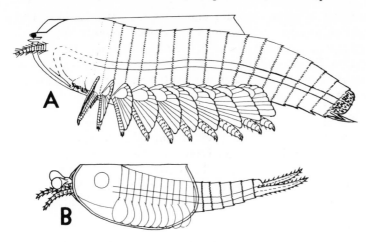

Fig. 26-6. (A) *Canadaspis perfecta* (from Briggs, 1978); (B) *Perspicaris dictynna* (from Briggs, 1977).

lobe arose laterally on each proximalmost segment, each lobe bearing a series of 10 broad rami, distally serrate, that formed a semielliptical fanlike exite.

These biramous appendages were rather unusual but can lend themselves to an interpretation as a highly modified phyllopodium. The basal seven segments could be considered a jointed specialization of the protopod, the basal joint bearing an extra large spinose gnathic lobe medially and a highly elaborate epipod laterally. The distinctive nature of these seven proximalmost segments is only detectable on the maxillule and maxilla. The seven distal segments of the telopod could then be modifications of the endopod. It would thus appear that the exopod may be absent.

The pleomeres, in terms of size and length, were barely distinguishable from the thoracomeres. The anterior abdominal segments lacked pleopods. The terminal segment was termed by Briggs (1977, 1978) a telson. Members of the genus *Perspicaris* (Fig. 26-5B) also had a set of rather large caudal rami on this unit. The so-called telson was a short rounded unit with a terminal anus; the rami were elaborate structures decorated with spines and setae. It is conceivable that interpretation of this region as an eighth pleomere with uropods (Bowman, 1971) might be as meaningful as the view of them as telson and caudal rami.

The mouth was under the labrum and directed posteriorly. The gut itself has been seen on a number of specimens and was composed of a simple alimentary canal extending through the thorax and abdomen. Little is known of the gut structure in the cephalon.

NATURAL HISTORY   *Canadaspis perfecta* is the best known and most completely preserved species (Briggs, 1978). The biramous appendages appear to have beat in a metachronal manner and seem structured around a mid-ventral food groove. The nature of the telopod seems to indicate an ambu-latory habit. Movement of exites probably facilitated feeding as well as respiration. Food would have moved forward in the food groove and been broken apart by the gnathic endites on the maxillules and maxillae before being passed to the mouth.

DEVELOPMENT   Nothing is known as to larval or juvenile stages. Briggs (1978) did record size-frequency distribution of the Burgess Shale *Cana-daspis* population. The material is bimodal. Several peaks are noticeable in the larger cohort, which may indicate molt stages within that supposedly older group. The smaller, presumably younger cohort, however, shows no apparent anatomical differences from members of the larger cohort.

TAXONOMY   The two species of the genus *Perspicaris* both bear uropods and for this reason are placed in a separate family, Perspicarididae, from the monotypic Canadispididae. Both of these Cambrian genera are better understood than almost any other fossil phyllocarid, with the possible exception of *Nahecaris stuertzi*, from the Devonian Hunsruck Slate.

REFERENCES

Bowman, T. E. 1971. The case of the nonubiquitous telson and the fraudulent furca. *Crusta-ceana* 21:165–75.

Briggs, D. E. G. 1977. Bivalved arthropods from the Cambrian Burgess Shale of British Col-umbia. *Palaeontol.* 20:595–621.

Briggs, D. E. G. 1978. The morphology, mode of life, and affinities of *Canadaspis perfecta*, Middle Cambrian, Burgess Shale, British Columbia. *Phil. Trans. Roy. Soc. Lond.* 281:439–87.

Briggs, D. E. G., and W. D. I. Rolfe. 1983. New Concavicarida (new order: ? Crustacea) from the Upper Devonian of Gogo, Western Australia, and the paleoecology and affinities of the group. *Sp. Pap. Palaeontol.* 30:249–76.

Chlupáč, I. 1960. Die Gattung *Montecaris jux* im älteren Paleozoicum der Tschechoslowakei. *Geol. Jahrb.* 9:638–49.

Glaessner, M. F. 1931. Eine Crustaceen fauna aus den Lunzer Schichten Niederosterreichs. *Jahrb. Kaiser. u. Konig. Geol. Bundesanst, Wien* 81:467–86.

Krestovnikov, V. N. 1961. Novyye rakoobraznie fillokaridy Paleozoya Russkoi platformy, Urala, Timana; Donbassa. *Akad. Nauk SSSR, Trudy, Geol. Inst.* 52: 1–67.

Rolfe, W. D. I. 1969. Phyllocarida. In *Treatise on Invertebrate Paleontology*, Part R, *Arthro-poda* 4(1) (R. C. Moore, ed.), pp. R296–331. Geol. Soc. Am. and Univ. Kansas Press, Lawrence.

Rolfe, W. D. I. 1981. Phyllocarida and the origin of the Malacostraca. *Geobios.* 14:17–27.

Rolfe, W. D. I., and V. A. Edwards. 1979. Devonian Arthropoda (Trilobita and Ostracoda excluded). *Spec. Pap. Palaeontol.* 23:325–29.

Schram, F. R. 1973. On some phyllocarids and the origin of Hoplocarida. *Fieldiana: Geol.* 26:77–94.

Schram, F. R. 1979. British Carboniferous Malacostraca. *Fieldiana: Geol.* 40:1–129.

Schram, F. R., and J. Horner. 1978. Crustacea of the Mississippian Bear Gulch Limestone of Central Montana. *J. Paleo.* 52:394–406.

27

LIPOSTRACA

DEFINITION   Carapace absent; trunk of 18 segments; antennules short and uniramous; antennae natatory and biramous; maxillules in male as a clasper; trunk limbs in ventrally directed series of two types, anterior two pairs with maxillae foliaceous and polyramous, posterior eight pairs biramous; caudal rami moderate in length and bladelike, preceded by a small pair of bladelike appendages on the elongate terminal somite.

HISTORY   The single species, *Lepidocaris rhyniensis* Scourfield, 1926, is known from abundant materials from a very unusual deposit. The Middle Devonian, Rhynie Chert, of Scotland preserves these tiny (3 mm) crustaceans in nodules and chips of silica. The preservation is superb and allows even details of setation to be determined. Recognized as significant at the time of its description, *L. rhyniensis* has subsequently become the focus of some stimulating phylogenetic speculation (e.g., Sanders, 1957; Sanders, 1963).

MORPHOLOGY   The cephalon (Fig. 27-1A) was broadly rounded anteriorly and somewhat horseshoe-shaped. About two-thirds of the length of the cephalon was a transverse groove just posterior to the level of the mandibles. No eyes were ever seen on any of the available specimens, and the animal was apparently 'blind.'

The antennules were uniramous and composed of three segments; the most distal was the longest and was armed with a terminal tuft of setae in two parallel rows (Fig. 27-2A).

The antennae were biramous and apparently natatory in function and are reminiscent in form to those of branchiopod larval antennae. The peduncle had three subequal segments. The exopod had five segments, of which the proximal two were much like the peduncular segments and armed with two and three stout spines, respectively, in females (Fig. 27-2B), and four and five spines in males (Fig. 27-2C). The distal three segments of the exopod were slender and equipped with long plumose setae. The endopod had two segments in the female, the distal one being the larger, with their margins armed with spinose setae. The male had three segments in the endopod, the basal segment had 13 spines, the middle segment without setae, and the distal segment with terminal papillae.

The mandibles were large and well sclerotized (Fig. 27-2D, E). There were no palps in the adults. The grinding surface was oval in form with

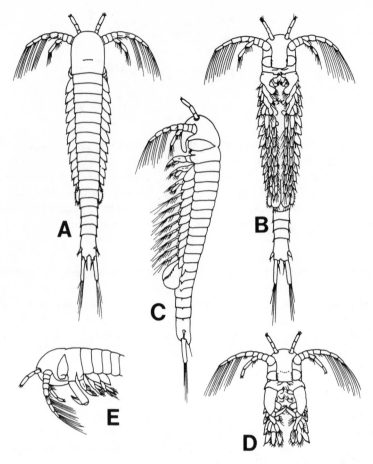

Fig. 27-1. *Lepidocaris rhyniensis* from the Devonian of Scotland. (A, B, C) Dorsal, ventral, and lateral reconstructions of female. (D, E) anterior ventral and lateral reconstructions of male. (From Scourfield, 1926)

minute denticles and a row of larger teeth. The right and left mandibles were asymmetrical in regard to size and tooth arrangement.

The female maxillules were small and apparently consisted of a simple flap armed with plumose setae on the inner margin (Figs. 27-1B, C, 27-2F). In the male this appendage was modified as a clasper (Figs. 27-1E, D, 27-2G). There was a basal setose flap on the clasper but, in addition, there was also a large ramus of three segments. The most proximal segment was the largest, the middle segment small, and the distal tapered segment about two-thirds the size of the proximalmost segment. The terminal segment was capable of being folded back and fitted into a scaled groove on the medial surface of the proximalmost segment.

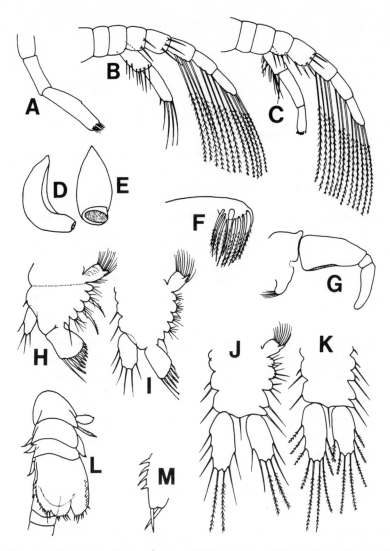

Fig. 27-2. *Lepidocaris rhyniensis* appendages. (A) antennules; (B) antennae of female; (C) antennae of male; (D, E) anterior and medial views of mandible; (F) maxillule of female; (G) maxillule of male as clasper; (H) maxilla; (I) thoracopods 1 and 2; (J) thoracopods 3 through 5; (K) thoracopods 6 through 10; (L) egg pouch cover; (M) rudimentary post egg pouch appendage. (From Scourfield, 1926)

The nature of the maxillae presents a problem. Scourfield (1926, p. 173) claimed only one specimen preserved a pair of tiny papillae just in front of the assumed first thoracopod (Fig. 27-3A) and suggested a possibility of these papillae as representing rudimentary maxillae. He was influenced in this interpretation in part by the fact that the so-called first thoracopod is similar to the next two succeeding pairs of appendages. However, these supposed first thoracopods arise from the ventral side of the posterior region of the cephalic shield (Fig. 27-3A). Given what we now know of the similarity of maxillae to first thoracopods in cephalocarids and the Cambrian phyllocarid *Canadaspis perfecta* (Briggs, 1978), we should not hesitate to identify the first thoracoform appendages posterior to the maxillules in *Lepidocaris* as in fact maxillae! The tiny papillations near these maxillae on the only specimen that preserves them are thus more likely to be openings for the maxillary glands. The maxillae (Fig. 27-2H) were composed of a central protopod from which several branches arise. There were six endites, the most proximal being a large subquadrangular lobe armed with plumose setae and a large pectinate tooth. The other five endites were small lobes each with only a few setae, and the third and fourth lobes had an additional long, comblike, toothed spine. The endopod was a palmate lobe with a series of terminal comblike spines directed mediad. The exopod was a single small oval segment with four short spines along its margin.

The first two trunk appendages were similar to the maxillae (Fig. 27-2I). They were, however, somewhat more slender than the latter; the endites were not armed with any comblike spines; the endopods were somewhat smaller with spines more ventrally directed; and the exopod was larger with six large setae on its margins.

The rest of the trunk appendage series presented a different form than that of the first two legs (Fig. 27-2J, K). The protopods were rectangular. There were endite lobes armed with setae as well as some slight exite lobation. The endopod and exopod were rectangular segments directed ventrally and armed with a few plumose setae. In addition, the fourth through sixth appendages had a large proximal endite lobe with a setal row and pectinate tooth, such as exists on the maxillae and first two trunk limbs.

Appendages more posterior to the above were noted by Scourfield. Females had a large flaplike 'egg pouch' on the segment behind the last trunk appendage; details are unclear and no eggs were seen in them. Behind the egg pouch apparently was a pair of rudimentary appendages bearing distal setae, but proximal portions of these appendages were not clearly preserved.

On the lateral surface of the last body segment were a pair of small one-segment appendages directed laterally and posteriorly with two terminal setae. They were something like miniature versions of the more posterior, larger caudal rami, which were moderate in length, bladelike, with five terminal setae. Scourfield termed these small one-segment appendages

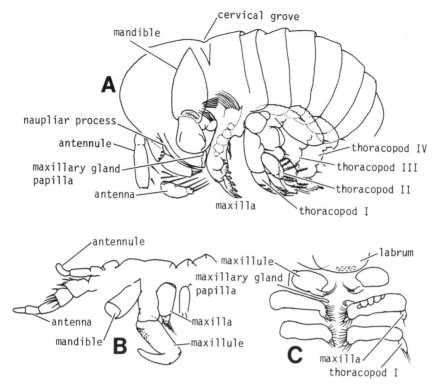

Fig. 27-3. *Lepidocaris rhyniensis* anterior structures. (A) Specimen that displays papilla of maxillary gland and associated appendages; (B) portions of a male specimen clearly elucidating the clasper and adjacent appendages; (C) ventral view of mouth region. (From Scourfield, 1926)

'lateral caudal processes.' It would thus appear that the last body segment was a fusion of the anal segment and last body somite.

The trunk was divisible into two regions. The first 10 to 12 segments, those with some type of appendages, had pleural lobes directed ventrolaterally and that decreased in size posteriorly. The next five segments were simple rings with lateral posteriorly directed spines that also decreased in size posteriorly. The last body segment had two knoblike processes (termed by Scourfield the 'primary furca') directed posteriorly and lying dorsal and medial to the large 'secondary furca.' These processes had one or two posterior spines and appear probably to represent a bilobed supra-anal plate above the anal opening.

The morphology of *L. rhyniensis*, with its ventrally directed foliaceous appendages adjacent to a food groove, would seem to indicate an animal that fed and swam simultaneously with metachronal beats that swept food forward to the mouth region (Fig. 27-3C). Several specimens are tightly doubled back on themselves and would seem to indicate the extreme flexibility of body associated with a possible grooming behavior. Both these features are reminiscent of the living brachypodans (cephalocarids).

NATURAL HISTORY   The paleoenvironment represented by the Devonian Rhynie Chert was a peculiar one indeed. Though plant remains are found in the chert, the only really abundant animal remains are those of *L. rhyniensis*. The other animal fossils in the chert nodules are those of a mite, a spiderlike form, and some Anthracomarti. Scourfield opted for an interpretation of the Rhynie habitat as that of a hot spring or geyser pool saturated with silica. The fossils in the chert thus represent those that were entombed in the precipitated silica gel from the saturated water. *Lepidocaris* would thus appear to have been adapted to a very special environment indeed, possibly of a rather temporary and transient nature.

DEVELOPMENT   The preservation of these fossils is so exceptional as to allow several developmental stages to be recognized. Several empty egg cases were noted, which Scourfield felt were probably those of *Lepidocaris*.

At least five metanaupliar stages (e.g., Fig. 27-4A, B) were recognized in the chert nodules (Scourfield, 1926; 1940). Morphological details remain unclear, but certain relevant features seen in *L. rhyniensis* can be mentioned here. The first antenna (Fig. 27-4C) has six or seven segments with the most distal armed with three plumose setae. Scourfield compared these to the antennules of juvenile *Chirocephalus diaphenous*, but they also bear comparison (Sanders, 1963) with the earliest first antennae of *Hutchinsoniella*. The antennae carry a bifid naupliar process that persists through all the stages seen and again resembles in its form that seen in various Branchiopoda and brachypodans. The mandibular palp closely resembles that of the anostracan *Chirocephalus* and is nearly as elaborate as that seen in cephalocarids. The early form of the maxillae and all the trunk appendages is biramous, the polyramous format of the maxillae and first two trunk appendages apparently appeared only as the adult condition was approached.

TAXONOMIC AFFINITIES   The speculations about phyletic relationships for *Lepidocaris* have been rather stable, if not particularly imaginative. Scourfield allied the Lipostraca with the Anostraca, though he noted a number of points of difference (1926, pp. 182–3). Without exception the connection of Lipostraca with Branchiopoda has remained through all subsequent exegetes. Even Sanders (1963) continued in this vein, although he noted several features of *L. rhyniensis* that were exceptions to the general branchiopod plan. Many features of the lipostracan larvae are as important in assessing possible relationships as those of the adults. No recent authors have offered an outline comparison with Brachypoda (Birshtein, 1960, ordinal name for cephalocarids) in the sense of Scourfield, despite the

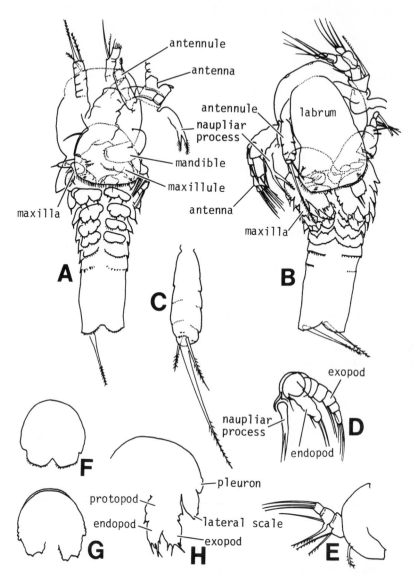

Fig. 27-4. *Lepidocaris rhyniensis* development. (A) Metanaupliar at eight postcephalic segment stage; (B) metanaupliar at nine postcephalic segment stage; (C) metanaupliar antennule; (D) metanaupliar antenna displaying the bifid naupliar process; (E) mandible with well-developed palp; (F, G, H) successive stages in the development of trunk appendages. (From Scourfield, 1926)

obvious importance of all of these groups in speculations on early crustacean evolution. Such is presented here:

Points of agreement of *L. rhyniensis* with Brachypoda

1. Horseshoe-shaped cephalon.
2. Ventrolaterally directed pleura on 'thorax.'
3. Short lateral spines on 'abdomen.'
4. Antennae natatory.
5. Maxillae 'thoracoform.'
6. Maxillae and first two thoracopods polyramous.
7. Extreme small size $\leqq$ 3 mm.
8. Long anamorphic development series.
9. Egg brooding on posterior 'thorax' by means of special appendages.
10. Absence of eyes.
11. Larval features:
    a. Large antennules with setae on several segments.
    b. Protopods of antennae $\leqq$ 1/2 total length, with spines on endopod other than just tip.

Points of agreement of *L. rhyniensis* with Branchiopoda

1. Reduced antennules of adult.
2. Apparent form of maxillules of adult.
3. Sexual dimorphism and use of some kind of clasper in male.
4. Proximal gnathic endite on trunk appendages.
5. Large labrum in larva.

Points unique to *L. rhyniensis*

1. Maxillules in male as claspers.
2. Trunk limbs in two distinct series; posterior trunk limbs copepoid, anterior limbs polyramous.
3. Specialized egg-carrying appendages.
4. Distinctive ramal and segment forms.
5. Larval mandibular palp incipiently biramous with small endopodal papilla.

The main affinities of *Lepidocaris* thus seem to be with the cephalocarids, though there are enough features in common with branchiopods to indicate that all these taxa seem to bear some relationship to each other.

The development and structure of the thoracopods of *Lepidocaris* are of special importance in this regard. J. P. Harding's discussion in Scourfield (1940) reinforced the significance of Scourfield's original observations, though they have gone largely unnoted in subsequent literature. Harding observed that it is usual for the posterior members of a limb series to be more primitive in form than the anterior members, and that the more phylogenetically primitive form of appendages frequently appear in the earlier ontogenetic stages. In both these instances, as applied to *L. rhyniensis*, the indication is that perhaps the thin, biramous copepodoid form of the thoracopods is more primitive than the polyramous 'cephalocaridoid' form. Besides hinting a possibly more ancestral position of lipostracans to cephalocarids than has heretofore been granted, it also would argue for a derived condition for Brachypoda and call further into question the emphasis that has been traditionally placed on cephalocarids in crustacean phylogenetic speculation.

REFERENCES

Birshtein, Ya. A. 1960. Podklass Cephalocarida. In *Osnovy Paleontologii: Chlenistongie, Trilobitoobraznie, i Rakoobraznie* (N. E. Chernysheva, ed.), pp. 421–22. Moscow.

Briggs, D. E. G. 1978. The morphology, mode of life, and affinities of *Canadaspis perfecta*, Middle Cambrian, Burgess Shale, British Columbia. *Phil. Trans. Roy. Soc. Lond.* (B)**281**:439–87.

Sanders, H. L. 1957. The Cephalocarida and crustacean phylogeny. *Syst. Zool.* **6**:112–29.

Sanders, H. L. 1963. The Cephalocarida functional morphology, larval development, comparative external anatomy. *Mem. Conn. Acad. Arts and Sci.* **15**:1–80.

Scourfield, D. J. 1926. On a new type of crustacean from the Old Red Sandstone—*Lepidocaris rhyniensis*. *Phil. Trans. Roy. Soc. Lond.* (B)**214**:153–87.

Scourfield, D. J. 1940. Two new nearly complete specimens of young stages of the Devonian fossil crustacean *Lepidocaris rhyniensis*. *Proc. Linn. Soc. Lond.* **152**:290–98.

28

BRACHYPODA

DEFINITION   Cephalon short, covered by head shield; no external eyes; antennules uniramous; mandibles without palp; large labrum forming atrium oris; maxillae pediform; thorax of eight segments, gonopores on sixth thoracomere, eighth thoracomere may lack appendage; thoracopods as polyramous phyllopods with several protopodal endites, ambulatory endopods, flaplike exopods, and a single flaplike epipod; abdomen without limbs except for remnants on first segment; anus terminal; anal segment with elongate caudal rami.

HISTORY   The group was first described from Long Island Sound by Sanders (1955). This species, *Hutchinsoniella macracantha*, has become the best known species of the group, with monographic descriptions of the morphology and development by Sanders (1963a) and skeletomusculature by Hessler (1964). A second genus and species, *Lightiella serendipita*, was described by Jones (1961) from material in San Francisco Bay, and then successively thereafter additional species were found in that genus, *L. incisa* Gooding, 1963, *L. monniotae* Cals and Delamare Deboutteville, 1970, and *L. floridana* McLaughlin, 1976. A third genus, *Sandersiella*, was erected for *S. acuminata* Shiino, 1965 and also includes *S. calmani* and *S. bathyalis* Hessler and Sanders, 1973. Finally a fourth genus was described from New Zealand, *Chiltoniella elongata* Knox and Fenwick, 1977. The group has been traditionally treated as a distinct subclass or class, depending on the viewpoint (Sanders, 1957), even to the point that when Harding and Ingle (1959) suggested in passing that only ordinal status be considered, Sanders (1961) vigorously defended his original evaluation. Birshtein (1960) erected an order name, Brachypoda, for the cephalocarids used herein. The cephalocarids had been considered a central group in speculations on crustacean phylogeny (e.g., Sanders, 1963b; Hessler and Newman, 1975), though some have questioned such an important status (e.g., Tiegs and Manton, 1958; Schram, 1982).

MORPHOLOGY   Cephalocarids are small animals, typically about three mm in length as adults. The cephalon is developed as a broad semicircular shield (Fig. 28-1). The trunk is composed of 20 segments. The first seven trunk segments have prominent pleural lobes directed ventrolaterally and bear more or less similar multiramous appendages directed ventrally. The eighth segment has small or reduced pleura and may or may not have

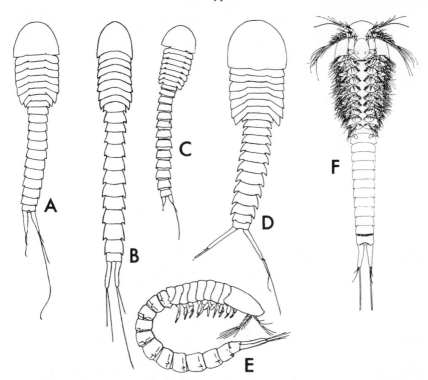

Fig. 28-1. Brachypodan body plan. (A, B, C, D. Dorsal views. (A) *Hutchinsoniella macracantha* (from Sanders, 1955); (B) *Chiltoniella elongata* (from Knox and Fenwick, 1977); (C) *Lightiella incisa* (from Gooding, 1963); (D) *Sandersiella bathyalis* (from Hessler and Sanders, 1973); (E) *Chiltoniella elongata*, lateral view in flexed grooming state (from Knox and Fenwick, 1977); (F) *H. macracantha*, ventral view (from Sanders, 1963a).

appendages (none being present on *Lightiella*). The tenth through nineteenth segments lack appendages and have variously developed lateral spines (small in *Hutchinsoniella* and *Lightiella* but large in *Chiltoniella* and *Sandersiella*).

The cephalocarid trunk has usually been divided into an anterior thorax of eight segments and a posterior abdomen. The common genital duct opens on the sixth segment (Hessler et al., 1970). The last segment, the anal segment (Bowman, 1971), has a pair of well-developed caudal rami. Sanders and Hessler, and Knox and Fenwick have used the term 'telson' in their papers, while Jones, Gooding, and Shiino tend to avoid use of that term. Bowman (1971) challenged the traditional concept of the crustacean telson and prefers the designation anal segment for cephalocarids used here.

The antennules (Fig. 28-3A) are uniramous with six segments and arise near the anterior end of the labrum.

The antennae (Fig. 28-3B) are biramous, large, and arise just lateral of the labrum. The exopods are rather robust and annulate, each segment bearing one or more stiff setae. The endopods are composed of two segments and are between one-half and one-third the length of the exopods. The rami arise from a two-segment protopod.

The labrum is large and forms a substantial atrium oris for the posteriorly directed mouth. The apparent eyes of cephalocarids have only been recently recognized (Burnett, 1981). Compound eye structures (Fig. 28-2A, B) have been seen in the anterior part of the labrum of *Hutchinsoniella*

Fig. 28-2. *Hutchinsoniella macrocantha* compound eye. (A) Photomicrograph of eye in anterior part of head; (B) schematic diagram of eye in relation to nervous system; (C) photomicrograph of ommatidia; (D) schematic diagram of ommatidium. (From Burnett, 1981)

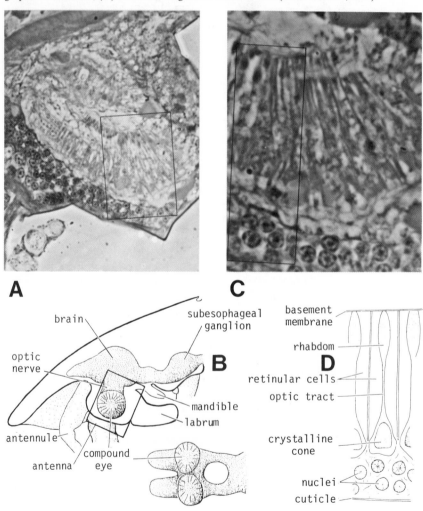

and, whether they are functional or not, individual ommatidia can be discerned (Fig. 28-2C, D).

The mandibles (Fig. 28-3C) are without a palp in the adult. Each consists of a molar process of fine teeth, an incisor process of two spines in *Hutchinsoniella*, and a spine flanked by a seta and denticles in *Lightiella*. The larval and juvenile mandibles, however, are rather maxilliform, with a six-segmented exopod and a lobate endopod. This extensive palp persists through several metanaupliar stages and only begins to disappear as the final instar is approached.

Fig. 28-3. Brachypodan limbs. (A–L) *Lightiella floridana*. (A) antennule; (B) antenna; (C) mandible; (D) maxillule; (E) maxilla; (F) first thoracopod; (G) second thoracopod; (H) third thoracopod; (I) fourth thoracopod; (J) fifth thoracopod; (K) sixth thoracopod; (L) seventh thoracopod (from McLaughlin, 1976). (M) *Sandersiella calmani*, eighth thoracopod (from Hessler and Sanders, 1973). (N) *S. acuminata*, sixth thoracopod with modified exopod (from Shiino, 1965).

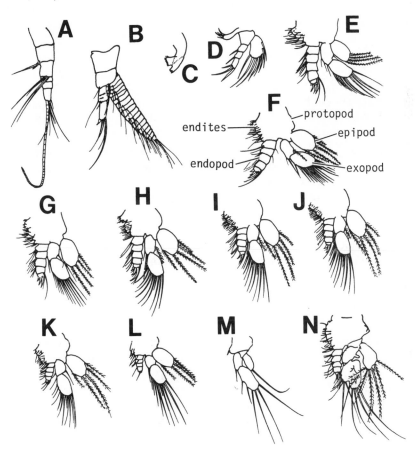

The maxillules (Fig. 28-3D) possess a relatively simple annulate endopod, a single-segment setose exopod, and a large endite that extends near the atrium oris. Successive stages in the larval development of the maxillules are directed toward the increase in the size of that endite.

The maxillae (Fig. 28-3E) are very similar in form to that seen in the first seven postcephalic appendages (Fig. 28-3F–28-3L). The protopod is a single flat segment with five to six medially directed, spinose endites. The somewhat ambulatory endopod has six segments with a terminal spine. These segments are somewhat developed distomedially where they are armed with one or more setae. The relative length of the endopods tends to decrease posteriorly through the series and also tends to vary between species. The exopod is two-segmented (some authors indicate an extra proximalmost segment, which is just probably a fold in the protopod, and the distal segment as two elements, which again are also due to folding or crinkling of that segment). Setation on the exopod varies, though the distal segment is typically armed with an abundance of long setae. The epipod, or pseudepipod, arises from the distal aspect of the protopod rather than more proximally, which accounts for the ambiguity in terms used by various authors. The epipods are a single flap with four or more simple or brush setae.

The sixth postcephalic appendages of *Hutchinsoniella* and *Lightiella* are similar to that of any other of thoracopods 1 through 7. However, in *Sandersiella* the sixth limb exopods are strongly modified (Fig. 28-3N). The proximal segment has a lateral process ranging from domelike (*S. calmani*) to fingerlike (*S. acuminata*). The distal segment is distinguished with a long lateral lobe (only slightly developed in *S. calmani*) and a shorter rounded medial lobe. The anterior face of the exopod has a fingerlike process (absent in *S. calmani*) and flanked in *S. acuminata* with a setose bump. The exact function of these specialized exopods is uncertain. Their location on segment 6 suggests a sexual connection, though none of the other genera have such structures.

Thoracopod 8 (Fig. 28-3M) lacks any endopod and the endite lobes are poorly developed. This appendage is totally lacking in the genus *Lightiella*.

The ninth postcephalic somite bears a reduced appendage. The basal segment is a broad, spherical proximal segment. The distal segment in *Hutchinsoniella* and *Lightiella* are short, lateral flaps, but in *Sandersiella* it is developed as a large central and two smaller flanking tubercles. The eggs are attached to these appendages while they are being brooded.

The pleura on the first seven postcephalic segments are well developed. The pleura of the eighth segment is generally smaller than those of the anterior segments but are markedly reduced in *Lightiella*. The ninth through nineteenth segments have generally been characterized in the literature as having pleura, but in fact these are armed with variously developed lateral and posteriorly directed spines (very large in *Sandersiella*). The nineteenth (or penultimate) segment bears a ventral comb of pectinate setae along its posterior margin (absent in *Lightiella*) (Fig. 28-4). The twentieth segment (telson or anal segment) also bears a ventral comb of very short setae, not quite the width of the anal segment in *Hutchinso-niella*, *Chiltoniella*, and *Sandersiella*, and as wide as the anal segment in *Lightiella*. The length of the caudal rami in *Lightiella* are generally equal to or less than the anal segment width; in *Hutchinsoniella* and *Chiltoniella* they are somewhat larger than but less than twice the anal segment width, and in *Sandersiella* about twice the anal segment width. Rami of all species are terminally surmounted by long setae.

Knowledge of brachypodan internal anatomy is incomplete. Hessler (1964) provided a most detailed analysis of the skeletomusculature, and Hessler et al. (1970) presented a short treatment of the reproductive system. However, no detailed presentations have ever been published on cephalocarid digestive, excretory, or nervous systems (with the exception of Burnett, 1981 on eyes). The lack of detailed treatment of the nervous system is especially vexing in that the massive compendium of Bullock and Horridge (1965) has placed great stock in nervous system comparative anatomy as an instrument in phylogenetic analysis in arthropods. Certainly a treatment of the development of the cephalocarid nervous system to the adult stage might be very interesting to crustacean phylogeneticists. For example, one interesting bit of information in this regard that has emerged has been recognition of the presence of apparent compound eyes just below the surface in the adult form (Burnett, 1981). The eyes have several ommatidial units and are connected to the brain. Whether these eyes function visually or serve some other purpose (e.g., neurosecretory) is not yet evident, but their presence in at least a reduced stage might lend some support to a notion of brachypodans as paedomorphs (Schram, 1982).

The skeletomusculature is known in tremendous detail. Certain important features can be pointed out here. The system exhibits serial homology and is basically simple (Fig. 28-5). A pair of dorsal and ventral tendons

Fig. 28-4. Abdominal combs. (A) *Sandersiella calmani* (from Hessler and Sanders, 1973); (B) *Lightiella floridana* (from McLaughlin, 1976).

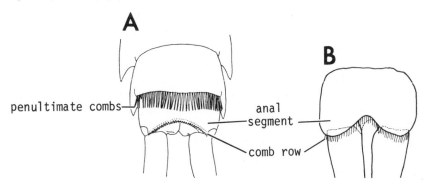

Crustacea

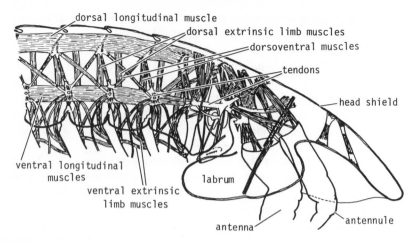

Fig. 28-5. Lateral view of thoracic skeletomusculature of *H. macracantha*. (Modified from Hessler, 1964)

marks the endoskeletal system and acts as a basic framework for the muscles. The paired dorsal and ventral longitudinal trunk muscles extend the length of the body. Three pairs of dorsoventral muscles establish a box truss structure in the thorax, and a lateral horizontal muscle extends from the ventral tendon to the thoracic pleura (which is a 'carapace adductor muscle' in the maxillary segment). All segments exhibit the same basic plan, though modified in the head region. The extrinsic limb muscles arise lateral to the dorsal muscles at the anterior and posterior ends of the segments and from the ventral tendon anterior and posterior to the limb. There are intrinsic limb muscles to operate the three major limb branches and the endites.

Brachypodans are functional hermaphrodites in the adult. These empty to the outside on the sixth thoracic segment by a common genital duct.

NATURAL HISTORY Cephalocarids tend to be found in benthic situations where generally organic-rich fine particulate matter prevails. The animals live in these flocculent oozes, and the metachronal movement of the limbs supposedly achieves locomotion and feeding simultaneously. The separation of the limbs on the forward stroke pulls fluid into the interlimb space. The backstroke pushes fluid out (propelling the animal forward), and a probable combination of backwash currents over the limb surface to sweep trapped particles off the setae and endite rotary movements propel the food forward in the food groove. A mucous gland is located in the protopod of each limb (Hessler, personal communication) and empties through a duct opening near the endites. The mucus supposedly helps form a food bolus. Another interpretation is possible, however. Due to their small size and the low Reynolds number at which brachypodans must operate, 'sticky' mucous secretions may be the last thing the animals need. It is possible that these glands produce surfactants or wetting agents; these would effectively mitigate the effects of 'sticky' water in the physics of a situation of laminar flow. Burnett (personal communication) in fact reports that the cuticle of *Hutchinsoniella* had distinct hydrophobic qualities that interfered with usual histochemical techniques. This certainly bears investigation with TEM in order to determine the nature of these glands. Although living cephalocarids have been watched in dishes of water, the actual mechanics of feeding has never been studied in detail.

Sanders (1963a) did clearly observe grooming behavior. The animal periodically doubles up and, starting at the thorax and progressing to the tips of the caudal rami, sweeps the body with its appendages. The combs of setae on the last segment and the telson in turn brush through the setae of the cephalic and thoracic appendages to remove debris from them.

Mating has never been observed in Brachypoda, but it is to be assumed that cross-fertilization takes place, at least when modest population densities prevail. However, it should not be ruled out that self-fertilization might also take place, especially in light of the apparent low population densities of some species.

Brachypoda are essentially worldwide in distribution (Fig. 28-6) and range in depth from intertidal to in excess of 1500 m (Table 28-1). Though the sediments associated with each of the collected species are generally very fine, not all have high organic flocculent content (Saloman, 1978).

Fig. 28-6. Biogeographic distribution of Brachypoda.

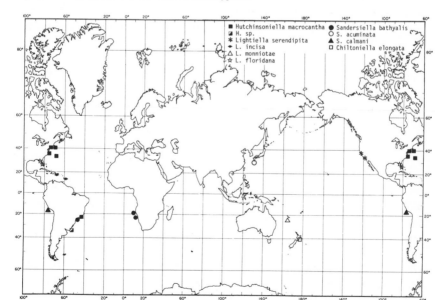

Table 28–1. Locality data for species of cephalocarids.

Species	Locality	Numbers	Depth
H. macracantha	Lond Island Sound	numerous	9–20 m
	Buzzards Bay	numerous	15 m
	Slope Station 3	4	300 m
	West of Rio	1	50 m
	Virginia	1	1 m
S. acuminata	Araike Bay, Japan	1	2–32 m
S. calmani	Peru Coast	2	85 m
S. bathyalis	Southwest Africa	1	1546–1559
	Walvis Ridge	1	1227 m
	Off Santos, Brazil	1	14.5 m
L. serendipita	San Franciso Bay	7	1–2 m
L. incisa	Barbados	2	Intertidal
	Maguey Island, P.R.	numerous (2 + 119)	1–2 m
L. monniotae	New Caledonia	1	4 m
L. floridana	Florida, gulf coast	1–50-site	0.75–6.10 m
L. sp.	Mobile Bay, Alabama	2	5 m
C. elongata	Hawke Bay, New Zealand	2	16 m

McLaughlin (personal communication) has a very large *Lightiella* from a rubble bottom in Biscayne Bay.

Of all the sites recorded in the literature only three (Long Island Sound, Buzzards Bay, and Maguey Island) can be said to have 'numerous' cephalocarids in the population; at the other sites the populations are for the most part quite sparse, see for example, Hessler and Sanders (1964) and Wakabara and Mizoguchi (1976). Hessler and Sanders (1964) record the density of *Hutchinsoniella* in Buzzards Bay at 176.5 per square meter, but only 7.1 per square meter at Slope Station 3 in the deep sea. Such wide fluctuations in population densities may be of considerable significance in the issues of cross- and/or self-fertilization.

DEVELOPMENT Only the larval and juvenile development has been described in great detail for *Hutchinsoniella macracantha* (Sanders, 1963a) and *Lightiella incisa* (Sanders and Hessler, 1963). A succession of meta-naupliar and juvenile stages were recorded (Fig. 28-7). In addition Jones (1961) noted a few larval stages for *L. serendipita*. The sequence in *Lightiella* seems to be shorter (12 preadult stages) than for *Hutchinsoniella* (18 preadult stages). Sanders and Hessler noted the scarcity of every other stage of their 12 stages in *Lightiella*. They postulated it was the 'normal' sequence of events to add two segments and a complete appendage per molt, and that this norm was capable of defects wherein occasionally only one segment was added at a molt. This would presumably result in an out-of-phase sequence of stages. Sanders also observed in passing that certain of his stages for *Hutchinsoniella* were more rare in his samples than others, though he did not elaborate. It may be a character of the group as a whole to exhibit such developmental variants.

Differences have been noted in ovulation and brooding patterns. *L. incisa* broods one egg in an ovisac attached to the ninth segment and also only prepares one egg at a time in the ovary on the side opposite the brooded ovisac. However, in *H. macracantha* there are two brooded ovisacs at a time, and both ovaries prepare eggs while brooding is going on. *L. incisa* apparently breeds all year round (Sanders and Hessler, 1963) while *H. macracantha* is a summer breeder only. Sanders and Hessler felt this was a climatic difference rather than a taxonomic one.

The developmental sequence of Brachypoda is anamorphic, and an example of the gradual changes that occur in structures is displayed in Figure 28-8. Several features are of special note. The naupliar process is retained on the second antenna until the transition from the metanaupliar to the juvenile series. The mandibular palp is lost soon after the naupliar process. Although the body segments are added quickly and steadily until the adult number of 20 is reached, appendages are added much more slowly and the metamorphosis of these to an adult condition lags behind

Fig. 28-7. Postembryonic mode of development. (A) *H. macracantha*; (B) *L. incisa*. ○ indicate rudimentary limb, ● indicate a more or less definitive form. (Modified from Sanders, 1963a; Sanders and Hessler, 1964)

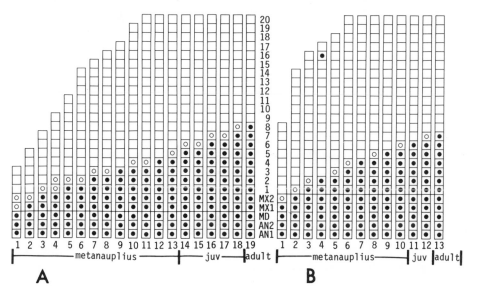

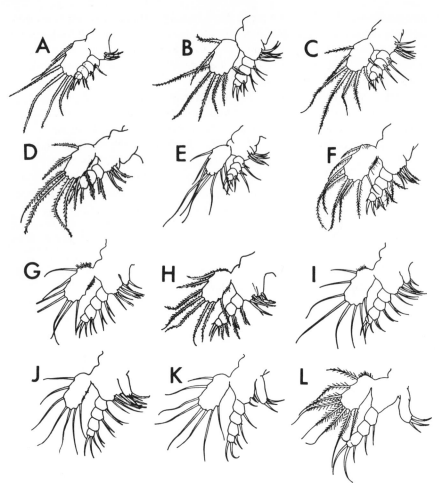

Fig. 28-8. Sequential changes during the anamorphic development of maxillules of *L. incisa* through the 12 recognized developmental stages. (Modified from Sanders and Hessler, 1963)

even that. The telson combs begin to appear early in development and are completely functional by about the middle of the larval sequence.

TAXONOMY   The genus *Lightiella* lacks an appendage on the eighth thoracomere, and Jones (1961) used this to erect a separate family. Hessler and Sanders (1972) preferred to place all genera within a single family, but Bowman and Abele (1982) maintained two families.

Family Hutchinsoniellidae Sanders, 1955
Family Lightiellidae Jones, 1961

REFERENCES

Birshtein, Ya. A. 1960. Podclass Cephalocarida. In *Osnovy Paleontologii, Chlenistongie, Trilobitoobraznie, i Rakoobraznie* (N.E. Chernysheva, ed.), pp. 421–22. Moskva.

Bowman, T. E. 1971. The case of the nonubiquitous telson and the fraudulent furca. *Crustaceana* 21:165–75.

Bowman, T. E., and L. G. Abele. 1982. Classification of the Recent Crustacea. In *The Biology of the Crustacea*, Vol. 1 (L. G. Abele, ed.) pp. 1–27. Academic Press, New York.

Bullock, T. H., and G. A. Horridge. 1965. *Structure and Function in the Nervous System of Invertebrates*. Vols. I and II. Freeman, San Francisco.

Burnett, B. 1981. Compound eyes in the cephalocarid crustacean *Hutchinsoniella macracantha. J. Crust. Biol.* **1**:11–15.

Calman, W. T. 1909. *Crustacea*. In *A Treatise on Zoology* (E. R. Lankester, ed.). Adam and Charles Black, London.

Cals, P., and C. Delamare Deboutteville. 1970. Une nouvelle espèce de crustacé céphalocaride de l'hemisphère austral. *C. R. Acad. Sci. Paris* (D)**270**:2444–46.

Gooding, R. U. 1963. *Lightiella incisa* from the West Indies. *Crustaceana* **5**:293–314.

Harding, J. P., and R. W. Ingle. 1959. *Crustacea. Zool. Rec.* **94**(10):1–99.

Hessler, A. Y., R. R. Hessler, and H. L. Sanders. 1970. Reproductive system of *Hutchinsoniella macracantha. Science* **168**:1464.

Hessler, R. R. 1964. The Cephalocarida comparative skeleto-musculature. *Mem. Conn. Acad. Arts Sci.* **16**:1–97.

Hessler, R. R., and W. A. Newman. 1975. A trilobitomorph origin for the Crustacea. *Fossils and Strata* **4**:437–459.

Hessler, R. R., and H. L. Sanders. 1964. The discovery of Cephalocarida at a depth of 300 meters. *Crustaceana* **7**:77–78.

Hessler, R. R., and H. L. Sanders. 1973. Two new species of *Sandersiella*, including one from the deep sea. *Crustaceana* **13**:181–96.

Jones, M. L. 1961. *Lightiella serendipita*, a cephalocarid from San Francisco Bay. *Crustaceana* **3**:31–46.

Knox, G. A., and G. D. Fenwick. 1977. *Chiltoniella elongata* n. gen. et sp. (Crustacea: Cephalocarida) from New Zealand. *J. Roy. Soc. N. Z.* **7**:425–32.

McLaughlin, P. A. 1976. A new species of *Lightiella* from the west coast of Florida. *Bull. Mar. Sci.* **26**:593–99.

Saloman, C. H. 1978. Occurrence of *Lightiella floridana* from the west coast of Florida. *Bull. Mar. Sci.* **28**:210–12.

Sanders, H. L. 1955. The Cephalocarida, a new subclass of Crustacea from Long Island Sound. *Proc. Nat. Acad. Sci.* **41**:61–66.

Sanders, H. L. 1957. The Cephalocarida and crustacean phylogeny. *Syst. Zool.* **6**:112–29.

Sanders, H. L. 1961. On the status of the Cephalocarida. *Crustaceana* **2**:251.

Sanders, H. L. 1963a. The Cephalocarida functional morphology, larval development, comparative external anatomy. *Mem. Conn. Acad. Arts and Sci.* **15**:1-80.

Sanders, H. L. 1963b. Significance of the Cephalocarida. In *Phylogeny and Evolution of Crustacea* (H. B. Whittington and W. D. I. Rolfe, eds.), pp. 163–75. Mus. Comp. Zool., Cambridge, Mass.

Sanders, H. L., and R. R. Hessler. 1963. The larval development of *Lightiella incisa. Crustaceana* **7**:81-97.

Schram, F. R. 1982. The fossil record and evolution of Crustacea. In *The Biology of Crustacea*, Vol. 1 (L. G. Abele, ed.), pp. 93–147. Academic Press, New York.

Shiino, S. M. 1965. *Sandersiella acuminata*, a cephalocarid from Japanese waters. *Crustaceana* **9**:181–91.

Tiegs, O. W., and S. M. Manton. 1958. The evolution of the Arthropoda. *Biol. Rev.* **33**:255–337.

Wakabara, Y., and S. M. Mizoguchi. 1976. Record of *Sandersiella bathyalis* from Brazil. *Crustaceana* **30**:220–21.

NOTOSTRACA AND
KAZACHARTHRA

These two groups are very closely related, so much so that it is conceivable that the Jurassic Kazacharthra may in fact be only a very specialized family of notostracans.

NOTOSTRACA

DEFINITION   Body of a variable but still rather high number of segments ($\approx$40); anterior two-thirds of body covered by carapace enveloping limbs; posterior body segments fused into 'rings' with which six or more pairs of appendages each can be associated; antennules and antennae range from reduced to vestigial or absent; maxillules small and spinescent, maxillae either greatly developed (*Lepidurus*) or virtually absent (*Triops*); first trunk appendage greatly modified while the second through eleventh trunk limbs are thin, polyramous, and distinctly lobate; posterior trunk limbs successively smaller and more flexible; telson may have a posteriorly extended supraanal plate and a pair of long annulate caudal rami.

MORPHOLOGY   Notostracans are moderate to large in size, ranging up to 10 cm in length. The body is distinctly characterized by a large concave shield-like carapace, which typically envelops most of the anterior two-thirds of the body (Fig. 29-1). The posterior somites are fused into 'rings' apparently formed by the complete or partial fusion of from two to six segments. The somites can be partially fused as half rings or peculiarly merged into spirals. The 11 'thoracic' segments each carry a single large pair of well-developed appendages; however, in the more posterior region ('abdomen') appendages are progressively reduced in size and in cuticular support and are massed several to a ring (Fig. 29-1C). All trunk appendages are so arranged as to form a midventral food groove. Only two genera are recognized, *Lepidurus* and *Triops*.

The paired, sessile, compound eyes are situated anteromedially on the carapace just posterior to a cluster of naupliar eyes variously recorded as from two to four in number. Posterior to the eyes on the carapace is an elliptical 'nuchal organ' of uncertain function.

The antennules (Fig. 29-2A) are short, uniramous, and not always

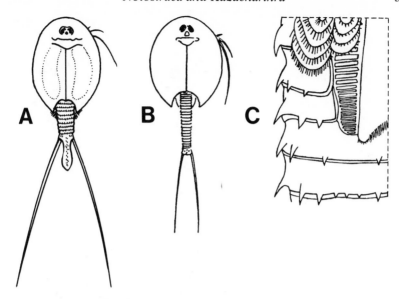

Fig. 29-1. Notostracan body plan. (A) *Lepidurus*; (B) *Triops*; (C) ventral portion of right side of *Lepidurus lynchi* with the posteriormost 25 legs removed clearly displaying the multisegmental nature of the body rings.

obviously jointed. The antennae (Fig. 29-2B), on the other hand, are very reduced or even completely absent.

The mandibles (Fig. 29-2C) have no palp and possess a toothed or spinous biting surface.

The maxillules (Fig. 29-2D) are simple in form, apparently bilobed, and spinescent on their medial margins. The maxillae (Fig. 29-2E) can have either a well-developed lobe with terminal setae (generally the case in *Lepidurus*) or are reduced to just the papillate opening of the maxillary gland (as in *Triops*).

The first trunk limb is very distinct in form. There are two somewhat elongate basal segments, and attached to the distal of these are three elongate flagella (Fig. 29-2F). The second trunk limb is similar to, though somewhat more lobate than, the first (Fig. 29-2G).

The third through eleventh limbs are all homonomous and well developed (Fig. 29-2H). The protopod is elongate and medially developed as six distinct, very setose endites. The endopod appears to be generally reduced, though a distal lobe on the protopod of some species may represent its remnant. The setose exopod or 'flabellum' is very large and extends dorsally as well as ventrally from its point of attachment. The epipod is an unadorned lobe extending dorsally from its point of attachment. The notostracan trunk limbs are the most distinctly lobate of all the phyllopods. The posterior trunk limbs (Fig. 29-2J) are similar to the anterior but are successively reduced in size as one proceeds posteriad.

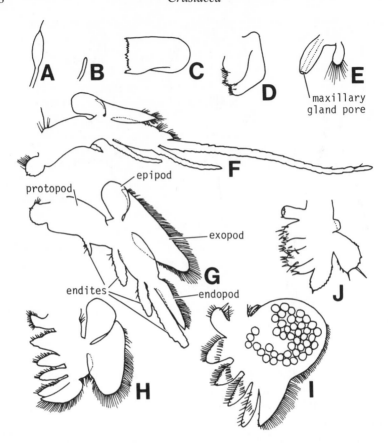

Fig. 29-2. Appendages of notostracans. (A, B, C, F, G, H, I, J) *Lepidurus*; (D, E) *Triops*. (A) antennule; (B) antenna; (C) mandible, ventral view; (D) maxillule; (E) maxilla; (F) first trunk limb; (G) second trunk limb; (H) eleventh trunk limb of male, indicative of the anterior trunk limb series; (I) eleventh trunk limb of female; (J) last (71st) limb in a series from *L. lynchi*.

The gut is a simple tube from which two ceca arise anteriorly. These cecae are extensively branched within the carapace. The maxillary glands have long loops that also extend into the carapace as well. Antennal glands function in the larvae.

The notostracan heart extends only throughout the 11 'thoracic' somites and has correspondingly 11 pairs of ostia. Blood is passed from the heart into the blood sinuses in the cephalon through three anterior openings.

The gonads are elaborately branched. The gonopores open on the eleventh trunk segment. There are both bisexual and unisexual populations known of individual species. Bisexual individuals are excluded from populations north of 45° to 50°N (Zaffagnini and Trentini, 1980), while in south temperate and tropical populations males are common (Kaestner, 1970).

Longhurst (1955c) studied the histology and cytology of the gonads of supposedly maleless populations and found clear evidence of both testicular
and ovarian tissues in the gonads. This was confirmed by Zaffagnini and
Trentini (1980), and while both sets of authors felt that these animals
reproduced by parthenogenesis, Longhurst also seemed to agree with von
Zograf (1906) that there was a potential of sex reversal in notostracans.

The nervous system of notostracans is basically a simple ladderlike system but exhibits some pecularities related to the distinct body form. The
antennular nerves arise from in front of the antennular ganglia, but these
nerves have nerve tracks that arise in the supraesophageal ganglion (Pelseneer, 1885). The atrophied maxillae have nerves that arise from no particular ganglion but take origin from the cord between the maxillular and first
thoracic ganglia (Fig. 29-3). Eloffson (1966) points out the naupliar eye
and frontal organs are identical with those of conchostraca and cladocera
but quite distinct from those of anostracans. Four cups form the naupliar
eyes, which have extensive variation in cell number between individuals.
The paired frontal organs touch the naupliar eyes, but the eye–organ complex is quite separate from the brain neutropile. Like anostracans there are
few pigment cells in the housing of the naupliar eyes, and there is an inver-
sion of the sensory cells with the rhabdomes located distally.

The skeletomusculature has been studied in some detail (Hessler,
1964). The box truss arrangement of dorsoventral trunk muscles, seen in
many phyllopodan groups, is present in *Triops longicaudatus*.

NATURAL HISTORY   These cosmopolitan crustaceans (except for Antarctica)
live in temporary and permanent fresh to brackish waters. A period of desiccation is necessary for the eggs of many species to complete development, although this is not a universal requirement (Tasch, 1969).

Fig. 29-3. Anterior central nervous system of *Triops*. Note lack of a ganglion for the nerves to
the maxillae. (From Calman, 1909)

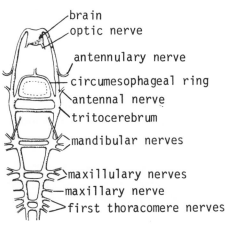

Notostracans are facultative detritus feeders, scavengers, or active predators. They are able to exploit any food resource, though typically they
live on pond bottoms and feed on the organic content of the muds. To this
end they can swim with metachronal beats of the appendages or crawl
across the bottom in the same way. However, they are also capable of
climbing over and through bottom debris in the waters they inhabit, and
they have been seen crawling out onto the banks of their ponds on rare
occasions. They can also burrow deep into the mud bottoms of the ponds
they inhabit.

Their chief adaptations for reproduction lie in their drought-resistant
eggs and ability to grow rapidly. These, combined with their versatile sexual habits, ensure an effective, though passive, mode of dispersal (Longhurst, 1955a, b). They occupy their habitat wherever it occurs on the planet
(north polar regions and isolated islands, as well as the temperate and tropical continents). Although separate males are known, hermaphroditism
seems to be possible and parthenogenesis is apparently frequent. The
advantages that this reproductive versatility might be for maximizing dispersal potential is obvious, since it takes only *one* egg to achieve colonization. Longhurst (1955a) felt that the great age of the group (Paleozoic
forms being known that are quite indistinguishable from modern types,
Tasch, 1969) and their dispersal potential ensured that notostracans are
now to be found almost everywhere.

Despite their universal availability, little is known about their early
development (Brusca, 1975). They apparently hatch in a nearly metanaupliar stage and then gradually but *rapidly* progress through many molts to
adulthood (Fig. 29-4). *Triops cancriformis* is reported by Kaestner (1970)
to molt 40 times in 14 days, that is, two to three times a day.

TAXONOMY   Only two genera are recognized *Triops* (= *Apus*), without an
anal plate, and *Lepidurus*, with an anal plate. Fossil forms are few, incompletely known, and essentially indistinguishable from the modern types.

KAZACHARTHRA

DEFINITION   Body of an indeterminate but high number of segments;
anterior one-third to one-half of body covered by a shieldlike carapace of
varying outline with optic tubercles; anterior part of thorax with apparently
six multiramous gnathal appendages; abdomen at least occasionally with
parallel rows of spines and segments varying from 32 to 40; terminus of
abdomen as an enlarged, clublike, spinous structure, which in turn may
bear a pair of distal spines or caudal rami.

HISTORY   These distinctive forms are known so far only from the Jurassic
of Kazachstan in the USSR. First described by Chernyshev (1940), Novozhilov (1959) felt they were sufficiently different to warrant a separate

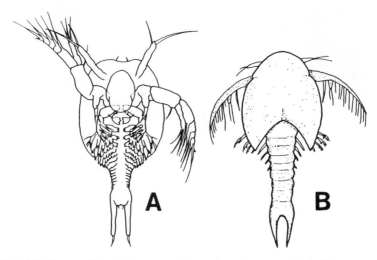

Fig. 29-4. (A) Metanauplius of *Triops cancriformis* (from Sanders, 1963a); (B) late larvae of notostracan (from Brusca, 1975).

order, though Sharov (in a footnote in Novozhilov, 1959) believed the evidence justified only a separate status for kazacharthrans within the Notostraca.

MORPHOLOGY These particular Jurassic fossils are blessed with a distinctive array of peculiar carapaces. They are quite large animals, their carapace lengths ranging from about 0.6 to over 5.0 cm. The apparently heavily sclerotized and mineralized carapaces were variable in outline ranging from inverted heart shapes (*Jeanrogerium*, Fig. 29-5A), ovals (*Kysyltamia*, Fig. 29-5F), and double ovals (*Iliella*, Fig. 29-5D). The carapace margins are armed with big spines. In those forms for which information exists, the carapace covers at least the anterior one-third, but generally less than one-half, of the body. An optic tubercle is borne on the anterior midline of the carapace.

The 'thoracic' appendages are quite distinctive but are only well known on *Jeanrogerium sornayi* Novazhilov, 1959 (Fig. 29-5B). The protopod or corm is subdivided into three segments, the most basal with a gnathobase. The intermediate and distal segments bear a series of setiferous bladelike lobes and a large, broad, flaplike, setose flabellum or exopod. The bulk of the appendage seems to have been extended laterally from the midline of the body.

The abdomen is composed of a large number of segments marked with longitudinal paired rows of spines. The 'segments' have the same general appearance of notostracan 'rings'; however, no appendages are preserved. The terminus of the abdomen is marked with an enlarged, spinose segment that is frequently oval in outline (Fig. 29-5E). The exact nature of this

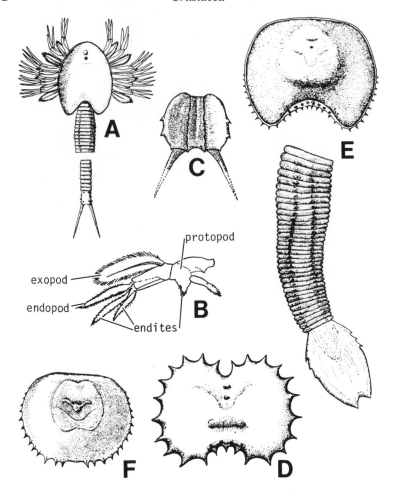

Fig. 29-5. Kazacharthran morphology. (A) *Jeanrogerium sornayi*; (B) trunk appendage of *J. sornayi*; (C) telson (?) and caudal furca (?) of *Panacanthocaris ketmenia*; (D) carapace, *Iliella spinosa*; (E) carapace and posterior half of *Ketmenia schultzi*; (F) carapace, *Kysyltamia tchiiliensis*. (Modified from Chernyshev, 1940; Novozhilov, 1957)

'segment' is not known. It could be a telson or an anal segment, it could be several fused abdominal segments, or it could be a supraanal plate. *Jeanrogerium* and *Panacanthocaris* (Fig. 29-5C) have what appear to be movable terminal spines or caudal rami of considerable length.

Nothing is known of the internal anatomy or development of these peculiar crustaceans. Tasch (1969) offers a review in English of available information.

REMARKS   Novozhilov (1957) erected a separate order to accommodate what he felt were the peculiar nature of these seven genera. Sharov (cited in footnotes in Novozhilov, 1959, p. 266), on the other hand, felt that the presence of caudal rami and the general form of the appendages warranted nothing else other than separate familial status within the Notostraca.

The general form of notostracan and kazacharthran bodies would indicate a close relationship, that is, shieldlike carapaces, certain general aspects of the thoracic appendages and probable presence of abdominal rings. However, there are distinct features for the Jurassic forms. The carapace is short, covering only the anterior parts of the body. The six thoracopods whose form is known in general resemble the first and second thoracopods found in notostracans, though they have flaplike vs. lobate endites, a gnathobase vs. gnathal lobe or flap, and distally enlarged and setose flabellum or exopod vs. a simple flabellum (Fig. 29-5B vs. Fig. 29-2F). Finally, the telson is sometimes rather distinctive but would seem to be somewhat reminiscent of its closest counterpart among the Notostraca, namely, the supraanal plate of *Lepidurus*. Certainly these differences may or may not be greater than those encountered between other families of crustaceans, though conceivably further knowledge of the appendage morphology of kazacharthrans might eventually definitively only justify a subordinal distinction within Notostraca, for the family Ketmeniidae Novozhilov, 1957.

REFERENCES

Brusca, G. J. 1975. *General Patterns of Invertebrate Development*. Mad River Press, Eureka, California.

Chernyshev, B. I. 1940. Mesozoic Branchiopoda from Turkestan and Transbaikal. *Akad. Nauk Ukrain. SSR., Inst. Geol. Nauk, J. Geol.* **7**(3):5–27 (in Russian).

Elofsson, R. 1966. The nauplius eye and frontal organs of the non-Malacostraca. *Sarsia* **25**:1–128.

Hessler, R. R. 1964. The Cephalocarida comparative skeleto-musculature. *Mem. Conn. Acad. Arts Sci.* **16**:1–97.

Kaester, A. 1970. *Invertebrate Zoology*, Vol. III. Interscience, New York.

Longhurst, A. R. 1955a. Evolution in the Notostraca. *Evol.* **9**:84–6.

Longhurst, A. R. 1955b. A review of the Notostraca. *Bull. B.M.(N.H.), Zool.* **3**:1–54.

Longhurst, A. R. 1955c. The reproduction and cytology of Notostraca. *Proc. Zool. Soc. Lond.* **125**:671–80.

Novozhilov, N. 1957. Un nouvel ordre d'Arthropodes particuliers: Kazacharthra du Lias des monts Ketmen. *Bull. Soc. Geol. Fr.* **7**(6):171–84.

Novozhilov, N. 1959. Position systématique de Kazacharthra d'après de nouveaux materiaux des monts Ketmen et Sajkan. *Bull. Soc. Geol. Fr.* **1**(7):265–69.

Pelseneer, P. 1885. Observations on the nervous system of *Apus*, 2. *J. Micr. Sci.* **25**:433–44.

Tasch, P. 1969. Branchiopoda. In *Treatise on Invertebrate Paleontology*, Part R, Arthropoda 4, Vol. 1, (R. C. Moore, ed.) pp. R128–191. Geol. Soc. Amer. and Univ. Kansas Press, Lawrence.

von Zograf, N. 1906. Hermaphroditismus bei dem Männchen von *Apus*. *Zool. Anz.* **30**:563–67.

Zaffagnini, F., and M. Trentini. 1980. The distribution and reproduction of *Triops cranciformis* in Europe. *Monitose Zool. Ital.* **14**:1–8.

30

ANOSTRACA

DEFINITION   Body with 19 to 27 postcephalic somites; carapace and head shield lacking; paired stalked compound eyes, and medium naupliar eye of a distinctive form; antennules short and uniramous; antennae with sexual dimorphism, usually reduced in females, as larger claspers in males in some cases with accessory structures; maxillae small to reduced or vestigial; 11 to 19 thoracomeres with broad, polyramous, leaf like appendages, each bearing several protopodal endites, broad thin endopod and exopod flaps, large epipodite, and one or more preepipodite flaps; all thoracopods generally homonomous; eight abdominal segments generally without appendages; anal segment with setose caudal rami.

HISTORY   This small group of branchiopods has had for the most part only scattered and isolated treatment in the literature. One of the earliest synthetic works was by the inimitable Claus (1873, 1886). Daday (1910a, b) originally monographed the group as known up to that time, and this was updated by Linder (1941). There are currently about 200 known species in the group.

MORPHOLOGY   Anostracans are small to medium in size for crustaceans, up to approximately 10 cm in maximum length (though more typically only about 1 cm). The body is circular in cross section, and does not possess any pleural lobes. The body is clearly defined into three regions: a shieldless cephalon with prominent stalked compound eyes (and with large antennal claspers in males), the thorax with its great foliaceous appendages, and the thin legless abdomen (Fig. 30-1).
   The antennules are (Fig. 30-2A, B) small, thin, and jointless appendages, which extend out from the anterior surface of the cephalon. These sometimes bear terminal setal tufts.
   The antennae are uniramous and in females are usually modest, unsegmented structures (Fig. 30-2B), though they can be relatively large, as in *Thamnocephalus* (Fig. 30-3B). In males the second antennae are developed as large claspers (Fig. 30-2A). The basal segment of the male is greatly inflated, and the distal segment is elongate and able to bend back against the basal segment. In effect a pair of sort of subchelate claspers are formed, capable of wrapping around a female. In many anostracans an 'antennal appendage' arises from the base of the antennae. These accessory structures seem to occur largely in males (Fig. 30-3A). These can be

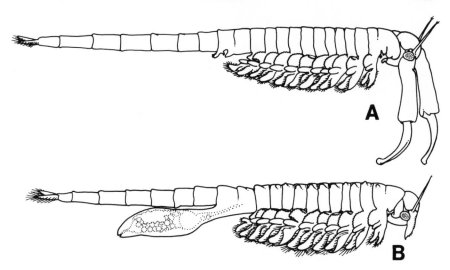

Fig. 30-1. Body plan of anostracan, with (A) male and (B) female of *Branchinecta campestris*. (From Lynch, 1960)

either rather elaborate structures and fused at the base to or inserted on the front of the cephalon separate from the antennae proper. The most elaborate development of these organs is seen in the Thamnocephalidae (see, e.g., Linder, 1941; Geddes, 1981; Belk and Pereira, 1982; Pereira, 1983.)

The mandibles (Fig. 30-2C) are simple, generally palpless appendages. A vestigial palp has been reported in the Polyartemiidae (Ekman, 1902) and the Branchinectidae (Linder, 1941; Belk, personal communication). The gnathobase of this appendage is broad and toothed.

The maxillules (Fig. 30-2D) are simple flaps with setae along the medial margin. The maxillae (Fig. 30-2E) are very reduced or vestigial. Both of these appendages frequently have better expression in the larvae, becoming reduced or almost absent in some adults.

The thoracopods (Fig. 30-2F, G) are generally homonomous, though the posterior appendages can be somewhat reduced over the form exhibited by the anterior. These limbs also can exhibit some subtle sexual dimorphism. The limb is centered around a large unjointed protopod medially developed into five endites, which variously may be either distinct or partially fused to each other. The endopod is an oval to rectangular, broad, single lobe, rather like the adjacent, similarly formed, articulated exopod. The endopod has frequently been termed a sixth endite; however, the early development of the limb seems to clearly reveal this as one of the distal rami (Fig. 30-3E, F, G). The various parts of this limb seem to be specialized for different functions: the endites and endopods for feeding, the exopod for locomotion, and the epipodites for respiration. The

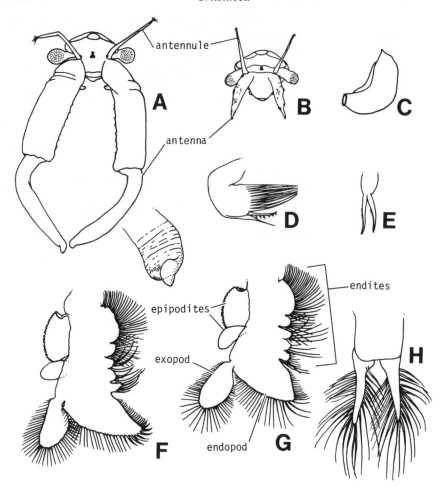

Fig. 30-2. Appendage morphology of anostracans. (A) anterior cephalon of male *Branchi-necta campestris*, with close-up of distal tip; (B) female of *B. campestris*; (C) mandible of *Streptocephalus seali*; (D, E) maxillule and maxilla, respectively, of 5 mm larva of *S. texanus*; (F, G) sixth thoracopod of male and female, respectively, of *B. campestris*; (H) caudal rami of *B. paludosa*. (A, B, E, F, G from Lynch, 1960; C from Baqai, 1963; D, E, H from Packard, 1883)

endopod, exopod, and endites are adorned with long, complex setae along the distal and median edges of the appendage. The lateral edge of the limb possesses an epipodite and one or more 'preepipodites'; this latter can have a somewhat serrated margin as in *Branchinella* or be simple unadorned lobes as in *Polyartemia*. The thoracomeres are somewhat concave ventrally and form a longitudinal food groove with the large appendages of the trunk.

The first two segments of the abdomen are fused and bear the gonopores. The males possess an eversible pair of penes in this region (Fig. 30-3C). The females carry a brood sac (Fig. 30-3D), which contains paired oviducal pouches, a median ovisac, and shell glands.

The anal segment is somewhat shorter than the other abdominal somites. The anus is terminal. The caudal rami (Fig. 30-2G) are long, setose, and are either broadly bladelike or developed as flaps.

The internal anatomy of anostracans is simple in its plan. Hessler (1964) described a body muscle arrangement of modest development not unexpected given the delicate membranous nature of the cuticle. Like other phyllopodans the dorsal and ventral longitudinal muscles are augmented with anteriorly and posteriorly descending dorsoventral muscles (Fig. 30-4). Unlike the other phyllopodans, however, these dorsoventral muscles are obliquely arranged. In addition the dorsal and ventral

Fig. 30-3. (A) frontal appendage of male *Thamocephalus platyuris*; (B) antenna of female *T. platyurus* (from Pennak, 1953); (C) ventral view of male genital segment with penes everted, *Branchinecta gigas*; (D) lateral view of female genital segment, *B. gigas* (from Lynch, 1937); (E, F, G) successive early developmental stages of a trunk limb, *Streptocephalus seali* (from Baqai, 1963).

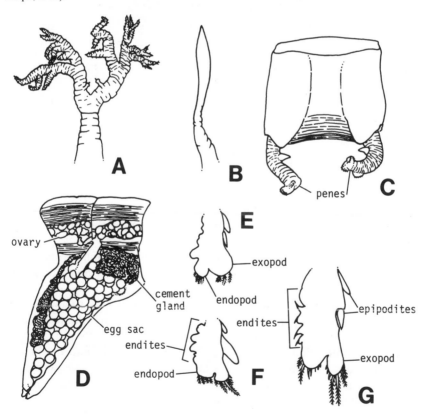

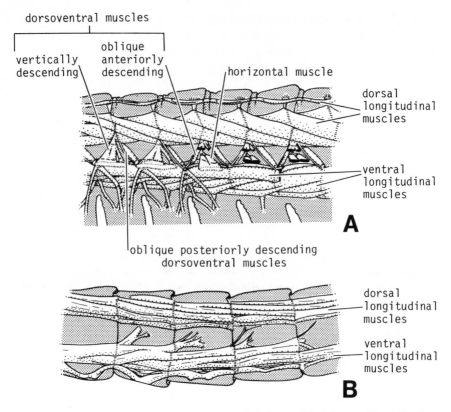

Fig 30-4. Body muscles of *Eubranchipus vernalis*. (A) thorax; (B) abdomen. (From Hessler, 1964)

longitudinal muscles not only contain straight intersegmental elements, there are also large bundles of muscles in the thorax that spiral around each other within the main muscle masses.

The digestive system is a simple tube, except for a pair of caeca that extend forward into the cephalon from the midgut in the region of the anterior thorax.

Excretion in adult anostracans is achieved through a pair of maxillary glands. Antennal glands are present in the larvae but degenerate as the adult condition is approached (Fryer, 1983).

The heart extends through the whole body and has 14 to 18 pairs of ostia. The system is open. Blood enters the heart through the ostia from the pericardial sinus. Contraction of the heart occurs along its entire length at the same time, with blood exiting the heart anteriorly into the hemocoele. Oval openings in the ventral wall of the pericardial sinus, the pericardial septum, allows blood from the body cavity to reenter the pericardial sinus.

The gonads are paired and generally located in the more posterior parts of the body. The oviducts combine into a single ovisac. The testes empty by means of a pair of vas deferens, which open on the distal tip of the penes. An enlargement of the vas deferens near its entry into the penes acts as a seminal vesicle. Small accessory glands open independently on the tip of the penes adjacent to the seminal vesicle and seem to produce a lubricant or sperm activator (Wolfe, 1971).

Elofsson (1966) studied the naupliar eye and frontal organs of branchiopods and noted some differences among groups. Anostracans have a naupliar eye (Fig. 30-5) composed of three cups with a stable number of cells. However, the distal rhabdome housed within a few pigment cells is basically like that of notostracans. In fairy shrimp, only paired ventral frontal organs are present and separate from the naupliar eye, and the whole structural complex of these organs is a projection from the brain neuropile. The central nervous system itself is marked by a pair of broadly spaced nerve cords (Fig. 30-6). The ganglia of the antennae and posterior segments form a ladderlike chain extending through the head and thorax, though the joined cords continue along the abdomen but without commissures. The antennal ganglia are postoral in location, and in this respect the anostracan nervous system seems to reflect a fairly plesiomorphic condition. The antennules and frontal organs appear to be the principal tactile and chemosensory structures.

NATURAL HISTORY Anostracans all live in such marginal freshwater habitats as brine pools, temporary melt waters, or flood water catchments. All these habitats are also characterized by their isolation and a lack of fish predators or major food competitors, though the anostracans are gorged upon when available by amphibians and predaceous insects. However,

Fig. 30-5. Semidiagrammatic view of anterior sensory organs of *Branchinecta paludosa*, showing the naupliar eye with three cups. (From Elofsson, 1966)

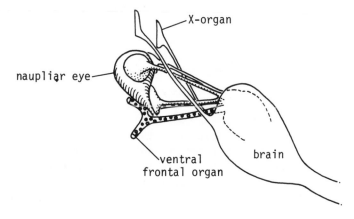

Fig. 30-6. Central nervous system of *Branchinecta*. (From Pennak, 1953)

their life cycle is so short and the reproductive potential so high, fairy shrimp are ideally suited to such transient habitats.

Anostracans typically swim with dorsal side down. They move with constant metachronal beats of the appendages, though beating of the limbs need not necessarily move the body (Lowndes, 1933). They generally feed by filtering suspended food particles, detritus, or microorganisms out of the water. There is a difference of opinion as to how feeding is achieved. Cannon (1928) and Cannon and Leak (1933) maintained that feeding in *Chirocephalus* was achieved simultaneously with locomotion. According to them, as the limb moves forward, water with suspended matter is drawn into the interlimb space and entangled on the marginal setae of the phyllopod. As the limb moves backward, the bulk of the water is expelled while pushing the animal forward, and simultaneously backwash currents along the limb surface sweep food particles into the food groove of the interlimb space. Material is then passed forward with successive limb beats to the mouth area for processing. Lowndes (1933) had a different view, claiming that the chief food source was detritus. *Chirocephalus* was noted to frequent the bottom, stirring up detritus when swimming away. Vortices are created by individual limbs, which serve to suck suspended matter in toward the limbs. Particles are pushed into the food groove between the thoracopods by the distal endites and endopod, and then pushed forward to the mouth by the basal endites. Lowndes maintained that the beating of the limbs was not strictly metachronal and that locomotion was probably achieved by the thoracic exopods acting as 'propellers,' that is, rotating on their bases. The conflict of issues here is important and deserves restudy to analyze the motion with modern high-speed cinematographic techniques.

Certain large species, such as *Branchinecta gigas*, are predators and feed on beetle larvae, copepods, anostracan eggs, and other smaller anostracans. However, rather than filtering setae on the appendages, *B. gigas* has spinelike setae that form a cage to trap prey. The mandibles in these predators act to crush and roll the victims into and through the mouth. Fryer (1966, 1983) discusses variations on the basic morphology and behavior of *B. gigas* as exhibited in related forms. Some *Branchinecta* species scrape the bottom with their spines and in that way collect food materials (reminiscent of Lowndes' observations in *Chirocephalus*). Though the details of the morphology of *B. gigas* are possibly a peculiar arrangement related in part to its large size (10 cm in length), it suggested to Fryer that the earliest anostracans *may not* have been filter feeders, since the basic pattern of anatomical form associated with predation in such carnivorous fairy shrimp is close to an assumed primitive type as seen in raptorial copepods.

Generally temperature has little effect on breeding (Wyman, 1981; Belk, 1984), though breeding behavior seems to be triggered by rising temperatures in *Eubranchipus moorei* (Moore and Ogren, 1962). Receptive females lay still at an angle in the water, allowing males to approach them. If the male should inadvertently grasp an unreceptive female, she quickly will shake him off. The beating of limbs becomes synchronous in coupling, the male acting as the pacemaker (Lent, 1971). The male grasps the middle of the female dorsum with his antennal claspers. In species with well-developed antennal or frontal appendages, those accessory structures stretch out anteriorly along the female's back and may function to identify the male to the female (Belk, 1984). Then periodically, sometimes over several hours in *Artemia* (Wolfe, 1973), the male bends his body around the abdomen of the female so their gonopores are opposed and inserts his penis into her ovisac, alternating on each side until one pene is inserted. Parthenogenic populations are encountered within *Artemia*.

As the eggs mature (Linder, 1959) they pass from the ovary to the lateral pouches of the oviduct and then to the median ovisac. The ovisac is surrounded by numerous cells comprising the shell gland, and these empty their secretions into the ovisac to form the thick, hard shell of the egg. Eggs are forced through the genital opening by contraction of the ovisac wall.

Anostracans of the same species are known to produce different-sized eggs and larvae depending on habitat of the populations involved. For example, Belk (1977) noted that *Streptocephalus seali* produced fewer larvae from large eggs on the Kaibab Plateau, Arizona, in spring melt-water ponds than in rainwater ponds in Louisiana. The Arizona ponds were 'predictable,' that is, yearly, pools of long duration, but the selective forces seemed to be due to pressures from well-established predators (in that case cyclopoid copepods) as well as natural limits on food resource availability. Thus widely separated but morphologically indistinguishable populations can evolve decidedly different reproductive strategies. The life cycle of fairy shrimp is very short, sexual maturity being reached, for example, in about a week after hatching in *Chirocephalus*, or two to three weeks in *Artemia*.

The anostracans are found worldwide, but, unlike notostracans, individual genera and species can be quite restricted. The genera *Artemia* and *Streptocephalus* are found on all continents; but their species are rather restricted in their individual distributions (see, e.g., Bowen, 1964; Browne and MacDonald, 1982). Some taxa are restricted to certain limited climatic conditions, such as the holarctic Polyartemiidae; or limited geographic areas, like the genus *Thamnocephalus* endemic to the western hemisphere (Belk, 1982) and the genus *Eubranchipus* found only in North America.

Some species are found only from one locality, for example, *Branchinella lithaca* (Creaser, 1940) from Stone Mountain, Georgia.

DEVELOPMENT   The most complete description of anostracan early ontogeny is given by Benesch (1969) for *Artemia salina*, which clarifies and supplements the work of Heath (1924). Cleavage is total (Fig. 30-7A) with the yolk being equally distributed among the blastomeres. Oddly, gastrulation in *Artemia* occurs in two distinct phases. The first (Fig. 30-7B) begins with the formation of a posteriorly located blastopore and results in the invagination of the mesoderm and primary germ cells (Fig. 30-7C), a process that begins about 32 hours after fertilization and is completed after about 50 hours. Naupliar anlagen then appear and are shortly followed about 60 hours later by the initiation of a distinct endodermal migration near the anterior part of the embryo (Fig. 30-7D, E).

The differentiation of the naupliar organs begins after the completion of gastrulation, when the stomodeum invaginates and the labrum begins to form (Fig. 30-7F, G). The paired naupliar anlagen give rise to the antennules, antennae, and mandibles, but anterior to these is an unpaired mass of preantennal mesoderm (Fig. 30-7D, F). This mass behaves like any of the more posterior metameres and produces a segmentally organized mass of muscles. This latter will eventually give rise to the labral, esophageal, and eye muscles, as well as lipid storage cells and connective tissue in the labrum (Fig. 30-7H). The teloblasts at the posterior end of the body (Fig. 30-7D, F) appear to be unpaired and indistinct and, as a result, the earliest stages of segment formation are not able to be determined in *Artemia*.

During the course of postnaupliar development, Benesch noted the appearance of cell masses derived from the somites in the mandibular and thoracic segments. These so-called 'blood glands' gave rise to hemocytes (Fig. 30-7J). Benesch maintained that these were homologous with the anlagen of the antennal and maxillary glands and were lacking only in the maxillulary segment. However, Anderson (1973) doubted whether these should be interpreted as segmental organs.

The eggs generally hatch as nauplii (Fig. 30-7L, M, N) in which the naupliar region is well developed and postnaupliar somites are internally delineated but not externally evident. Belk (personal communication) observed *Eubranchipus serratus* hatches as a metanauplius. There are 10 to 22 larval stages (Figs. 30-8, 30-9). The animal lives on yolk reserves through the first three stages and commences naupliar feeding at stage 4. However, at the third molt a series of postnaupliar appendage buds suddenly appear. The trunk limbs continue to develop either essentially in blocks or in small gradual steps (see discussion below). Through these larval molts, the animal subsists by 'naupliar' feeding, that is, the antennae and mandibles sweep food into the sublabral space, facilitated in part by a suction created by the very large labrum as the animal moves forward, as

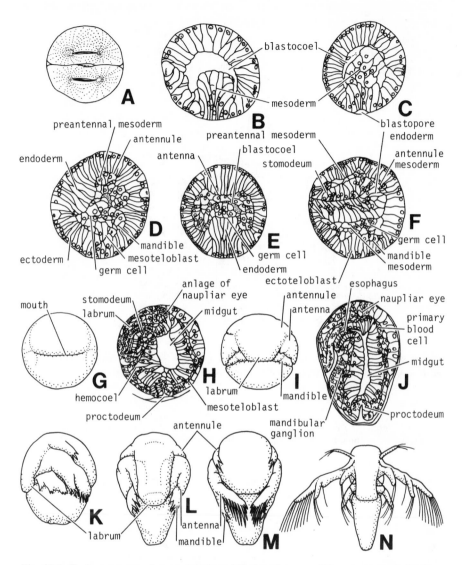

Fig. 30-7. Early ontogeny of *Artemia salina*. (A) first cleavage; (B) early stage of the first phase of gastrulation, involution of mesoderm; (C) advanced stage of mesoderm formation; (D) longitudinal section of second phase of gastrulation, inward migration of endoderm cells; (E) cross section of D; (F) initiation of stomodeum, begining of naupliar differentiation; (G) ventral view of F; (H) anterior gut and labrum complete, initiation of protodeum; (I) ventral view of H; (J) advanced naupliar differentiation; (K) lateral view of J; (L) ventral; (M) dorsal views of nauplius near to hatching (modified from Benesch, 1969). (N) hatched and free-swimming nauplius (from Heath, 1924).

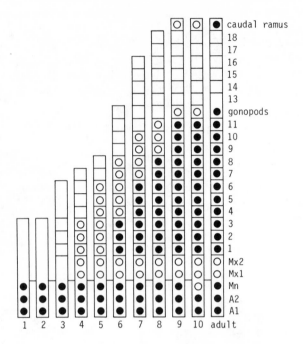

Fig. 30-8. Postembryonic development in *Artemia salina*. ● mature and functioning limbs, ○ immature and/or reduced limbs. (Modified from Anderson, 1967)

Fig. 30-9. Sequence of appendage development in *Branchinecta ferox*. ● mature and functioning limbs, ○ immature and/or reduced limbs. (Modified from Fryer, 1983)

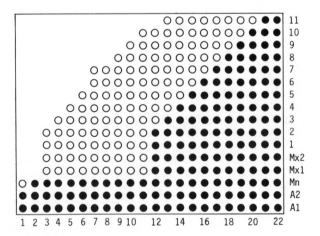

mechanical manipulation by the endite spines. The maxillules and maxillae are reduced in size to accommodate this sweeping action of the naupliar appendages.

Anderson (1967) recognized 10 instars in *Artemia salina*. Heath's stage 4 corresponds to Anderson's stage 2; however, Heath felt he could delineate more instar stages at the terminal end of this sequence. For example, Heath recognized 17 instars in *Branchinecta occidentalis*, though he admits that after stage 10 the differences are at best subtle. Baquai (1963) recognized 18 instars in *Streptocephalus seali*. Fryer (1983) delineated 22 instars in *B. ferox*.

The outline of this developmental sequence according to Anderson (1967) is summarized in Figure 30-8, but the pattern exhibited by *A. salina* may not be typical for anostracans as a whole. Anderson recorded distinct 'metamorphic' stages in *Artemia* the developmental sequence that occurred as groups of thoracopods were incorporated into the feeding behavior. In contrast, Fryer (1983) observed a gradual and sequential incorporation of trunk limbs into the feeding cycle in *Branchinecta* (Fig. 30-9). Fryer also noted a distinct overlap of naupliar feeding with thoracic feeding; in fact, the antennae continued to help in food manipulation to the end of the larval instars. This pattern of Fryer's is even more primitive than that noted for *Hutchinsoniella* (Sanders, 1963), heretofore considered to display the most primitive of crustacean developmental sequences, in that it is initiated with a true naupliar stage. Since Fryer only noted changes in the development of the limbs and did not record specifics as to segment addition, a complete comparison of Anderson's and Fryer's work is not possible. In addition, Anderson and Fryer deal with animals in different families; but clearly enough information is available from other, less complete studies (for a review, see Fryer, 1983, pl.333) to indicate that a major reevaluation of branchiopod, and particularly anostracan, development is needed.

FOSSIL RECORD The history of such delicately structured forms as fairy shrimp is rather poor, not too surprisingly. Some of the fossil material placed within the Anostraca undoubtedly belongs elsewhere (Tasch, 1969); for example, *Brachipusites* from the German Carboniferous is probably an insect nymph, and the lower Devonian *Gilsonicaris* Van Straelen, 1943 may not be anything at all. The problem with anostracan fossils are that they are generally so poorly preserved they don't look like much of anything. An example of this occurs in a specimen from the Silurian of Indiana (Fig. 30-10), which is reminiscent of an anostracan form but may or may not be so. On the other hand, a Miocene silicified anostracan described by Palmer (1957) is so beautifully preserved that the details afforded on that specimen indicate that the material undoubtedly represents a new family.

The distribution in space and time combine to reveal that anostracans are an old group, with both stable long-lived generalist taxa with

Fig. 30-10. Possible fossil anostracan (specimen no. 10975—Indiana University fossil invertebrate collection) from the Silurian Kokomo Limestone, Indiana. (c) cephalon, (t) trunk.

ubiquitous distributions, and highly speciose (probably actively evolving) taxa composed of endemic specialists.

TAXONOMY   There are eight living families whose differences are largely based on structures of the male claspers and associated structures (Belk, 1982). As indicated above, fossil forms may pose some problems in accommodating to this scheme.

> Family Artemiidae Grochowski, 1896 Pleistocene–Recent
> Family Branchinectidae Daday, 1910 Recent
> Family Branchipodidae Simon, 1886 (?Eocene) Recent
> Family Chirocephalidae Daday, 1910 Recent
> Family Linderiellidae Brtek, 1964 Recent
> Family Polyartemiidae Simon, 1886 Recent
> Family Streptocephalidae Daday, 1910 Recent
> Family Thamnocephalidae Simon, 1886 Recent

REFERENCES

Anderson, D. T. 1967. Larval development and segment formation in the branchiopod crustaceans *Limnadia stanleyana* and *Artemia salina*. *Aust. J. Zool.* **15**:47–91.

Anderson, D. T. 1973. *Embryology and Phylogeny in Annelids and Arthropods.* Pergamon Press, Oxford.

Baqai, I. U. 1963. Studies on the postembryonic development of the fairy shrimp *Streptocephalus seali*. *Tulane Stud. Zool.* **10**:91–120.

Belk, D. 1977. Evolution of egg size strategies in fairy shrimp. *Southw. Nat.* **22**:99–105.

Belk, 1982. Branchiopoda. In *Synopsis and Classification of Living Organisms*, Vol. 2 (S. P. Parker, ed.), pp. 174–80. McGraw-Hill, New York.

Belk, D. 1984. Antennal appendages and reproductive success in the Anostraca. *J. Crust. Biol.* **4**:66–71.

Belk, D., and G. Pereira. 1982. *Thamnocephalus venezuelensis*, n. sp., first report of *Thamnocephalus* in South America. *J. Crust. Biol.* **2**:223–26.

Benesch, R. 1969. Zur Ontogenie und Morphologie von *Artemia salina*. *Zool. Jahrb. Abt. Anat.* **86**:307–458.

Bowen, S. 1964. The genetics of *Artemia salina*. IV. Hybridization of wild populations with mutant stocks. *Biol. Bull.* **126**:333–344.

Browne, R. A., and G. H. MacDonald. 1982. Biogeography of the brine shrimp, *Artemia*: distribution of parthenogenetic and sexual populations. *J. Biogeogr.* **9**:331–38.

Cannon, H. G. 1928. On the feeding mechanisms of the fairy shrimp, *Chirocephalus diaphanus*. *Trans. Roy. Soc. Edinb.* **55**:807–22.

Cannon, H. G., and F. M. C. Leak. 1933. On the feeding mechanisms of the Branchiopoda. *Phil. Trans. Roy. Soc. Lond.* (B)**222**:267–352.

Claus, C. 1873. Zur Kenntniss des Baues und der Entwicklung von *Branchipus stagnalis* und *Apus cancriformis*. *Abhl. König. Gesell. wissen. Gött.* **18**:93–136.

Claus, C. 1886. Untersuchungen ober die Organization und Entwicklung von *Branchipus* und *Artemia* nebst vergleichenden Bemerkungen über andere Phyllopoden. *Arb. zool. Inst. Wien* **6**:267–370, 12 pls.

Creaser, E. P. 1940. A new species of phyllopod from Georgia. *J. Wash. Acad. Sci.* **30**:435–37.

Daday de Deés, E. 1910a. Monographie systématique des Phyllopodes Anostracés. *Ann. Sci. Nat. Zool.* (9)**11**:91–492.

Daday de Deés, E. 1910b. Quelques Phyllopodes Anostracés nouveaux. *Ann. Sci. Nat. Zool.* (9)**12**:241–64.

Ekman, S. 1902. Beiträge zur Kenntniss die phyllopoden Familie Polyartemiidae. *Bihang till R. Svenska Vet. Akad. Handl.* **28**:1–4.

Elofsson, R. 1966. The nauplius eye and frontal organs of the non-Malacostraca. *Sarsia* **25**:1–128.

Fryer, G. 1966. *Branchinecta gigas*, a nonfilter feeding raptatory anostracan, with notes on the feeding habits of certain other anostracans. *Proc. Linn. Soc. Lond.* **177**:19–34.

Fryer, G. 1983. Functional ontogenetic changes in *Branchinecta ferox*. *Phil. Trans. Roy. Soc. Lond.* (B)**303**:229–343.

Geddes, M. C. 1981. Revision of Australian species of *Branchinella*. Aust *J. Mar. Freshwater Res.* **32**:253–95.

Heath, H. 1924. External development of certain phyllopods. *J. Morph.* **38**:453–83.

Hessler, R. R. 1964. The Cephalocarida comparative skeletomusculature. *Mem. Conn. Acad. Sci.* **16**:1–97.

Lent, C. M. 1971. Metachronal limb movements by *Artemia salina*: synchrony of male and female during copulation. *Science* **173**:1247–48.

Linder, F. 1941. Contributions to the morphology and the taxonomy of the Branchiopoda Anostraca. *Zool. Bidrag. Uppsala* **20**:101–302, 1 pl.

Linder, H. J. 1959. Studies on the freshwater fairy shrimp *Chirocephalus bundyi*. *J. Morph.* **104**:1–47.

Lowndes, A. G. 1933. The feeding mechanism of *Chirocephalus diaphanus*, the fairy shrimp. *Proc. Zool. Soc. Lond.* **1933**:1093–118, 7 pls.

Lynch, J. E. 1937. A giant new species of fairy shrimp of the genus *Branchinecta* from the state of Washington. *Proc. U.S. Nat. Mus.* **84**:555–62, pls. 77–80.

Lynch, J. E. 1960. The fairy shrimp *Branchinecta campestris* from northwestern United States. *Proc. U.S. Nat. Mus.* **112**:549–61.

Moore, W. G., and L. H. Ogren. 1962. Notes on the breeding behavior of *Eubranchipus holmani*. *Tulane Stud. Zool.* **9**:315–18.

Packard, A. S. 1883. A monograph of the phyllopod Crustacea of North America with remarks on the order Phyllocarida. *Twelfth Ann. Rept. U.S. Geol. Geogr. Surv. Terr., 1878* **1**:295–592.

Palmer, A. R. 1957. Miocene arthropods from the Mojave Desert, California. *U.S.G.S. Prof. Paper* **294**:237–80.

Pennak, R. W. 1953. *Fresh Water Invertebrates of the United States*. Ronald Press, New York.

Pereira, G. 1983. Taxonomic importance of the frontal appendage in the genus *Dendrocephalus*. *J. Crust. Biol.* **3**:293–305.

Sanders, H. L. 1963. The Cephalocarida functional morphology, larval development, comparative external anatomy. *Mem. Conn. Acad. Arts and Sci.* **15**:1–80.

Tasch, P. 1969. Branchiopoda. In *Treatise on Invertebrate Paleontology*, Part R, *Arthropoda* **4**(1) (R. C. Moore, ed.), pp. R128–91. Geol. Soc. Am. and Univ. Kansas Press, Lawrence.

Wolfe, A. F. 1971. A histological and histochemical study of the male reproductive system of *Artemia*. *J. Morph.* **135**:51–69.

Wolfe, A. F. 1973. Observations on the clasping behavior of *Artemia salina*. *Amer. Zool.* **13**:472.

Wyman, F. H. 1981. Mating behavior in the *Streptocephalus* fairy shrimps. *Southwest. Nat.* **25**:541–46.

31

CONCHOSTRACA

DEFINITION   Carapace bivalved, laterally compressed, completely enveloping body and limbs; sessile compound eyes; antennules small to moderate,
uniramous, sometimes unsegmented; antennae biramous, flagelliform;
mandibles without palps; maxillules small; maxillae vestigial or absent;
trunk limbs as phyllopods, 10 to 32 pairs; telson with spikelike terminus,
caudal rami styliform or clawlike.

HISTORY   Attention in this uncommonly encountered group has been largely
restricted to basic systematics. The work of Sars (1896) and Daday (1915–27)
served as benchmarks. However, most advances in the knowledge of conchostracans has come about as various isolated studies from restricted
regions about the world. About 200 species are currently recognized.

MORPHOLOGY   The all-enveloping carapace (Fig. 31-1A) is generally,
though not always, marked by concentric lines of growth. The carapace is
bivalved, though the dorsal hinge is considered by some to be not a true
one, in that it's merely a flexible portion of an otherwise dorsally continuous shield. It is closed by an adductor muscle in the head region. The carapace is connected to the dorsum of the head by a ligament. The body is
thus suspended in the carapace chamber from the ligament and is able to
move freely, even when the carapace is tightly closed.

The head proper is relatively small (Fig. 31-1B). It is somewhat subtriangular in shape when looking at it dorsally, with the anterior apex bearing the compound eyes. Posterior to the eyes on the dorsal surface is a
clublike stalked organ, which Sars (1896) originally interpreted (?) as 'an
organ of attachment,' but it may be actually the homolog of the head pores
seen on cladocerans. A prominent cervical groove marks the level of the
mandibles. Ventrally, the head bears a 'rostrum,' a subtriangular structure
anterior to the labrum, which is actually more akin to an epistome than a
rostrum. The naupliar eye is embedded in tissue beneath the surface of the
epistome. The epistome is separated from the labrum by a groove.

The antennules (Fig. 31-2A) flank the epistome and are relatively well
developed for a branchiopod. A short basal segment bears a long process,
which in turn is distally equipped with a number of lobes covered by chemosensory papillae.

The biramous antennae (Fig. 31-2B) are large. The basal part proximally bears a small setose lobe. The distal portion is marked by a series of

379

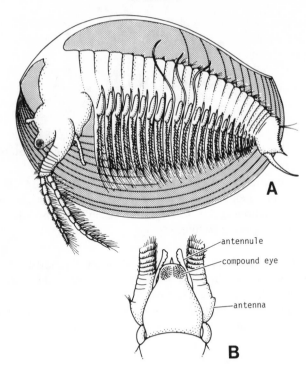

Fig. 31-1. Body plan of *Limnadia lenticularis*. (A) lateral view of body; (B) dorsal view of cephalon. (From Sars, 1896)

annulae, each of which is equipped with a row of spinelike setae. The two flagella have fewer than 10 or up to 25 segments, each bearing a set of brushlike setae posteriorly and a set of simple spinelike setae anteriorly.

The labrum is a subrectangular structure whose free edge is marked by a prominent spinelike process. The labrum (Fig. 31-2C) forms a large atrium oris in which the molar processes of the mandible are situated. The mandibles (Fig. 31-2D) are simple lobate gnathobases with ridged molar processes. Sars could not discern any paragnaths, but others have reported apparently small, densely setose, pointed lobes.

The maxillules (Fig. 31-2E) are simple lobes. Anteriorly there is a small setose accessory lobe. The main lobe bears long setae along the margin, parallel to which on the posterior extent are a row of special comblike setae.

The maxillae (Fig. 31-2F) are tiny setose lobes.

The trunk is composed of generally 10 to 32 segments, each bearing a phyllopodous limb. The appendages are more or less homonomous, though posterior to the eleventh pair they begin to decrease in size and development. The limbs are rather narrow and elongate (Fig. 31-2G). The protopod has five rather broad setose endites, the most proximal of which is rather more distinct and complex (Fig. 31-2H). The fifth endite on the trunk limbs is modified as a 'palp' or 'tactile process' in males. The endopod is a distinctly articulated paddlelike lobe and is often referred to as a sixth endite. The exopod is an elongate plate, sometimes referred to as a 'flabellum.' It extends distally from the lateral side of the protopod and appears to be developed proximally as a long, dorsally directed lobe. This lobe may actually represent a distal epipod. There is an additional respiratory epipod of rather fleshy character arising near the base of the protopod. In males the first one or two trunk limbs lose their phyllopodous lobes and bear hypertrophied endites to assist in grappling females.

Fig. 31-2. Appendages of *Limnadia lenticularis*. (A) antennule; (B) antenna; (C) labrum; (D) mandible; (E) maxillule; (F) maxilla; (G) trunk limb; (H) proximalmost endite of G; (I) body terminus with caudal rami. (From Sars, 1896)

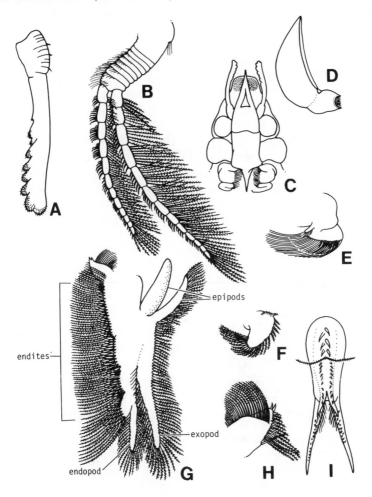

The anal segment (Fig. 31-2I) is large and directed ventrally. It dorsally bears a double row of stout spines. Terminally, this unit has a pair of blade-like serrate anal spines, and a pair of slender serrate and spinose caudal rami flanking the terminal anus.

The gut is generally a simple tube extending from mouth to anus. In the cephalon, however, a pair of gut caeca arises laterally from the anterior of the midgut and extends anastomosing rami ventrally and dorsally into the anterior part of the head.

The circulatory system centers on the heart, which extends from the fourth trunk segment into the head. The heart has four sets of ostia, one in the head, and the others in the three anterior trunk segments. Blood enters the heart through the ostia and posterior channel. Three blood paths can be recognized. Blood is supplied anteriorly to the head, ventroposteriorly to the viscera and trunk limbs, and laterally over the adductor muscles into the carapace and wherein it flows through a complex system of channels. The carapace circulation is a principal locus for respiration, in addition to the fleshy epipods on the trunk limbs.

Excretion is achieved by a maxillary gland.

The sexes are separate when there are more than one, but some species are apparently parthenogenetic or both parthenogenetic and sexual (Belk, 1972; Strenth, 1977). The gonads extend throughout the trunk adjacent to the midgut. They connect to the outside by means of oviducts or vasa deferentia that open on the eleventh trunk segment.

Little seems to have been recorded concerning the nervous system. The antennal nerves arise from the circumesophageal ganglion. The nerve cord is ladderlike with two widely separated trunks and connected in each segment with delicate commissures. The ganglia along its length, however, are only slightly developed (Fig. 31-3).

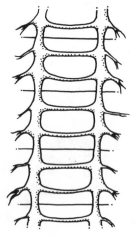

Fig. 31-3. Ventral nerve cord of *Limnadia lenticularis*. (From Sars, 1896)

NATURAL HISTORY   Although conchostracans are denizens of fresh water all over the world, they have not been the subject of intense study. For example, few detailed investigations have been carried out concerning locomotion or feeding (Cannon, 1933; Eriksson, 1935), and only a few works have focused on other aspects of their biology (e.g., Belk, 1972; Eriksen and Brown, 1980).

Details concerning reproductive habits are vague. During copulation the male grasps the ventral edge of the female shell and extends his abdomen into the carapace space to deposit a spermatophore in the area of the tenth or eleventh trunk limbs (Strenth, 1977). Eggs are brooded inside the carapace on or around the epipodites. The eggs are shed at the time of a female molt. Breeding is constant throughout the adult stage of the life cycle, the female shedding eggs with each molt (Tasch, 1969).

The eggs are equipped with thick shells. However, quite contrary to popular opinion (e.g., Pennak, 1953) it seems that this shell is not so much effective to prevent desiccation as it is at fending off the effects of the ultraviolet radiation of sunlight or protecting the egg and embryo from mechanical abrasion (Belk, 1970). Actual distributions of at least some species, for example, *Caenestheriella setosa*, seem to be determined by the effect of temperature on the egg-hatching response (Belk, 1975).

DEVELOPMENT   Cannon (1924) investigated the internal development from the naupliar stage onward and observed some early aspects of organogenesis (Fig. 31-4). At the naupliar hatching, the space between the ecto- and endodermis is completely filled by mesoderm and germ cells. First a split in the mesodermal mass occurs dorsally, forming the cardiac cavity, then the ventral mesoderm separates from the gut, forming the perivisceral cavity. Both of these cavities are continuous in the region of the proctodeum. The ventral mesoderm develops some transitory segmentation, while the dorsal part of the mesoderm delineates seven coelomic pouches. The heart develops from the first four of these pouches, after which the sacs diminish in size. In the anterior part of the ventral mesoderm, a coelomic cavity appears that becomes the end sac of the maxillary gland.

The external postnaupliar development is reviewed by Anderson (1967) and Monakov and Dobrynina (1977). The pattern is distinctive, with the addition of segments and limbs occurring in bursts through six stages before an adult condition is achieved (Fig. 31-5). The initial nauplius is laden with yolk, and feeding does not begin until stage 3.

FOSSIL RECORD   Tasch (1969) reviewed the record of fossil conchostracans. It is a modest one, extending from the Devonian. For the most part, however, it is a disappointing history, largely characterized by only fossils of carapaces. Because the major diagnostic features of the higher taxa of clam shrimp are currently related to body shape, these fossils can generally be fit fairly well into the conchostracan classification scheme. However, they are not particularly helpful concerning details of evolution within the group.

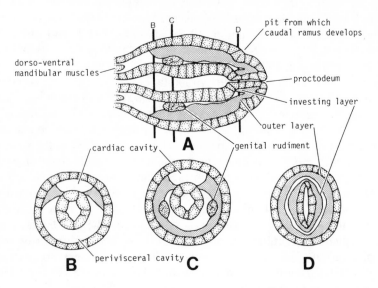

Fig. 31-4. Diagrams of posterior part of early estheriod larvae. (A) frontal section; B, C, -D) cross sections at designated levels of A. (Modified from Cannon, 1924)

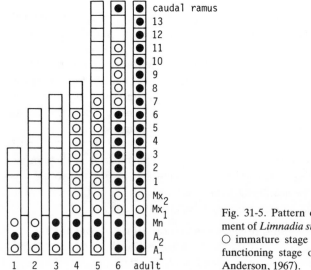

Fig. 31-5. Pattern of posthatching development of *Limnadia stanleyana*.
○ immature stage of limb, ● mature and functioning stage of limb. (Modified from Anderson, 1967).

TAXONOMY Though the conchostracans are not a large group. Linder (1945) did see fit to divide them into several higher taxa based on body form. Whether or not some of these features are convergent remains to be seen. The lyncaeids have a globular carapace with the cephalon completely exposed. The limnadioids have prominent umbones on the carapace, while the cyzicoids have a more discrete umbone.

Suborder Laevicauda Linder, 1945 Lower Cretaceous–Recent
    Family Lynceidae Stebbing, 1902
Suborder Spinicauda Linder, 1945 Lower Devonian–Recent
   Superfamily Limnadioidea Baird, 1849 Carboniferous–Recent
    Family Cyclestheriidae Sars, 1899 Recent
    Family Leptestheriidae Daday, 1923 Recent
    Family Limnadiidae Baird, 1849 Carboniferous–Recent
   Superfamily Cyzicoidea Stebbing, 1910 Lower Devonian–Recent
    Family Cyzicidae Stebbing, 1910 Lower Devonian–Recent
    Family Aslmussiidae Kobayashi, II 1954 Lower Devonian–Upper Cretaceous
   Superfamily Estherielloidea Kobayashi, 1954 Upper Carboniferous–Upper Cretaceous
    Family Estheriellidae Kobayashi, 1954
   Superfamily Leaioidea Raymond, 1946 Middle Devonian–Lower Cretaceous
    Family Leaiidae Raymond, 1946
   Superfamily Vertexioidea Kobayashi, 1954 Devonian–Recent
    Family Vertexiidae Kobayashi, 1954 Lower Carboniferous–Triassic
    Family Limnadopsidae Tasch, 1969 Lower Carboniferous–Recent
    Family Pemphilimnadiopsidae Tasch, 1961 Pennsylvanian
    Family Ipsiloniidae Novozhilov, 1958 Devonian–Lower Cretaceous

REFERENCES

Anderson, D. T. 1967. Larval development and segment formation in the branchiopod crustaceans *Limnadia stanleyana* and *Artemia salina*. *Aust. J. Zool.* **15**:47–91.
Belk, D. 1970. Functions of the conchostracan egg shell. *Crustaceana* **19**:105–106.
Belk, D. 1972. The biology and ecology of *Eulimnadia antlei*. *Southwest. Nat.* **16**:297–305.
Belk, D. 1975. Hatching temperatures and new distributional records for *Caenestheriella setosa*. *Southwest. Nat.* **20**:409–20.
Cannon, H. G. 1924. On the development of an estheriid crustacean. *Phil. Trans. Roy. Soc. Lond.* (B)**212**:395–430.
Cannon, H. G. 1933. On the feeding mechanism of the Branchiopoda. *Phil. Trans. Roy. Soc. Lond.* (B)**222**:267–352.
Daday, E. 1915, 1927. Monographie systématique des phyllopodes conchostracés. *Ann. Sci. rNat. Zool.* (9)**20**:39–330, (10)**10**:1–112.

Eriksen, C. H., and R. J. Brown. 1980. Comparative respiratory physiology and ecology of phyllopod Crustacea. I. Conchostraca. *Crustaceana* **39**:3–9.

Eriksson, S. 1935. Studien über die Fangapparate der Branchiopoden. *Zool. Bidr. Uppsala* **15**:23–287.

Linder, F. 1945. Affinities within the Branchiopoda, with notes on some dubious fossils. *Ark. Zool.* **37**(4):1–28.

Monakov, A. V., and T. I. Dobrynina. 1977. Postembryonic development of Lynceus brachyurus. *Zool. Zhur.* **56**:1877–79.

Pennak, R. W. 1953. *Fresh-water Invertebrates of the United States.* Ronald Press, New York.

Sars, G. O. 1896. *Fauna Norvegiae.* Bd. I. *Phyllocarida og Phyllopoda.* Christiania.

Strenth, N. E. 1977. Successful variation of sex ratios in *Eulimnadia texana. Southwest. Nat.* **22**:205–12.

Tasch, P. 1969. Branchiopoda. In *Treatise on Invertebrate Paleontology*, Part R, *Arthropoda* **4**(1) R. C. Moore, ed.), pp. R128–91. Geol. Soc. Am. and Univ. Kansas Press, Lawrence.

32

CLADOCERA

DEFINITION   Carapace laterally compressed, not hinged, usually enclosing trunk and appendages, sometimes atrophied, cephalon always exposed; compound eyes sessile when present; antennules modest to large in size, uniramous; antennae biramous; mandibles lacking palps; maxillules small; maxillae vestigial or absent; thoracopods of a reduced phyllopod type, four to six pairs; trunk somites obscure, posterior trunk recurved under body, terminating with clawlike caudal rami.

HISTORY   Serious study of cladocerans began as far back as 1776 when O. F. Müller published a monograph on the group. A great deal of systematic work has been done since then, but works by Lilljeborg (1901) and Birge (1918) were pivotal in cladoceran studies. The former is still a benchmark for current work, and the latter was unfortunately largely responsible for the recurrent but inaccurate concept of cladoceran species being cosmopolitan. Study of the group has always attracted a steady stream of monographers who have devoted detailed attention to particular taxa or geographic regions, for example, Brooks (1957), Goulden (1968), or Smirnov and Timms (1983). Most recently, taxonomic observations have been most effectively and creatively combined with functional morphologic analysis to enlarge our understanding of the overall biology of the group (e.g., Fryer 1963, 1968, 1974).

MORPHOLOGY   The 600 or so species of water fleas are among the most distinctive of crustacean body plans (Fig. 32-1). The body is reduced to, at most, a half-dozen limb-bearing trunk segments and a recurved postabdomen. The carapace is bilobed and generally envelopes the trunk and all the legs, though in the polyphemids it is reduced to a small brood structure (Fig. 32-1G, F). The cephalon is exposed, but covered with a head shield of variable development (Fig. 32-1A, D). The head is marked by a set of sessile compound eyes that are typically fused together at the midline. The naupliar eye is small and often lacks pigment. Frey (1959) also noted the structure and taxonomic importance of head pores. These are minute pores on the posterior median aspect of the head shield (thought once to perhaps be respiratory in nature) whose exact function is unknown.

One of the more confusing aspects to cladoceran studies is the phenomenon of cyclomorphosis. This is a within-species variation in shape with season. Brooks (1946) demonstrated that this was related to differential

387

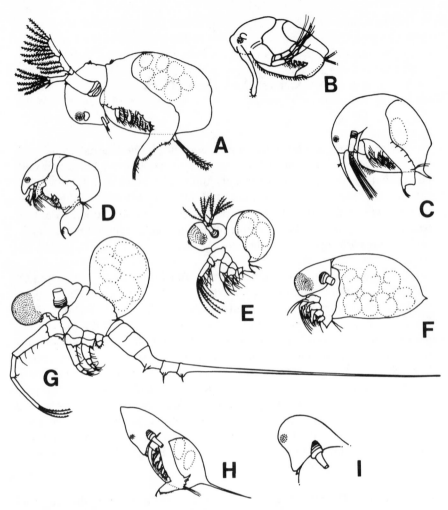

Fig. 32-1. Body plans of cladocerans. (A) *Sida crystallina*; (B) *Moina brachycephala*; (C) *Bosmina longispina*; (D) *Eurycercus glacialis*; (E) *Podon intermedius*; (F) *Evadne nordmanni*; (G) *Bythotrephes longimanus*; (H, I) some cyclomorphs of *Daphnia galeata*. (A, C–I from Lilljeborg, 1901; B from Goulden, 1968)

rates of growth between the cephalon and trunk regulated by temperature and food availability. For example, in *Daphnia recurvata* and *D. galeata* Brooks showed that temperatures less than 7°C resulted in animals with round heads, while temperatures in excess of 7°C produced spiked-head creatures (Fig. 32-1H, I).

The antennules (Fig. 32-2A) are typically modest in size and distally marked by a tuft of terminal setae. These can be rather large structures as, for example, in the males of *Moina brachycephala* (Fig. 32-1B) or *Bosmina longispina* (Fig. 32-1C).

The antennae (Fig. 32-2B) are prominent. The peduncles are usually rather large and distally equipped with two modest rami of two to four segments bearing prominent setae. On occasion, as in *Latona*, there is a variation to effectively produce an almost triramose antenna (Fig. 32-3C).

The mandibles (Fig. 32-2C) are generally rather simple, that is, no palps and massive molar processes. However, specialized variants occur (Fig. 32-3A, B) with either more complex gnathobases, as in *Bythotrephes*, or styliform blades, as in carnivores like *Leptodora*.

The maxillules (Fig. 32-2D) are reduced to small, flat lobes with long, stiff setae. The maxillae are either absent or developed as tiny lobes, which are little more than vehicles for an opening of the maxillary glands. Fryer (1963) raised the issue of whether the first trunk limb is possibly the

Fig. 32-2. Appendages of *Sida crystallina*. (A) antennule; (B) antenna; (C) mandible; (D) maxillule; (E) first trunk limb; (F) sixth trunk limb; (G) caudal end. (From Lilljeborg, 1901)

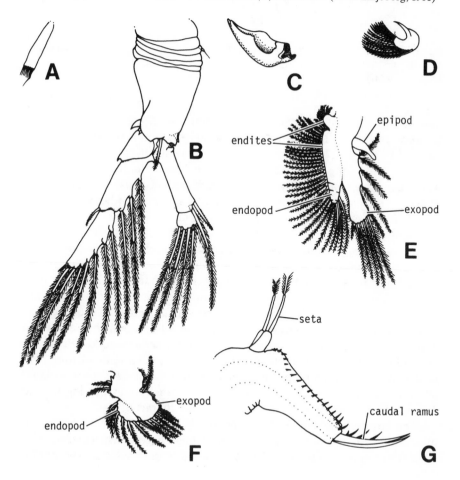

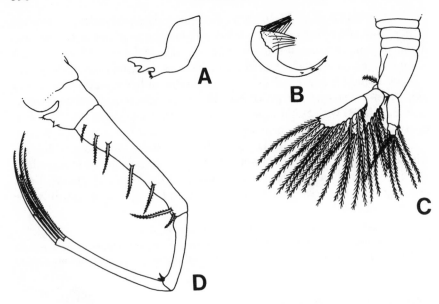

Fig. 32-3. Mandible of (A) *Bythotrephes* sp.; (B) *Leptodora* sp.; (C) antenna of *Latona seti-fera*; (D) first trunk limb *Bythotrephes longimanus*. (A, B from Wagler, 1927; C, D from Lilljeborg, 1901)

maxilla, since he felt the arguments of Grobben (1879, p. 245) on this matter were unconvincing (a position somewhat akin to those expressed by this author in Chapter 27 concerning lipostracan maxillae).

The trunk limbs (Fig. 32-2E, F) range from four to six in number. These can vary considerably along the series, in response to slight differences in function (see discussion below under feeding); or between species with different modes of feeding. In a filtering and/or grazing form like *Sida*, the phyllopodous nature of the limbs is preserved. Generally in these forms almost all traces of segmentation disappear, although the broad outlines of the primitive rami can be more or less discerned. A setose endite, or gnathobase, may be seen, but distinctions of other endites typically become obscure, and the remaining setal row becomes continuous with that of the remnant endopod. This latter ramus, however, as well as that of the exopod, is continuous with the protopodal region. Only a branchial epipodite retains an articulation. The posterior limbs in the series tend to be reduced and indeed may be absent altogether in examples of extreme oligomerization of the cladoceran body. In carnivorous forms, like *Polyphemus* (Fig. 32-3D), the trunk limbs are segmented and rather robust, losing for the most part traces of their original phyllopodous character.

The posterior part of the trunk generally loses traces of the body segmentation. The terminus recurves ventrally and becomes specialized (Fig. 32-2G) as the postabdomen. This is modified to both assist in locomotion and ensure that feces get voided outside the carapace chamber. The post-abdomen can be variously adorned with spine rows, elaborate setae, and the distal, bladelike, caudal rami.

Internal anatomy is relatively simple, as one might suspect in such creatures that display such distinct oligomerization (Fig. 32-4). The digestive system, for example, typically is a simple tube. If digestive caeca occur, they are simple paired pockets. In some forms, the midgut coils once or twice before entering the hindgut, as in *Drepanothrix* or *Eurycercus*.

The circulatory system is reduced to a single bulbous heart in the region of the first trunk segment. It has a single pair of ostia, and blood is pumped anteriad directly into the head sinus. However, Wagler (1927) reports in *Leptodora* a bulbous arteriosus and short aorta anterior to the heart proper.

Excretion is achieved with maxillary glands.

The reproductive system consists of simple single or paired gonads. In females the oviducts open along the lateral or dorsal surface of the

Fig. 32-4. Internal anatomy of *Daphnia pulex*. (From Kükenthal and Matthes, 1953)

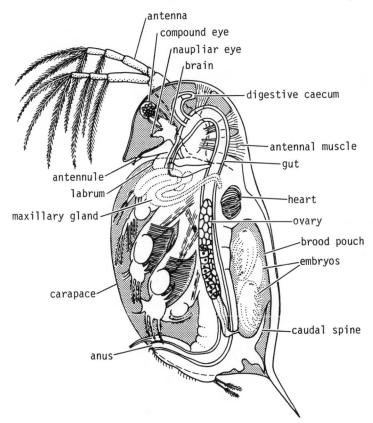

posterior trunk. In males the openings are lateral or ventral and typically placed near the body terminus. The males may fuse the openings to a single pore, and some groups, for example, the Sididae, have the terminal portions of the vas deferens everted to form a penis.

The nervous system can show signs of reduction. In some forms the ganglia of the cord fuse to form a single postoral mass. There seems to be no remnant of a maxillary ganglion, even in those forms that do not fuse the ganglia. The small naupliar eyes sometimes lack pigment, but the compound eyes can have up to 500 ommatidia, especially in carnivores like *Leptodora*.

NATURAL HISTORY   Cladocerans sort themselves into two general types, swimmers and crawlers. The former typically uses the antennae for locomotion, assisted to a greater or lesser degree by the buoyancy of the body. The crawlers live on or in the bottom or on the surface of macrophytes. The trunk limbs and postabdomen push the animal along. Sometimes the ventral surface of the carapace can have specially arranged setae on a recurved flange that helps form a seal such that a negative pressure within the carapace chamber can facilitate movement up vertical surfaces and even under overhangs.

Fryer (1963, 1968, 1974) carried out detailed investigations on cladoceran feeding, with other important studies done by Cannon (1933), Eriksson (1935), Lockhead (1936), Smirnov (1968), and Crittenden (1981). The details noted in Fryer's work, however, reveals that generalizations are difficult since specific animals use particular structures in different ways depending on individual needs. The first trunk limb is often mainly a locomotory organ, but in macrothricids it also serves to help collect food. Food collection proper is more often a function of the second trunk limb, but the intermediary and posterior trunk limbs serve to set up water currents that sweep food particles into the carapace chamber from outside (filter feeding) or bring scraped particles from the distal parts of the anterior limbs up into the food groove along the limb bases (grazing). One common theme, however, is the lack of an anteriorly directed current in the food groove. Particles there are manipulated by the gnathobases and passed forward mechanically.

Fryer (especially 1962) carefully documented the secretory nature of labral and trunk limb glands. Taking his cue from Cannon (1922), these were assumed to produce mucuslike secretions to agglutinate food particles. However, as Koehl and Strickler have shown for copepods (see Chapter 37), the problem with animals the size of cladocerans (on the average about 1 mm long, but ranging down to 0.26 mm as in *Alonella nana*) is 'sticky' water around food particles, and limb and setal surfaces. Indeed, Crittenden (1981) confirms that cladocerans function in a regime of laminar flow. It would seem a profitable line of investigation would be to determine whether these secretions in cladocerans might not actually be surfactants or 'wetting agents.' Such might be suspected from consideration of forms like *Pseudochydorus globosus*, a scavenger, that produces copious labral gland secretions even though they are not obviously needed to 'glue' particles together. Fryer (1974) observed in some species, such as in *Macrothrix triserialis*, *Onchobunops tuberculatus*, and *Streblocerus serricaudatus*, enormous reservoirs for the labral glands that sometimes extended up into the head. The shear volume of these secretions relative to the size of the animal and the particles being fed upon might suggest a function other than agglutination.

Other feeding modes are known in cladocerans. The leptodorids and polyphemoids are carnivores. Their robust and spinose limbs grapple victims and bring them to the mouth for their serrate and styliform mandibles to render. *Pseudochydorus globosus* is a scavenger, with the spinose second thoracopod being used to grab food while the proximal elements of the first through third limbs push the food up into the food groove. *Anchistropus emarginatus* is an ectoparasite on *Hydra*. Hooks on the first thoracopod serve alternately for attachment, movement on the host, and rendering of flesh.

Reproductive biology of cladocerans has been the subject of considerable experimental manipulation (see Kaestner, 1970, for a summary). Although bisexual, cladocerans alternate between asexual and sexual phases. Asexual phases predominate in spring and summer; sexual phases in the fall produce resting eggs that overwinter.

The number of eggs produced during the life of a female varies (Green, 1954, 1956). Generally in nature, the number of eggs per brood increases with the age of *Daphnia* females; with increasing age of the mother, the proportion of young maturing at the fifth instar, rather than the sixth, increases. These can be seen as modes of reproduction meant to increase the probability of overwintering success in the population. Green (1954), however, found that under laboratory conditions the number of eggs decreased with the age of the female, which is a caveat to be careful when interpreting laboratory data relevant to crustacean reproduction.

Cladocerans have evolved a useful mechanism, like aphids, to induce some genetic variation even during asexual phases of reproduction (Bacci et al., 1961). At least some populations of *Daphnia pulex* utilize endomeiosis, whereby bivalent chromosomes in summer eggs engage in crossover before anaphase. The genetic variation is said to be as great as that of sexual (amphigonic) reproduction. This seems to be especially useful in modifying sex genotypes and, thus, sex determination is not entirely environmentally determined.

Resting eggs are rich in yolk and develop to a gastrula stage before going into a diapause. In some cladocerans, the resting eggs are released to float or sink as the case may be; in others they are attached to the bottom or to surrounding macrophytes. Even the oceanic *Evadne* produces resting eggs, but their fate is currently not known (Bainbridge, 1958). Some cladoceran females molt during brooding and leave the eggs enclosed within the exuvia, forming ephippia. These afford some added protection during the diapause. Some ephippia sink; others, like daphniids, float; thus, like floating resting eggs, they become vehicles of dispersal, while still others, like chydorids, fix them to macrophytes (Frey, 1982).

Cladocerans inhabit lakes, ponds, and vernal pools. Their preference seems to be toward more permanent bodies of water rather than the more transient pools frequented by other branchiopods. They also are the only branchiopods known from marine situations, with forms like *Evadne* and *Penilia* being not uncommon denizens of the plankton. *Evadne*, in fact, can in some areas and some seasons be very abundant and occur in a high proportion of samples (Longhurst and Seibert, 1972).

The life history of cladocerans seems directed, at least in part, toward minimizing the effects of invertebrate predation (Lynch, 1980). The eggs of cladocerans are relatively large and well equippped with yolk. The larvae hatched are quite large relative to the size of the adult and have a high rate of growth. Such a strategy does encounter other problems. Large size may minimize invertebrate predation but increases the probabilities of vertebrate predation (Lynch, 1980). In addition, larger animals run the risk of outstripping resources, and cladocerans do often exhibit seasonal succession with smaller forms following the larger forms as food resources shift (Lynch, 1978).

Cladocerans were thought to be cosmopolitan forms, that is, having few species but those being rather ubiquitous. Frey (1982) seriously questions such a generalization. This misconception stems largely from the work of Birge (1918) who began the practice of superficially comparing material from various parts of the world with the excellent descriptions and illustrations of Lilljeborg (1901). For example, it had been widely assumed there were only two species of *Eurycercus* in the world: the northern hemisphere *E. lamellatus* and the north temperate *E. glacialis*. Frey (1973) revealed that this simple picture was not the case when he described a third species, and so effectively has the old bispecific concept been overturned that the total number of *Eurycercus* species 'cannot even be guessed at the moment' (Frey, 1982, p. 492). *Chydorus sphaericus* was once thought to be the most ubiquitous of cladocerans. It is now known to be a complicated species complex in which *C. sphaericus* (sensu stricto) is not even present in North America.

Even the marine genus *Evadne* was thought to be restricted to neritic habitats. However, Longhurst and Seibert (1972) showed *E. spinifera* to be a denizen of central gyral waters in the eastern Pacific, while *E. tergestina* avoids gyres and frequents warm waters in low latitudes.

As far as cladoceran biogeography is concerned, the issues were thought mistakenly to be simple and long settled. Now, until cladoceran systematics can restabilize at some point in the future, and the field biology of the group is studied (separate from laboratory analysis), we simply cannot make any reliable observations on cladoceran distribution.

DEVELOPMENT   Detailed knowledge of early ontogeny in cladocerans is largely restricted to the earliest stages (Grobben, 1879; Kühn, 1908, 1911, 1913). Cleavage is total (Fig. 32-5A) and of a spiral type, and presumptive cells are able to be recognized already at the 16-cell stage (Fig. 32-5B, C). Eggs undergo a dormancy period.

Gastrulation is initiated by invagination of cells to form the endo- and mesodermal cells (Fig. 32-5D) while the primordial germ cells serve to mark the site of the blastopore. Epiboly of the ectoderm eventually serves

Fig. 32-5. Development of cladocerans. (A) second cleavage; (B) ventral view of embryo, 16-cell stage; (C) blastopore region as gastrulation is about to commence; (D) longitudinal section during gastrulation; (E) external ventral view of naupliar stage; (F) longitudinal section of E; (G) advanced embryo in metanaupliar stage. (From Grobben, 1879)

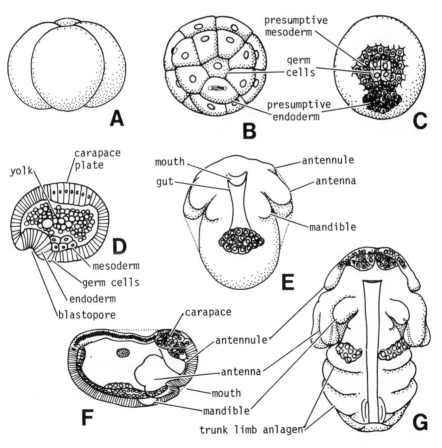

to close the site of the blastopore, but it is unclear just how the internal germ layers organize into their traditional positions.

At this point, the embryo begins to elongate (Fig. 32-5E, F), the naupliar anlagen appear (Fig. 32-5E), and the carapace begins to develop. Apparently the teloblasts become active quite soon as well, since the development of the posterior part of the body accelerates with trunk limb anlagen differentiating even before the naupliar limb buds are formed (Fig. 32-5G). The embryo continues to maintain an elongate linear form, and no ventral flexion occurs until the postabdomen and anus are differentiated. By the time the embryos hatch within the mother's brood pouch, all limbs are present in some degree of development.

FOSSIL RECORD   Because of the small size and generally delicate nature of the cuticle, the fossil record of cladocerans is limited. The earliest known forms are Oligocene in age (Tasch, 1969), but this is not necessarily any indicator of the real age of the group. Fossils consist of ephippia, fragments of carapaces, postabdomens, and caudal rami, and tell us very little about evolution within the group.

TAXONOMY   The systematics of cladocerans has been a constant focus of study since the days of Müller in the late eighteenth century. Until relatively recently it was thought to be stabilizing, but now it appears as if the work is just beginning (e.g., see the comments of Frey, 1982 and Christie, 1983). The higher taxonomy has remained stable for some time, but the family level taxa may begin to proliferate as generic groups and species complexes become better known. The old names of some of these groups are still encountered in the literature and are included here for completeness.

Suborder Haplopoda Sars, 1865 Recent
    Family Leptodoridae Lilljeborg, 1900
Suborder Eucladocera Eriksson, 1932 Oligocene–Recent
    Superfamily Sidoidea Baird, 1850 (= Ctenopoda) Recent
        Family Sididae Baird, 1850
        Family Holopedidae Sars, 1865
    Superfamily Daphnoidea Straus, 1820 (= Anomopoda) Oligocene–Recent
        Family Daphnidae Straus, 1820 Oligocene–Recent
        Family Bosminidae Baird, 1845 Recent
        Family Chydoridae Stebbing, 1902 Recent
        Family Macrothricidae Norman and Brady, 1867 Recent
        Family Moinidae Goulden, 1968 Recent
    Superfamily Polyphemoidea Baird, 1845 (= Onychopoda) Recent
        Family Polyphemidae Baird, 1845
        Family Cercopagidae Mordukhai–Boltovskoi, 1968
        Family Podonidae Mordukhai–Boltovskoi, 1968

REFERENCES

Bacci, G., G. Cognetti, and A. M. Vaccari. 1961. Endomeiosis and sex determination in *Daphnia pulex*. *Experimentia* **17**:505–6.

Bainbridge, V. 1958. Some observations on *Evadne nordmanni*. *J. Mar. Biol. Assc. U.K.* **37**:349–70.

Birge, B. E. A. 1918. The water fleas (Cladocera). In *Freshwater Biology* (H. B. Ward and G. C. Whipple, eds.), pp. 676–740. Wiley, New York.

Brooks, J. L. 1946. Cyclomorphosis in *Daphnia*. *Ecol. Monogr.* **16**:409–447.

Brooks, J. L. 1957. Systematics of North American *Daphnia*. *Mem. Conn. Acad. Arts and Sci.* **13**:1–180.

Cannon, H. G. 1922. On the labral glands of a cladoceran (*Simnocephalus vetulus*) with a description of its mode of feeding. *Quart. J. Micro. Sci.* **66**:213–34.

Cannon, H. G. 1933. On the feeding mechanism of the Branchiopoda. *Phil. Trans. Roy. Soc. Lond.* (B) **222**:267–352.

Christie, P. 1983. A taxonomic reappraisal of the *Daphnia hyalina* complex: an experimental and ecological approach. *J. Zool.* **199**:75–100.

Crittenden, R. N. 1981. Morphological characteristics and dimensions of the filter structures from three species of *Daphnia*. *Crustaceana* **41**:231–48.

Eriksson, S. 1935. Studien über die Fangapparate der Branchiopoden. *Zool. Bidr. Uppsala* **15**:23–287.

Frey, D. G. 1959. Phylogenetic significance of head pores of the Chydoridae. *Int. Rev. ges. Hydrobiol.* **44**:27–50.

Frey, D. G. 1973. Comparative morphology and biology of three species of *Eurycercus* with a description of *Eurycercus macracanthus*. *Int. Rev. ges. Hydrobiol.* **58**:221–67.

Frey, D. G. 1982. Questions concerning cosmopolitanism in Cladocera. *Arch. Hydrobiol.* **93**:484–502.

Fryer, G. 1962. Secretions of the labral and trunk limb glands in the cladoceran *Eurycercus lamellatus*. *Nature* **195**:97.

Fryer, G. 1963. The functional morphology and feeding mechanism of the chydorid cladoceran *Eurycercus lamellatus*. *Trans. Roy. Soc. Edinb.* **65**:335–81.

Fryer, G. 1968. Evolution and adaptive radiation in the Chydoridae: a study in comparative morphology and ecology. *Phil. Trans. Roy. Soc. Lond.* (B)**254**:221–385.

Fryer, G. 1974. Evolution and adaptive radiation in the Macrothricidae: a study in comparative morphology and ecology. *Phil. Trans. Roy. Soc. Lond.* (B)**269**:137–274.

Goulden, C. E. 1968. The systematics and evolution of the Moinidae. *Trans. Am. Phil. Soc. Philadel.* **58**:1–101.

Green, J. 1954. Size and reproduction in *Daphnia magna*. *Proc. Zool. Soc. Lond.* **124**:535–45.

Green, J. 1956. Growth, size and reproduction in *Daphnia*. *Proc. Zool. Soc. Lond.* **126**:173–204.

Grobben, C. 1879. Die Entwicklungsgeschichte der *Moira rectirostris*. *Arb. zool. Inst. Wien* **2**:203–68, 7 pls.

Kaestner, A. 1970. *Invertebrate Zoology*, Vol. III, *Crustacea*. Wiley, New York.

Kühn, A. 1908. Die Entwicklung der Keimzellen in den parthenogenetischen Generationen der Cladoceren *Daphnia pulex* and *Polyphemus pediculus*. *Arch. Zellforsch.* **1**:538–86, pls. 18–21.

Kühn, A. 1911. Über determinierte Entwicklung bei Cladoceren. *Zool. Anz.* **38**:345–57.

Kühn, A. 1913. Die Sonderung der Keimesbezirke in der Entwicklung der Sommierer von *Polyphemus pediculus*. *Zool. Jahrb. Anat.* **35**:243–340, pls. 11–17.

Kükenthal, W., and E. Matthes. 1953. *Zoologishes Practicum*. Gustav Fischer Verlag, Jena.

Lilljeborg, W. 1901. Cladocera Sueciae. *Nova Acta Regiae Soc. Sci. Uppsala* **(3)19**:1–701, 87 pls.

Lockhead, J. H. 1936. On the feeding mechanism of a ctenopod cladoceran, *Penilia avirostris*. *Proc. Zool. Soc. Lond.* **1936**:335–55.

Longhurst, A. R., and D. L. R. Seibert. 1972. Oceanic distribution of *Evadne* in the eastern Pacific. *Crustaceana* **22**:239–48.

Lynch, M. 1978. Complex interactions between natural exploiters—*Daphnia* and *Ceriodaphnia*. *Ecol.* **59**:552–64.

Lynch. M. 1980. The evolution of cladoceran life histories. *Quart. Rev. Biol.* **55**:23–42.

Smirnov, N. N. 1968. On comparative functional morphology of limbs of Chydoridae. *Crustaceana* **14**:76–96.

Smirnov, N. N., and B. V. Timms. 1983. A revision of the Australian Cladocera. *Records Aust. Mus. Suppl.* **1**:1–132.

Tasch, P. 1969. Branchiopoda. In *Treatise on Invertebrate Paleontology*, Part R, *Arthropoda* **4**(1) (R. C. Moore, ed.), pp. R128–91. Geol. Soc. Am. and Univ. Kansas Press, Lawrence.

Wagler, E. 1927. Branchiopoda, Phyllopoda. In *Handbuch der Zoologie*, **3**(1) (W. Kükenthal and T. Krumbach, eds.), pp. 309–98. Der Gruyter, Berlin.

33

OSTRACODA

DEFINITION   Bivalved carapace completely enclosing body, typically calcareous, hinged dorsally; antennules and antennae typically locomotory; postcephalic trunk region greatly reduced, with 0 to 2 postcephalic appendages, body segmentation typically absent or obscure; well-developed caudal rami. All larval stages with a carapace.

HISTORY   Latreille was the first to use the term Ostracoda in 1802 (originally written as Ostrachoda); however, G. O. Sars (1866) erected the first effective suprafamilial taxonomy of the ostracodes, recognizing four groups. This system enjoyed wide acceptance until G. W. Müller (1894) erected the two main groups generally recognized today, Myodocopa and Podocopa. The Müller system was objected to in the monographic treatment of Skogsberg (1920), but the Müller taxonomy has generally prevailed in its essential form, modified by the addition of the myriad fossil types (see Benson et al., 1961 or McKenzie et al., 1983). Claus wrote exhaustively on the taxonomy of the group, yet most of our knowledge of internal anatomy and functional morphology dates from the work of Cannon (1931, 1933, 1940). In considering the modern literature of the group, one is overwhelmed with the sheer mass of taxonomic descriptions. McKenzie et al. (1983) estimated that about 40,000 species had been described, including both fossil and Recent forms. Hartmann (1975) provides a review of the group

MORPHOLOGY   The most obvious external feature of the ostracodes is the all-enveloping carapace. Two major types (Fig. 33-1) are recognized among living forms: one with smooth or convex borders and an anterior antennal notch, the myodocopidan condition, and one with no notch and the ventral border straight or concave, the podocopidan form. Consideration of fossil types complicates the picture. Because the carapace is typically the only source of knowledge about extinct ostracodes, it has been studied in great detail with the concomitant development of a complex terminology to deal with these studies (see, e.g., Scott, 1961).

The carapace originates from the naupliar head. In this respect it would not correspond to the traditional definition of a carapace as an outgrowth of the maxillary segment (as given in Chapter 1) but is rather more akin to a specialized head shield. The carapace is typically calcareous, and although bivalved it is not symmetrical, that is, one valve fits inside the

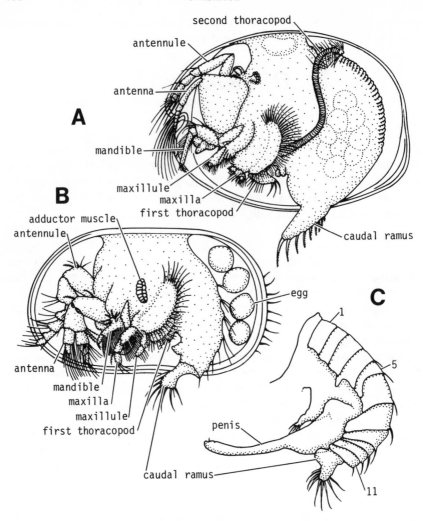

Fig. 33-1. Modern ostracode body types. (A) myodocopidan *Cypridina norvegica*; (B) podo-copidan *Cytherella abyssorum* (A, B modified from Howe et al., 1961); (C) detail of trunk segmentation of *Cytherella pari* (C modified from Schulz, 1976).

other along the margins. The valves are joined by a flexible cuticle, which has traditionally been viewed as an antagonist to the adductor muscles to serve in opening the shell. Harding (1964) suggested that the hinge cuticle is no different from any other arthrodial membrane, and hydrostatic pressure on the mandible apex opens the shell. The body and appendages for the most part are completely enclosed within the carapace.

The shell is typically calcareous; however, in purely planktonic forms calcium carbonate may be all but absent, and the fossil phosphatocopines are thought by many to have had a primary mineral phase of calcium apatite. The inner wall of the carapace fold, termed a duplicature, can also be partially calcified. Externally the entire array of possible frills, lobes, knobs, spines, denticles, ridges, grooves, and pits have been analyzed in great detail (Benson, 1981), and seen as solutions to a diverse series of structural problems related to demands placed on the shell in different ecologic situations.

The carapace of these tiny animals, complex as it is, is only part of an elaborate skeletal system. The other elements are the endoskeletal framework supporting the limbs and caudal rami (Cannon, 1931; Schulz, 1976). The rest of the body covering joined to these skeletal elements was characterized by Cannon (1940) as 'the flimsiest of cuticles.'

The antennules (Fig. 33-2A) are uniramous and typically composed of a total of five to eight segments. They can serve a variety of functions other than sensation, including climbing, digging, swimming, and assisting in copulation. In copulation the antennules are sexually dimorphic in myodocopidans with some distal setae equipped with suctorial structures in males (Fig. 33-3A).

The antennae (Fig. 33-2B) are biramous, locomotory structures. The myodocopidans have the basal segment greatly inflated (Fig. 33-3C). In the podocopidans the protopodal joints are subequal but with a distinct knee between the coxa and basis. The rami are variously developed, being subequal as in the platycopines, the exopod being reduced as in podocopines (Fig. 33-3B), or the endopod being either reduced or developed as a clasper in the males in the myodocopidans (Fig. 33-3C).

The mandibles (Fig. 33-2C) are basically similar in all groups. The coxa forms the gnathobase. The distal elements are biramous, with the basis and endopod forming the principal components of the palp. The condition of an elongate and setose basis is a specialization, with a basis subequal to the endopod being considered the more primitive state (Fig. 33-3D).

There is a difference of opinion as to terminology applicable to the branches of the postmandibular limbs. The setose vibratory plate is variously termed an epipod or an exopod. This structure is believed herein to be an exopod, since to use the term epipod would imply an independent evolution of such a specialized exite in the ostracodes.

The maxillules (Fig. 33-2D) exhibit a wide variety of form and possess important characters used in the taxonomy of the living ostracodes. The protopod is typically marked with two or three well-developed setose endites. The exopod is developed as a large plate or lobe equipped with long plumose setae, while the endopod in most forms is a robust, setose palp. The proximal segment of the endopod in platycopines is rather unusual in that it is distinctively setose. A more primitive overall arrangement of this limb is seen in the cladocopines (Fig. 33-3E) in which the protopodal segments are distinct and the exopod is not particularly distinguished or absent.

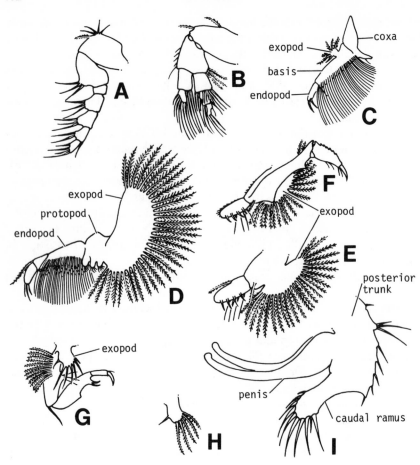

Fig. 33-2. Appendages of the platycopine ostracode *Cytherella abyssorum*. (A) antennule; (B) antenna; (C) mandible; (D) maxillule; (E) female maxilla; (F) male maxilla; (G) male first trunk limb; (H) female first trunk limb; (I) posterior trunk of male with caudal rami. (Modified from Howe et al., 1961)

The identity of the ostracode maxillae (Fig. 33-2E, F) has been the subject of considerable debate. This limb has been also variously termed the first thoracic leg, or the fifth limb, or the maxillipede. The dispute here arises from the fact that during the course of ostracode development, limbs appear sequentially at each molt, except in the second molt. It is thus thought among some ostracode workers that the maxillae should have appeared at that molt and that the usual lack of an appendage may mean that the maxillae are supposedly suppressed. The appendage that appears at the third molt is thus thought to be the first thoracopod. However, in *Cyprideis* the anlagen of the maxillae appear with the second molt, and coincidently these limbs most resemble the maxillules when the adult stage is reached (Weygoldt, 1960). In addition, however, there is no extra set of ganglia between that of the maxillules and those of the 'maxillae,' further negating the idea that these are trunk limbs. Furthermore, consideration of the endoskeletal association of these appendages (Cannon, 1931; Schulz, 1976) reveals that these limbs are 'cephalic' in association and not 'thoracic' (Fig. 33-4). The issue, of course, is complicated by the variation in structure that this limb can exhibit from being very 'maxilliform' (Fig. 33-3F) to very 'pediform' (Fig. 33-3G), depending to what extent the setose exopod is developed as a plate or lobe. Sexual dimorphism is also noted in these limbs (Fig. 33-2F), as well as left–right asymmetry.

The maxillae and sometimes the maxillules exhibit a posterior orientation of the limb axis (Fig. 33-4). This is to allow the exopods of these

Fig. 33-3. (A) suctorial setules on antennule of male *Cypridina norvegica*; (B) antenna of podocopine *Bairdia frequens*; (C) antenna of myodocopine *Conchoecia elegans*; (D) mandible of cladocopine *Polycope orbicularis*; (E) maxillule of *P. orbicularis*; (F) maxilla of myodocopine *Cypridina norvegica*; (G) maxilla of podocopine *Macrocypris minna*; (H) first trunk limb of myodocopine *C. norvegica*; (I) second trunk limb of podocopine *Limnocythere sanctipatricii*; (J) asymmetrical brush-shaped organs of *B. frequens*. (Modified from Howe et al., 1961)

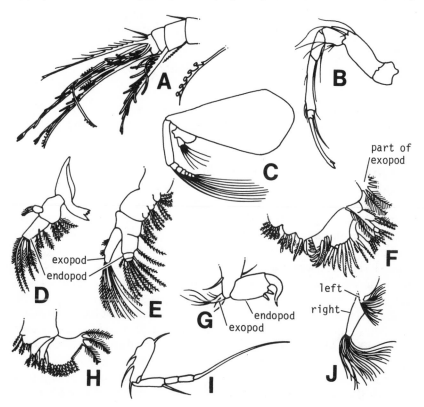

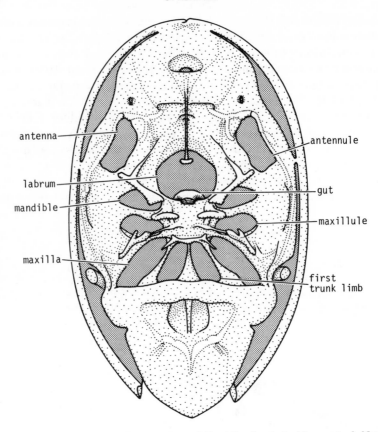

Fig. 33-4. Inner view of body endoskeleton of *Cypridina laevis*, looking ventrad. Note the progressively posteriad orientation of maxilla and first trunk limb. (From Cannon, 1931)

limbs to vibrate in opposition to the first thoracopods and thus achieve the anterior-to-posterior current flow (discussed below).

The first thoracopods, absent from the cladocopines, also display a great variation in form. It is a locomotory limb in the podocopidans. In the platycopines there is strong sexual dimorphism, with the males displaying the basic locomotory form (Fig. 33-2G), while females bear a vestigial lobe (Fig. 33-2H). In the myodocopidans this limb ranges from being very similar to the maxillae to being a nonlocomotory lamelliform plate (Fig. 33-3H).

The second thoracopods are absent in the cladocopines and platycopines. In the podocopines this appendage is pediform (Fig. 33-3I), but in the cypridinoid myodocopines it is developed as a distinctive, unique, vermiform, terminally setose structure.

Brushlike structures occasionally found in some male podocopidans (Fig. 33-3J) are thought by some authorities to represent a vestigial set of third thoracopods.

Schulz (1976) has nicely documented and illustrated postcephalic segmentation in some of the podocopidans (Fig. 33-1C, D). The platycopines display the most primitive condition in which 11 such segments, including the telson with its caudal rami, can be demonstrated. The penis can thus be seen to be associated with the sixth or seventh postcephalic segments, a typical maxillopodan condition.

The body terminus is marked by the development of caudal rami (Fig. 33-2I). These are typically clawlike lamellae, but in the podocopines they can be variously developed, or even absent as in the darwinulids.

The narrow esophagus passes dorsally to the enlarged midgut. Cannon (1931) recorded a bulging of the anterior wall of the esophagus into the lumen to form a semilunar cross section. In the podocopidans the upper end of the esophagus, as it projects into the stomach, is armed with ridges, spines, and setae (which in the bairdiids is developed as a full-fledged gastric mill). The esophagus is equipped with circular and dilator muscles. The midgut may have a pair of caeca extending laterally into the valves, but these caeca are sometimes absent. The walls of the midgut are enveloped with a mass of parenchymous tissue, especially in forms lacking the caeca. The midgut opens into a short hindgut. The anus is typically ventral to the caudal rami in the myodocopidans and dorsal to the rami in the podocopidans.

For such generally small animals, myodocopidans have a well-developed circulatory system (Fig. 33-5) (Cannon, 1931, 1940). The pericardium is divided into a distinct posterior portion, paired in *Gigantocypris* (Cannon, 1940), that drains the trunk, and an anterior region around the heart that drains the posterior pericardium and the carapace valves. Blood enters the heart through a posterodorsal pair of ostia, and exits the heart either anteroventrally through the aortic valve or posteroventrally through a pair of hepatic valves. These latter deliver blood to the gut parenchyme. The former delivers blood into an aorta, which descends in a ventroposterior direction toward the circumesophageal nerve ring. There the aorta splits into a lateral antennal branch, and a posterior branch, which in turn splits into a mandibular branch and a branch that joins its mate from the opposite side to form a supraneural arterial ring. The supraneural ring opens posteriorly into the body cavity.

The various portions of the pericardium and the arteries are anchored by a complex of muscles (Cannon, 1931, 1940). Cannon felt that these muscles all contracted in synchrony with the muscles of the heart itself in order to achieve the flow of blood in the body. Such a system, he concluded, would move hemolymph even in the absence of a heart proper (as, in fact, occurs in the podocopidans).

Specialized respiratory structures generally seem to be absent, which is not too surprising in such tiny animals. An apparent exception to this occurs in the cylindroleberidids (except *Bruniella*) and some species of *Cypridina*. In these, lamellar branchiae are found on the dorsum of the trunk (Kornicker, 1981).

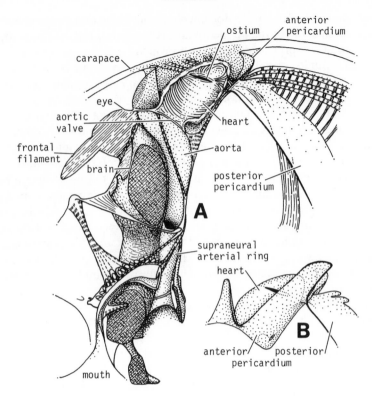

Fig. 33-5. Lateral view of central circulatory system in *Cypridina laevis*. (A) cutaway diagram; (B) outer view of heart and pericardium. (From Cannon, 1931)

Excretion is achieved by antennal and maxillary glands. In the podoco-pidans both antennal and maxillary glands function in the adult phase, while adult myodocopidans have only functional antennal glands.

The ostracode nervous system (Fig. 33-6) is very well developed, though markedly compact because of the constraints of body size. The center of the system is a circumesophageal nerve ring, in front of which are fused the optic and protocerebral lobes of the brain. The antennulary nerve arises from the anterior part of the nerve ring, while mandibular, maxillulary, and maxillary nerves arise on the posterior part of the ring. The trunk limb nerves arise from a compact mass behind the ring, though in *Gigantocypris* this mass is somewhat elongate and suggests its original segmental character (Fig. 33-6). Cannon (1931) reports that the central nervous system of *Cypridina* is unique in that the segmental nerves swell into ganglia within the base of the limb supplied. Visceral nerves arising with the mandibular nerves form a large ganglion at the junction of the eso-phagus and stomach.

Both naupliar and compound eyes are known among ostracodes,

though many forms are blind. The naupliar eye has three cups with both tapetal and lens cells. Many authors record frontal organs, but Elofsson (1966) maintains these are not such *sensu stricto* but are rather frontal 'tentacles' or filaments.

Sexes are separate, and gonads are usually paired and may extend into the carapace valves. The female system has the oviducts exiting to the outside in front of the caudal rami. Seminal receptacles are present and open separately in front of the oviducts, and these frequently have a separate internal opening into the oviducts. Only the cyprids lack the separate external opening of the seminal receptacles.

In the male system the testes can be rather variable in form, but the vasa deferentia exit posteriorly in front of the rami. In the cypridinids they open by a single median pore between the penes; in all other ostracodes, however, the vasa exit at the tip of the penis. In halocyprids this is a single organ on the right-hand side; in the other families the penes are paired. The studies of Schulz (1976) indicate that the copulatory organs are associated with the sixth or seventh postcephalic segment.

Fig. 33-6. Central nervous system of *Gigantocypris muelleri*, with nerves to relevant major organs and appendages so labeled. (From Cannon, 1940)

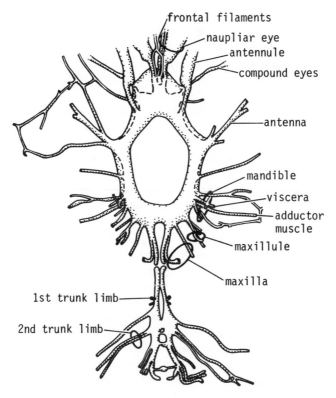

NATURAL HISTORY   Ostracodes are generally marine forms, but the podo-copines are found in fresh water as well and have several terrestrial genera (Shornikov, 1980).

The myodocopidans have limbs that are well developed for swimming with the antennae, but occasionally with the antennules as well, which are the principal organs of locomotion. The podocopidans generally walk by means of the tips of the antennae and first trunk limbs. However, platyco-pines move by means of thrusting–digging movements of the antennules, antennae, and caudal rami. Generally the animals are either pushed or pulled along the bottom by the relevant limbs, though cytherids can climb with the assistance of silk draglines produced by glands exiting on the antennae.

The basic mode of feeding in ostracodes is supposedly that of the maxil-lary filterer, and other forms have been suggested as derived from that. Cannon (1924, 1933) and Storch (1933) elucidated the basic mechanism. A vibrating exopodal plate moves a current of water from anterior to poster-ior. This revolves around the maxillae and first trunk limb arranged so they are parallel to such a flow of water. A filter screen is arranged at right angles to and across the flow. Particles then trapped in the flow are trans-ferred to the mouth. The exact elements employed differ in the two ostra-code orders (see Table 33-1) in response to the prevailing modes of locomotion. Myodocopidans, being for the most part antennary swimmers, use the exopods of the maxillae to vibrate against the first trunk limb and so generate the current flow over the maxillules. Podocopidans, being trunk limb walkers, shift everything forward and use the exopods of the maxillules to generate the current through a filter plate on the mandible. The filter is a double one; each side is quite separate and capable of inde-pendent action. Separation of right and left halves is achieved by the first trunk limb and the flexed elements of the caudal rami.

The hypothesis of filter feeding as the basic mode of feeding in ostra-codes is perhaps too simplistic. Certainly, different limbs are used in differ-ent groups to achieve filtration (Table 33-1). Most ostracodes living today appear to be detritus feeders. Adamczak (1969) thought the podocopines and metacopines of the Paleozoic were detrital feeders, though conceded that palaeocopids may have been filter feeders. Furthermore, no workers to date have considered the effect of body size and scaling on the physics of fluid flow around ostracode limbs. Certainly the small size of ostracodes ensures that they live in a world of very low Reynolds numbers and vis-cous, laminar fluid flow, and these must have profound effects on how their limbs actually function (see the discussion of feeding in copepods in Chapter 37). Most studies of ostracode feeding have not dealt with living animals, but rather are based on inferences concerning the functional mor-phology of the dead, that is, preserved material.

Scavenger and carnivorous ostracodes seem to rely largely on the mobility of the mandibles to acquire and manipulate food. The maxillules

Table 33–1. Comparison of filtration mechanism between ostracode orders. (Modified from Cannon, 1933 and Hartman, 1975)

	Myodocopida (*Asterope*)	Podocopida (*Cytherella*)	Podocopida (*Notodromas*)
Current generation	maxillary exopod	maxillulary exopod	maxillulary exopod
Working against	first trunk limb	maxillary exopod, first trunk limb, and posterior part of body	—
Filter	maxillulary basis	mandibular endopod	mandibular palps
Filter cleaned by	comb setae on maxillary protopod and exopod	comb setae on maxillulary endites and endopod	maxillulary endites
Food transferred to gnathobase by	maxillulary endites and spinose lobe on mandibular base	comb setae on maxillulary endites and endopod	maxillulary endites
Food pushed into esophagus by	mandibular gnathobase	maxillulary and mandibular endites	labral and paragnathal teeth

and maxillae are used to assist in this, but their largely reciprocating action is used for the most part to position food for processing by the mandibles (Lockhead, 1968). This in turn is facilitated by secretions of the labral glands. Scavengers like *Conchoecia* continue to use the vibrating maxillary plate to generate a respiratory current; however, there is no filtration of particles. Indeed, in experimental situations (Lochhead, 1968) the setae of the vibratory plate could become clogged, with no structural or behavioral modifications in *Conchoecia* for cleaning such setae.

The most pressing problem mitigating against presenting an effective overview of the evaluation of feeding in ostracodes is the scarcity of published studies of feeding based on detailed examination of living forms. Until that is remedied, most statements about ostracode feeding are merely suppositions.

Reproduction within ostracodes generally involves two sexes. Copulation can occur in any number of ways: venter to venter, dorsal mounts, posterodorsal mounts. In all cases, copulation is achieved within minutes and such copious amounts of sperm are placed by a single liaison into the seminal receptacles to suffice the female for life. Females may produce one or more broods, depending on species and ecologic conditions. Frequently, males are lacking in some freshwater species, but this too can vary. Some species are purely parthenogenetic; others are so only in part of their range, still others are only functionally so and react to environmental factors, such as reduced food availability, to trigger a bisexual phase.

Brood care varies. Most podocopines (except most darwinulids) lay their eggs freely, or attach them to something, and then go their own way. All other ostracodes tend to brood the eggs between the dorsal part of the body and the carapace, and one or more molts of the developing juveniles typically occur before the larvae are shed.

Ostracodes live in all manner of habitats and seem to prefer benthic or epibenthic life-styles, though pelagic forms are not unknown. What is especially important, however, is their almost ubiquitous occurrence in aquatic habitats of all kinds combined with their exacting environmental requirements. These make ostracodes important monitors and indicators of ecologic conditions, not only among the Recent forms but also among the fossils (see, e.g., Benson, 1961). This interest in ostracode ecology is well established in the discipline and stems from the pioneering work of G. W. Müller (1894). Not only do ostracodes closely respond to depth, temperature, and associated biotas, but they are also indicators of environmental factors of salinity, sediment type, pH, and oxygen level. Shallower water forms thus can be highly regional in their distributions, reflecting localized prevailing conditions (see Hazel, 1970 for an analysis of North Atlantic ostracode biogeography).

Beyond factors that influence localized distributions, the great wealth of living and especially fossil species can allow for some considerable insight into ostracode history. Several interesting phenomena have been recorded. One is the importance of the ancient Tethys as a dispersal corridor (McKenzie, 1967, 1973a, b). For example, the distribution through time from the Cretaceous to Recent of the genus *Saida* (Fig. 33-7A) reveals a clear track wherein this ostracode spread from the western Tethys in the Cretaceous, to include an eastern Tethys component in the early Cenozoic, while in more recent times the genus has come to be restricted to the remnants of the eastern Tethys in and around Australia.

In general the ostracodes have been important in demonstrating the effectiveness of the Tethys in dominating the biogeography of the planet until its obliteration in the middle of the Cenozoic. Besides influencing distribution patterns, the presence of that ancient open equatorial ocean allowed vertical water exchanges. The world's oceans prior to 40 million years ago were probably largely thermospheric, that is, were composed of fairly uniformly warm waters. The deep-sea ostracode faunas of that time were distinctly different and more diverse than those after 40 million years ago (Benson, 1975). Apparently, the disappearance of the Tethys cut off the circulation of warm, dense, and more saline water from the surface tropics into the deep sea and, in turn, allowed a pattern to develop that prevails today wherein cold dense water from the poles supplies the deep sea. This was facilitated by the geographic isolation of Antarctica, the

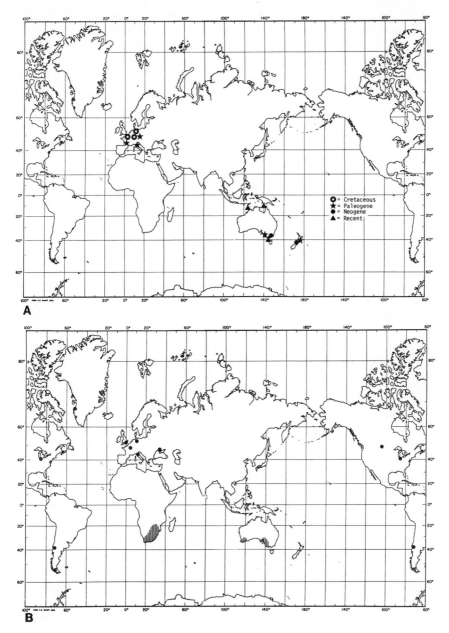

Fig. 33-7. Some biogeographic distributions of ostracodes. (A) Tethyan track through time species of *Saida*: (B) distribution of modern isocypridines: hatched areas bisexual populations, ● parthenogenetic populations. (From McKenzie, 1973b)

411

establishment of Antarctic circumpolar currents, and the isolation of the Arctic basin, all of which contributed to the refrigeration of the poles. The deep-sea habitat today is characterized by a not very diverse, but rather ubiquitous, ostracode fauna adapted to the cold but uniform conditions now prevailing at that depth.

Finally, McKenzie (1971, 1973b) claims that consideration of reproductive patterns coupled with distributions can pinpoint centers of origin. For example, there are both bisexual and parthenogenetic forms in the freshwater isocypridines. Apparently, as in many such forms, they reproduce bisexually in the original source area and are parthenogenetic in areas to which they have subsequently dispersed. Thus, in South Africa the genera *Isocypris* and *Amphibilocypris*, and in Australia *Platycypris*, are bisexual, but species of *Isocypris* in the Americas and Europe are apparently parthenogenetic only (Fig. 33-7B). This in turn would seem to imply an original southern Gondwanan origin of this subfamily with subsequent distribution elsewhere.

DEVELOPMENT   The most recent effective treatment of ostracode development is that of Weygoldt (1960). Cleavage is total and equal (Fig. 33-8A). A remnant of its spiral derivation is shown by the orientation of the spindle apparatuses during these divisions. Gastrulation occurs in two distinct phases. The first (Fig. 33-8C, D) is an invagination of cells through a blastopore, which eventually narrows to form a ventral groove (Fig. 33-8E). This invagination delineates the presumptive mesoderm on the inside of the embryo. The second phase is a delamination of endodermal cells along the ventral groove (Fig. 33-8F).

As the ventral groove closes, the anteriormost portion remains to form the mouth (Fig. 33-8G) as the stomodeum grows inward (Fig. 33-8H). At this stage the naupliar anlagen appear laterally. Those of the antennules form just lateral to and somewhat posterior to the mouth. As the limb anlagen grow toward the midline, those of the antennules shift anteriad (Fig. 33-8I). Meanwhile the carapace folds grow ventrad to envelope the naupliar limbs (Fig. 33-8J). This is unique within the crustaceans, where the carapace usually arises later in development as a fold of the maxillary segment.

The protocerebrum arises as a pair of lobes anterodorsal to the deutocerebrum. The deuto- and tritocerebra along with the mandibular ganglion arise as ectodermal masses near their respective limbs. Thus the deutocerebrum is more or less postoral in origin. With the anterior migration of the antennules and caudad growth of the stomodeum and labrum the proto- and deutocerebra fuse to form the anterior portion of the nerve ring.

The midgut is delineated into distinct anterior and posterior regions (Fig. 33-8K). The caeca arise as simple lateral evaginations of the anterior region. The lumina of the two regions join eventually rather late in devel-

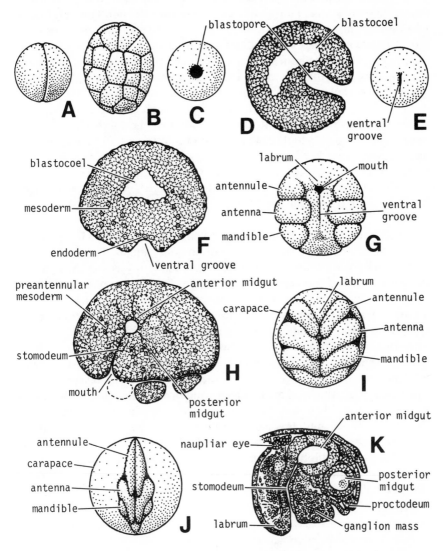

Fig. 33-8. Development of *Cyprideis torosa*. (A) first cleavage; (B) 64-cell stage blastula; (C) initiation of gastrulation; (D) sagittal section of C; (E) late gastrulation, blastopore closing to form ventral groove; (F) cross section of E, with delamination of presumptive endoderm; (G) initiation of naupliar anlagen; (H) sagittal section of G; (I, J) successive stages of carapace formation; (K) sagittal section of nauplius. (Modified from Weygoldt, 1960)

413

opment. The lumina are free of yolk, which remains essentially intracellular throughout development.

The mesoderm organizes into vague, segmental units, though no clear segment boundaries are formed. A portion of the mesoderm migrates early to a position in front of the stomodeum (Fig. 33-8H). This develops later into stomodeal and labral mesoderm and, in this regard, is homologous to the preantennal mesoderm seen in some other crustaceans. At no time are any coelomic sacs formed within the mesoderm. The postnaupliar region arises from sets of teloblasts in the posterior part of the body.

'Larval' development is initiated early, though for the most part the young remain brooded within the protection of the mother's carapace. Essentially, a single nauplius stage is followed by a series of metanaupliar molts (Fox, 1964; McKenzie et al., 1983). A typical pattern (Fig. 33-9) reveals a set limb anlagen added for each molt, except for the second molt in which sequential development is delayed. Anywhere from five to eight stages lead up to the adult condition, after which ecdysis typically ceases.

FOSSIL RECORD   The magnitude of the fossil record of ostracodes is almost to a point beyond the comprehension of any one person. The most recent synthesis (Benson, 1961) was out of date as soon as it was published, and a current revision of the *Treatise* for ostracodes (only in the planning stages) will entail several volumes. A problem with the fossil ostracodes is that for the most part they are only carapaces (with some rare exceptions, e.g., Dzik, 1978), and thus difficult, though not impossible, to compare with the living orders. The problem is very marked for taxa without living representatives. Thus the Cambrian bradoriidans (which Jones and McKenzie, 1980, believe are polyphyletic) are characterized by a chitinous, poorly calcified carapace, which is finely punctate and/or wrinkled. The early Paleozoic leperditicopidans had thick, well-calcified shells, long hinges, and complex adductor muscle scars. The Paleozoic palaeocopidans had thick shells lacking a calcified inner lamella or duplicature, frequently possessed complex surface sculpturing, and commonly had ventral structures on the valves. The Cambrian phosphatocopidans are distinguished by their phosphatized cuticles; some would unite these with bradoriids (Andres, 1969).

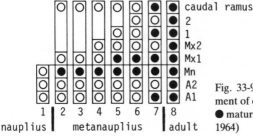

Fig. 33-9. Postembryonic mode of development of cypridids. ○ immature form of limb, ● mature form of limb. (Modified from Fox, 1964)

Frederick Sundberg (personal communication) has collected silicified phosphate copines from the Cambrian of the Great Basin in North America.

By far the most interesting of these groups are the phosphatocopidans (Fig. 33-10) in that they have been preserved with great details of their soft anatomy (K. Müller, 1979, 1982; McKenzie et al., 1983). There is some confusion concerning the identity of limbs (e.g., Briggs, 1983), which apparently arose between the initial studies of *Falites* and *Vestrogothia* (K. Müller, 1979) and subsequent work (K. Müller in McKenzie et al., 1983). *Hesslandona* (Fig. 33-10A) has small uniramous antennules and large mandiblelike antennae. It was thought that the antennules of the falitids (Fig. 33-10C) were antennalike (see, e.g., K. Müller, 1979, Figs. 34, 35), but these are now understood to have been reduced or rudimentary (McKenzie et al., 1983, p. 36).

The phosphatocopines have unique combinations of both primitive and apparently derived features. They can retain several sets of trunk limbs. *Hesslandona* (Fig. 33-10A) can have up to four trunk limbs (McKenzie et al., 1983), though *Vestrogothia* (Fig. 33-10B) has only one or possibly two. It appears that the phosphatocopidans retain the typical larval capacity of the antennae to assist in feeding (e.g., McKenzie et al., 1983, Fig. 6). However, the maxillules and maxillae have a characteristic ostracode anteroposterior orientation (Fig. 33-10A), which would seem to imply they might have been used to generate a feeding current. Whereas modern ostracodes use maxillary or maxillulary exopods as vibratory plates, phosphatocopidans seem to have used setose and highly annulate exopods to generate the current (Fig. 33-10E, F). The stout and robust character of the setae and spines on the endites and endopods of all the appendages from the antennae to the maxillae (Fig. 33-10C, D, E, F) would seem to be best interpreted as indicating a more carnivorous or at least large-particle mode of feeding rather than a filter-feeding mode. Such an interpretation would be seconded by the large labrum (Fig. 33-10B), probably related to well-developed labral glands, that are also noted in living ostracodes with such feeding habits.

TAXONOMY   One of the frustrating aspects of ostracode studies is that no two papers, or books, or authorities seem to agree on the details of higher classification of the group. In part, this might be related to the difficulty of comparing characters of fossils with living forms. It is also in part related to the difficulty of rigid character analysis of any of the living groups when dealing with such highly derived and oligomerized animals and the resultant almost complete lack of such studies (for an all too rare analysis in this regard, see Maddocks, 1976). In addition, there seems to be an all too ready tendency to inflate the status of new taxa. One ostracode worker jokingly remarked on this penchant for taxonomic inflation by observing that one rule of thumb seems to be that if you can tell two ostracodes apart they

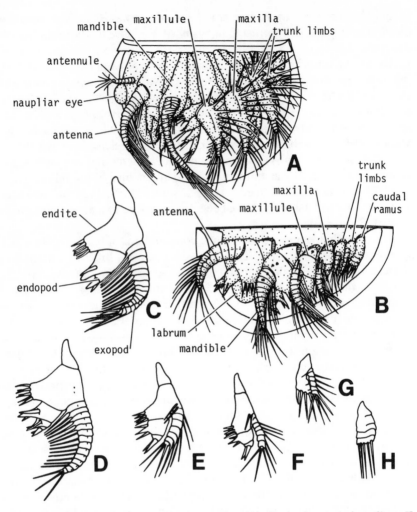

Fig. 33-10. Cambrian phosphatocopine ostracodes. (A) *Hesslandona unisulcata* (from K. Müller, 1982). (B–H) *Vestrogothia spinata*, (B) somewhat distorted ventral oblique view of body; (C) antenna; (D) mandible; (E) maxillule; (F) maxilla; (G) first trunk limb; (H) ? caudal ramus (from K. Müller, 1979, 1982). (Labels and limb identifications my own and not necessarily those of Müller)

are different genera; if you can't, they are different species. The higher classification of the ostracodes may not stabilize until careful analysis of characters is attempted across all groups. What follows here is a 'hybrid' between the recent classifications of Bowman and Abele (1982) and McKenzie et al. (1983) with the addition of fossil families. Endings used here follow Bowman and Abele (1982) and the ICZN recommendation that =*oidea* be the ending of choice for superfamilies. Use of the ending =*acea* for ostracode superfamilies causes untold confusion among all but ostracode workers. The ending has some difficulties, for when perusing lists of Carboniferous biota one may sometimes be misled into positing the presence of some obscure family of plants only to find that the item in question was, instead, some equally obscure superfamily of ostracode. In addition =*acea* is also an ordinal ending among some crustaceans. The classic orders Myodocopida and Podocopida have been retained here. To go along with the tendency of some modern authors to inflate the suborders and superfamilies to ordinal and subordinal status would blur some real similarities within the classic orders.

Order Bradoriida Raymond, 1935 Cambrian (?Ordovician)
    Family Bradoriidae Matthew, 1902 Cambrian
    Family Beyrichionidae Ulrich and Bassler, 1931 Cambrian (?Ordovician)
    Family Comptalutidae Öpik, 1968 Cambrian
    Family Hipponicharionidae Sylvester–Bradley, 1961 Cambrian
    Family Indianidae Ulrich and Bossler, 1931 Cambrian
    Family Svealutidae Öpik, 1968 Cambrian
Order Phosphatocopida K. J. Müller, 1964 Cambrian
  Suborder Vestrogothicopina K. J. Müller, 1982
    Family Falitidae K. J. Müller, 1982
    Family Oepikalutidae Jones and McKenzie, 1980
    Family Monasteridae Jones and McKenzie, 1980
    Family Vestrogothiidae K. J. Müller, 1964
  Suborder Hesslandonocopina K. J. Müller, 1982
    Family Hesslandonidae K. J. Müller, 1982
Order Leperditicopida Scott, 1961 (?Cambrian) Ordovician–Devonian
    Family Leperditiidae Jones, 1856 (?Cambrian) Ordovician–Devonian
    Family Isochilinidae Swartz, 1949 Ordovician–Devonian
Order Palaeocopida Henningsmoen, 1953 Ordovician–Recent
  Suborder Kirkbyocopina Gründel, 1969 Devonian–Recent
   Superfamily Kirkbyoidea Ulrich and Bassler, 1906 Devonian–Permian
    Family Kirkbyidae Ulrich and Bassler, 1906 Mississippian–Permian
    Family Amphissellidae Kesling and Chilman, 1978 Devonian
    Family Amphissitidae Knight, 1928 Devonian–Permian
    Family Arcyzonidae Kesling, 1961 Devonian
    Family Cardiniferellidae Sohn, 1953 Mississippian
    Family Kellettinidae Sohn, 1954 Carboniferous–Permian
    Family Placideidae Schneider, 1956 Devonian–Permian
   Superfamily Puncioidea Hornibrook, 1949 Recent
    Family Punciidae Hornibrook, 1949
  Suborder Beyrichicopina Scott, 1961 Ordovician–Permian

Superfamily Beyrichioidea Matthew, 1886 Ordovician–Permian
Family Beyrichiidae Matthew, 1886 Ordovician–Permian
Family Zygobolbidae Ulrich and Bassler, 1923 (?Ordovician) Silurian (?Devonian)
Superfamily Tribolbinoidea Sohn, 1978 Carboniferous–Permian
Family Tribolbinidae Sohn, 1978
Suborder Hollinocopina Henningsmoen, 1965 Ordovician–Permian
Superfamily Drepanelloidea Ulrich and Bassler, 1923 Ordovician–Permian
Family Drepanellidae Ulrich and Bassler, 1923 Ordovician–Devonian
Family Dolborellidae Melnikova, 1976 Ordovician
Family Aechminellidae Sohn, 1961 Devonian–Permian
Family Aechminidae Bouček, 1936 Ordovician–Mississippian
Family Bolliidae Bouček, 1936 Ordovician–Mississippian
Family Kirkbyellidae Sohn, 1961 Silurian–Pennsylvanian
Family Richinidae Scott, 1961 Ordovician–Devonian
Superfamily Eurychilinoidea Ulrich and Bassler, 1923 Ordovician–Devonian
Family Eurychilinidae Ulrich and Bassler, 1923 Ordovician–Devonian
Family Bolbinidae Ivanova, 1979 Ordovician
Superfamily Hollinacea Jaanusson, 1957 Ordovician–Permian
Family Hollinidae Schwartz, 1936 Ordovician–Permian
Family Bassleratiidae Schmidt, 1941 Ordovician
Family Chilobolbinidae Jaanusson, 1957 Ordovician–Silurian
Family Piretellidae Öpik, 1937 Ordovician
Family Quadrijugatoridae Kesling and Hussey, 1953 Ordovician
Family Sigmoopsidae Herningsmoen, 1953 Ordovician (?Silurian)
Family Tetradellidae Swartz, 1936 Ordovician–Silurian
Family Tvaerenellidae Jaanusson, 1957 Ordovician
Superfamily Nodelloidea Zaspelova, 1952 Devonian
Family Nodellidae Zaspelova, 1952
Superfamily Oepikelloidea Jaanusson, 1957 Ordovician–Pennsylvanian
Family Oepikellidae Jaanusson, 1957 Ordovician
Family Aparchitidae Jones, 1901 Ordovician–Pennsylvanian
Family Prybylitidae Pokornỹ, 1958 Silurian–Devonian
Superfamily Primitiopsoidea Schwartz, 1936 Ordovician–Devonian
Family Primitiopsidae Schwartz, 1936
Order Myodocopida Sars, 1866 Ordovician-Recent
Suborder Myodocopina Sars, 1866 Devonian–Recent
Superfamily Cypridinoidea Baird, 1850 Devonian–Recent

Family Cypridinidae Baird, 1850 (?Carboniferous) Recent
Family Cypridinellidae Sylvester–Bradley, 1961 Devonian–Carboniferous
Family Philomedidae G. W. Müller, 1906 Carboniferous
Family Rhombinidae Sylvester–Bradley, 1951 Carboniferous
Superfamily Cylindroleberidoidea G. W. Müller, 1906
    Family Cylindroleberididae G. W. Müller, 1906 Recent
    Family Cyprellidae Sylvester–Bradley, 1961 Carboniferous
Superfamily Sarsielloidea Sylvester–Bradley, 1961
    Family Sarsiellidae Brady and Norman, 1896 (?Devonian) Recent
    Family Rutidermatidae Brady and Norman, 1896 Recent
Suborder Halocyprina Dana, 1853
  Infraorder Halocypridina Dana, 1853 Devonian–Recent
    Superfamily Halocypridoidea Dana, 1852 Recent
      Family Halocyprididae Dana, 1852
    Superfamily Entomoconchoidea Sylvester–Bradley, 1953 Devonian–Carboniferous
      Family Entomoconchidae Brady, 1868
    Superfamily Thaumatocypridoidea G. W. Müller, 1906 Jurassic–Recent
      Family Thaumatocyprididae G. W. Müller, 1906
  Infraorder Cladocopidina Sars, 1866 Ordovician–Recent
    Superfamily Entomozooidea Přibyl, 1951 Ordovician–Recent
      Family Entomozoidae Přibyl, 1951 Ordovician–Recent
      Family Bolbozoidae Bouček, 1936 Silurian–Devonian
    Superfamily Polycopoidea Sars, 1866 (?Devonian) Jurassic–Recent
      Family Polycopidae Sars, 1865
Order Podocopida Sars, 1866 Ordovician–Recent
  Suborder Platycopina Sars, 1866 (?Ordovician) Silurian–Recent
    Superfamily Kloedenelloidea Ulrich and Bassler, 1908 (?Ordovician) Silurian–Permian
      Family Kloedenellidae Ulrich and Bassler, 1908 (?Ordovician) Silurian–Pennsylvanian (?Permian)
      Family Beyrichiopsidae Henningsmoen, 1953 Devonian–Permian
      Family Glyptopleuridae Guty, 1910 (?Devonian) Mississippian–Permian
      Family Lichviniidae Posner, 1950 Devonian–Permian
      Family Miltonellidae Sohn, 1950 (?Mississippian) Pennsylvanian
      Family Monotiopleuridae Gruber and Jaanusson, 1964 Ordovician
    Superfamily Cytherelloidea Sars, 1866 Jurassic–Recent
      Family Cytherellidae Sars, 1866

Superfamily Cavellinoidea Egorov, 1950 Pennsylvanian–Triassic
    Family Cavellinidae Egorov, 1950
Superfamily unknown
    Family Geisinidae Sohn, 1961 Devonian–Permian
    Family Suchonellidae Mishina, 1972
Suborder Metacopina Sylvester–Bradley, 1961 Ordovician–Recent
    Superfamily Darwinuloidea Brady and Norman, 1889
        Family Darwinulidae Brady and Norman, 1889 (?Ordovician)
            Pennsylvanian–Recent
        Family Panxianiidae Wang, 1980 Carboniferous–Triassic
    Superfamily Healdioidea Harlton, 1933 Ordovician–Cretaceous
        Family Healdiidae Harlton, 1933 Devonian–Cretaceous
        Family Bairdiocyprididae Shaver, 1961 (?Ordovician) Silurian–
            Permian (?Jurassic)
        Family Barychilinidae Ulrich, 1894 Devonian (?Mississippian)
        Family Krausellidae Berdan, 1961 Ordovician–Devonian
        Family Pachydomellidae Berdan and Sohn, 1961 (?Ordovician)
            Silurian–Devonian
    Superfamily Quasillitoidea Coryell and Malkin, 1936 Devonian–
        Carboniferous
        Family Quasillitidae Coryell and Malkin, 1936 Devonian–Car-
            boniferous
        Family Bufinidae Sohn and Stover, 1961 Devonian (?Pennsylva-
            nian)
        Family Ropolonellidae Coryell and Malkin, 1936 Devonian
    Superfamily Thlipsuroidea Ulrich, 1894 (?Ordovician) Silurian–
        Devonian
        Family Thlipsuridae Ulrich, 1894
    Superfamily Sigilioidea Mandelstam, 1960 Paleogene–Recent
        Family Sigiliidae Mandelstam, 1960 Paleogene
        Family Saipanettidae McKenzie, 1968 Recent
Suborder Podocopina Sars, 1866 Ordovician–Recent
    Superfamily Bairdioidea Sars, 1888 Ordovician–Recent
        Family Bairdiidae Sars, 1888 Ordovician–Recent
        †Family Beecherellidae Ulrich, 1894 Silurian–Devonian
    Superfamily Cypridoidea Baird, 1845 (?Ordovician) Triassic–
        Recent
        Family Cyprididae Baird, 1845 (?Permian) Jurassic–Recent
        Family Candoniidae Kaufmann, 1900 Tertiary–Recent
        Family Cypridopsidae Kaufmann, 1910 (?Permian) Cretaceous–
            Recent
        Family Ilyocyprididae Kaufmann, 1900 Triassic–Recent
        Family Macrocyprididae G. W. Müller, 1912 (?Ordovician)
            (?Miocene) Pliocene–Recent
        Family Notodromadidae Kaufmann, 1900 Paleocene–Recent

Family Paracyprididae Sars, 1923 (?Silurian) Jurassic–Recent
Family Pontocyprididae G. W. Müller, 1894 (?Devonian) Trias-
sic–Recent
Family Terrestricypridae Shornikov, 1980 Recent
Superfamily Cytheroidea Baird, 1850
Family Cytheridae Baird, 1850 Jurassic–Recent
Family Acronotellidae Schwartz, 1936 Ordovician–Silurian
Family Australocytheridae McKenzie, 1977 Recent
Family Brachycytheridae Puri, 1954 Jurassic–Recent
Family Bythocytheridae Sars, 1866 Devonian–Recent
Family Cobanocytheridae Shornikov, 1975 Recent
Family Cushmanideidae Puri, 1973 Eocene–Recent
Family Cytherettidae Triebel, 1952 Cretaceous–Recent
Family Cytherideidae Sars, 1925 Permian–Recent
Family Cytherissinellidae Kashevanova, 1958 Triassic
Family Cytheromatidae Elofsson, 1939 Recent
Family Cytheruridae Müller, 1894 Jurassic–Recent
Family Dryelbidae Sohn, 1982
Family Entocytheridae Hoff, 1942 Recent
Family Eucytheridae Puri, 1954 Jurassic–Recent
Family Hemicytheridae Puri, 1953 Eocene–Recent
Family Kliellidae Schäfer, 1945 Recent
Family Krithidae Mandelstam, 1958 Recent
Family Leguminocythereididae Howe, 1961 Eocene–Recent
Family Leptocytheridae Hanai, 1957 Recent
Family Limnocytheridae Klie, 1938 Jurassic–Recent
Family Loxoconchidae Sars, 1925 Cretaceous–Recent
Family Microcytheridae Klie, 1938 Miocene–Recent
Family Neocytherideidae Puri, 1957 Cretaceous
Family Osticytheridae Hartman, 1980
Family Paracytheridae Puri, 1973 Recent
Family Paracytherideidae Puri, 1957 (?Pennsylvanian) Creta-
ceous–Recent
Family Paradoxostomatidae Brady and Norman, 1889 (?Creta-
ceous) Eocene–Recent
Family Parvocytheridae Hartmann, 1959 Recent
Family Pectocytheridae Hanai, 1957 Cretaceous–Recent
Family Permianidae Schneider, 1947 Permian
Family Progonocytheridae Sylvester–Bradley, 1948 (?Pennsyl-
vanian) Jurassic–Recent
Family Psammocytheridae Klie, 1938 Recent
Family Pseudolimnocytheridae Hartmann and Puri, 1974
Recent
Family Schizocytheridae Howe, 1961 Cretaceous–Recent
Family Sinusuellidae Kashevarovna, 1958 Permian

Family Terrestricytheridae Schornikov, 1969
Family Tomiellidae Mandelstam, 1956 Permian
Family Trachyleberididae Sylvester–Bradley, 1948 Jurassic–Recent
Family Xestoleberididae Sars, 1928 Cretaceous–Recent
Superfamily Paraparchitoidea Scott, 1959 Devonian–Permian
Family Paraparchitidae Scott, 1959
Suborder uncertain
Superfamily uncertain
Family Sansabellidae Sohn, 1961 Mississippian–Pennsylvanian
Superfamily Paraparchitoidea Scott, 1959 Devonian–Permian
Family Paraparchitidae Scott, 1959

REFERENCES

Adamczak, F. 1969. On the question of whether the paleocope ostracodes were filter feeders. In *The Taxonomy, Morphology, and Ecology of Recent Ostracoda* (J. Neale, ed.), pp. 93–98. Oliver & Boyd, Edinburgh.

Andres, D. 1969. Ostracoden aus dem mittleren Kambrium von Öland. *Lethaia* 2:165–80.

Benson, R. H. 1961. Ecology of ostracode assemblages. In *Treatise on Invertebrate Paleontology*, Part Q, *Arthropoda* 3 (R. C. Moore, ed.), pp. Q56–Q63. Geol. Soc. Am. and Univ. Kansas Press, Lawrence.

Benson, R. H. 1975. The origin of the psychrosphere as recorded in changes of deep-sea ostracode assemblages. *Lethaia* 8:69–83.

Benson, R. H. 1981. Form, function, and architecture of ostracode shells. *Ann. Rev. Earth Planetary Sci.* 9:59–80.

Benson, R. H. (+ 16 other authors). 1961. *Treatise on Invertebrate Paleontology*, Part Q, *Arthropoda* 3 (R. C. Moore, ed.). Geol. Soc. Am. and Univ. Kansas Press, Lawrence.

Bowman, T. E., and L. G. Abele. 1982. Classification of the Recent Crustacea. In *The Biology of Crustacea*, Vol. I (L. G. Abele, ed.), pp. 1–27. Academic Press, New York.

Briggs, D. E. G. 1983. Affinities and early evolution of Crustacea: the evidence of the Cambrian. *Crust. Issues* 1:22.

Cannon. H. G. 1924. On the feeding mechanism of a freshwater ostracode, *Pionocypris vidus*. *J. Linn. Soc. Lond. Zool.* 36:325–35.

Cannon, H. G. 1931. On the anatomy of a marine ostracod, *Cypridina (Doloria) laevis*. *Disc. Rpts.* 2:435–82, pls. 6, 7.

Cannon, H. G. 1933. On the feeding mechanism of certain marine ostracods. *Trans. Roy. Soc. Edinb.* 57:739–64.

Cannon, H. G. 1940. Anatomy of *Gigantocypris muelleri*. *Disc. Rpts.* 19:185–244.

Dzik, J. 1978. A myodocopid ostracode with preserved appendages from the Upper Jurassic of the Volga River region. *N. Jahrb. Geol. Paläontol. Mh.* 1978:393–99.

Elofsson, R. 1966. The nauplius eye and frontal organs of the non-Malacostraca. *Sarsia* 25:1–128.

Fox, H. M. 1964. On the larval stages of cyprids and on *Sipholocandona*. *Proc. Zool. Soc. Lond.* 142:165-76.

Harding, J. P. 1964. Crustacean cuticle with reference to the ostracod carapace. In *Ostracods as Ecological and Palaeoecological Indicators* (H. S. Puri, ed.). *Pubbl. Staz. Zool. Napoli* 33 (suppl.):9–31.

Hartmann, G. 1975. Ostracoda. *Bronn's Klassen und Ordnungen des Tierreichs* 5(4):569–786.

Hazel, J. E. 1970. Atlantic continental shelf and slope of the United States—ostracode zoogeography in the southern Nova Scotian and northern Virginian faunal provinces. *U.S.G.S. Prof. Pap.* 529-E:1–21.

Howe, H. V., R. V. Kesling, and H. W. Scott. 1961. Morphology of living ostracodes. In *Treatise on Invertebrate Paleontology*, Part Q, *Arthropoda* 3 (R. C. Moore, ed.), pp. Q3–Q17. Geol. Soc. Am. and Univ. Kansas Press, Lawrence.

Jones, P. J., and K. G. McKenzie. 1980. Queensland Middle Cambrian Bradoriida: new taxa, paleobiogeography, and biological affinities. *Alcheringa* 4:203–25.

Kornicker, L. S. 1981. Revision, distribution, ecology, and ontogeny of the ostracode subfamily Cyclasteropinae. *Smith. Contr. Zool.* 319:1–548.

Lockhead, J. H. 1968. The swimming and feeding of *Conchoecia*. *Biol. Bull.* 134:456–64.

Maddocks, R. F. 1976. Quest for the ancestral podocopid: numerical cladistics analysis of ostracode appendages, a preliminary report. *Abh. Verh. naturwiss. Ver. Hamburg* 18/19:39-53.

McKenzie, K. G. 1967. The distribution of Cainozoic marine Ostracoda from the Gulf of Mexico to Australasia. In *Aspects of Tethyan Biogeography* (G. A. Adams and O. V. Ager, eds.), pp. 219–38. Syst. Assc., London.

McKenzie, K. G. 1971. Paleozoogeography of freshwater Ostracoda. In *Paléoécologie des Ostracodes* (H. J. Oertli, ed.). *Bull. Centre Rech. Pau-SNPA* 5 (suppl.):207–37.

McKenzie, K. G. 1973a. The biogeography of some Cainozoic Ostracoda. *Sp. Pap. Palaeontol.* 12:137–53.

McKenzie, K. G. 1973b. Cenozoic Ostracoda. In *Atlas of Paleobiogeography* (A. Hallam, ed.), pp. 477–87. Elsevier, Amsterdam.

McKenzie, K. G., K. J. Müller, and M. N. Gramm. 1983. Phylogeny of Ostracoda. *Crust. Issues* 1:29–46.

Müller, G. W. 1894. Die Ostrakoden des Golfes von Neapel und der angrenzenden Meeres-Abschnitte. *Fauna Flora Golf. Neapel* 21:1–404.

Müller, K. J. 1979. Phosphatocopine ostracodes with preserved appendages from the Upper Cambrian of Sweden. *Lethaia* 12:1–27.

Müller, K. J. 1982. *Hesslandona unisulcata* with phosphatized appendages from Upper Cambrian 'orsten' of Sweden. In *Fossil and Recent Ostracods* (R. H. Bate, E. Robinson, and L. M. Sheppard, eds.), pp. 276–304. Ellis Horwood, Ltd., Chichester.

Sars, G. O. 1866. Oversigt af Norges mariner Ostracoder. *Fork. Vid= Selsk. Christiania* 1965:1–130.

Schulz, K. 1976. Das Chitinskelett der Podocopida und die Frage der Metamerie dieser Gruppe. Unpublished doctoral dissertation, Univ. Hamburg.

Scott, H. W. 1961. Shell Morphology of Ostracoda. In *Treatise on Invertebrate Paleontology*, Part Q, *Arthropoda* 3 (R. C. Moore, ed.), pp. Q21–Q37. Geol. Soc. Am. and Univ. Kansas Press, Lawrence.

Shornikov, E. I. 1980. Ostrakody v nazemnykh biotopakh. *Zool. Zh.* 59:1306–19.

Skogsberg, T. 1920. Studies on marine ostracods, Part I. *Zool. Bidr. Uppsala Suppl.* 1:1–784.

Storch, O. 1933. Morphologie und Physiologie des Fangapparates eines Ostrakoden (*Notodromas monacha*) *Biol. Gen.* 9(1,2):151–98, 355–94; 9(2,3):299–330.

Weygoldt, P. 1960. Embryologische Untersuchungen an Ostrakoden, die Entwicklung von *Cyprideis littoralis*. *Zool. Jahrb. Abt. Anat.* 78:369–426.

34

MYSTACOCARIDA

DEFINITION   Small meiofaunal forms with body divided into an elaborate two-region cephalon and long trunk; the latter with a short maxillipedal segment and 10 large and subequal thoracoabdominal somites; antennules uniramous; antennae and mandibles biramous and similar; maxillules and maxillae uniramous and multisegmented; maxillipedes with large, spinose proximal segments; thoracopods few in number and reduced to simple flaps; telson with ventral combs and dorsal supraanal plate; large clawlike caudal rami.

HISTORY   These interesting little creatures were described originally from a meiofaunal habitat in Nobska Beach near Woods Hole (Pennak and Zinn, 1943). These strange little animals soon attracted a great deal of phylogenetic speculation; for example, Armstrong (1949) viewed them as a separate, albeit peculiar, order of copepods. They were soon relegated by Dahl (1952) to a perplexing, but secure, affiliation with copepods amid other maxillopodan subclasses. So similar are all the species, that only two genera are recognized, *Derocheilocaris* and *Ctenocheilocaris*. Since their original discovery in New England, the range of mystacocarids has been extended throughout the Atlantic basin and around the southern tips of Africa and South America. Despite diligent search elsewhere in the world in suitable habitats, the range of mystacocarids has not been extended. However, it must be pointed out that their distribution in any particular occupied beach is patchy (Hessler, 1971). A review of the literature was compiled by Zinn et al. (1982).

MORPHOLOGY   Mystacocarids are tiny (typically 0.5—1.0 mm in length) forms with a very conservative morphology (Fig. 34-1). They possess a distinct body plan, and all species are very similar to each other. The most distinctive feature of the anatomy is the cephalon. This head region occupies fully one-third of the total body length and is divisible into two distinct regions, the anterior antennulary portion being delineated from the posterior region by a distinct groove called the cephalic constriction. This anterior area carries the antennules and four dorsal ocelli. The margins of this region are developed into rather prominent paired extensions, the anteromedian and anterolateral lobes. Differences noted in these lobes are subtle but can be used to partially differentiate species. The posterior portion of

424

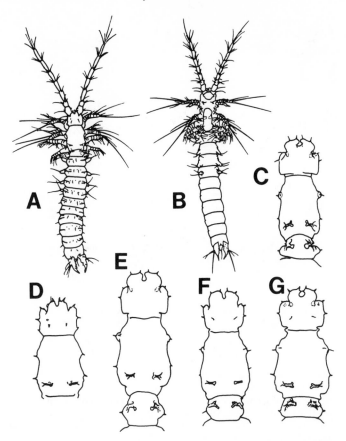

Fig. 34-1. Mystacocarid body plan. (A) dorsal view of *D. ingens*; (B) ventral view of *D. typicus*; note great development of labrum; (C) cephalon and first trunk (maxillipede) segment, *D. typicus*; (D) cephalon of *D. ingens*; (E) cephalon and first trunk segment, *D. remanei*; (F) cephalon and first trunk segment, *D. angolensis*; (G) cephalon and first trunk segment, *D. delamarei*. (Modified from Hessler, 1969, 1972)

the cephalon ventrally bears all the other head appendages and a very elongate labrum (Fig. 34-1B) that extends beyond the posterior edge of the cephalon to underlay part of the maxillipedal segment. Thus the maxillipede is a functional cephalic appendage even though the segment itself is not fused to the cephalon.

The antennules (Fig. 34-2A) have eight setose segments and are uniramous. The appendage arises from the anterior surface of the cephalon and in life remains rather inflexibly extended anteriorly and somewhat laterally.

The antennae (Fig. 34-2B) are biramous. The exopod (the longer of the two branches) has nine segments armed with setae, the length and

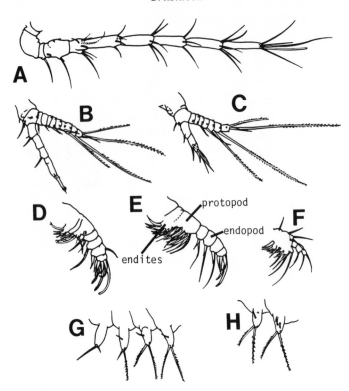

Fig. 34-2. Appendages of *Derocheilocaris ingens*. (A) antennule; (B) antenna; (C) mandible; (D) maxillule; (E) maxilla; (F) maxillipede; (G) the four thoracoabdominal rudimentary appendages of a female; (H) the third and fourth thoracoabdominal limbs of a male; note 'copulatory' spines on fourth limb. (From Hessler, 1969)

complexity of which increases from proximal simple types to distal brush setae. The shorter four-segmented endopod is armed with stiff spines. The protopod is fused to the cephalon in a manner that obscures possible divisions of that proximal area.

The mandibles (Fig. 34-2C) resemble the antennae with the added feature of a prominent gnathal element projecting under the labrum. The exopod also differs slightly from that of the antennae by having only seven segments, and typically bears only the distal brush setae and some long simple setae on the third and fourth segments.

The maxillules (Fig. 34-2D) are uniramous with a long protopod and four-segmented shaft. This appendage is armed with simple stout setae concentrated both at the base and on the terminal segment.

The maxillae (Fig. 34-2E) are similar to the maxillules except that the protopods are more heavily setose. All these posterior cephalic appendages (antennae, mandibles, maxillules, and maxillae) are apparently used equally in both locomotion as well as feeding.

The maxillipedes (Fig. 34-2F) in most species are biramous with a three-segment endopod and a small single-segment exopod. The protopod is equipped with a series of small endites and well-armed with spinous setae. The exopod and endopod are not present on all species.

The next four trunk somites bear reduced, flaplike, and virtually immobile appendages (Fig. 34-2G). The fourth appendage in the series is modified in the male (Fig. 34-2H) with spinous setae that apparently facilitate copulation. These limbs are located just posterior to the single genital pore generally on the right side located on the third somite (Dahl, 1952). The segments posterior to these four trunk appendages do not possess appendages of any kind. The dorsal surfaces of the posterior cephalon and all trunk segments are decorated with paired, toothed furrows of unknown function.

The last segment, the anal somite, is as large as any of the preceding segments. Dorsal to the terminal anus is a supraanal plate, and flanking this are large, serrate, and spinous caudal rami. The ventral surface of the anal segment bears two sets of combs, most likely useful in grooming. The variations in anal plate, setal length, rami, and ventral combs have proven useful in species delineation (Fig. 34-3).

Internal anatomy has been observed by only a few workers. Brown and Metz (1967) studied sperm morphology in *D. typicus* and observed that the gonads occupy the entire posterior portion of the trunk. The vas deferens was found to contain masses of spermatophores, each with two sperm. The sperm themselves were generally of a primitive form (unusual for crustaceans) with an acrosome, nucleus, mitochondria, and flagellum. Dahl (1952) observed a slightly different arrangement of the gonads in *Ctenocheilocaris galvarina* from Isla Guafo, Chile, in that the gonads were predominately on the right side. The left side was occupied by the gut. Dahl also noted some peculiarities of the digestive system: the great labrum contains a large gland, the stomodeum is quite long (related to the large size of the cephalon), and the proctodeum extends through the last three to four segments.

The nervous system is marked by very large ganglia, and the nerve cords are a 'rope ladder' type with relatively widely spaced tracks separated by the large ganglia. Dahl (1952) was able to trace some nerves, especially those to and around the eyes. He reported some tissues around the eyes, and the setae on the anteromedian lobes represent the frontal organs.

NATURAL HISTORY   Many of the broad aspects of mystacocarid biology are known, but detailed analyses seem to be lacking. These meiofaunal animals seem to prefer beaches that are exposed to ocean waves (Hall, 1972) and that are composed of well-sorted clean sand with grain sizes from 0.25to 0.5 mm (Delamare–Deboutteville, 1960). Hessler (1971) recorded that the animals are likely to be found intertidally where wave wash is

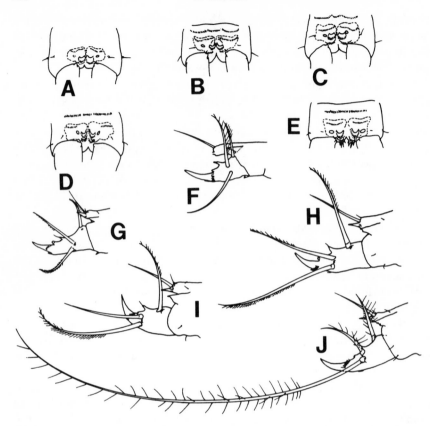

Fig. 34-3. Subtle taxonomic differences among mystacocarids. (A–E) ventral view of telson displaying telsonic combs. (A) *D. ingens*; (B) *D. typicus*; (C) *D. remanei*; (D) *D. angolensis*; (E) *D. delamarei*. (F, G, H, I, J) lateral view of telson and furca. (F) *D. typicus*; (G) *D. ingens*; (H) *D. remanei*; (I) *D. angolensis*; (J) *D. delamarei*. (From Hessler, 1972)

allowed to sink into the sand. Exceptions to these general patterns are known; Dahl (1952) and Delamare–Deboutteville (1953) collected subtidal forms, and Grimaldi (1963) found *Derocheilocaris* in poorly sorted beach sands with high amounts of calcium carbonate. Jansson (1966) established some environmental parameters that can restrict mystacocarid distributions in a beach: their inner limit is established by salinities not falling below five parts per thousand and the temperature not exceeding 27°C, while the outer limit seems to be constrained by the amount of turbulence.

Exact feeding modes are obscure in *Derocheilocaris* since its size and habitat make observation difficult. Buchholz (1953), Hessler (1971), and Jansson (1966) deduced that mystacocarids graze off bacterial films as they crawl in among the sand grains in their habitat. The motion of appendages while crawling then also serves to gather food under the labrum for processing. Lombardi and Ruppert (1982) observed that *Derocheilocaris typica* employs the antennae and mandibles as the motive organs during locomotion. The animals require both a dorsal *and* a ventral surface to move. The exopods of the antennae and mandibles are directed dorsolaterally from the body against the dorsal substrate. This presses the body down against the ventral substrate, allowing the antennal and mandibular endopods to gain traction. The locomotory abilities are quite effective, producing a maximum rate of movement of about 420 μm/s, but only so long as there are two substrates for the limbs to work against. One substrate alone results in the animal's thrashing about ineffectively.

The life cycle of mystacocarids is known in some detail. Reproductive patterns vary with climate from seasonal (Hall and Hessler, 1971) to continuous (Delamare–Deboutteville, 1960). Sexes are separate. Fizi (1963) reported very large eggs inside the females, and Brown and Betz (1967) noted spermatophores in the males. However, the exact manner of copulation is unknown. Hessler (1971) remarks that eggs are freely shed into the environment, though males of all species apparently have the fourth thoracoabdominal limb modified to assist in copulation. Eggs have never been observed free in the environment.

In light of the limited dispersal capabilities of their life-style, several aspects of their biogeographic distribution take on special interest. First, species are strangely morphologically conservative, though some of them have rather long ranges. Since mystacocarids are apparently restricted entirely to their infaunal habitat through all developmental stages with little presumed gene flow outside of populations, it is surprising that they retain their morphologic integrity over such long distances (Fig. 34-4). *Derocheilocaris typicus*, for example, extends from Massachusetts on the north to Miami Beach, Florida on the south. Another long range is exhibited by *D. remanei*, which extends though with several subspecies from the French Atlantic coast down into Africa with an extension into the easternmost Mediterranean. With restricted gene flow and dispersal capacity assumed for mystacocarids, unless there is some unknown naupliar dispersal phase that we do not yet recognize, these wide species' distributional patterns are difficult to explain in traditional terms.

Though individual *Derocheilocaris* species can have long ranges, the order itself is apparently rather restricted. Mystacocarids are essentially circum-Atlantic with dispersal salients out of the South Atlantic into the Pacific up the coast of Chile and into the Indian Ocean up the east coast of South Africa (Fig. 34-4). Such a distribution suggests that the evolution of mystacocarids is linked to the relatively recent opening of the Atlantic Ocean since the Jurassic and Cretaceous (Friauf and Bennett, 1974). Subsequent migration must have occurred out of the European Atlantic into the Mediterranean after the Miocene salinity crisis there. The time frame implied by their biogeography indicates essentially a post-Cretaceous dispersal and evolution.

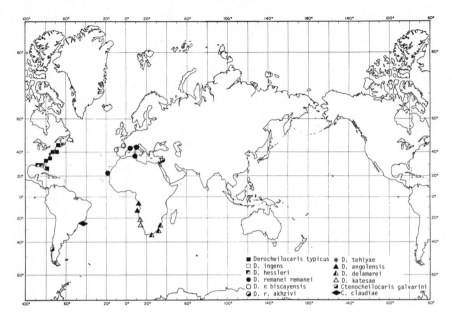

Fig. 34-4. Biogeographic distribution of mystacocarids.

DEVELOPMENT   Ontogenetic sequences are well known for *D. remanei* (Delamare–Deboutteville, 1954), *D. typicus* (Hessler and Sanders, 1966), and *D. katesae* (McLachlin, 1977). The developmental patterns display some differences between the species. The development of segments is gradual, with the addition of somites being completed before the achievement of the definitive appendage form. However, the addition of postnaupliar appendages and the acquisition of their definitive adult form tends to be episodic (Fig. 34-5). The earliest recognized stages are either an advanced nauplius or metanauplius, with nine metanaupliar instars in *D. remanei* and six in *D. typicus* and *D. katesae*. Whether the first stage that is recognized in any of the known cases is the actual hatchling stage, we cannot be sure. All larvae and adults appear to be entirely infaunal. The changes that occur through successive instars are very slight (Fig. 34-6).

TAXONOMY   The most striking aspect of mystacocarid biology that all authors never fail to comment on is their great anatomical similarity to each other, usually termed 'conservativeness.' Ten species are recognized, all within two very similar genera, *Derocheilocaris* and *Ctenocheilocaris* (Renaud-Mornant, 1976). One species, *D. remanei*, has three subspecies, but the differences between mystacocarid taxa are generally so subtle (Hessler, 1972) as to have slowed the sorting and recognition of separate taxa. Though the essential limits of the range of mystacocarids (Dahl, 1952; Noodt, 1954) were established within a decade of the erection of the order (Pennak and Zinn, 1943), species and subspecies within that range were delineated only very slowly over the subsequent 20 years, the latest species having been described by Friauf and Bennett (1974).

Slight differences in ventral telsonic combs, the median lobes of the antennulary portion of the cephalon (referred to by some authors as the rostral plate), terminus of the supraanal plate, and setation on the caudal rami are the major features that are useful taxonomically. However, no one set of features can be used across the board to separate species.

The small size, generalized and spinose nature of the anterior appendages, reduced anterior and lack of posterior trunk limbs, and the general 'metanaupliar' form of the body bespeak the importance of progenetic paedomorphosis in the evolution of this group.

Fig. 34-5. Postembryonic mode of development in *Derocheilocaris*. (A) *D. remanei*, with the earliest stage an advanced nauplius; (B, C) *D. typicus* and *D. katesae*, with the earliest stage a metanauplius and relatively fewer instars than *D. remanei*. ● appendage in definitive form, ○ appendage in an immature state.

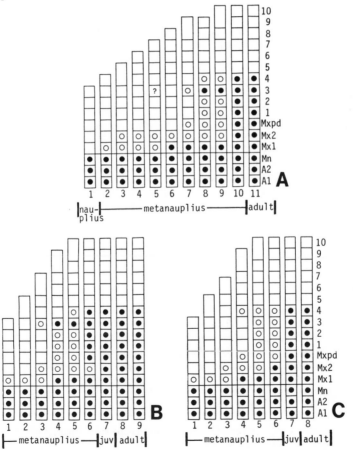

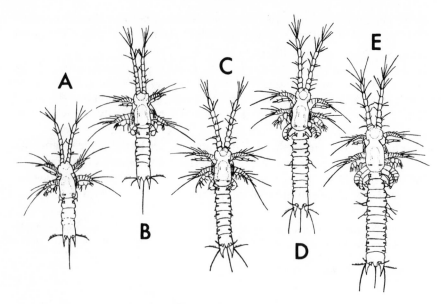

Fig. 34-6. Representative metanauplii of *D. katesae*. (From McLachlan, 1977)

REFERENCES

Armstrong, J. C. 1949. The systematic position of the crustacean genus *Derocheilocaris* and the status of the subclass Mystacocarida. *Am. Mus. Nov.* **1413**:1–6.
Brown, G. G., and C. B. Metz. 1967. Ultrastructure studies on the spermatozoa of two primitive crustaceans, *Hutchinsoniella macracantha* and *Derocheilocaris typicus*. *Zeit. f. Zellforsch. mikroak. anat.* **80**:78–92.
Buchholz, H. A. 1953. Die Mystacocarida. Eine neue Crustaceenordnung aus dem Lüchensystem der Meerssande. *Mikrokosmas* **43**:13–16.
Dahl, E. 1952. Mystacocarida. Reports of the Lund University Chile Expedition 1948–49. *Lunds Univ. Arssk.* **48**(2):1–41.
Delamare–Deboutteville, C. 1953. Revision des Mystacocarides du genre *Derocheilocaris*. *Vie et Milieu* **4**:459–69.
Delamare–Deboutteville, C. 1954. Le développement postembryonaise des Mystacocarides. *Arch. Zool. Exp. et Gen.* **91**:25–34.
Delamare–Deboutteville, C. 1960. *Biologie des Eaux Souterraines Littorales et Continentales.* Hermann, Paris.
Fizi, A. 1963. Contribution a l'etude de la microfaune des sables littoraux du Golfe d'Aigues-Mortes. *Vie et Milieu* **14**:669–774.
Friauf, J. J., and L. Bennett. 1974. *Derocheilocaris hessleri*: a new mystacocarid from the Gulf of Mexico. *Vie et Milieu* **24**:487–96.
Grimaldi, P. 1963. Primo rinvenimento di *Derocheilocaris remanei* del Mediterraneo orientale. *Ann. Inst. Mus. Zool. Univ. Napoli* **15**:1–7.
Hall, J. R. 1972. Aspects of the biology of *Derocheilocaris typica* II. Distribution. *Mar. Biol.* **12**:42–52.
Hall, J. R., and R. R. Hessler. 1971. Aspects in the population dynamics of *Derocheilocaris typica*. *Vie et Milieu* **22**:305–26.

Hessler, R. R. 1969. A new species of Mystococarida from Maine. *Vie et Milieu* 22:105–116.

Hessler, R. R. 1971. Biology of the Mystacocarida: a prospectus. *Smith. Contr. Zool.* 76:87–90.

Hessler, R. R. 1972. New species of Mystacocarida from Africa. *Crustaceana* 22:259–73.

Hessler, R. R., and H. L. Sanders. 1966. *Derocheilocaris typicus* revisited. *Crustaceana* 11:141–55.

Jansson, B. A. 1966. The ecology of *Derocheilocaris remanei*. *Vie et Milieu* 17:143–86.

Lombardi, J., and E. Ruppert. 1982. Functional morphology of locomotion in *Derocheilocaris typica*. *Zoomorph.* 100:1–10.

McLachlan, A. 1977. The larval development and population dynamics of *Derocheilocaris algoensis*. *Zod. Afr.* 12:1–14.

Noodt, W. 1954. Crustacea Mystacocarida von Süd-Africa. *Kieler Meerlsforsch* 10:243–46.

Pennak, R. W., and D. J. Zinn. 1943. Mystacocarida, a new order of Crustacea from intertidal beaches in Massachusetts and Connecticut. *Smith. Misc. Coll.* 103:1–11.

Renaud-Mornant, J. 1976. Un nouveau genre de Crustacé Mystacocaride de la zone néotropicale: *Ctenocheilocaris claudiae*. *C. R. Acad. Sci. Paris* 282:863–66.

Zinn, O. J., B. W. Found, and M. G. Kraus. 1982. A bibliography of the Mystacocarida. *Crustaceana* 42:270–74.

35

BRANCHIURA

DEFINITION   Carapace as a bilobed dorsal shield; paired compound eyes, naupliar eyes persistent in adult; antennules and antennae modified with hooks for attachment; labrum, mandibles, and paragnaths reduced to serrate blades and enclosed in a suctorial proboscis; maxillules highly modified either as suckers or with hooks for attachment; four trunk somites, each with a biramous, setose, somewhat cirriform pair of appendages, first two trunk appendages with accessory flabellum; 'abdomen' unsegmented, terminal anus flanked with tiny caudal rami.

HISTORY   Throughout their history, branchiurans have been linked to a variety of different groups. Until the 1850s they were treated as probable siphonostome copepods. In 1854 Zenker removed them from the Copepoda and placed them within the Branchiopoda, coequal in status with the phyllopodans. Thorell in 1864 agreed with the branchiopod placement and was the first to use the name Branchiura. The classic work of Claus (1875) resulted in the group's reassignment to the Copepoda but as a distinct and separate taxon; the comprehensive review of Wilson (1903) seconded this placement. Martin (1932) finally established the completely separate status of the group. However, as to which of the other entomostracan groups the Branchiura might be closely related continues to be the basis of considerable debate (see, e.g., Grygier, 1983). Presently four genera (Fig. 35-1) are recognized (*Argulus*, *Chonopeltis*, *Dolops*, and *Dipteropeltis*) with some 150 species.

MORPHOLOGY   The branchiuran body plan is distinctly characterized by its great flattening. The head shield is laterally extended and together with the two folds of the carapace (which are partially fused to the first trunk somite) form an oval shield on which the posterior cleft is variously developed (Fig. 35-1). The dorsal surface of this shield is marked by three naupliar eyes and a pair of compound eyes. The compound eyes project into a blood sinus below the surface of the cuticle and are suspended by a stalk formed by the optic nerves. The eyes are freely movable within this sinus by means of muscles. Martin (1932) documents the fine detail of this unique arrangement and concludes that it is not homologous to the similar water sac of some branchiopods. The ventral surface of the carapace shield has two paired respiratory areas marked by very thin cuticle, one flanking the maxillary region and the other flanking the thoracopods.

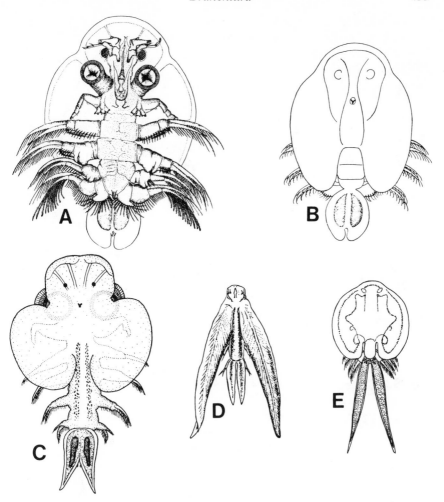

Fig. 35-1. Branchiuran body types. (A, B) *Argulus laticauda* (from Wilson, 1903); (C) *Chonopeltis brevis* (from Fryer, 1961a); (D) *Dipteropeltis hirundo* (from Kaestner, 1970); (E) *Dolops longicauda* (from Wilson, 1903).

The antennules (Fig. 35-2A) are short and typically bear spines or hooks developed on the basal segments. Within the anterior antennular spine a large gland or sense organ has been reported by several workers, but the function of this organ is uncertain. A small palp medially flanks the anterior antennal spine. Antennules are absent in *Chonopeltis*.

The antennae (Fig. 35-2A) are also small, closely associated with the antennules and armed with terminal spines and basal hooks. The basal segment contains a large mass of glandular tissue that extends into the cephalon, but no duct to the outside has been noted.

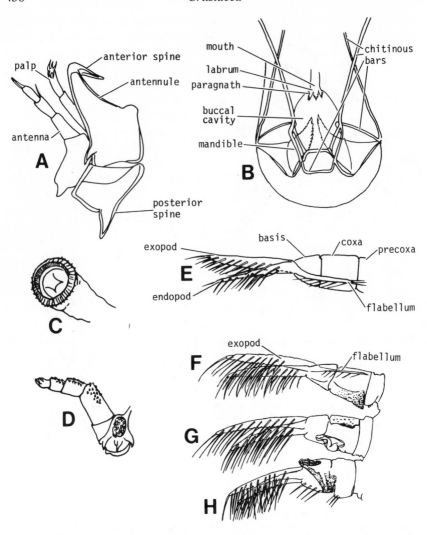

Fig. 35-2. Appendages of *Argulus foliaceus*. (A) antennules and antennae; (B) distal proboscis details with mouthparts; (C) maxillule; (D) maxilla; (E, F, G, H) first through fourth thoracopods of male. (From Martin, 1932)

Anterior to the proboscis is a delicate, needlelike, preoral spine (see Fig. 35-4) whose distal portion is retractable into a basal socket. This spine is apparently capable of injecting a substance from a 'poison gland.' The spine is variously developed depending on species.

The mouth and its surrounding labrum, mandibles, and paragnaths are enclosed in a large suctorial proboscis (Figs. 35-2B, 35-3). The proboscis is supported distally by a complex series of chitinous bars (Fig. 35-2B). The fleshy lips of the distal proboscis form a buccal cavity. The outer portion of the cavity is filled with a pair of spinose swellings associated with glands (Fig. 35-3). Proximal to these swellings, within the buccal cavity, are the serrate palpless mandibles (Fig. 35-2B). The actual mouth is guarded by serrate lamellate flaps, homologized with the labrum and paragnaths. When not feeding, the entire proboscis lies at rest in a groove along the venter of the body.

The maxillules (Fig. 35-2C) are one of the most characteristic features of the branchiurans. They are typically developed in adults as large suckers. The size is variable, but in some species (e.g., *Argulus funduli*, or *A. latus*) the suckers are almost one-third the width of the carapace. In the genus *Dolops*, the maxillules have large hooks and are not developed as suckers.

The maxillae (Fig. 35-2D) serve largely as grooming appendages. The basal joints bear large spines on their posterior margins and also possess a patch of smaller spines toward the anterior margin. The distal joints of the limb also bear batteries of spines as well as a terminal claw. Individuals have been observed to contort the body to draw the thoracopods across the spinose surface of the maxillae in order to remove debris. There may also be hooklike processes on the sternite between the maxillae to further assist in this grooming process.

The thoracopods (Fig. 35-2E, F, G, H) are basically biramous. The protopod appears to consist of three joints, including a small precoxal ring, coxa, and basis. The rami are setose, and the exopods on the first two appendages also bear a medially directed flabellum (Fig. 35-2E, F). The protopodal joints on the last thoracopod are typically equipped with accessory lobes and spines over that seen on the anterior thoracic limbs (Fig.

Fig. 35-3. Longitudinal section of the proboscis of *Argulus foliaceus*. (From Martin, 1932)

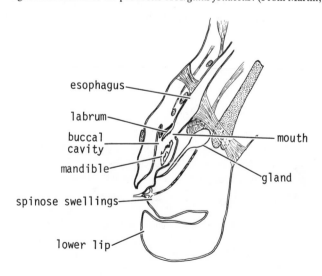

35-2H). Additional slight modifications of the second through fourth thoracopods (Fig. 35-2F, G, H) occur on males for copulation (detailed descriptions of these can be found in Martin, 1932).

The posterior part of the body or abdomen is a thin, flexible, double-lobed, unsegmented structure. The anus lies in the cleft between the two lobes and is flanked by a pair of tiny flaps variously termed anal papillae, anal furcae, or caudal rami.

Claus (1875) and Wilson (1903) still remain the best sources on the internal anatomy of the group. The digestive system (Fig. 35-4) is basically a simple tube, except for the elaborate caeca. These arise from the anterior part of the midgut and are greatly branched, becoming increasingly so with age. The caeca lie inside the carapace. At the time of feeding the caeca become engorged with food and effectively fill the entire volume of the carapace.

The circulatory system (Fig. 35-5) is very simple, consisting of heart and anterior aorta. The heart is triangular and is located in the last thoracic segment (Fig. 35-4). In *A. americanus*, blood enters the heart through paired lateral ostia at the posterior angles of the heart. The blood is then pumped anteriad through the aorta to supply the cephalic region, ventrad through a large ventral opening to supply the thoracic viscera, and posteriad through a large posterior opening into the abdomen. The blood moves about freely in body sinuses facilitated by the beating heart and ordinary body movements. Respiration is achieved when the blood passes over the thin cuticle of the respiratory areas on the ventral surface of the carapace.

Excretion is achieved with maxillary glands.

The female reproductive system (Fig. 35-6A) consists of a large median ovary that opens to the outside by means of a short oviduct on the last thoracic sternite. The vaginal opening is flanked by two seminal papillae that are connected by ducts to a pair of seminal receptacles in the anterior part of the abdomen (Fig. 35-6B). When the papillae are extruded they curve

Fig. 35-4. Longitudinal section to illustrate internal anatomy of *Argulus foliaceus* male. (From Martin, 1932)

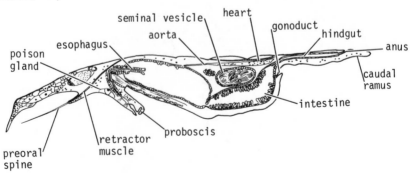

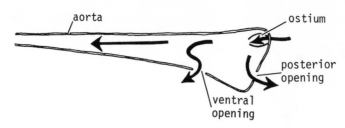

Fig. 35-5. Diagrammatic view of heart, and circulatory pattern within it. (From Wilson, 1904)

inward toward the vaginal opening. The male reproductive system (Fig. 35-6C) is more complex. The paired testes lie in the abdomen. These are connected by vasa efferentia to a median seminal vesicle in the thorax. A pair of vasa deferentia exits from the anterior end of the vesicle and proceeds posteriad, where they are joined by the ducts draining blind capsules or glands. The vasa then turn mediad to fuse into a single ejaculatory duct opening on the last thoracic sternite.

The nervous system is highly centralized (Fig. 35-7). A large brain

Fig. 35-6. Reproductive system of branchiurans. (A) female system; (B) detail of seminal receptacles and papillae; (C) male system. (From Wilson, 1903)

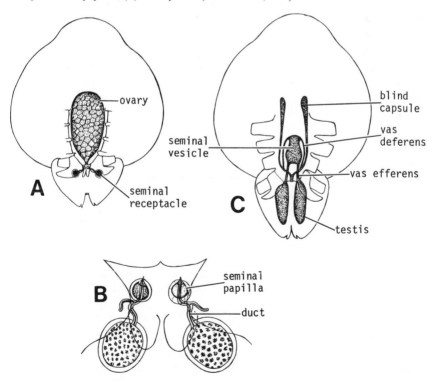

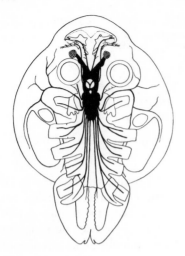

Fig. 35-7. Nervous system of branchiurans. (From Wilson, 1904).

supplies the antennules and eyes. The antennal nerves arise from the thick circumesophageal commissures. The posterior ganglia are all fused, but six lobes are still recognizable. Nerves to the proboscis leave the front of this ganglion mass. The six ganglionic lobes then each supply the remaining appendages, with the maxillary ganglion also supplying the carapace and the last ganglion also innervating the abdomen.

NATURAL HISTORY   Branchiurans are ectoparasites of fish and occasionally amphibians. They seem to prefer the gill chamber but can be found on the skin or fins of their victims. They fasten to the host by the antennal hooks and the maxillules (be they developed as suckers or hooks), and are capable of scuttling across the surface of the host. However, they do leave the host to find mates, to lay eggs, and to locate new hosts. As a result of their mobility, branchiuran species generally do not exhibit host specificity. The swimming abilities of *Chonopeltis* and *Dipteropeltis* seem to be more limited than those of *Argulus* or *Dolops*.

The parasites generally feed by rasping away at the host's integument with the serrated mandibles and burying the proboscis into a blood vessel to suck blood. Other species feed primarily on mucus, epithelial cells, and extracellular fluids. With feeding, the caeca become engorged, allowing the branchiuran to go two or three weeks between meals.

During mating the male mounts the female dorsally, and then wraps his abdomen around that of the female on alternate left and right sides to inseminate the right and left receptacles. Spermatophores have been noted in *Dolops*, which impales the sperm packets on the seminal papillae. When it comes time to lay her eggs, the female branchiuran leaves the host to find a suitable plant or rock surface on which to deposit her brood. As each egg emerges from the oviduct, it is pricked by one of the papillae and fertilized from the sperm reserve. The female will lay several broods in the course of her life, returning to a host between egg-laying sessions. It is not known whether the female may endure multiple inseminations from different males in the course of her life.

Branchiurans are found worldwide in both marine and fresh water habitats. Wilson (1903) reports great tolerance of *Argulus* species for alternating from salt- to freshwater conditions. Some endemicity is recorded (McLaughlin, 1980). *Argulus* is ubiquitous, but the freshwater genera seem to be endemic to certain regions: *Chonopeltis* is African, *Dipteropeltis* is South American, and *Dolops* is Gondwanan with species in South America, Tasmania, as well as Africa. It would appear from this that the Branchiura is an ancient group with strong connections to Gondwanaland.

DEVELOPMENT   Studies of branchiuran development have been almost exclusively restricted to the sequence of events after the first larva hatches from the egg. Only the most general obvservations have been offered on ontogeny within the egg (e.g., Tokioka, 1936). This is especially vexing because early ontogenetic studies might teach us much concerning branchiuran relationships and possibly lend some insight into the homology of peculiar structural features. Collection of egg masses in nature or in laboratory-rearing situations has been frequently reported in the literature since the days of Claus and Wilson, and Clark (1903) even reported culturing the entire life cycle in the laboratory. This aspect of branchiuran biology certainly bears some investigative attention.

The embryonic stages take three to five weeks to transpire if laid in spring when the temperature is relatively high. Fall eggs may overwinter. The emergent larvae soon attach themselves to a host. Two kinds of larvae are recognized (Wilson, 1903): those that swim by means of the antennae and mandibular palps at the base of the developing proboscis, and those that swim with a complete set of thoracopods. The number of larval stages varies from seven to nine (Fig. 35-8C, D, E, F, G, H) and takes four to five weeks to pass through (Fryer, 1961b; Tokioka, 1936; Stammer, 1959; Shimura, 1981). In the initial larval stages the head appendages are different in form from those of the adult. The antennules are somewhat adultlike. However, the antennae are separated from the antennules and bear large setose palps (Fig. 35-8A). The mandibles have long setose palps articulating to the base of the proboscis, and the gnathobase only gradually develops the serrate form of the adult (Fig. 35-8A, B). The maxillules are jointed and equipped with hooks (Fig. 35-8C, D, E); the suckers do not begin to appear until the fourth larval stage as a specialization of the protopod (Fig. 35-8F), the palp persisting in some form (Fig. 35-8G) until the final molt to an adult conditon. (*Dolops*, of course, never develops suckers.) The maxillae resemble the adult condition but become more robust and spinose with each molt until their final form is achieved. In species like *A. japonicus* that hatch with rudimentary thoracopods, an adult

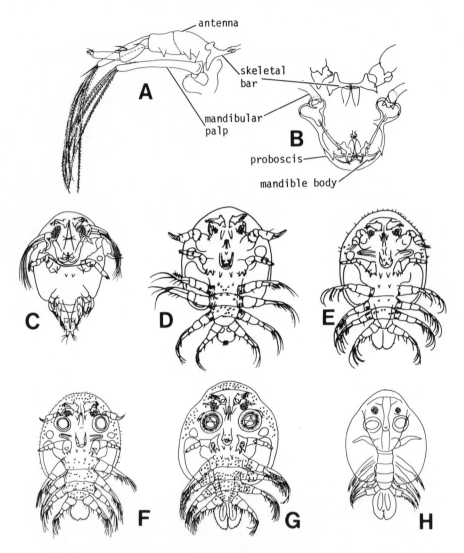

Fig. 35-8. *Argulus japonicus* development. (A) first stage antenna and mandibular palp; (B) detail of first stage proboscis, and antenna and palp bases; (C) first stage larva; (D) second stage; (E) third stage; (F) fourth stage; (G) fifth stage; (H) seventh stage. (From Tokioka, 1936)

442

segmentation of these limbs is soon achieved by the second stage larva. The abdomen is initially only a small lobe with terminal rami. This gradually enlarges and by the fifth to sixth stages begins to resemble an adult form.

TAXONOMY   The four genera of branchiurans are placed in a single family, the Argulidae Leach, 1819. No fossils are known. Grygier (1983) discusses reasons for possibly placing the problematic Pentastomida in relationship to the Branchiura.

REFERENCES

Clark, F. N. 1903. *Argulus foliaceus*, a contribution to the life history. *Proc. South London Ent. & Nat. Hist. Soc.* **1902**:12–21.
Claus, C. 1875. Über die Entwicklung, Organization, und systemische Stellung der Arguliden. *Zeit. wiss. Zool.* **25**:217–84, pls. 14–18.
Fryer, G. 1961a. The parasitic Copepoda and Branchiura of the fishes of Lake Victoria and the Victoria Nile. *Proc. Zool. Soc. Lond.* **137**:41–60.
Fryer, G. 1961. Larval development in the genus *Chonopeltis*. *Proc. Zool. Soc. Lond.* **137**:61–69.
Grygier, M. J. 1983. Ascothoracida and the unity of the Maxillopoda. *Crust. Issues* **1**:73–104.
Kaestner, A. 1970. *Invertebrate Zoology*, Vol. III. Interscience, New York.
Martin, M. F. 1932. On the morphology and classification of *Argulus*. *Proc. Zool. Soc. Lond.* **1932**:771–806, pls. 1–5.
McLaughlin, P. A. 1980. *Comparative Morphology of Recent Crustacea*. Freeman, San Francisco.
Shimura, S. 1981. The larval development of *Argulus coregoni. J. Nat. Hist.* **15**:331–48.
Stammer, J. 1959. Morphologie, Biologie, und Bekämpfung der Karpfenläuse. *Zeit. Parasitenk* **19**:135–208.
Tokioka, T. 1936. Larval development and metamorphosis of *Argulus japonicus. Mem. Coll. Sci. Kyoto Imp. Univ.* (B)**12**:93–114.
Wilson, C. B. 1903. North American parasitic copepods of the family Argulidae, with a bibliography of the group and a systematic review of all known species. *Proc. U.S. Nat. Mus.* **25**:635–742, pls. 8–27.
Wilson, C. B. 1904. A new species of *Argulus*, with a more complete account of two species already described. *Proc. U.S. Nat. Mus.* **27**:627–55.

36

TANTULOCARIDA

DEFINITION   Juvenile with head shield; cephalon with anterior oral disc and median ventral stylet, apparently no other recognizable appendages; thorax of six segments, thoracopods 1 to 5 biramous with lobate endites, thoracopod 6 uniramous without endite, median genital papilla on fifth thoracic sternite; abdomen with up to six limbless segments. Adult with cephalon as in juvenile; trunk saclike, abdominal segments and appendages apparently lost in terminal molt; median female gonopore on anteroventral portion of trunk.

HISTORY   Becker (1975) described *Basipodella harpacticola*, a parasite on deep-sea harpacticoids, and placed it within the Copepoda. A second genus and species, *Deoterthron dentatum*, was described by Bradford and Hewitt (1980), who removed both species from the copepods and suggested that their closest relatives were probably among the cirripedes. Grygier (1983) felt that their affinities were with no specific group but did lie somewhere within the Maxillopoda. Boxshall and Lincoln (1983) concluded that the exact affinity of these species was still a problem but that the group deserved separate status as Tantulocarida.

MORPHOLOGY   Understanding of these microscopic parasites is complicated by the fact that only females have been recognized so far. In addition, juveniles and adults are not known for all species, *Deoterthron* (Fig. 36-1A) being based on juveniles, while *Basipodella* (Fig. 36-1B) is known from both juveniles and adults. However, commonalities of the cephalic anatomy seem to ensure that there is some relationship between these two genera. The interpretations of body form of Boxshall and Lincoln (1983) are followed here.

The cephalon is covered dorsally by a head shield, which may bear some cuticular ornament or longitudinal ridges as well as a rostrum. On the ventral side of the head (Fig. 36-2A) is an oral hood with a central mouth. The mouth appears to be continuous with a tube whose walls are composed in part of chitinous bars with prominent basal swellings. Extending into the mouth tube from the posterior is a tubular longitudinal organ. This is overlain by a longitudinal striated organ, which extends from the midlength of the cephalic stylet. Apparently, a bulbous organ, which extends into the host tissues, is everted from the parasite's mouth. A long cephalic stylet

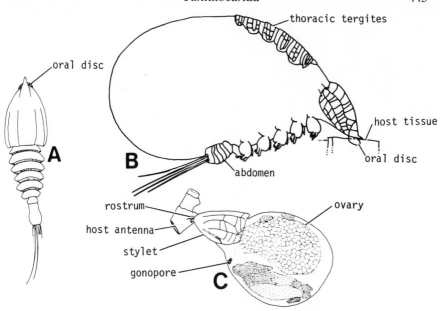

Fig. 36-1. Tantulocarid types. (A) *Deoterthron aselloticola*; (B) juvenile, *Basipodella harpacticola*; (C) adult, *B. atlantica*. Scale 50 μm. (From Lincoln and Boxshall, 1983)

arises behind the oral area; it is hollow at its base, but solid distally. Additional paired clear and dense areas of tissue are noted in the head, but details of their structure or possible function are yet unknown.

The thoracic segments in the juveniles are clearly delineated with tergal plates. The thoracopods (Fig. 36-2B) have large, flat protopods, which on legs 1 to 5 bear oval, lobate endites with a terminal spine. Usually the small exopods on the anterior limbs each bear two long and two short setae (Fig. 36-2C, D). The endopods on *Deoterthron* are long and typically bear two terminal spines (Fig. 36-2C), and both genera on legs two to five (Fig. 36-2D) also have a pair of long setae arising midlength on the ramus. The sixth thoracopod (Fig. 36-2B) lacks an endite and bears two long setae distally on the protopod. A median genital papilla is found on the fifth thoracic sternite.

The adult trunk (Fig. 36-1C) is saclike and lacks all traces of segmentation. The tergites and thoracopods are lost at the time of the final molt to the ultimate reproductive stage. A large median genital pore is located on the anteroventral aspect of the saclike trunk.

The abdomen in juveniles has two segments in *Deoterthron* and six in *Basipodella*. There are no abdominal appendages. There are tiny terminal caudal rami that bear long stiff setae. The abdomen appears to be lost in the course of metamorphosis to the adult condition.

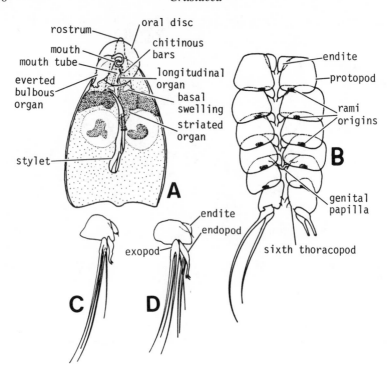

Fig. 36-2. Appendages of *Deoterthron aselloticola*. (A) ventral view of cephalon; (B) ventral view of thoracopod series, with rami of thoracopods 1 to 5 removed; (C) first thoracopod; (D) second thoracopod. (From Boxshall and Lincoln, 1983)

NATURAL HISTORY   These minute, generally deep-sea crustaceans are ecto-parasites on other crustaceans. *Basipodella* has been reported on harpacticoid copepods, while *Deoterthron* has heen found on the isopods *Haploniscus tangaroae* and *Hydroniscus lobocephalus*, and the benthic myodocopid ostracode *Metavargula mazeri*. It appears that the animals attach to the host by means of the oral disc (perhaps assisted by the stylet) early in the life cycle, and evert through the mouth some digestive or absorptive structure, the bulbous organ, into the tissues of the host.

Many fascinating aspects of the life cycle of these creatures remain to be described, especially their mode of development, the existence of males, the possible manner of copulation, the disposition of the eggs, and details of the internal anatomy (Boxshall, personal communication).

The known species are widely distributed (Table 36-1): *Deoterthron aselloticola* in the Tasman Sea (Boxshall and Lincoln, 1983), *D. dentatum* from New Zealand (Bradford and Hewitt, 1980), *D. megacephala* also from New Zealand (Lincoln and Boxshall, 1983), *Basipodella atlantica* from southwest of the Azores (Boxshall and Lincoln, 1983), and *B. harpac-*

Table 36–1. Distribution of tantulocarids.

	Host	Locality	Depth
Basipodella *harpacticola*	harpacticoid copepods (3 species in 2 families)	12°3′S 78°45′W 12°4′S 78°5′W 12°54′S 80°41′W	4100 m 2000 m 5000 m
B. atlantica	tisbid harpacticoid	34°57′N 32°55′W	3000 m
D. dentatum	*Metavargula mazeri*	43°35′S 178°3′W	384 m
Deoterthron *aselloticola*	*Hydroniscus* *lobocephalus*	41°9′S 166°28′E	3250–3340 m
D. megacephala	*Haploniscus* *tangaroae*	45°21′S 173°36′W	1386 m

ticola from several sites off the coast of Peru (Becker, 1975). So few species precludes any meaningful comments on their distribution at this time.

TAXONOMY   Though the crustacean diagnostic characters of the head are completely lacking in tantulocarids, all workers have felt that their general facies and the form of the thoracic legs mark them as Crustacea. Boxshall and Lincoln (1983) formalized their separate status, but placed both genera within a single family Basipodellidae.

REFERENCES

Becker, K.-H. 1975. *Basipodella harpacticola* n. gen., n. sp. *Helg. wiss. Meers.* **27**:96–100.
Boxshall, G. A., and R. J. Lincoln. 1983. Tantulocarida, a new class of Crustacea ectoparasitic on other crustaceans. *J. Crust. Biol.* **3**:1–16.
Bradford, J. M., and G. C. Hewitt. 1980. A new maxillopodan crustacean, parasitic on a myodocopid ostracod. *Crustaceana* **38**:67–72.
Grygier, M. J. 1983. Ascothoracida and the unity of the Maxillopoda. *Crust. Issues* **1**:73–104.
Lincoln, R. J., and G. A. Boxshall. 1983. A new species of *Deoterthron* ectoparasitic on a deep-sea asellote from New Zealand. *J. Nat. Hist.* **17**:881–89.

37

COPEPODA

DEFINITION   Cephalon with well-developed head shield; one or more thoracic somites fused to head to form cephalosome, first thoracopod as a maxilliped; thorax with total of six segments, including above; abdomen of five segments, including anal somite; antennules uniramous, antennae bi- or uniramous, mandible usually with palp that can be biramous, mouthparts generally raptorial or grappling in some way, posterior thoracopods biramous, occasionally a reduced pair of limbs on the first abdominal somite associated with gonopores, well-developed caudal rami. Development with six 'nauplii,' five copepodid larvae.

HISTORY   Copepods have probably been noted since ancient times, and the first named forms appeared in the compendia of Linnaeus. The higher classification of the group, however, has been subject to changing fashions. Thorell (1859) divided copepods into groups based on mouthpart anatomy; he recognized gnathostomes, poecilostomes, and siphonostomes. Claus during the course of his career contributed much to our basic understanding of the copepods, such that by 1892 Giesbrecht was able to suggest a new system of higher taxonomy based on division of the body: gymnopleans and podopleans. Sars (1901–11) suggested an alternative system based on certain 'type' genera (*Calanus*, *Caligus*, *Cyclops*, *Harpacticus*, *Lernaea*, *Notodelphys*, and *Monstrilla*) with each higher taxon based on these of equal status to any of the others. The Sars system came to be recognized as containing polyphyletic groups, though various 'hybrid' taxonomies containing elements of both Giesbrecht and Sars were proposed by different authors over the years. Kabata (1979), returning to certain aspects of Thorell's system, has advanced a higher classification that seems to constitute a step toward eventually arriving at a natural system of classification for the copepods (see Chapter 43).

MORPHOLOGY   A central focus of study in the 8 to 10,000 species of copepods has been on body tagmosis. Copepods are herein considered to conform to a maxillopodan body plan (five cephalic, six thoracic, and five abdominal segments). Calanoids, or gymnopleans, have the principal point of flexure between the last thoracic and first abdominal segments, while the other groups, collectively known as the podopleans, have it between the fifth and sixth thoracic segments. The location of this articulation has been used as a dividing point between an anterior prosome and posterior urosome regions employed by many authors. It is generally assumed that the ancestral 'precopepod' lacked such an articulation. Indeed some forms, for example, many harpacticoids, virtually lack this flexion point and move rather with lateral wormlike gyrations. In addition, the location of this flex point can be variable. It was Calman (1909) who originally documented that the flex point appears at the fourth and fifth segment margins in early development and moves posteriad with subsequent molts. However, Gurney (1931) pointed out exceptions to this among podopleans where, in *Thespesiopsyllus*, artotrogids, and cancerillids, the adult flex point remains at the ontogenetic flex point between the fourth and fifth thoracic segments, and in *Caligidium* the flex point even falls between the third and fourth thoracic segments. The location of the podoplean flexure can be subject to even earlier paedomorphic arrest in individual cases.

The genital pores are on the first abdominal segment. The possession of rudimentary appendages on the genital segment has been the basis by many workers of consideration of a seven-segment abdomen in copepods. However, the fusions, enlargements, and reductions of segments in both the thorax and abdomen, especially in parasitic forms, makes strict generalizations about copepod tagmosis impossible (Fig. 37-1). As a rule, however, infaunal types (Fig. 37-1C) are frequently characterized by a vermiform body plan wherein the typical copepod dorsoventral flexure is co-empted by lateral flexion. The epibenthic and phytal forms (Fig. 37-1D) have some regionalization of the trunk apparent, but the caudal rami are not particularly specialized. Pelagic copepods (Fig. 37-1A, B) have well-developed antennules, strong body tagmosis, and large and setose caudal rami. Parasites can range from forms like some cyclopoids, with recognizable copepod plans, to ones that are hardly recognizable as arthropods, let alone copepods (Fig. 37-1E, F, G).

The cephalosome is dorsally covered by an enveloping head shield that wraps around the head laterally and frequently anteriorly; this latter is usually referred to as a 'rostrum' (see Fig. 37-3A). Frontal filaments and typically maxillopodan naupliar eyes (Elofsson, 1966) mark the principal special sensory organs in copepods. In addition, Parker (1891) reported the rudiments of compound eyes in pontellids, though these may have been a mistaken impression of the characteristically large naupliar eye ocelli of pontellids (see, e.g., Park, 1966). The first thoracopod is developed as a maxillipede, but the fusion of the relevant thoracomere is not always complete; for example, in some longipediid harpacticoids the first thoracomere is ventrally articulated with the head.

The antennules (Fig. 37-2A) are uniramous and are variable in regard to size and number of segments. Marcotte (1982) felt that an intermediate number of 13 articles (as in the misophrioids or some cyclopoids) possibly gave rise alternately to forms with long antennules and high numbers of segments (such as calanoids) and to forms with short antennules and low numbers of articles (like harpacticoids, mormonilloids, some cyclopoids,

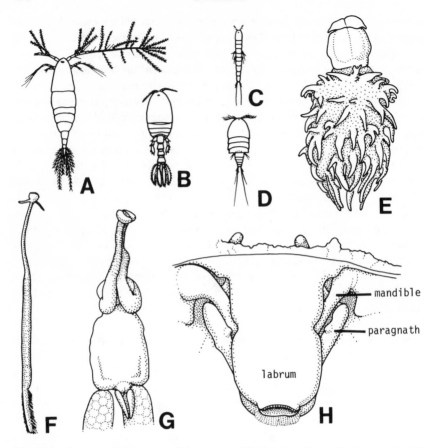

Fig. 37-1. Copepod body forms. (A) *Acartia*; (B) *Pseudocyclops*; (C) *Leptocaris*; (D) *Tegastes*; (E) *Chondracanthus*; (F) *Pennella*; (G) *Albionella*, (H) mouth of *Caligus* illustrating gap in base between labrum and paragnaths to accommodate the insertion of the mandible into the siphon. (A, B, C, D, from Noodt, 1972; E, F, G, H, from Kabata, 1979).

and many of the parasitic groups). Boxshall et al. (1984) felt that long antennules were primitive. This appendage is frequently geniculate in males, that is, equipped with a hinge joint to facilitate grasping the female during mating. Antennules can bear aesthetascs, as well as be modified for grasping (Fig. 37-3B).

The antennae (Fig. 37-2B) are frequently biramous. However, the exopod can be reduced as in harpacticoids and some cyclopoids or absent as in some cyclopoids and the parasitic groups. In the monstrilloids the antennae are absent in the adults. Parasitic groups like the siphonostomes and poecilostomes typically have the antennae equipped with claws to facilitate attachment to the host (Fig. 37-3C).

The mandibles (Fig. 37-2C) can exhibit great variation of form. In the primitive condition the gnathobase is well developed and surmounted by a setose, biramous palp. The endopod can have one or two segments, while the exopod can have up to six articles. Parasitic types have a wide variety of mandibles, and the distinct styliform jaws of siphonostomes (Fig. 37-3D) and falcate or sicklelike gnathobases of poecilostomes (Fig. 37-3E) are one of the principal bases upon which these two groups are recognized. Mouthfields of most groups are generally open (gnathostomous), but in the siphonostomes the labrum and paragnaths are formed into a tube or siphon that is not completely closed off at the base in order to allow the mandibles entry to the buccal chamber (Fig. 37-1H).

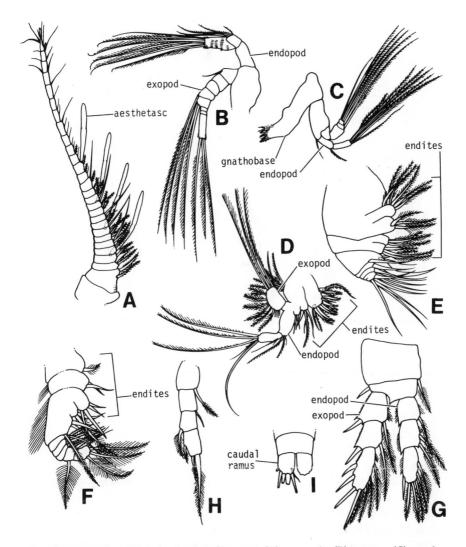

Fig. 37-2. Appendages of *Archimisophria discoveryi*. (A) antennule; (B) antenna; (C) mandible; (D) maxillule; (E) maxilla; (F) maxillipede; (G) third thoracopod; (H) sixth thoracopod; (I) body terminus with caudal rami. (From Boxshall, 1983)

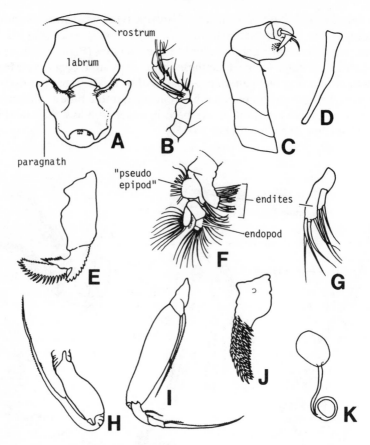

Fig. 37-3. (A) mouth region, *Misophriopsis dichotoma* (modified from Boxshall, 1983); (B) antennule, *Thalestris rufoviolascens* (from Gurney, 1931); (C, D) antenna and mandible *Ceuthoecetes aliger* (from Humes and Dojiri, 1980); (E) mandible, *Ergasilus nanus* (from Kabata, 1979); (F) maxillule, *Pseudocalanus* sp. (from Gurney, 1931); (G, H, I) maxillule, maxilla, and maxillipede, *C. aliger* (from Humes and Dojiri, 1980); (J) maxilla, *E. sieboldi* (from Kabata, 1979); (K) spermatophore, *C. aliger* (from Humes and Dojiri, 1980).

The maxillules (Fig. 37-2D) are generally setose and/or spinose appendages. Calanoids (Fig. 37-3F) have the most elaborate form, bearing not only well-developed endites and a biramous palp, but also having an epipoditelike exite on the coxa. (Unlike a true epipodite, however, this lobe is not articulated.) The exite is absent or incorporated into the protopod, and the palps are reduced in free-living podopleans. However, extreme reductions can take place (Fig. 37-3G) wherein only a small, narrow limb is present. Much has been made of protopodal segmentation in this limb (as indeed has been the case of protopodal segmentation in general), but the admonitions of Borradaile (1926, p.206) is appropriate:

'The recognition of a three-segmented protopodite in Crustacea is purely empirical...'

The maxillae (Fig. 37-2E) seem to fall into two basic types. The first is lamellate setose limb with prominent endites. In its most elaborate form, as seen in calanoids, the setae are quite plumose. The second type displays a marked reduction or absence of setae, and the development of terminal spines (Fig. 37-3J) to produce a raptorial organ, frequently subchelate (Fig. 37-3H). A distinctively unique set of maxillae occurs in the parasitic lernaeopodids (Fig. 37-1G), wherein the maxillae are clublike organs that extend in front of the head and *fuse* distally to form an attachment organ.

The maxillipedes (Fig. 37-2F) also display the dichotomy of form seen in the maxillae, that is, a lamellate and setose limb, or a more robust, raptorial, frequently subchelate appendage (Fig. 37-3I). Though both forms of the maxillipedes are generally more robust than the maxillae, in some calanoids the maxillipedes are reduced to a size smaller than that of the maxillae.

The second through fifth thoracopods (Fig. 37-2G) conform typically to a basic biramous paddlelike form. The coxa and basis are flat and rather broad, and frequently the opposed coxae are united by a fold of the sternite, the intercoxal plate, which ensures simultaneous strokes of a pair of legs during fast swimming. The rami are well developed, marginally setose, and can have up to three segments each. (Gurney, 1931–33 makes much of the prominent inner plumose seta on the coxa and an outer robust spine of the basis; the ultimate consequence of speculations along these lines are phylogenetic scenarios like that of Itô, 1982).

The sixth thoracopod of gymnopleans is often modified in the male as a copulatory organ, whereas in females it is reduced or absent. In podopleans, these limbs are present but reduced (Fig. 37-2H) and never modified for copulation. Some harpacticoids and cyclopoids enlarge this last appendage in the females to facilitate egg brooding. Thoracopods are more or less reduced in the parasitic forms.

The abdomen typically has no well-developed appendages. The first segment bears the gonopores. Typically the genital openings of podopleans are covered by movable plates that, because they are operated by muscles, are decorated with spines and/or setae and have a special nerve supply (Fig. 37-8), which some workers have surmised represent a reduced set of appendages on the first abdominal segment. There are usually five segments in the abdomen, with the last bearing the terminal anus flanked by a pair of caudal rami (Fig. 37-2I).

Detailed knowledge of internal anatomy presently arises largely from work on *Calanus finmarchicus* (Lowe, 1935), *Epilabidocera amphitrites* (Park, 1966), *Diarthrodes cystoecus* (Fahrenbach, 1962) and *Benthomisophria palliata* (Boxshall, 1982). The digestive system is relatively simple. The foregut is short and consists of an esophagus with well-developed circular and dilator muscles (Fig. 37-4). The midgut has an anterior, median

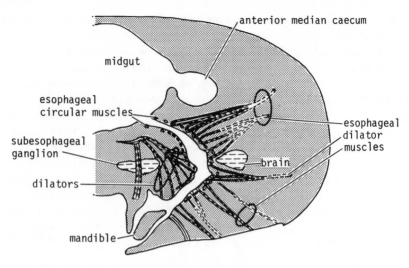

Fig. 37-4. Anterior digestive system of *Benthomisophria palliata*. (Modified from Boxshall, 1982)

caecum, which in some instances at least can be completely sealed off from the main gut lumen by sphincter muscles. Frequently there are also pairs of lateral diverticula that may be either simple, bifurcate, or multilobed sacs. The anterior midgut can be capable of considerable distension, as in misophroids, and has walls of columnar vacuolar cells. The midgut rather abruptly narrows in the anterior part of the thorax and continues posteriad toward the hindgut as a narrow tube. The proctodeum is short and occupies the last two abdominal segments. Like the foregut, the hindgut too is equipped with circular and dilator muscles.

The work of the digestive system is aided by labral glands and possibly (at least in *Benthomisophria*) cone organs. These latter consist of clusters of glands on the ventrolateral aspect of the head shield. They are placed such that their secretions seem to be available to the antennae and mandibles as these limbs sweep across the edge of the shield. It is also possible that the cone organs may serve in grooming behavior (Boxshall, personal communication)—their exact function is still unknown.

In close proximity to the gut within the visceral cavity is an oilsac (see Fig. 37-7A). This lipid-filled bag can vary in size, helps to regulate buoyancy in calanoids, but generally serves as a food reserve in most forms.

The circulatory system is lacking in all podopleans, except misophrioids in which it is poorly developed; but it is well developed in the calanoids. In the higher podopleans blood is kept in motion by the gyrations of the gut and body musculature. When present, the circulatory system centers on a small ovoid heart (Fig. 37-5) located in the second and third thoracomeres.

It has a single posterior and two lateral ostia draining a pericardial chamber, which is separated from the main visceral cavity by a broad pericardial floor (Fig. 37-7A). An aorta extends forward to the extreme anterior portion of the body where it splits to form a pair of large tracks encircling the head and joining ventrally into a supraneural sinus (Fig. 37-6). Blood flows in this sinus posteriad with the sinus splitting temporarily around the esophagus and anterior endosternite (with an opening into the visceral chamber). The sinus joins again behind the anterior endosternite to resplit around the posterior endosternite and then opens into the ventral sinus around the nerve cord and into the main visceral chamber. Blood passes from the visceral chamber into the pericardium by passing through dorsoventral afferent canals between the cuticle and dorsoventral muscles. The broad outlines of this system are not too unlike that seen in ostracodes.

Fig. 37-5. Heart of *Calanus finmarchicus*. (Modified from Lowe, 1935)

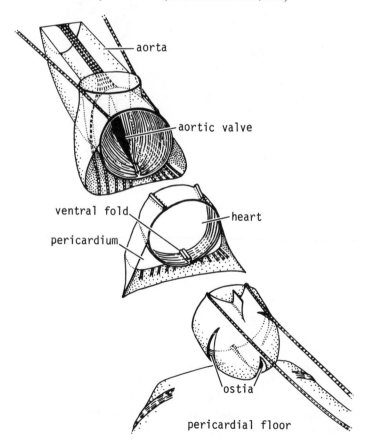

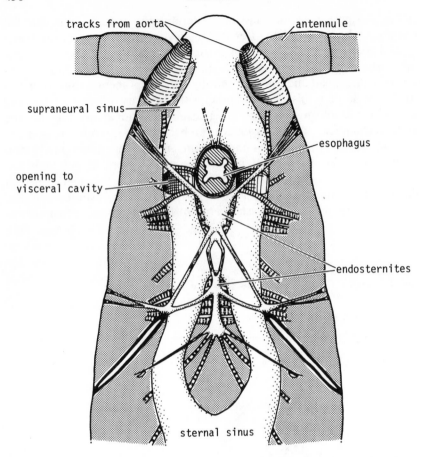

Fig. 37-6. Anterioventral circulatory system of *Calanus finmarchicus*. (Modified from Lowe, 1935)

Excretion is achieved in adult copepods by means of paired maxillary glands. However, no such organs have been noted in adult *Benthomisophria*, which does have functional antennal glands. Antennal glands also occur in the nauplii of copepods (Chapter 33).

Sexes are separate, and the gonads located in the cephalosome are typically median in position and unpaired in free-living forms (Fig. 37-7A). However, the systems are paired in the parasitic forms and at least in some of the primitive types such as the misophrioids. In females (Fig. 37-7B), whether the gonads are single or paired, the oviducts arise from the anterior portion of the ovary and extend back along the side of the body to the genital somite (Fig. 37-7A). The ducts typically open into a genital antrum. These antra are paired invaginations of the genital sternite that meet at the midline to form a transverse bilobed structure. The antra are 'floored' by the movable plates on the first abdominal segment. Also opening into the antra are seminal receptacles, sometimes with paired openings, sometimes with a single median opening. A variety of other glands opens into the antra (Boxshall, 1982); their exact functions are not yet known.

The testes (Fig. 37-7C) anteriorly give rise to a vas deferens, which in its posterior region is somewhat dilated to act as a seminal vesicle and ejaculatory duct. The glandular nature of the posterior portion of the duct helps produce the spermatophores (Fig. 37-3K). Each genital pore opens onto the surface of the cuticle, and the opening can be covered by a plate formed by the sixth thoracopods.

The central nervous system is rather condensed in its anterior aspect. The brain, circumesophageal connectives and anterior ganglia form a solid ring with batches of nerves arising to supply the appendages of the cephalosome (Fig. 37-8). The thoracic portion of the nerve cord is not paired nor are ganglia readily visible. In most copepods the cord does not extend beyond the thorax; however, in misophrioids the cord becomes paired at this point but bears no ganglia or segmental nerves in the abdominal region except to the genital flap.

Copepods have an array of chemosensory aesthetascs on the antennules. A well-developed naupliar eye is present (Elofsson, 1966), but no frontal organs are recognized. Naupliar eyes can be rather elaborate, as in

Fig. 37-7. (A) Diagrammatic cross section of copepod thorax at about the level of third and fourth thoracomeres (modified from Lowe, 1935); (B, C) lateral views of reproductive systems in *Benthomisophria palliata* (from Boxshall, 1982).

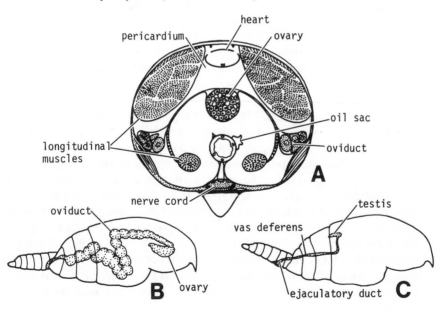

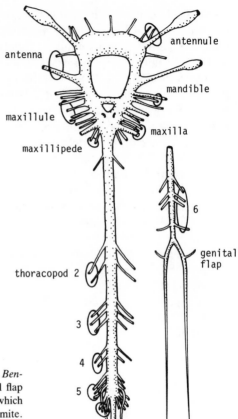

Fig. 37-8. Central nervous system of *Ben-thomisophria palliata*. Nerve to genital flap supplies 'rudimentary appendage' which covers gonopores on first abdominal somite. (Modified from Boxshall, 1982)

pontellids where the three ocelli become widely separated and are specialized with large lens and 'retinal surfaces' (Park, 1966). Misophrioids lack naupliar eyes.

NATURAL HISTORY   Copepods have been reported to have two basic locomotory modes: a slow speed achieved largely by the antennae and other mouthparts, and a fast jumping mode by means of the thoracopods (Cannon, 1928). This latter appears to be an apomorphic behavior among copepods and is facilitated by the median coupler plate that synchronizes thoracopodal movements. Some dispute exists as to the exact nature of this movement. Park (1966), and Boxshall et al. (1984) suppose the limbs beat simultaneously through the intermediary of giant neuronal tracks originating in the antennules. Boxshall (1982) and Boxshall et al. (1984) observed that there is a sequential two millisecond gap between the initiation of the remotor swings of each of the thoracopods beginning with the fifth thoracopod and passing forward to the second. Jumping movement is thus achieved by the extreme abbreviation of a basically metachronal sequence. This rapid remotion (only four milliseconds in duration for each leg) is facilitated by the extreme flexibility of the sternal cuticle and the insertion of the remotor muscles on the sternite just posterior to the limb rather into the leg itself. The total arc of swing is about 110°, almost twice the usual 50 to 60° typical for most crustaceans (Boxshall, 1982; Manton, 1977). The functional rationale for this unique method of locomotion may be related to the physics of movement in viscous media (discussed below for feeding) where 'sticky' water has to be literally thrown off the limbs in order to project the animal forward.

Feeding in copepods has been the subject of considerable debate. Cannon (1928) postulated a basic maxillary filter; however, soon afterward Gurney (1931–33) and Lowndes (1935) questioned whether this was indeed the basic feeding mode. They felt raptorial feeding was primitive, since gut contents and observation of actual animals seemed to indicate that copepods are omnivorous in their diet, and that strict filter feeding is relatively rare. Marcotte (1977) initially concluded raptorial feeding was primitive for at least podopleans and probably primitive for all copepods (Fryer, 1957; Marcotte, 1982). Cushing (1959) felt that copepods were encounter feeders, that is, 'catching' items when they actually touched them, whereas Fryer (1957) and Beklemishev (1961) believed that copepods can also detect food several millimeters away and actively seek it out.

The physics of this situation was elegantly analyzed by Koehl and Strickler (1981). Because of their size, copepods live in a world of low Reynolds numbers, wherein the medium is viscous and flow is laminar (rather than the medium being inertial with the flow turbulent as it would be with larger animals). Water is moved past the animal by the flapping of the antennae, mandibular palps, maxillules, and maxillipeds (Fig. 37-9B, C, D). It is the maxillae that actually capture food particles but without actually 'touching' them (Fig. 37-9EG'). In a viscous, laminar world the setae on the limbs behave as solid paddles rather than sieves, with the water layer immediately near the limb surface adhering to it and acting as part of the paddle. As the maxillae are flung apart (Fig. 37-9FF'), particles with *their* adhering surface water are sucked into the intermaxillary space (Fig. 37-9GG'). As the maxillae adduct, the particles are pushed inward and forward to the maxillulary endites, which in turn push them into the mouth.

The physics of this system provides an explanation for the earlier conclusions as to the importance of raptorial feeding in copepods. Details of this system were verified in calanoids (Paffenhöfer et al., 1982) and were directly applicable to the cyclopoids as well (Fryer, 1957). Fryer pointed out that the radial orientation of limbs around the mouth, not unlike that seen in ostracodes and cirripedes, was important. When combined with the reciprocating abductor–adductor action of the limbs, this configuration allowed food to be stuffed down the mouth and also served well in either an encounter-feeding habit or a grazing-herbivorous one.

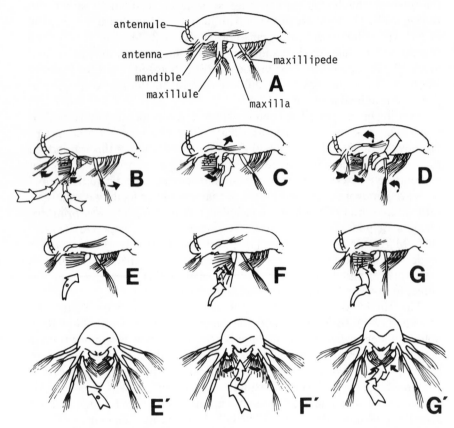

Fig. 37-9. Feeding in *Eucalanus pileatus*. (A) diagrammatic outline of body; (B, C, D) diagrams of feeding appendage movements (black arrows) and surrounding water currents they produce (white arrows) (an arrow with a narrow shaft and wide head indicates lateral movement out of the plane of the page toward the reader; an arrow with a wide shaft and narrow head indicates medial movement away from the reader); (B) outward movements of second antennae and maxillipedes suck water toward copepod's maxillae; (C) posteromedial movement of the first maxillae and dorsolateral movement of mandibular palps suck water laterally; (D) inward movements of second antennae and maxillipedes coupled with dorsolateral movement of mandibular palps shove water posterolaterally; (EE'FF', GG') particle capture (white arrows are water currents; black arrows indicate movements of second maxillae and of a first maxilla in G'); (E, F, G) copepod is viewed from its left side, and first maxilla has been left off for clarity; (E', F', G') animal is viewed from its anterior end. Feeding currents bypass second maxillae (EE') until an alga nears them. Alga is captured by an outward fling (FF') and an inward sweep (GG') of second maxillae. (Modified from Koehl and Strickler, 1981)

460

Fryer felt the raptorial mode was the only method of feeding from which herbivorous as well as parasitic types could evolve. As to the latter, Fryer observed that free-living cyclopoids immediately enter a quiescent period after feeding. Fryer postulated a scenario in which a raptorial cyclopoid 'attacks' a vertebrate rather than an easily devoured small invertebrate. When recovered from its postfeeding quiescent period, such an animal would find itself still attached to a food source and could easily feed again. All that would then remain for the cyclopoid to complete the adaptation would be to evolve an obligate relationship with its vertebrate 'prey–host,' and a parasitic state would be achieved. Further specialization of mouthparts would then lead to poecilostomous and siphonostomous conditions.

Though sexual reproduction is the norm in copepods, parthenogenesis is known among harpacticoids (Borutskii, 1964). The basic mode of copulation (Gauld, 1957; Blades, 1977; Blades and Youngbluth, 1977; Jacoby and Youngbluth, 1983) is fairly uniform among free-living forms with some variations noted. The male grabs the female by the urosome or caudal rami using his geniculate antennules. This initial grip may last for minutes or even days and is eventually exchanged for a second grip by the sixth thoracopod, grasping the female genital somite and coupling the pair head-to-tail. The spermatophore is transferred to the antra before the sperm are injected into a seminal receptacle, whereupon the male either drops off or reverts to the first position. Some females have been observed to have two or three spermatophores. It is thought that males, if they retain contact, can repeat the sequence more than once, or it is conceivable that more than one male can copulate with a female. This last might seem unlikely in terms of a possible premium to be placed on selecting systems whereby males might seek to minimize competing copulations by other males. However, few studies have been made on reproductive strategies in copepods (e.g., Hopkins, 1982).

Fertilization occurs when an egg is extruded into the antrum and sperm are released. The antrum is only large enough to accommodate one egg at a time. Females commonly retain eggs in an egg sac produced by the distal oviduct or by glands associated with the genital antra. However, some planktonic forms void the eggs freely with fertilization. If the eggs are brooded, they are retained typically only until the nauplii hatch, although some parasites hatch at the first copepodid stage.

Copepods inhabit all manner of aquatic habits. Some harpacticoids are known to occupy moist terrestrial environs such as moist moss, humus, and leaves. Some groups are restricted to particular habitats, for example, misophrioids are essentially deep-sea forms (Boxshall, 1982, 1983), as are the spinocalanids (Damkaer, 1975). Many Copepoda are parasites and infest hosts ranging from sponges to mammals. The fish parasites have been reviewed by Kabata (1981) and those of invertebrates by Gotto (1979).

Because several copepod species tend to be found in broad

environments (e.g., the *Siboga* expedition recorded over 100 species from one station), the group has had to develop adaptive mechanisms for such sympatric species to coexist. It has been postulated that species segregate themselves by (1) differing in size (implying differing in food resources), (2) vertical segregation, and (3) seasonal separation. Actual studies of resource partitioning have not been common. For example, Sandercock (1967) found that in coexisting species of *Diaptomus* in Clarke Lake, Ontario all three mechanisms were observable. *D. minutus* reached a spring population maximum as a small form, whereas *D. oregonensis* was a large species with a summer population peak. *D. minutus* and *D. sanguinensis* were about the same size and both peaked in spring, but *D. minutus* lived in the upper 5 m of the lake while *D. sanguinensis* is found below 6 m. Marine copepods are well known for their daily vertical migrations. Though numerous explanations for this have been advanced (e.g., Kaestner, 1970), it seems possible this is in part a mechanism to partition resources in the open ocean. Recent work suggests that, at least in part, vertical migration is also a predator avoidance strategy (Boxshall, 1980). However, little concerted effort at understanding these mechanisms in terms of reproductive needs and strategies has been undertaken. For instance, vertical migrations have been shown to have some effects on body growth and egg productivity; where warmer, light-intensive waters have greater food resources, cooler temperatures produce slow but maximum growth.

Though copepods occur widely and there is no shortage of copepod workers, synthetic analyses of biogeography are rare. Borutskii (1964) surveys the patterns of biogeographic distribution in harpacticoids, but does not offer many conclusions as to what these may tell us about the history of the group. However, much of the basic information *is* available in the literature, for example, Gurney (1931–33), in treating the British fauna, also mentions occurrences of taxa in other parts of the world, and the same can be said of Kabata (1979). Some more detailed work in this field will prove enlightening and holds great potential (van der Spoel and Pierrot-Boults, 1979; van der Spoel, 1983).

DEVELOPMENT   Aspects of copepod development have not been effectively examined since the efforts of Grobben (1881) and Fuchs (1914), with some additional observations on eggs by Harding (in Marshall and Orr, 1955). The eggs are laid near the surface and, in *Calanus* at least, develop over a 24-hour period. Since they are denser than water, this means freely laid eggs hatch into nauplii at some depth. The rate of development and the speed of descent are both related to temperature of the water (Table 37-1). Some forms, especially those inhabiting shallow water, produce resting eggs that sink and remain dormant in the sediments for extended periods.

Table 37–1. Variations in hours of time for development of *Calanus finmarchicus* under varying temperature. (Modified from Marshall and Orr, 1955)

	0°	5°	10°	15°	20°
First cleavage	2.50	—	1.33	1	0.75
Blastocoel	10–25	10–18	—	2–3	2–3
Nauplius form	90	27–48	—	16–20	14–18
Hatching	116–120	40–65	25–30	20–26	19–22

The cleavage is total (Fig. 37-10A); however, the blastomeres are pretty much the same size. Fuchs (1914) did attempt to trace cell lineages in *Cyclops*, but Grobben (1881) did not do so in *Calanus*. The presumptive germ layers are all delineated in the blastula stage when about the 32-cell stage the primary endoderm and mesoderm cells differentiate. In *Calanus* a mass of endodermal cells lies anterior to a pair of primary mesodermal cells, whereas in *Cyclops* primary germ and endodermal cells are surrounded by a ring of presumptive mesoderm (Fig. 37-10C, D, E).

In all known cases gastrulation (Fig. 37-10F, G, H, I) is by involution. The endo- and mesodermal cells move inwards more or less together, with the germ and/or mesodermal cells migrating somewhat earlier than the endodermal cells. In *Calanus* the endoderm tissues differentiate before the more slowly developing mesoderm, whereas in *Cyclops* the mesoderm divides more rapidly than the endoderm. In either case a stage is reached that is a three-layered ball, with ectoderm on the outside, endoderm and primary germ cells within, and mesoderm in between (Fig. 37-10J).

Knowledge of differentiation after this stage is somewhat vague. The naupliar limb anlagen appear, while a labrum and esophagus differentiate. Hatching, in *Calanus* at least, is achieved largely by changes in the egg membrane chemistry rather than any mechanical effort of the twitching larva (Marshall and Orr, 1955). Osmotic forces appear to rupture the outer egg membrane, which then allows the inner membrane to swell to five times its original volume before it splits.

The first one or two naupliar stages do not feed, the completion of gut development not being typically achieved until nauplius III. The copepod nauplii (Dietrich, 1915) can be sorted as ortho- and metanauplii (Fig. 37-11). The first two naupliar stages are orthonauplii, possessing only the naupliar appendages while the next four stages are metanauplii that successively add postnaupliar limbs and segments. The metamorphosis to the first of five copepodid stages involves several changes: separation of the nervous system and the naupliar eyes from the body wall, and appearance of recognizable gonadal tissues, heart if present, maxillary glands, and caudal rami. Development through the five copepodite stages involves progressive delineation of body segments and thoracic appendages until the final molt produces a sexually mature adult.

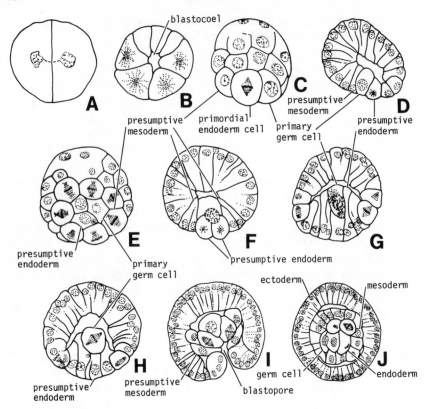

Fig. 37-10. Early ontogeny in *Cyclops viridis*. (A) first cleavage; (B) 16-cell stage blastula; (C) lateral surface view at about 64-cell stage with primordial cells indicated; (D) section through C; (E) vegetal pole surface of C; (F, G, H) early stages of gastrulation; (I) advanced gastrulation; (J) gastrulation complete, germ layers delineated. (From Fuchs, 1914)

FOSSIL RECORD   Fossils of copepods are virtually nonexistent. Some unnamed cyclopoids and cletodid harpacticoids fossils are known to occur in the western hemisphere (Palmer, 1969) in situations of unusual preservation, and an elegantly preserved specimen of dichelesthiid siphonostome was found in the gill chamber of a Lower Cretaceous fish (Cressy and Patterson, 1973).

TAXONOMY   The classification of Kabata (1979) seems for the time being to have stabilized the higher taxonomy of the group. However, some reservations still exist over the status of some groups, for example, Marcotte (1982) feels siphonostomes may yet prove to be polyphyletic. The major groups of copepods and their definitions are given here, followed by an arrangement of families.

Calanoida. Antennules long, 16 to 26 articles, geniculate in males; antennae, mandibles, and maxillules biramous; last thoracopod as copulatory organ in male, reduced in female; single median genital pore; flexure point between last thoracic and first abdominal segments, abdomen narrow; gnathostomous mouth; heart.

Misophrioida. Antennules moderate in size, 13 to 27 articles, geniculate in males; antennae, mandibles, and maxillules biramous but without exites; sixth thoracopod biramous; body flexure between fifth and sixth thoracomeres; gnathostomous mouth; heart; carapace fold on maxillipede segment.

Harpacticoida. Antennules short, 10 or fewer articles, geniculate in males; antennae with exopod reduced; mandibles biramous; maxillules with exites, bi- or uniramous; first two abdominal segments fused; sixth thoracopod with endopod reduced and fused to basis; body flexure between fifth and sixth thoracomeres; thorax and abdomen not always sharply differentiated; gnathostomous mouth; no heart.

Mormonilloida. Antennules long, three to four articles, not geniculate; antennae biramous with eight-segmented exopod; mandible and maxillule biramous, without exites; single median genital pore; complete fusion of first two abdominal segments; sixth thoracopod absent; body flexure between fifth and sixth thoracomeres; abdomen narrow; mouth gnathostomous; no heart.

Monstrilloida. Antennules long; no antennae; mouth and mouthparts absent in adult; sixth thoracopod with endopod reduced; body flexure between fifth and sixth thoracomeres; abdomen narrow; no heart; larvae parasitic.

Cyclopoida. Antennules moderate in size, 10 to 16 articles, geniculate in males; antennae uniramous; mandible with palp frequently reduced or

Fig. 37-11. Postembryonic mode of development in *Eucalanus elongatus*. (Based on Johnson, 1937)

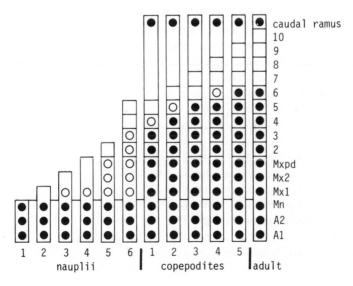

absent; maxillules usually biramous without exites; sixth thoracopod with endopod reduced; body flexure between fifth and sixth thoracomeres (lacking in lernaeids); abdomen narrow, first two segments fused; gnathostomous mouth; no heart; dorsal or dorsolateral genital pores.

Poecilostomatoida. Antennules short, not geniculate; antennae clawed and uniramous; mandible uniramous and falcate; maxillules reduced, without exites; maxillae reduced and clawed; maxillipede subchelate in males, may be absent in females; sixth thoracopod with endopod reduced, sometimes modified or vestigial; mouth slitlike; no heart; dorsal or dorsolateral genital pores; first two abdominal segments fused.

Siphonostomatoida. Antennules variable in length, geniculate; antennae clawed with exopod reduced to no more than two segments, or absent; mandible enclosed in siphon, styliform, palp reduced; maxillule biramous but reduced, without exites; maxillae subchelate; maxillipede subchelate or absent; sixth thoracopod with endopod lost, sometimes modified or vestigial; body flexure variable (sometimes between fifth and sixth, or fourth and fifth thoracomeres, or lacking altogether); first two abdominal segments fused; mouth with tubular siphon formed by labrum and paragnaths; no heart.

Order Calanoida Sars, 1903 Recent
    Superfamily Augaptiloidea Sars, 1905
        Family Arietellidae Sars, 1902
        Family Augaptilidae Sars, 1905
        Family Discoidae Gordejeva, 1975
        Family Epacteriscidae Fosshagen, 1973
        Family Heterorhabdidae Sars, 1902
        Family Lucicutiidae Sars, 1902
        Family Metridinidae Sars, 1902
        Family Phyllopididae Brodsky, 1950
    Superfamily Bathypontioidea Brodsky, 1950
        Family Bathypontiidae Brodsky, 1950
    Superfamily Centropagoidea Giesbrecht, 1892
        Family Acartiidae Sars, 1903
        Family Candaciidae Giesbrecht, 1892
        Family Centropagidae Giesbrecht, 1892
        Family Diaptomidae Baird, 1850
        Family Parapontellidae Giesbrecht, 1892
        Family Pontellidae Dana, 1853
        Family Pseudodiaptomidae Sars, 1902
        Family Sulcanidae Nicholls, 1945
        Family Temoridae Giesbrecht, 1892
        Family Tortanidae Sars, 1902
    Superfamily Clausocalanoidea Giesbrecht, 1892
        Family Aetideidae Giesbrecht, 1892

Family Clausocalanidae Giesbrecht, 1892
Family Diaixidae Sars, 1902
Family Euchaetidae Giesbrecht, 1892
Family Mesaiokeratidae Matthews, 1961
Family Phaennidae Sars, 1902
Family Pseudocyclopiidae Sars, 1902
Family Scolecithricidae Giesbrecht, 1892
Family Spinocalanidae Vervoort, 1951
Family Stephidae Sars, 1902
Family Tharybidae Sars, 1902
Superfamily Eucalanoidea Giesbrecht, 1892
Family Eucalanidae Giesbrecht, 1892
Superfamily Megacalanoidea Sewell, 1947
Family Calanidae Dana, 1849
Family Calocalanidae Bernard, 1958
Family Mecynoceridae Andronov, 1973
Family Megacalanidae Sewell, 1947
Family Paracalanidae Giesbrecht, 1892
Superfamily Platycopioidea Sars, 1911
Family Platycopiidae Sars, 1911
Superfamily Pseudocyclopoidea Giesbrecht, 1893
Family Pseudocyclopidae Giesbrecht, 1893
Family Ridgewayiidae M. S. Wilson, 1958
Superfamily Ryocalanoidea Andronov, 1974
Family Ryocalanidae Andronov, 1974
Order Misophrioida Gurney, 1933 Recent
Family Misophriidae Brady, 1878
Order Harpacticoida Sars, 1903 Miocene–Recent
Suborder Polyarthra Lang, 1948 Recent
Family Canuellidae Lang, 1948
Family Longipediidae Sars, 1903
Suborder Oligarthra Lang, 1948 Miocene–Recent
Infraorder Maxillipedasphalea Lang, 1948 Recent
Superfamily Cervinioidea Lang, 1948
Family Aegisthidae Giesbrecht, 1892
Family Cerviniidae Sars, 1903
Superfamily Ectinosomatoidea Lang, 1948
Family Ectinosomatidae Sars, 1903
Superfamily Neobradyoidea Lang, 1948
Family Chappuisiidae Chappuis, 1940
Family Darcythomsoniidae Lang, 1936
Family Neobradyidae Oloffson, 1917
Family Phyllognathopodidae Gurney, 1932
Infraorder Exanechentera Lang, 1948 Recent
Superfamily Tachidioidea Lang, 1948

Family Harpacticidae Dana, 1846
Family Tachidiidae Sars, 1909
Superfamily Tisboidea Lang, 1948
Family Peltidiidae Sars, 1904
Family Porcellidiidae Sars, 1904
Family Clytemnestridae Scott, 1909
Family Tegastidae Sars, 1904
Family Tisbidae Stebbing, 1910
Infraorder Podogennonta Lang, 1948 Miocene–Recent
Superfamily Ameiroidea Lang, 1948 Recent
Family Ameiridae Monard, 1927
Family Canthocamptidae Sars, 1906
Family Cylindropsyllidae Sars, 1909
Family Louriniidae Monard, 1927
Family Paramesochridae Lang, 1948
Family Parastenocarididae Chappuis, 1933
Family Tetragonicipitidae Lang, 1944
Superfamily Cletodoidea Lang, 1948 Miocene–Recent
Family Ancorabolidae Sars, 1909 Recent
Family Cletodidae T. Scott, 1904 Miocene–Recent
Family Laophontidae T. Scott, 1904 Recent
Superfamily Metoidea Lang, 1948 Recent
Family Metidae Sars, 1910
Superfamily Thalestroidea Lang, 1948 Recent
Family Balaenophilidae Sars, 1910
Family Diosaccidae Sars, 1906
Family Miraciidae Dana, 1846
Family Parastenheliidae Lang, 1936
Family Thalestridae Sars, 1905
Infraorder incertae sedis Recent
Family Gelyellidae Rouch and Lescher–Moutoué, 1977
Suborder Incertae Sedis Recent
Family Latiremidae Bozic, 1969
Order Mormonilloida Boxshall, 1979 Recent
Family Mormonillidae Giesbrecht, 1892
Order Monstrilloida Sars, 1903 Recent
Family Monstrillidae Giesbrecht, 1892
Family Thaumatopsyllidae Sars, 1913
Order Cyclopoida Burmeister, 1834 (?Miocene) Recent
Family Archinotodelphyidae Lang, 1949
Family Ascidicolidae Thorell, 1860
Family Botryllophyllidae Sars, 1921
Family Buproridae Thorell, 1859
Family Cyclopidae Dana, 1853
Family Cyclopinidae Sars, 1913

Family Doropygidae Brady, 1878
Family Enterocolidae Sars, 1921
Family Enteropsidae Aurivillius, 1887
Family Lernaeidae Cobbold, 1879
Family Namakosiramiidae Ho and Perkins, 1977
Family Notodelphyidae Dana, 1853
Family Oithonidae Dana, 1853
Family Schizoproctidae Aurivillius, 1887
Order Poecilostomatoida Thorell, 1859 Recent
Family Anomoclausiidae Gotto, 1964
Family Anomopsyllidae Sars, 1921
Family Bomolochidae Sumpf, 1871
Family Catiniidae Bocquet and Stock, 1957
Family Chondracanthidae Milne Edwards, 1840
Family Clausidiidae Embleton, 1901
Family Clausiidae Giesbrecht, 1895
Family Corallovexiidae Stock, 1975
Family Corycaeidae Dana, 1852
Family Cucumaricolidae Bouligand and Delamare–Deboutte-
ville, 1959
Family Echiurophilidae Delamare–Deboutteville and Nunes–
Ruivo, 1955
Family Ergasilidae Nordmann, 1832
Family Eunicicolidae Sars, 1918
Family Gastrodelphyidae List, 1890
Family Lamippidae Joliet, 1882
Family Lichomolgidae Kossmann, 1877
Family Mantridae Leigh–Sharpe, 1934
Family Myicolidae Yamaguti, 1936
Family Mytilicolidae Bocquet and Stock, 1957
Family Nereicolidae Claus, 1875
Family Oncaeidae Giesbrecht, 1892
Family Pharodidae Illg, 1948
Family Philoblennidae Izawa, 1976
Family Philichthyidae Bassett–Smith, 1899
Family Phyllodicolidae Delamare–Deboutteville, 1960
Family Pseudanthessiidae Humes and Stock, 1972
Family Rhynchomolgidae Humes and Ho, 1967
Family Sabelliphilidae Gurney, 1927
Family Sapphirinidae Thorell, 1859
Family Sarcotacidae Yamaguti, 1963
Family Serpulidicolidae Stock, 1979
Family Shiinoidae Cressey, 1975
Family Splanchnotrophidae Norman and Scott, 1906
Family Staurosomatidae Ardeev and Ardeev, 1975

Family Synaptiphilidae Bocquet and Stock, 1957
Family Taeniacanthidae Wilson, 1911
Family Telsidae Ho, 1967
Family Tuccidae Vervoort, 1962
Family Urocopiidae Humes and Stock, 1972
Family Vahiniidae Humes, 1966
Family Ventriculinidae Leigh–Sharpe, 1934
Family Xarifidae Humes, 1960
Order Siphonostomatoida Thorell, 1859 Lower Cretaceous–Recent
Family Atrotrogidae Brady, 1880 Recent
Family Asterocheridae Giesbrecht, 1899 Recent
Family Brychiopontiidae Humes, 1974 Recent
Family Caligidae Burmeister, 1835 Recent
Family Calvocheridae Stock, 1968 Recent
Family Cancerillidae Giesbrecht, 1897 Recent
Family Cecropidae Dana, 1852 Recent
Family Dichelesthiidae Dana, 1853 Lower Cretaceous–Recent
Family Dinopontiidae Murnane, 1967 Recent
Family Dirivultidae Humes and Dojiri, 1980 Recent
Family Dissonidae Yamaguti, 1963 Recent
Family Dyspontiidae Giesbrecht, 1895 Recent
Family Entomolepidae Brady, 1899 Recent
Family Eudactylinidae Yamaguti, 1963 Recent
Family Euryphoridae Wilson, 1905 Recent
Family Hatschekiidae Kabata, 1979 Recent
Family Herpyllobiidae Hansen, 1892 Recent
Family Hyponeoidae Heegaard, 1962 Recent
Family Kroyeriidae Kabata, 1979 Recent
Family Lernaeopodidae Olsson, 1869 Recent
Family Lernanthropidae Kabata, 1979 Recent
Family Megapontiidae Heptner, 1968 Recent
Family Micropontiidae Gooding, 1957 Recent
Family Myzopontiidae Sars, 1915 Recent
Family Nanaspididae Humes and Cressey, 1959 Recent
Family Naobranchiidae Yamaguti, 1939 Recent
Family Nicothoidae Dana, 1852 Recent
Family Pandaridae Milne Edwards, 1840 Recent
Family Pennellidae Burmeister, 1835 Recent
Family Pontoeciellidae Giesbrecht, 1895 Recent
Family Pseudocycnidae Wilson, 1922 Recent
Family Rataniidae Giesbrecht, 1897 Recent
Family Saccopsidae Lützen, 1964 Recent
Family Sphyriidae Wilson, 1915 Recent
Family Spongiocnizontidae Stock and Kleeton, 1964 Recent
Family Stellicomitidae Humes and Cressey, 1958 Recent
Family Tanypleuridae Kabata, 1969 Recent

Family Trebiidae Wilson, 1932 Recent
Family Xenocoelomatidae Bresciani and Lützen, 1966 Recent
Order uncertain
Family Antheacheridae M. Sars, 1870
Family Mesoglicolidae Zulueta, 1911
Family Sponginticolidae Topsent, 1928
Family Staurosomitidae Zulueta, 1911

REFERENCES

Beklemishev, C. W. 1961. Superfluous feeding in marine herviborous zooplankton. *Int. Council Expl. Sea, Symp. Zooplankton*, no. 8 (manifold).

Blades, P. I. 1977. Mating behavior of *Centropages typicus. Mar. Biol.* **40**:57-64.

Blades, P. I., and M. J. Youngbluth. 1977. Mating behavior of *Labidocera aestiva. Mar. Biol.* **51**:339–55.

Borradaile, L. A. 1926. Notes upon crustacean limbs. *Ann. Mag. Nat. Hist.* (9)**17**:193–213.

Borutskii, E. V. 1964. *Fauna of USSR, Crustacea, Freshwater Harpacticoida.* Israel Prog. Sci. Transl., Jerusalem.

Boxshall, G. A. 1980. Community structure and resource partitioning—the plankton. In *The Evolving Biosphere* (P. L. Forey, ed.), pp. 143–56. Cambridge Univ. Press, Cambridge.

Boxshall, G. A. 1982. On the anatomy of the misophrioid copepods, with special reference to *Benthomisophria palliata. Phil. Trans. Roy. Soc. Lond.* (B)**297**:125–81.

Boxshall, G. A. 1983. Three new genera of misophrioid copepods from the near-bottom plankton community in the North Atlantic Ocean. *Bull. B.M.(N.H.) Zool.* **44**:103–24.

Boxshall, G. A., F. D. Ferrari, and H. Tiemann. 1984. The ancestral copepod: towards a consensus of opinion at the First International Conference on Copepods. *Crustaceana*, Suppl. **7**:68–84.

Calman. W. T. 1909. Crustacea. In *Treatise on Zoology* (E. R. Lankester, ed.), Vol. 7. Adam & Charles Black, London.

Cannon, H. G. 1928. On the feeding mechanism of the copepods *Calanus finmarchicus* and *Diaptomus gracilis. Brit. J. Exp. Biol.* **6**:131–44.

Cressy, R., and C. Patterson. 1973. Fossil parasitic copepods from a Lower Cretaceous fish. *Science* **180**:1203–85.

Cushing, D. H. 1959. On the nature of production in the sea. *Fish. Invest. Lond.* (2)**22**:1–40.

Damkaer, D. M. 1975. Calanoid copepods of the genera *Spinocalanus* and *Minocalanus* from the Central Arctic Ocean, with a review of the Spinocalanidae. NOAA Tech. Rept. NMFS Circ. 391.

Dietrich, W. 1915. Die Metamorphose der frielebenden Süsswassencopepoden. *Zeit. f. wiss. Zool.* **113**:252–323.

Elofsson, R. 1966. The nauplius eye and frontal organs of the non-Malacostraca. *Sarsia* **25**:1–128.

Fahrenbach, W. H. 1962. The biology of a harpacticoid copepod. *La Cellule* **62**:302–76.

Fryer, G. 1957. The feeding mechanism of some freshwater cyclopoid copepods. *Proc. Zool. Soc., Lond.* **1957**:1–25.

Fuchs, K. 1914. Die Keimblätterentwicklung von *Cyclops viridis. Zool. Jahrb. Anat.* **38**:103–56, 39 figs.

Gauld, D. T. 1957. Copulation in calanoid copepods. *Nature* **180**:510.

Giesbrecht, W. 1892. Systematik und Faunistik der pelagischen Copepoden des Golfes von Neapel und der angrenzenden Meeres-Abschnitte. *Fauna u. Flora Neapel Monog.* **19**:1–831.

Gotto, R. V. 1979. The association of copepods with marine invertebrates. *Adv. Mar. Biol.* **16**:1–109.

Grobben, C. 1881. Die Entwicklungsgeschichte von *Cetochilus septemtrionalis*. *Arb. Zool. Inst. Wien* **3**:243–82, 4 pls.

Gurney, R. 1931–33. British Freshwater Copepoda, Vols. I–III. Ray Society, London.

Hopkins, C. 1982. The breeding biology of *Euchaeta norvegica* in Loch Etive, Scotland. *J. Exp. Mar. Biol. Ecol.* **60**:71–102.

Humes, A. G., and M. Dojiri. 1980. A siphonostome copepod associated with a vestimentiferan from the Galapagos Rift and the East Pacific Rise. *Proc. Biol. Soc. Wash.* **93**:697–707.

Itô, T. 1982. The origin of biramous copepod legs. *J. Nat. Hist.* **16**:715–26.

Jacoby, C. A., and M. J. Youngbluth. 1983. Mating behavior in three species of *Pseudodiaptomus*. *Mar. Biol.* **76**:77–86.

Johnson, M. W. 1937. The developmental stages of the copepod *Eucalanus elongtus* var. *bungii*. *Trans. Am. Micro. Soc.* **56**:79–98.

Kabata, Z. 1979. *Parasitic Copepoda of British Fishes*. Ray Society, London.

Kabata, Z. 1981. Copepoda parasitic on fishes: problems and perspectives. *Adv. Parasitol.* **19**:1–71.

Kaestner, A. 1970. *Invertebrate Zoology*, Vol. III. Interscience, New York.

Koehl, M. A. R., and J. R. Strickler. 1981. Copepod feeding currents: food capture at low Reynolds number. *Limnol. Oceanogr.* **26**:1062–73.

Lowe, E. 1935. On the anatomy of a marine copepod, *Calanus finmarchicus*. *Trans. Roy. Soc. Edin.* **58**:561–603.

Lowndes, A. G. 1935. The swimming and feeding of certain calanoid copepods. *Proc. Zool. Soc. Lond.* **1935**:687–715.

Manton, S. M. 1977. *The Arthropoda*. Oxford Univ. Press, London.

Marcotte, B. M. 1977. An introduction to the architecture and kinematics of harpacticoid feeding: *Tisbe furcata*, *Mikrof. Meersb.* **61**:183–96.

Marcotte, B. M. 1982. Part 2: Copepods. In *The Biology of Crustacea*, Vol. I (L. G. Abele, ed.), pp. 185–97. Academic Press, New York.

Marshall, S. M., and A. P. Orr. 1955. *The Biology of a Marine Copepod*. Oliver & Boyd, Edinburgh.

Noodt, W. 1972. Ecology of the Copepoda. *Smith. Contr. Zool.* **76**:97–102.

Paffenhöfer, G. A., J. R. Strickler, and M. Alcaraz. 1982. Suspension feeding by herbivorous calanoid copepods: a cinematographic study. *Mar. Biol.* **67**:193–99.

Palmer, A. R. 1969. Copepoda. In *Treatise on Invertebrate Paleontology*, Part R, *Arthropoda* **4**(1) (R. C. Moore, ed.), pp. R200–R203. Geol. Soc. Am. and Univ. Kansas Press, Lawrence.

Park, T. S. 1966. The biology of a calanoid copepod *Epilabidocera amphitrites*. *Cellule* **66**:129–251, pls. 1–10.

Parker, G. H. 1891. The compound eyes of crustaceans. *Bull. Mus. Comp. Zool.* **21**:45–141, 10 pls.

Sandercock, G. A. 1967. A study of selected mechanisms for the coexistence of *Diaptomus* species in Clarke Lake, Ontario. *Limnol. Oceanogr.* **12**:97–112.

Sars, G. O. 1901–11. An Account of the Crustacea of Norway. Bergen.

Thorell, T. T. T. 1859. Bidrag till kannedomen om krustaceer som levfa i arter of slägtet *Ascidia*. *K. svenska Vetensk-Akad. Handl.* **3**:1–84.

van der Spoel, S. 1983. Patterns in plankton distribution and their relation to speciation. The dawn of pelagic biogeography. In *Evolution in Time and Space: The Emergence of the Biosphere* (R. W. Simms, J. H. Price, and P. E. S. Whalley, eds.). Academic Press, London.

van der Spoel, S., and A. C. Pierrot–Boults. 1979. *Zoogeography and Diversity of Plankton*. Bunge, Utrecht.

38

RHIZOCEPHALA

DEFINITION   Parasites; lacking as adults any appendages, traces of segmentation, and all internal organs except for gonads and degenerate remnants of the nervous system; four nonfeeding naupliar larval stages, with caudal rami, frontolateral horns, and frontal filaments; cyprid larva lacking mouthparts and segmented abdomen.

HISTORY   Though as adults rhizocephalans bear no similarities at all to any arthropods, they were nonetheless typed quite early as crustaceans when Cavolini recognized in 1787 their nauplius larvae. And Thompson astutely identified them as highly modified cirripedes in 1836. However, detailed knowledge of the biology of rhizocephalans has only slowly come to light. The first, and very effective, review of the group was by Delage (1884), who at that time elucidated the complete life cycle and internal anatomy, confirmed the male phase, and described the kentrogon of the forms he studied. The exact nature of sexuality in the group, however, was not understood until Ichikawa and Yanagimachi (1958). Though the dichotomy between kentrogonids and akentrogonids was formally recognized by Häfele (1911), the general acceptance of such an arrangement has only slowly come to prevail and considerable uncertainty still prevails as to whether all akentrogonids truly lack a kentrogon.

LIFE CYCLE   The kentrogonid rhizocephalans have one of the most distinctive life cycles of any animal parasite (Fig. 38-1). Both Yanagamachi (1961) and Ritchie and Høeg (1981) report that a single externa can produce either large eggs or small eggs, which in turn develop into large larvae or small larvae. The large larval forms settle as males on juvenile externa, while the small larvae settle on a host and develop into the endoparasitic females. Though the rhizocephalan nauplii have frontolateral horns like other cirripedes, they also have a pair of terminal caudal rami rather than a long stylet (Fig. 38-2A).

In the Kentrogonida, the small-bodied female cyprid (Fig. 38-2D) attaches to a host (Fig. 38-2C, D, E) and develops into a kentrogon. This is an attachment stage with a chitinous sac surrounding the remnants of the cyprid tissue. The kentrogon develops a stylet in the region of the mouth-field within 20 to 40 hours after attachment (Ritchie and Høeg, 1981). Delage (1884) observed that the stylet in *Sacculina* was arranged to pierce the host through the 'cyprid antennule,' but Ritchie and Høeg (1981)

473

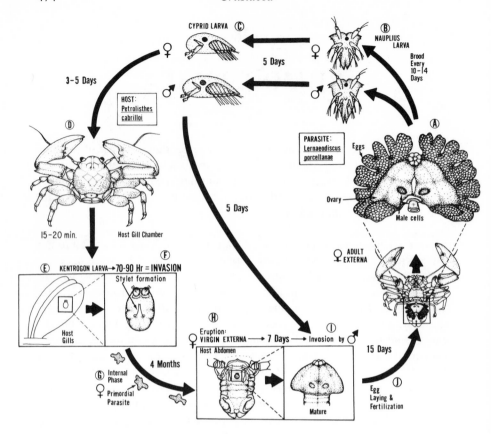

Fig. 38-1. Life cycle of the kentrogonid rhizocephalon, *Lernaeodiscus porcellanae*, parasitic on the porcelain crab *Petrolisthes cabrilloi*. For details see text. (From Ritchie and Høeg, 1981)

recorded that the stylet of *Lernaeodiscus* penetrated the host tissue directly. In 70 to 90 hours after attachment the tissue mass within the kentrogon sac is injected into the host (Fig. 38-2F) to form the interna, a root-like system centering on the gut of the host.

What happens immediately thereafter is not clear. Eventually a bud or 'tumor' is formed within the abdomen beneath where the reproductive sac will eventually erupt. This bud is the externa in rudimentary form (Fig. 38-3A, B). The host is effectively taken over 'body and soul' by the rhizocephalan, such that when the externa erupts it will be perceived as 'self' and not be subject to other than normal host-grooming attention. The internal phase seems to persist for at least two to four months before the eruption of the externa on the surface of the host.

Eventually the rudimentary bud begins to grow (Fig. 38-3C) and presses up against the host cuticle. This causes necrosis of the host epidermis, such that after the molt that soon follows, the parasite erupts onto the surface of the host as an externa (Fig. 38-3D). This initial external stage remains in an immature condition until a large-bodied male cyprid attaches to the externa near its mantle opening. The tissues of the male are then injected into the externa (Fig. 38-4) and migrate to a receptacle wherein they undergo spermatogenesis. The 'hyperparasitized' externa then matures and within two weeks produces eggs. The host behaves as if the egg mass were its own to the extent that even male hosts become feminized and neatly tend to the brooding egg mass. A single externa can produce several broods, apparently all fertilized by the output from a single male. However, Delage (1884) reported that as a result of sperm depletion the externa of *Sacculina* is periodically shed, and a new immature externa erupts to start the reproductive phase anew. The host is unaffected by all this, except for its castration, and is parasitized for life. Ritchie and Høeg (1981) report that infected *Petrolisthes cabrilloi* were kept in the laboratory for over two years with both host and parasite doing well in all that time.

The akentrogonids have a more direct life cycle. The female cyprid attaches to the host at the site of infection and develops directly into the parasitic stage. An internal root system is developed prior to the complete formation of the externa at the site of infection. However, this pattern has only been confirmed for one rhizocephalon, *Chthamalophilus delagei* (Bocquet–Védrine, 1961). Such a 'direct' life cycle makes the parasite highly

Fig. 38-2. Larval stages of *Sacculina carcini*. (A) nauplius; (B) cyprid; (C) attachment of cyprid and molt; (D) initiation of kentrogon; (E) kentrogon; (F) injection of kentrogon tissue into the host. (From Delage, 1884)

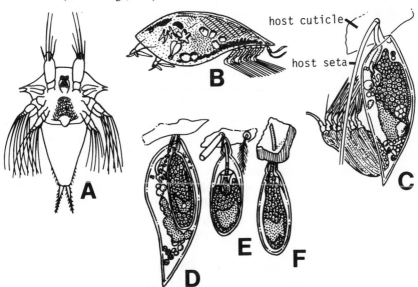

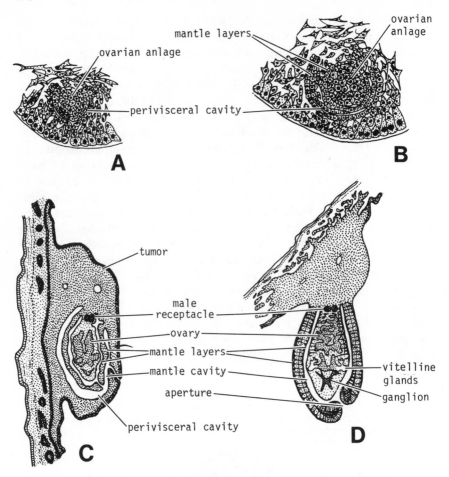

Fig. 38-3. Development of interna of *Sacculina carcini*. (A) initiation of interna bud or tumor; (B) older tumor, dormant under site of future external eruption; (C) interna preparing to erupt; (D) early external phase. (From Delage, 1884)

susceptible to removal by host grooming until the host is effectively 'taken over.' As a result akentrogonids are known largely from groups with poorly developed grooming behaviors, such as barnacles, isopods, and some carideans in addition to the more typical reptantians. The kentrogonids, on the other hand, have been able to exploit infestations of reptantians, because the small, smooth, low profile of the kentrogon resists removal by grooming, early development is internal, and the externa mimics the egg mass of the host and is 'cared for' as such.

TAXONOMY   The classification of the rhizocephalans is not particularly pelucid. The kentrogonids have been divided into a series of families, some

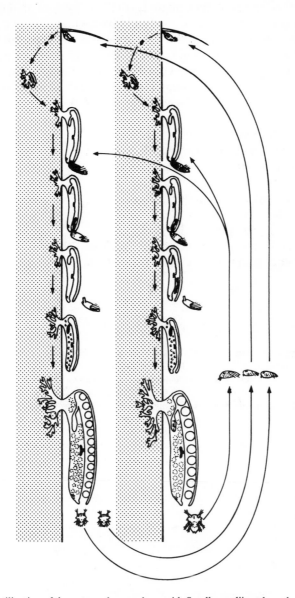

Fig. 38-4. Fertilization of the externa by a male cyprid. Small nauplii settle and eventually give rise to erupted female externa of two kinds; one (on the left) produces small nauplii; the other (on the right) produces large male nauplii. (Modified from Yanagimachi, 1961)

of them monotypic. Most of the genera placed within the akentrogonids are there only tentatively, awaiting confirmation that they do in fact lack a kentrogon. This apparent lack might seem likely for genera like *Duplorbis* and *Microgaster*, parasites of isopods and barnacles respectively, less certain for *Mycetomorpha*, a parasite on carideans. The apparent preference of *Thompsonia* for parasitizing reptantians, however, might indicate that this genus may yet prove to have a kentrogon. Complicating the whole issue of rhizocephalan classification is the general consensus that the group is possibly polyphyletic (e.g., Newman et al., 1969).

Suborder Kentrogonida Delage, 1884
    Family Clistosaccidae Boschma, 1928
    Family Lernaeodiscidae Boschma, 1928
    Family Peltogastridae Lilljeborg, 1861
    Family Sacculinidae Lilljeborg, 1861
    Family Sylonidae Boschma, 1928
Suborder Akentrogonida Häfele, 1911
    Family Chthamalophilidae Bocquet–Védrine, 1957
    A number of genera of uncertain familial affinities

REFERENCES

Bocquet–Védrine, J. 1961. Monographie de *Chthamalophilus delagei*, rhizocephale parasitede *Chthamalus stellatus*. *Cah. Biol. Mar.* **2**:455–593.
Delage, Y. 1884. Evolution de la sacculine (*Sacculina carcini*) Crustacé endoparasite de l'ordre nouveau. *Arch. Zool. Exp. Gen.* (2)**2**:417–736, pls. 22–30.
Häfele, F. 1911. Anatomie und Entwicklung eines neuen Rhizocephalen, *Thompsonia japonica*. *Beiträge zur Naturgesch. Ostasiens, Abhl. math.-phys. (K.K.) Bayer. Akad. Sci.* Suppl. 2. **7**:1–25.
Ichikawa, A., and R. Yanagimachi. 1958. Studies on the sexual organization of the Rhizocephala, I. The nature of the 'testes' of *Peltogasterella socialis*. *Annot. Zool. Japon.* **31**:82–96.
Newman, W. A., V. Zullo, and T. H. Withers. 1969. Cirripedia. In *Treatise on Invertebrate Paleontology,* Part R, *Arthropoda* **4**(1) (R. C. Moore, ed.), pp. R206–95. Geol. Soc. Am. and Univ. Kansas Press, Lawrence.
Ritchie, L. E., and J. T. Høeg. 1981. The life history of *Lernaeodiscus porcellanae* and coevolution with its porcellanid host. *J. Crust. Biol.* **1**:334–47.
Yanagimachi, R. 1961. The life cycle of *Peltogasterella*. *Crustaceana* **2**:183–96.

39

ASCOTHORACIDA

DEFINITION Carapace bivalved, typically developed as a firm or fleshy mantle enveloping most if not all of body; antennules prehensile in adult and cyprid larva, frequently raptorial or subchelate; antennae absent or greatly reduced beyond nauplius stages; with oral pyramid or cone formed from labrum, mouthparts modified for piercing and sucking; up to six pairs of thoracopods, primitively biramous, paddlelike, often uniramous or absent; abdomen with at least four or five segments, terminal anal somite with well-developed caudal rami; female gonopores on first thoracopods, penis or rudiment on first abdominal segment.

HISTORY The first ascothoracid, *Laura gerardiae*, was described by Lacaze–Duthiers in 1865, who also established the distinct status of Ascothoracida in 1880. Through most of their history, most workers have regarded the group as part of the cirripedes. However, Vagin (1937, 1947a, 1976) has consistently advocated an independent position for ascothoracids outside of the Cirripedia *sensu stricto*. Such an arrangement has been supported by Grygier (1982, 1983c, 1984) as result of formal analyses of characters, though Boxshall (1983) opposes their separation from Cirripedia. Grygier (1984) is the most recent complete review of the group's morphology, development, and taxonomy.

MORPHOLOGY These animals are ecto- and endoparasites of echinoderms and anthozoan coelenterates. The body is enveloped by a carapace mantle (Fig. 39-1A, B). In males the carapace encloses the thorax and anterior parts of the abdomen, while in females it envelops the entire body and also acts as a brood chamber. In the laurids, some synagogids, and *Ulophysema*, the margins of the mantle fuse, except for a small opening, to completely encapsulate the female; in *Dendrogaster* the mantle is greatly elaborated for respiratory purposes (Fig. 39-1C). Adults are blind, but a few species have been observed to have either a functional nauplius eye in the nauplius or a compound eye in the cyprid (Grygier, 1983a).

The prehensile antennules (Fig. 39-2A) of the cyprid and adult can have as few as two or as many as six segments, though four is the most common. The joints are highly sclerotized, with the terminal segments typically rather spinose and forming subchelae. The muscular system within the limb is highly developed to facilitate prehensility.

The antennae were generally thought to be absent after the nauplius

479

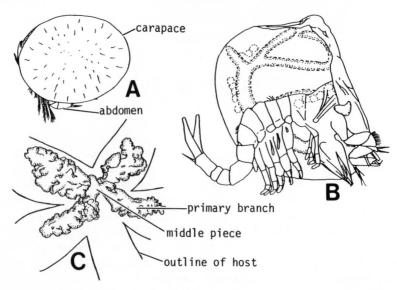

Fig. 39-1. Body forms of ascothoracicans. (A) *Synagoga mira*, external view of whole animal; (B) *S. normani*, whole animal with right mantle–carapace valve removed and setae omitted; (C) *Dendrogaster fisheri*, external view of female in host starfish showing great elaboration of mantle. (A, B from Grygier, 1983b; C from Grygier, 1981c)

stage. Grygier (1981b) felt that a greatly reduced second antenna could have been present in at least some ascothoracids, though he now feels (Grygier 1984, personal communication) that what he thought were antennae are really frontal filaments between the antennules and labrum. In adult females (well developed in protanders), the antennae (Fig. 39-2B) are simple biramous structures with little or no jointing evident in the rami. A basal appendix arises from the ventral ramus. In the protander stages of *Waginella* and *Gorgonolaureus muzikae* the rami are highly setose and possibly jointed.

The oral pyramid or oral cone is a specialized arrangement of the mouthparts. The labrum (Fig. 39-2C) typically forms an enveloping sheath, which may or may not be fused posteriorly. Within the sheath are the simple suctorial mandibles (Fig. 39-2D) and the bladelike maxillules (Fig. 39-2E). These former are reported by Vagin to be broad at the base but are folded and formed into a slender, grooved tube distally. Grygier (1983d) illustrated setose, brushlike mandibles in *Ascothorax*. The mandible and maxillules are capable of being protruded outside the labral sheath. The maxillae (Fig. 39-2C) are long and extend beyond the sheath. They are usually fused along most of their length and, thus, form a posterior wall to the oral cone. Their distal ends remain free, are almost harpoonlike, and bifid.

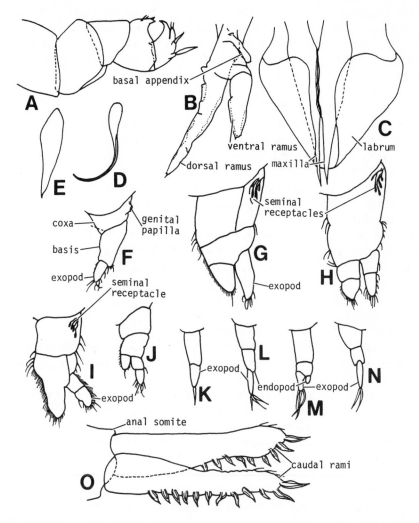

Fig. 39-2. Appendages of *Ascothorax ophioctenis*. (A) antennule; (B) antenna (from *Gorgonolaureus muzikae*); (C) oral pyramid; (D) mandible; (E) maxillule; (F, G, H, I, J) female thoracopods; (F) first; (G) second; (H) third and fourth; (I) fifth; (J) sixth; (K, L, M, N) male thoracods; (K) first; (L) second and fifth; (M) third and fourth; (N) sixth; (O) caudal rami. (From Vagin, 1947, except B from Grygier, 1981b)

481

The thoracic appendages, except for the first, are biramous paddles. They can be better formed in females (Fig. 39-2F, G, H, I, J) while in males they can be reduced (Fig. 39-2K, L, M, N), but more often than not they are generally similar between sexes. The first thoracopod is sometimes uniramous and in females bears a prominent genital papilla laterally (Fig. 39-2F). The second through fifth limbs have rami that may be either virtually vestigial, as in *Ascothorax*, or be well developed and composed of one to three joints. The sixth limb can be uniramous (Fig. 39-2N) or weakly biramous (Fig. 39-2J).

The segments of the abdomen lack limbs, except for the penis in males. The terminal segment bears a dorsal or terminal anus and a pair of posteriorly directed, well-developed, typically setose caudal rami.

The digestive system has some distinctive specializations related to the parasitic life-style. The foregut is developed as a very muscular esophagus with circular and longitudinal muscles as well as dilator muscles. The esophagus protrudes into the large midgut. Arising laterally from the midgut are the distinctive digestive caeca, up to three pairs in females and two pairs in males. These are highly branched, especially in females, and extend out into the folds of the mantle. The hindgut extends throughout the abdomen and terminates in the dorsal or terminal anus.

There is no organized circulatory system. Blood sinuses are scattered about the body but are especially prominent around the gut and gonads. Hemocytes have been noted in these sinuses, but no organized circulatory pattern of fluid has ever been observed.

Excretion is achieved with maxillary glands. These are well-developed organs with large saclike reservoirs emptying to the outside by a pair of ducts on the base of the maxillae.

Ascothoracicans are dioecious, with distinct sexual dimorphism, except for the petrarcids, which are simultaneous hermaphrodites. The smaller males may range from exhibiting complete independence of the female, though living in some proximity to her, to being much modified and living as a parasite in the female mantle cavity.

The testes are paired, though Vagin (1947a) reported a single unpaired organ in *Ascothorax ophioctenis*. In either case extensive testicular lobes extend into the mantle from the main organ. Vasa efferentia arise near the lobe bases and merge into vasa deferentia. These latter ducts open on the sternite of the first abdominal segment by means of a penis, which in the genus *Petrarca* is larger than the abdomen itself. The flagellate sperm is apparently of a rather generalized, primative type (Grygier, 1981c; 1982).

The ovaries are paired and sometimes united anteriorly and, like the testes, are laterally extended into the mantle, occupying the entire posterior portion in *Ascothorax*. The sinuous oviducts extend anteriad and empty to the outside on conical genital papillae most often right above the coxa but occasionally on the posterolateral surfaces of the first thoracopod coxae. Seminal receptacles can be found usually in the coxae of thoraco-pods 2 through 5, though they are lacking in *Dendrogaster*.

The nervous system is quite reduced. There is a large brain or supraeso-phageal ganglion, which gives rise anteriorly to a set of large antennular nerves. These nerves have a large ganglion along their length in the basal-most segments of the antennules. Circumesophageal commissures connect the brain to a large subesophageal ganglionic mass from which mantle, buccal, and thoracic nerves arise. The subesophageal mass is continuous posteriorly with a thoracic mass with no evident ganglia along its length.

The muscular system is well developed in ascothoracicans (Grygier, 1984). Of special note, however, is the large carapace or mantle adductor muscle located posterior to the esophagus and ventral to the midgut just over the subesophageal ganglion (Fig. 39-3).

NATURAL HISTORY   Except for the free-swimming species *Synagoga mira*, all ascothoracids are obligate parasites permanently resident on other invertebrates. The synagogoidids are generally found in or on echinoderms (though *S. mira* and *S. normani* occur on anthozoans) while the lauroidids infest anthozoan coelenterates. Although it is widely assumed that the oral pyramid and muscular esophagus serve together to actively suck in host tis-sues and fluids, this may not always be the case. The normal modes of feed-ing are variations on a single theme. The synagogids insert their oral pyramids into the permanent host (or temporary host as is the case in *S. mira*) and suck fluids into the gut. Grygier (1981b) once suggested the possibility of filter feeding for *Gorgonolaureus*, but now no longer feels this is so (Grygier, personal communication). Dendrogastrids live in echino-derm visceral cavities and may subsist, at least in part, on cells of the host that adhere to the mantle opening, which are apparently collected by the antennules (Vagin, 1976). The laurid females are endoparasites in antho-zoans while the males are ectoparasites of the host. Petrarcids live in galls

Fig. 39-3. Internal anatomy of *Ascothorax ophioctenis*. (From Vagin, 1947a)

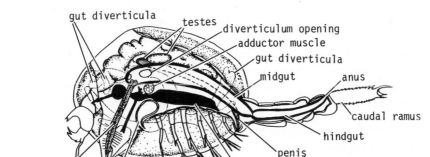

or cysts in scleractinian corals and feed on the surrounding host tissues, and there must be at least two individuals to a gall in order to achieve reproduction.

Locomotion, except in the larval states, is limited once the host is attained. Only *S. mira* and *S. normani* are free-swimming throughout their life cycle, functioning as an ectoparasite only intermittently.

Modes of reproduction in ascothoracidans are not entirely clear. In most forms, copulation occurs once with storage of sperm in the seminal receptacles. In *Dendrogaster*, because of the symbiotic nature of the male with the female, there is no need apparently for seminal receptacles.

All variations on the normal gonochoristic sexual theme exist among ascothoracidans. *Petrarca* and *Introcornia* species are functional hermaphrodites (Grygier, 1983e). The greatly enlarged penes are incapable of achieving self-fertilization, however, and the inevitable existence of individual pairs of *Petrarca* in a host is probably related to obligate cross-fertilization (Gryger, 1981a). Protandric hermaphroditism is indicated in *Gorgonolaureus muzikae* (Grygier, 1981b) and *Isidascus bassindalei* (Moyse, 1983), where distinct 'protander' stages appear to precede a functional female condition. Protandry has also been suggested for species of *Waginella* (Newman, 1974; Grygier, 1983b). In *Ascothorax* and *Parascothorax* (Vagin, 1976) sex is environmentally determined. The first individuals in a bursal cavity of the host are always females, while subsequent arrivals become males.

Currently, over 58 species in 14 genera of ascothoracidans are recognized. The group is ubiquitous as a whole, though because it is currently subject to intense study and revision (Grygier, 1984), no meaningful comments can yet be made on the biogeography of the group. Vagin (1976) presents a summary of then current knowledge with some general speculations as to the group's biogeographic history, but as shown by Grygier (1983d) Vagin's predictions concerning distributions have their limits.

DEVELOPMENT   Knowledge of ascothoracidan ontogeny is uneven. Some attention has been paid to early cleavage patterns and the sequence of events in the larvae, but nothing is known of what goes on in between.

Vagin (1947b, 1976) elucidated and summarized early cleavage in ascothoracids. Early divisions are complete but unequal, eventually resulting in macro- and micromeres. It is a modified type of spiral cleavage and is similar to that seen in barnacles as well as some copepod groups.

Patterns of larval development vary. Eggs are shed, fertilized as they pass the seminal receptacles, and are brooded in the mantle chamber. Nauplii lack the frontolateral horns seen in the rhizocephalans and cirripedes *sensu stricto*. Hatching can occur at any stage in development; for example in *Ulophysema* eggs hatch as nauplii (Brattström, 1948), in *Ascothorax* they hatch as metanauplii (Vagin, 1976), while in *Dendrogaster* they sometimes hatch as an ascothoracid or cyprid larva. In addition, the

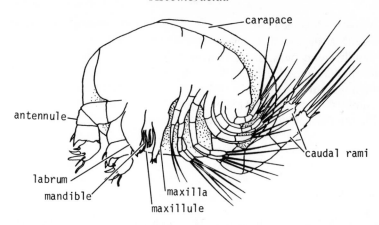

Fig. 39-4. Diagram of the ascothoracid or cyprid larva of *Ascothorax ophioctenis*. (From Vagin, 1947a)

number of molts within each stage is quite variable (Vagin, 1976), but a hypothetically complete sequence would contain two naupliar, six meta-naupliar, and two ascothoracid or cyprid stages. However, Grygier (1984) maintains that he has been able to identify up to five naupliar instars in some species. All stages can pass inside the mother's mantle cavity, even the ascothoracids, which remain with the parent for some time before swimming off in search of a host. However, some forms release larvae: laurids shed nauplii, and *Parascothorax* never harbors ascothoracids. Vagin (beginning in 1937 and continuing from then on) argued against applying the term cyprid to the ascothoracid larva, but except in particulars of antennules, mouthparts, and abdomen the general body form of the larva is cypridlike (Fig. 39-4).

FOSSIL RECORD   Traces have been found in Cretaceous echinoderms and anthozoans of what may be ascothoracidans. Voigt (1959, 1967) named some cysts, *Endosacculus moltkiae* and *E. najdini*, on the Maastrichtian octocorals *Moltkia minuta* and *Isis* sp., respectively. Similarly, holes similar to those made by *Ulophysema* have been noted (Madsen and Wolff, 1965) on the Upper Cretaceous irregular urchin *Echinocorys*.

TAXONOMY   Gruvel (1905) established four families and Thiel (1925) added another, but Vagin (1976) grouped these into two suborders. Relationships among these taxa, and even placement of specifc genera within families, are still in some state of flux (Grygier, 1983b, d, c), and Grygier (1984) indicates that some extensive revision on the higher level taxonomy of the group will be necessary.

Synagogoida Vagin, 1976
  Family Synagogidae Gruvel, 1905
  Family Dendrogasteridae Gruvel, 1905
Lauroidida Vagin, 1976
  Family Lauridae Gruvel, 1905
  Family Ctenosculidae Thiele, 1925
  Family Petracidae Gruvel, 1905

REFERENCES

Boxshall, G. A. 1983. A comparative functional analysis of the major maxillopodan groups. *Crust. Issues* **1**:121–44.
Brattström, H. 1948. Undersökningar över Öresund 33. On the larval development of the ascothoracid *Ulophysema öresundense*. Studies on *Olophysema öresundense* 2. *K. Fysiogr. Sällsk Handl. N.F.* **59**(5):1–10.
Gruvel, A. 1905. *Monographie des Cirrhipèdes ou Thécostracés*. Masson et Cie, Paris.
Grygier, M. J. 1981a. *Petrarca okadi*, a new crustacean from the Great Barrier Reef, the first shallow water record of the genus. *J. Crust. Biol.* 1:183–89.
Grygier, M. J. 1981b. *Gorgonolaureus muzikae* sp. nov. parasitic on a Hawaiian gorgonian, with special reference to its protandric hermaphroditism. *J. Nat. Hist.* **15**:1019–45.
Grygier, M. J. 1981c. Sperm of the ascothoracican parasite *Dendrogaster*, the most primitive found in Crustacea. *Int. J. Invert. Reprod.* **3**:65–73.
Grygier, M. J. 1982. Sperm morphology in Ascothoracida: confirmation of generalized nature and phylogenetic importance. *Int. J. Invert. Reprod.* **4**:323–32.
Grygier, M. J. 1983a. A novel, planktonic ascothoracid larva from St. Croix. *J. Plank. Res.* **5**:197–202.
Grygier, M. J. 1983b. Revision of *Synagoga*. *J. Nat. Hist.* **17**:213–39.
Grygier, M. J. 1983c. Ascothoracida and the unity of the Maxillopoda. *Crust. Issues* **1**:73–104.
Grygier, M. J. 1983d. *Ascothorax*, a review with descriptions of new species and remarks on larval development, biogeography, and ecology. *Sarsia* **68**:103–26.
Grygier, M. J. 1983e. *Introcornia conjugans*, parasitic in a Japanese hermatypic coral. *Senck. Biol.* **63**:419–26.
Grygier, M. J. 1984. Comparative Morphology and Ontogeny of the Ascothoracida, a Step Toward a Phylogeny of the Maxillopoda. Ph.D. dissertation. Scripps Inst. Oceanogr., Univ. California, San Diego.
Madsen, N. and T. Wolff. 1965. Evidence of the occurrence of Ascothoracica in the Upper Cretaceous. *Medd. Dansk. Geol. Foren.* **15**:556–58, pl. 1.
Moyse, J. 1983. *Isidascus bassindalei*, from the north-east Atlantic with a note on the origin of barnacles. *J. Mar. Biol. Assc. U.K.* **63**:161–80.
Newman, W. A. 1974. Two new deep-sea Cirripedia from the Atlantic. *J. Mar. Biol. Assc. U.K.* **54**:437–56.
Thiel, J. 1925. Prosobranchia. In *Handbuch der Zoologie* 5(1,1), (W. Kükenthal and T. Krumbach, eds.), pp. 40–94. de Gruyter, Berlin.
Vagin, V. L. 1937. Die Stellung der Ascothoracida ord. nov. im System der Entomostraca. *Dokl. Akad. Nauk SSSR* **15**:273–78.
Vagin, V. L. 1947a. *Ascothorax ophioctenis* and the position of Ascothoracida in the system of the Entomostraca. *Acta Zool. Stoch.* **27**:155–267.
Vagin, V. L. 1947b. Cleavage in Ascothoracida and its connection with the original type of cleavage in Arthropoda. *Dokl. Akad. Nauk SSSR* **55**:363–66.

Vagin, V. L. 1976. *Meshkogrudye Raki Ascothoracida.* Izdatelstvo Kazanskogo Univ., Kazan.

Voigt, E. 1959. *Endosacculus moltkiae* n.g. n.sp., ein vermutlicher fossiler Ascothoracide als Cystenbildner bei der Oktokoralle *Moltkia minuta. Palaönt. Zeit.* **33**:211–23, 2 pls.

Voigt, E. 1967. Ein vermutlicher Ascothoracide (*Endosacculus* ? *najdini*) als Bewohner einer kretazischen *Isis* aus der USSR. *Palaöntol. Zeit.* **41**:86–90.

40

THORACICA

DEFINITION   Adults highly metamorphosed, attached more or less permanently to substrate; carapace as mantle enclosing body and generally armored with calcareous plates; antennules (of cyprid larva) the organs of attachment; antennae typically absent except in nauplius; mandible as serrated blade, palp as setose lobe on labrum; trunk limbs biramous, annulate, setose cirri; posterior part of body vestigial; typically hermaphrodites, female opening on first thoracic segment, male intromittant organ (penis) on rudiment of abdomen. Larvae consist of six naupliar stages, nauplii with frontolateral horns and frontal filament; one postnaupliar cyprid stage.

HISTORY   Originally placed by Linnaeus among the Mollusca, the barnacles continued to be the property of malacologists long after Lamarck gave them their name, 'Cirrhipedes,' and Vaughan Thompson (1830) described the metamorphosis from the cyprid and placed them among crustaceans. It was not until Darwin's series of monographs on fossil and recent barnacles (1851a, b, 1854a, b) that the study of these animals was placed on a sound foundation and within the province of carcinologists. So effective were Darwin's volumes that they are still standard reference works today. Taxonomic knowledge of the group has advanced steadily since Darwin, with classic works being achieved by Aurivillius, Gruvel, Hoek, Newman, Nilsson-Cantell, Pilsbry, Withers, and Zevina.

MORPHOLOGY   Thoracicans come in essentially two flavors (Fig. 40-1): stalked (lepadomorph, Fig. 40-1A, B, C, D, E), and unstalked (balanomorph, Fig. 40-1G, H, and verrucomorph, Fig. 40-1F) forms. All are noted for the more or less extreme metamorphosis they undergo from the larval condition. The stalk or peduncle may or may not bear an armature of generally scalelike imbricating plates. The body or capitulum of the animals is usually protected by a well-armored mantle or carapace in which the discrete plates have characteristic arrangements. The primary plates include a single carina and paired scuta and terga (Fig. 40-1A). To these primary plates, a single rostrum and various paired lateral plates may also be added. The primary plates are derived from calcification centers in the carapace; the additonal plates are derived phylogenetically from peduncular plates shifted into the capitulum. As lateral plates became incorporated into the capitulum, the scuta and terga became restricted to functioning as guards to the opercular opening of the mantle (Fig. 40-1F, G, H). Not all

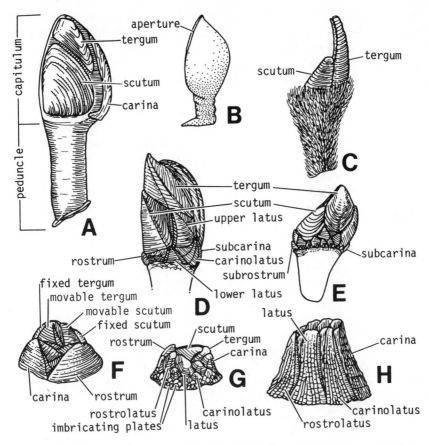

Fig. 40-1. Body forms of Thoracica. (A, B, C, D, E) Lepadomorpha; (A) *Lepas*, illustrating basic plate plan; (B) *Malacolepas*, an unplated form; (C) *Ibla*, (D) *Arcoscalpellum*; (E) *Pollicipes*. (F) *Verruca*, a verrucomorph. (G, H) Balanomorpha; (G) *Catomerus*, note imbricating plates around base; (H) *Semibalanus*. (Modified from Newman et al., 1969)

barnacles bear plates on the capitulum; several families, namely, the heterolepadids, koleolepadids, and malacolepadids (Fig. 40-1B) lack them. Nor is the line between stalked and unstalked forms a clear one, since some unstalked groups, for example, some chthamalids like *Catomerus* (Fig. 40-1G) or the extinct brachylepadids, have several rows of imbricating peduncular plates arranged around the base of the capitulum. Finally, though the barnacle body is generally more or less symmetrical, one group, the verrucomorphs (Fig. 40-1F), have a distinctly asymmetrical shell in which only a single scutum and tergum act as opercular valves, the other set being incorporated with the rostrum and carina to form the body wall.

The body of the barnacle is enclosed within the mantle cavity. The antennules may persist embedded in the cement at the point of attachment.

The mouthparts, or trophi (Fig. 40-2A), of the various barnacles, as Petriconi (1969) nicely demonstrated, are all very much alike. The labrum is large, frequently bullate, and bears a pair of setose lobes (Fig. 40-2B), the mandibular palps, which become dissociated in the course of metamorphosis from the jaws proper. The mandibles (Fig. 40-2C) are simple serrate blades.

The maxillules (Fig. 40-2D) are subrectangular lobes, their margins typically being armed with spinose setae and their sides variously setose.

The maxillae (Fig. 40-2E) are located along the midline and are oriented anteroposteriorly. They form a sort of 'lower lip,' and act to push food forward toward the mandibles and maxillules.

The thoracic limbs, or cirri, have two-segmented protopods and rami that are annulate and very setose. The cirral setae are specialized to assist various modes of feeding. Ctenopodous setae (Fig. 40-2H) are characteris-

Fig. 40-2. Appendages of *Calantica villosa*. (A) mouthfield; (B) mandibular palp; (C) mandibular gnathobase; (D) maxillule; (E) maxilla; (F) first cirrus; (G) fourth cirrus (from Anderson, 1983). (H, I, J) cirral setal types; (H) ctenopod; (I) lasiopod; (J) acanthopod (from Newman and Ross, 1971).

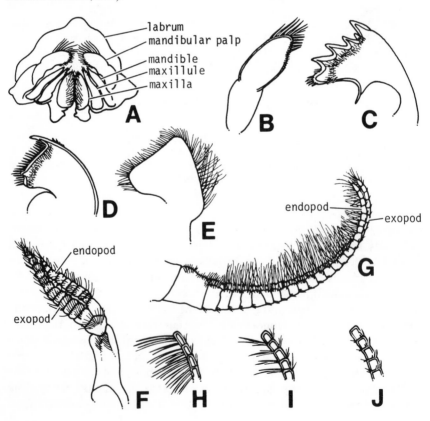

tic of filter feeders, lasiopodous setae (Fig. 40-2I) are found typically on large particle feeders, while acanthopodous setae (Fig. 40-2J) are a carnivorous adaptation. The first through third cirri (Fig. 40-2F) may be modified as maxillipedes, that is, they can be shorter than the posterior cirri, may have special setae, and be more closely associated with the trophi. Such maxillipedes serve to assist in the transfer of food from the posterior cirri to the mouthparts. The posterior cirri (Fig. 40-2G) when unfurled and extended through the aperture are food collectors.

In hermaphrodites and males, a penis is located posterior to the last pair of cirri. Darwin originally thought this structure an atrophied abdomen, but Newman et al. (1969) indicate probable homology with the penis cone on the first abdominal segment of ascothoracidans. The abdominal rudiment may support a pair of uni- or multiarticulate caudal appendages.

The gut is a sinuous tube. The esophagus, lined with cuticle, opens into the U-shaped midgut. The midgut is somewhat larger anteriorly than posteriorly, and a pair of caeca are given off laterally from the anterior portion. Absorption, however, occurs all along the midgut (Rainbow and Walker, 1978). The midgut empties into the narrow, cuticular lined hindgut.

The circulatory system of barnacles is among the most complex of any crustacean (Burnett, 1972, 1975, 1977). The system is in good part closed. The rostral vessel appears to be a vestige of the heart. Actually pumping, however, may be achieved by the rostral sinus (a remnant of the pericardium) in lepadids and balanomorphs, or three pairs of body muscles that contract the dorsolateral channels as in *Pollicipes*. The basic pattern of circulation can be illustrated by that of *Lepas*. Blood leaves the rostral sinus and goes into the peduncular vessel (Fig. 40-3A). From the pedicle plexus blood enters the mantle sinus, and then collects in one of the paired scutal vessels. Blood can then flow (Fig. 40-3B) either to the adductor muscle, gut, caeca, or maxillary glands. The gut circulation supplies blood to the cirri, penis, and oral cone by means of the epineural sinuses. Blood is collected (Fig. 40-3C) in a prosomal sinus, which receives blood from the gut, the cirri by way of the superior gastric vessel, and the testicular plexus. The prosomal sinus empties into the rostral vessel, which in turn empties into the rostral sinus along with blood from the adductor muscle and oral cone. In this entire system, blood rarely leaves true vessels or membrane-lined sinuses. The pattern noted in *Megabalanus* (Burnett, 1977) is similar to the above except that the latter possesses a large thoracic sinus (instead of a prosomal sinus), which receives blood from the epineural sinuses, the cirri, and penis (Fig. 40-3D). The rostral sinus is also larger in *Megabalanus* than that seen in the lepadomorphs and receives blood directly from the peripheral circulation, while the prosomal sinus is absent.

Excretion is achieved primarily by means of maxillary glands. Calman (1909) also records supplementary excretion being achieved by the gut caeca and somatic nephrocytes. Broch (1927) indicated that the cement

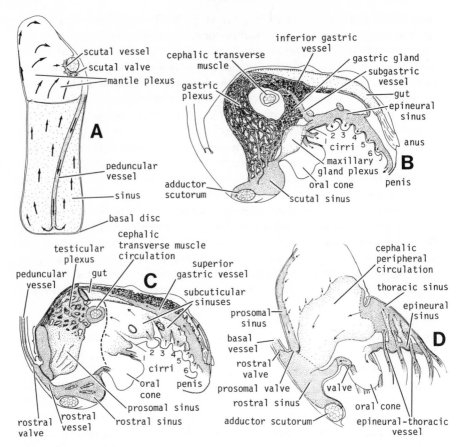

Fig. 40-3. Circulatory system of thoracicans. Direction of flow shown by arrows. (A) circulation of the peduncle–basis and the mantle of lepadids; (B) distributive circulation of *Lepas anatifera*; (C) peripheral and collecting circulation of *L. anatifera*; (D) peripheral and collecting circulation of *Megabalanus californicus*.

glands may serve an excretory function in the larvae and, in that respect, might be analogous to antennal glands of other crustaceans.

The reproductive system is for the most part hermaphroditic. The ovaries are located in the peduncle in the stalked forms, but otherwise are placed in the anterior portion of the body cavity. The oviducts exit near the bases of the first pair of cirri. The testes are found in the thorax. Each vas deferens forms a seminal receptacle near the base of the penis; however, the distal vasa unite within the penis, a long whiplike organ capable of extending beyond the mantle cavity.

Though generally hermaphrodites, there are several instances of the occurrence of separate complemental males within the thoracicans (Darwin, 1851a; Foster, 1983). These males exhibit successive degrees of degeneracy that range from complete but miniature versions of the host hermaphrodite, to those with degenerate mouthparts and living near the aperture of the host and those with degenerate or absent mouthparts living within the host mantle cavity. Some species, especially in the deep sea, have separate sexes, the female being accompanied by a dwarf male.

The nervous system of thoracicans exhibits great variety of form (Cornwall, 1953) ranging from a primitive ladderlike arrangement of the nerve cord, as seen in *Lepas hilli*, to an advanced state in which all ganglia are fused into a single postoral mass, as in *Balanus nubilis*. Barnacles are equipped with special sense organs around the margins of the aperture (Foster and Nott, 1969). The cuticle forms a membrane around the margin of the opening. This membrane is supplied with sensory hairs that project toward the outside when the scutal and tergal valves are closed. These hairs seem to be chemosensory and are well placed in order to monitor 'outside' conditions without having to open the valves and expose the mantle cavity.

Naupliar and compound eyes are functional in the larvae and, although the former may persist in the adult, they remained buried in tissue in the mature stages. The naupliar eyes have the distinctive maxillopodan features of tapetal and pigment cells (Elofsson, 1966; Hallberg and Elofsson, 1983). In addition, the nauplii bear sensory frontal filaments near the antennules.

NATURAL HISTORY   Adult barnacles are sessile, but the larvae are motile. This is so especially for the cyprids, which swim with jerking movements as all thoracopods beat synchronously. The cyprid larvae alight frequently to test the bottom in their search for an appropriate place to permanently settle (Barnes, 1970). Once that task is achieved, the cyprid undergoes metamorphosis to the adult.

The manner of barnacle feeding, once thought to be fairly straightforward, is now perceived as a rather complex issue in which the evolution of external shell form was concurrent with repeated experimentation with particular feeding strategies (Anderson, 1981). These strategies are postulated to involve behavioral variation in neurological evolution, without alteration in any major way of the basic trophal and cirral anatomy (Anderson, 1978, 1981). The basic thoracican feeding pattern is illustrated by that of the scalpellid *Pollicipes polymerus* (Barnes and Reese, 1959). In this instance, cirri are extended. They are left so until some tactile stimulation triggers the cirri to enfold and withdraw back into the mantle and, thus, capture the particle stimulant. The maxillipedes then remove the particle from the cirri and transfer it to the trophi. The maxillae push the food anteriad and downward, while the alternatingly reciprocating action of the maxillules and mandibles push food into the esophagus. The mandibular palps and enfolded cirri act to screen the trophi anteriorly and, thus, prevent food from being pushed too far foward. This pattern is widespread and is seen generally in lepadomorph, verrucomorph, and chthamaloid and coronuloid balanomorphs. A variant on this basic theme is seen in the coral barnacle *Megatrema anglicum* (Fig. 40-4A), where pulsating beats of the cirri occur after a long quiet extension; but these beats only continue until the tactile stimulant can be swept in close enough to be captured by the anteriormost cirrus (Anderson, 1978).

Another form of feeding (Fig. 40-4B) is like that seen in shallow-water verrucomorphs (Anderson, 1980a) and *Lepas pectinata* (Anderson, 1980b). Here the cirri are quietly extended but are then periodically withdrawn into the mantle. The prolonged pause phase of the cycle is in the extended position, and the time duration of the extended phase does not seem to be related to external conditions. This type of feeding seems directed toward utilizing microparticulate matter (Stone and Barnes, 1973) as well as large particle and carnivorous items.

The advanced balanomorphs exhibit a distinctly different mode. These animals exhibit a wide variety of cirral actions (Crisp and Southward, 1961; Anderson and Buckle, 1983): (1) testing, where the valves are hardly open with no movement of body or cirri; (2) pumping, where the valves open rhythmically and the body moves, but the cirri are hardly projected; (3) normal beating, where the cirri and valves beat in unison; (4) fast beating, where the valves remain open while the cirri beat; and (5) prolonged extension. However, the beating pattern of balanomorphs is just the opposite of that seen in other barnacles (Anderson, 1981); the prolonged pause in the cycle occurs in the flexed position within the mantle cavity, the extension phase being very short. This versatility of cirral activity has been responsible for the great radiation noted within balanomorphs, since feeding behaviors can be directed toward exploiting a variety of specific food resources (Barnes, 1959; Ota, 1957; Crisp and Southward, 1961; Anderson, 1981).

Cyprid larvae do not feed, but barnacle nauplii do and in a distinctive manner (Lochhead, 1936; Rainbow and Walker, 1976; Moyse, 1984). The naupliar limbs beat metachronally. On the backstroke, particles are swept posteriad and upward, converging just behind and beneath the body. These are then sucked forward with the forestroke of the antennae and mandibles. The particulate matter is retained just posterior to the labrum by the limb setae, and body and labral spines. On the backstroke, the limb setae (especially those of the antennal naupliar process) push the food into the mouth. Because of their size, the limbs cannot be viewed as rakes or leaky paddles (Moyse, 1983). Rather, the low Reynolds numbers and laminar flow of the medium must ensure that the limbs act as solid paddles (see the discussion of feeding in Chapter 37). The larvae seem to have food preferences depending on latitude (Moyse and Knight–Jones, 1965; Rainbow and Walker, 1976), with high latitude species requiring diatoms, while oceanic and low latitude types prefer flagellates.

Thoracicans are typically hermaphroditic, and are generally noted for

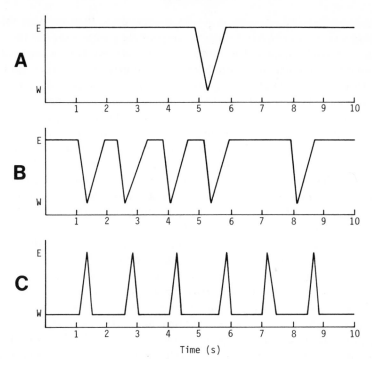

Fig. 40-4. Feeding rhythms in thoracicans showing extended (E) and withdrawn (W) phases through units of time. (A) prolonged extension with occasional withdrawal, such as in *Megatrema anglicum* (Anderson, 1978); (B) pulsating beats of cirri with the pause occurring in the extended phase, for example, *Verruca stroemia* (Anderson, 1980a); (C) pulsating beats of cirri with the pause occurring in the withdrawn phase, such as in, *Balanus perforatus* (Anderson, 1981).

their tendency to clump together when settling. The latter is necessitated to facilitate copulation between individuals. Collier et al. (1956) and Barnes (1962) demonstrated that the presence of vitamin C in the seminal fluid initiates mating receptivity. Not all individuals in a particular locality will respond equally at any one time to a 'groping' penis for, as Barnes and Barnes (1956) pointed out, in *Semibalanus balanoides* three to four 'males' will end up copulating at once with one 'female.' There is some evidence that self-fertilization (or perhaps parthenogenesis) can take place (Barnes and Crisp, 1956), since isolated individuals of several species have been recorded brooding eggs. The percentages of normal and viable embryos in such situations, however, is significantly less than in those in which cross-fertilization can take place.

Though thoracicans are generally hermaphrodites, several times in the history of barnacles complemental males have evolved (Newman, 1980; Foster, 1983). It was generally assumed that larvae of these species are 'ambisexual'; that is, if they settle on the substrate they become hermaphrodites and if they settle on a conspecific they become males. However, Gomez (1975) has suggested a possible genetic basis for determination of sex in barnacles. The complemental males are all small and exhibit various degrees of degeneration. Some have mouthparts and are essentially miniature versions of the hermaphrodite, for example, males of *Bathylasma alearum*, and in the genera *Calantica*, *Euscalpellum*, *Scillaelepas*, and *Smilium*. Other such males have degenerate trophi and live in special pockets near the host aperture, for example, inside the scuta of scalpellids, as in *Graviscalpellum pedunculatum*, or with females in *Arcoscalpellum*, and inside the rostra or the terga of some species of *Conopea* or *Solidobalanus*. Still others have very degenerate mouthparts or lack them altogether and are virtually parasites inside the mantle cavity of the host, for example, species of *Ibla*. Foster (1983) believes the phenomenon of complemental males arose independently at least seven times within Cirripedia.

Patterns of breeding themselves are quite variable among thoracicans and are very responsive to environmental conditions (Barnes, 1963; Hines, 1979). Some species breed over wide ranges of temperature, like *Elminius modestus*. Others have very narrow temperature requirements and are limited by food resources. Animals like *Chthamalus fissus*, *C. stellatus*, *Balanus amphitrite*, and *B. perforatus* produce several broods one after the other, their total fecundity regulated only by food availability. Copulation under such circumstances follows the release of the previous brood. Predation pressures can also affect reproductive strategies. *B. glandula* stores nutrients through the summer and breeds in the winter, allowing the larvae to settle in the spring, when food resources are highest and quick growth can get the young barnacles beyond the range of predator size preference. *Tetraclita squamosa* delays its breeding until it is at least 22 months. All energy in this period is directed into growth in order to quickly grow out of potential predator preference ranges. Though reproductive age is delayed in *T. squamosa*, fecundity is high and settlement takes place in fall when early winter growth stages can occur with minimal predator pressures. The patchy settlement in *Tetraclita* also serves to thwart predation, serving to 'frustrate' potential predators in their search for dinner.

Thoracican barnacles, despite their aberrancy of form, or perhaps because of it, inhabit almost every conceivable marine habitat from the deep sea to the intertidal, from pole to pole, and from rock solid substrates, to mud bottoms, and even to living tissue.

The distribution of barnacles in nature is only now coming to be understood as faunas of regions become well known, for example, Antarctica (Newman and Ross, 1971), the eastern Pacific (Zullo, 1966), the California transition zone (Newman, 1979a; Newman and Abbott, 1980), the Gulf of Mexico (Spivey, 1981), or New Zealand (Foster, 1978); or as closely related species come to be revised, for example, *Scillaelepas* spp. (Newman, 1980). As might be expected under such circumstance, understanding is spotty, and patterns applicable to some future grand synthesis of barnacle history are only just beginning to be recognized. However, patterns that are thought to be clear can change abruptly, as was evidenced when the classic endemic Indo-Pacific distribution of the genus *Tesseropora* was nullified with the description of a North Atlantic form (Newman and Ross, 1977).

Nevertheless, certain broad patterns of biogeographic significance have emerged. One such pattern is the nature of the barnacle fauna of the southern oceans (Newman, 1979b). Balanomorphs appear in the fossil record in the Late Mesozoic and the basic groups evolved rapidly through the Paleogene after the breakup of Pangaea. The effects of vicariance would thus seem to play a minimal role in explaining the wide distribution of balanomorphs. Yet there are twice as many endemics among the southern hemisphere balanomorphs than in the north (even with an almost exclusively lepadomorph fauna in Antarctica). The endemic southern balanomorph taxa for the most part have affinities with Tertiary forms, and many of these endemics occur on islands of Miocene age or younger. The distribution of recent southern balanomorphs would thus seem better explained (Newman, 1979b) by scenarios of long-range dispersal coupled with subsequent speciation rather than reliance on vicariance. However, in this same vein, Zullo (1966) explained the endemism in the eastern Pacific (60–80%) as related to its long-term isolation from the western Pacific. Eastern Pacific forms seem to be predominately derived from a pantropical, pre-Pliocene, American track with Atlantic and Mediterranean affinities. Vicariance from the parent fauna occurred when Central America intervened to isolate the eastern Pacific from the Caribbean in the late Tertiary.

A second pattern is the importance of species interactions in determining global and habitat distributions within the balanomorphs (Stanley and Newman, 1980; Newman and Stanley, 1981). The more primitive Chthamaloidea have been in decline since their origin in the Cretaceous, apparently in response to the evolution of first the Coronuloidea, in the Paleocene, and then Balanoidea, in the Eocene. The small number of chthamaloid species, which are largely characterized today by a relict and disjunct pattern of distribution, is indicative of a group which had much greater global diversity in the past. A prime cause for the chthamaloid decline are apparently the balanoids, who are more effective than chthamalids in the competition for space (Connell, 1961; Dayton, 1971). A prime element in this success of the balanoids is their tubiferous plates, which fosters rapid growth without sacrificing plate rigidity (Stanley and Newman, 1980). Furthermore, the bathymetric gap in the distribution of modern chthamaloids (Fig. 40-5) corresponds to the species diversity peak among the balanoids (Newman and Stanley, 1981). The modern distribution of chthamaloids is thus most effectively explained by realizing it is a group in decline that now occupies geographic and ecologic refugia.

In those few species investigated, peculiarities in the distribution of

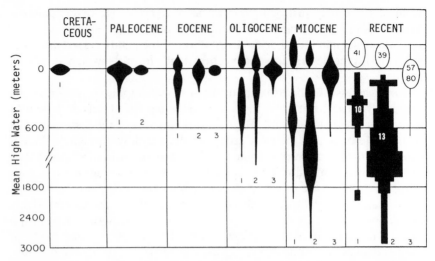

Fig. 40-5. Time of appearance in fossil record and inferred generic diversity with depth of Balanomorpha. (1) Chthamaloidea; (2) Coronuloidea, (3) Balanoidea, numbers in Recent of known species. (From Newman and Stanley, 1981)

individual taxa have been shown frequently to have causal bases, though little work has been done in this regard. For example, the only balanomorph south of the Antarctic convergence is *Bathylasma corolliforme*, which has a circumpolar, bathyal distribution (100–1500 m). Transplant experiments (Dayton et al., 1982) have shown that this species, which sustains itself by a passive feeding mode, is especially sensitive to the lack of sufficient currents. It is not that larvae do not penetrate outside the observed depths or that recruitment does not occur in colonies transplanted to shallower depths. It is just that the animals effectively starve where there are insufficient currents.

Biogeographic generalizations are difficult to make for a complex group, such as the thoracicans that continues to undergo rapid radiation. However, more attention to such matters in the future will undoubtedly increase our understanding of the evolution of the group.

DEVELOPMENT   The most recent and detailed study of thoracican development is that of Anderson (1969) utilizing *Tesseropora rosea*, *Tetraclitella purpurascens*, *Chthamalus antennatus*, and *Chamaesipho columna*. The study of this early ontogeny has revealed that not only is cleavage total, it is uniquely primitive in that it is of a modified spiral type. The egg is oval and drawn out to a papilla. The pointed end will be posterior in the developing embryo. The yolk initially concentrates on one side of the egg and then moves posteriad as well. The yolk–cytoplasm border then is obliquely oriented, the first spindle apparatus is perpendicular to this plane (Fig. 40-6A). The first cleavage thus segregates an anterior cytoplasm-rich (AB)

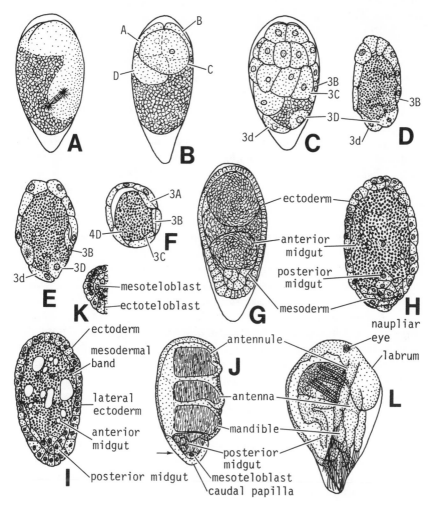

Fig. 40-6. Early development of *Tetraclita rosea*, cells labeled to show spiral cleavage derivatives. (A) egg; (B) after second cleavage; (C) advanced gastrulation at 28-cell stage; (D) section through C; (E) sagittal section of 28- to 30-cell stage with shifting of micromeres to begin formation of presumptive mesoderm; (F) sagittal section of 31- to 33-cell stage with presumptive mesoderm (3A, 3B, 3C); (G, H) dorsal view and frontal section, respectively, of embryo with presumptive areas delineated; (I) section, with mesodermal proliferation underway; (J) naupliar somites indicated; (K) cross section through J at point indicated by arrow; (L) an embryo just before hatching with development complete. (From Anderson, 1969)

cell and a posterior yolk-rich (CD) cell. The anterior cell then divides again and forms the A and B cells. The posterior cell divides more slowly, first segregating a yolk-free region, and then producing a yolk-free C cell and a yolk-rich D cell (Fig. 40-6B).

The third cleavage is sequential, with first B, then C, then A, and finally D dividing. The 1d cell is again yolk-free and divides off in the anterior end of the embryo. Each of these early cleavage stages are distinct, with interphase resting periods to set them apart, so that distinct 2, 4, 8, 16, and 28 cell stages are recognized. The net result of early cleavage is to effectively enclose the yolk-rich 3D cell within the interior of the embryo (Fig. 40-6C, D).

A portion of the sixth cleavage that occurs in the posterior part of the embryo in the 2d lineage results in a complex shift of micromeres (Fig. 40-6E) to displace the 3A, 3B, and 3C cells inward. These cells come to lie between the surface ectoderm layer and the endodermal yolk cell (Fig. 40-6F). These three cells form the presumptive mesoderm. The 3D cell then divides to form a yolky 4D cell, which is presumptive endoderm and a yolk-free 4d. The remaining surface cells become ectoderm.

The 4D yolk cell divides and forms large yolky anterior midgut cells and smaller less yolky posterior midgut cells (Fig. 40-6H).

The mesodermal cells divide and migrate posteriad until all mesoderm is concentrated in the caudal papilla at the terminus of the embryo (Fig. 40-6G). The mesoderm then begins to proliferate up each side of the embryo (Fig. 40-6I). These paired bands spread anteriad and ventrad until they meet at the anterior end and along the ventral midline. At that point the bands begin to thicken into the three paired masses of the naupliar somites. Formation of these somites then induces the ectoderm to form the anlagen of the naupliar appendages (Fig. 40-6J). The mesodermal mass in the caudal papilla is separated as postnaupliar mesoderm and two of the posterior cells in this mass enlarge to become the mesoteloblasts (Fig. 40-6J, K). These divide twice to produce eight mesoteloblast cells, which remain dormant until after hatching. The cells of the anterodorsal mesoderm above the antennulary somite divide and spread through the hemocoel and for the most part become associated with the midgut. No coelomic spaces appear in any of the naupliar somites.

The growth of the appendage anlagen proceeds rapidly in a posterodorsal direction so that the limbs come to occupy the space between the posterior embryo and the posterior egg membrane (Fig. 40-6L). The antennal and mandibular limb buds then become biramous. However, Kaufmann (1965) observed in *Scalpellum scalpellum* that the anlagen of the antennules were biramous as well. At this point, an invagination occurs between the antennae to form the stomodeum. The mesoderm carried along with this invagination forms the stomodeal musculature. The initiation of the stomodeum induces the ectoderm between the antennules to evaginate and grow posteriad as the labrum. Adjacent mesoderm proliferates to fill the labrum and eventually gives rise to the labral muscles. The stomodeal invagination also marks the proliferation of cells in the ectoderm that will eventually form the central nervous system, but inward movement of these cells does not occur until cuticular segmentation has begun.

The anterior midgut cells proliferate to form a dense mass. By the time the naupliar limbs begin to twitch the anterior mesoderm reorganizes to form a central cavity. This lumen joins that of the stomodeum and posterior midgut and forms the naupliar stomach (Fig. 40-6L). The posterior midgut cells begin to divide when the naupliar anlagen are formed, and these cells grow forward beneath the anterior yolk cells. A tube is formed that is closed at its posterior end and joined anteriorly with the stomach wall with which it eventually opens into to form the naupliar intestine. Connection to the proctodeum is not achieved until after hatching and the first naupliar molt.

Developing eggs are brooded in the mantle cavity. Hatching is achieved when the adults produce a diffusable factor (probably a lactone or lactan) that stimulates the embryo (Crisp and Spencer, 1958) to gyrate and break the egg membranes. The mantle is emptied within 24 hours. The escaping individuals are typically stage I nauplii, but occasionally some as yet unhatched eggs are also voided (Barnes, 1955).

There are six naupliar stages (see, Barnes and Barnes, 1959; Barnes and Costlow, 1961; or Bainbridge and Roskell, 1966). The first stage cannot feed but, after the first molt and the development of the gut are completed, the nauplii take up active feeding. The naupliar form is quite distinctive, characterized by a prolonged caudal process, paired anterolateral horns on the head shield, and frontal filaments anterior to the antennules (Fig. 40-7A-F). The anterolateral horns have glands that produce a water-insoluble secretion (Taylor, 1970). This secretion may serve a protective function or be a wetting agent to prevent the nauplius from getting trapped in surface films. The frontal filaments have complex innervation (Walker, 1974) and have been noted in some instances to be 'segmented' (Bassindale, 1936). Their function is still unknown; Walker suggests a possible ion receptor that in turn might affect phototaxis. Finally, the nauplii bear well-developed naupliar eyes, and in stages V and VI they also have compound eyes (Hallberg and Elofsson, 1983). Both types of eyes persist into the cyprid stage. Also by stages V and VI, the teloblasts are activated, so these 'naupliar' stages are effectively metanauplii.

The cyprid is a nonfeeding stage, characterized by a large, bivalved carapace and the presence of all the adult body limbs (Fig. 40-7G). The cyprid is a settlement phase and spends its time searching for a suitable substrate. To this end the antennules (Fig. 40-7H) have become highly modified (Nott and Foster, 1969). The third segment of the limb is the actual attachment point, while the fourth segment is retained and equipped with sensory setae and processes. The antennule is not a suctorial device, but rather is equipped with glandular tissue that is functional and adhesive even in the exploratory phase.

Metamorphosis is rather profound when it occurs, especially in the

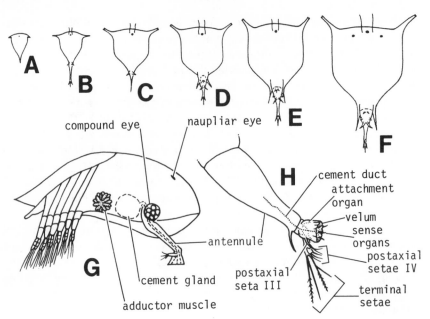

Fig. 40-7. Larval development. (A, B, C, D, E, F) body outline of six naupliar stages of *Balanus nubilis* (from Barnes and Barnes, 1959); (G) cyprid larva; (H) cyprid antennule of *Semibalanus balanoides* modified for settlement (from Nott and Foster, 1969).

balanomorphs. The carapace develops five symmetrical areas of calcification, which develop into the five primary plates, a single dorsal carina, and paired lateral terga and scuta (Runnström, 1925). In lepadomorphs the cyprid attaches by the antennule, and the anterior portion of the head becomes the peduncle. The body of the animal somewhat reorients itself within the mantle cavity and the thoracopods become cirri. In balanomorphs, the cyprid settles and casts off its cuticle. An amorphous cytoplasmic mass remains on the substrate and develops a chitinous ring, within which the barnacle body is reconstituted. The ring calcifies into carina, rostrum, and lateral plates, and opercular plates develop to cover the aperture. The balanomorph metamorphosis is so profound that it tells us little if anything as to how the balanomorph form evolved.

FOSSIL RECORD   Of all the crustaceans, the barnacles have one of the most abundant of fossil records. The broad outlines are dealt with here but more detailed reviews can be found in Newman et al. (1969), Alekseev (1981), Buckeridge (1983), Schram (1982), and Newman (1982).

The oldest suggested barnacle is *Priscansermarinus* from the Middle Cambrian Burgess Shale (Collins and Rudkin, 1981). It would appear to be a lepadomorph with a globular carapace lacking plates and, thus, possibly a heteralepadoid of some kind. However, Briggs (1983) has suggested caution in forcing that interpretation too far.

The oldest generally conceded barnacle is *Cyprilepas holmi* (Fig. 40-8A) from the Silurian (Wills, 1963). These fossils seem to be of lepadomorphs with a clearly bivalved, uncalcified carapace covering the capitulum. *Cyprilepas* occurs attached to the paddles of the sea scorpion *Baltieurypterus fisheri*.

The next oldest barnacles, from the Carboniferous, have been referred to a single family Praelepadidae. They are characterized by chitinous plates. The first known species, *Praelepas jaworskii* Chernyshev (1931) has five well-developed plates tightly arranged to cover the capitulum and act to close the aperture (Fig. 40-8B). It seems to fit well as a prelepadid stock. The other species *Praelepas damrowi* Schram (1975) is so distinctly different (Fig. 40-8C) that it should be referred to its own genus *Illilepas* (hereby proposed). The scuta and terga are large and cover the entire subrectangular capitulum; the aperture is along the distal margin; and the margin is marked by a row of fine spines and a large platelike spine (originally thought to be a reduced tergum). Though possessing chitinous plates, the affinities of *Illilepas* may actually lie with the iblids rather than lepadids. Another Carboniferous lepadamorph is *Pabulum spathiforme* Whyte (1976). Reexamination of the type series indicates that this is probably a scalpellid with a long peduncle covered with imbricating plates. Such peduncular plates are frequently found in the Late Paleozoic in association with corals (Schram et al., in prep.).

Fig. 40-8. Fossil thoracicans. (A) *Cyprilepas holmi*, Silurian; (B) *Praelepas jaworskii*, Carboniferous; (C) *Illilepas damrowi*, Carboniferous; (D) *Brachylepas cretacea*, Jurassic; (E) *Archaeolepas redtenbacheri*, Jurassic. (From Schram, 1982).

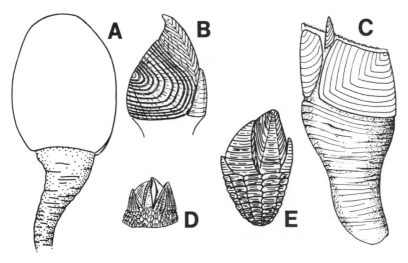

The Mesozoic and Cenozoic record seems to indicate a steady increase in diversity of thoracican types with successive groups reaching peaks and yielding to 'more advanced' types. Scalpellids extend from the Carboniferous (because of *Pabulum*), but otherwise have a good record along with lepadids from the Triassic on. The verrucomorphs first appear in the Cretaceous, but the record as yet is insufficient to understand the course of that radiation. The balanomorphs first appear in the Jurassic (Fig. 40-8D) as do the brachylepadoids (Newman, in press); subsequently chthamaloids appear in the Cretaceous, coronuloids in the Paleocene, and balanoids in the Eocene.

An interesting correlation has been noted between the inferred sequence of evolutionary events derived from consideration of adult morphology and ontogeny and those seen in the fossil record. For example, the evolution of scalpellids inferred from the ontogeny of *Pollicipes*; the sequential comparative anatomy of living lepadid–oxynaspids, *Neolepas*, and calanticine scalpellids; and the fossil sequence of Carboniferous *Praelepas*, Triassic *Eolepas/Archaeolepas* (Fig. 40-8F), early Cretaceous *Blastolepas* (Drushits and Zevina, 1969), and late Cretaceous *Scillaelepas* (Newman, 1979c) all coincide into a common pattern of increasing capitular plate complexity.

TAXONOMY The discovery of new barnacles in the deep sea, new fossil finds, and reassessment of old fossil records are resulting in constant revision of old taxonomies. Zevina (1981, 1982) has suggested a sweeping revision of the lepadomorphs, though whether her new genera will stand the test of time remains to be seen. Newman and Ross (1976) revised the balanomorphs, but more recent information (Newman, in press) concerning the brachylepadoids has suggested that these are better treated as balanomorphs rather than as a separate suborder. However, no across-the-board analysis of barnacle relationships has been done using rigorous character analysis, and when this occurs it will undoubtedly result in further modifications of thoracican classification. The following taxonomy is thus not definitive. It is based largely on that of Newman (in press) and Zevina.

Suborder Lepadomorpha Pilsbry, 1916 (?Cambrian) Silurian–Recent
    Superfamily Cyprilepadoidea Newman et al., 1969 Silurian
        Family Cyprilepadidae Newman et al., 1969
    Superfamily Heteralepadoidea Nilsson–Cantrell, 1921 (?Cambrian) Recent
        Family Heteralepadidae Nilsson–Cantell, 1921
        Family Koleolepadidae Hiro, 1937
        Family Malacolepadidae Hiro, 1937
        Family Microlepadidae Zevina, 1980
        Family Pagurolepadidae Zevina, 1980

Family Rhizolepadidae Zevina, 1980
Family Anelasmatidae Zevina, 1980
Superfamily Praelepadoidea Chernyshev, 1930 Carboniferous
 Family Praelepadidae Chernyshev, 1930
Superfamily Lepadoidea Darwin, 1851 (?Triassic) Eocene–Recent
 Family Lepadidae Darwin, 1851 (?Triassic) Eocene–Recent
 Family Oxynaspididae Pilsbry, 1907 Eocene–Recent
 Family Poecilasmatidae Annandale, 1909 Eocene–Recent
Superfamily Ibloidea Leach, 1825 Carboniferous–Recent
 Family Iblidae Leach, 1825
Superfamily Scalpelloidea Pilsbry, 1916 (?Carboniferous) Triassic–Recent
 Family Scalpellidae Pilsbry, 1916 Triassic–Recent
 Family Stramentidae Withers, 1920 Cretaceous
Suborder Verrucomorpha Pilsbry, 1916 Cretaceous–Recent
 Family Verrucidae Darwin, 1854
Suborder Balanomorpha Pilsbry, 1916 Jurassic–Recent
 Superfamily Brachylepadoidea Woodward, 1901 Jurassic–Miocene
  Family Brachylepadidae Woodward, 1901
 Superfamily Chthamaloidea Darwin, 1854 Cretaceous–Recent
  Family Catophragmidae Utinomi, 1968 Cretaceous–Recent
  Family Chthamalidae Darwin, 1854 Miocene–Recent
 Superfamily Coronuloidea Newman and Ross, 1976 Paleocene–Recent
  Family Coronulidae Leach, 1817 Pliocene–Recent
  Family Bathylasmatidae Newman and Ross, 1971 Paleocene–Recent
  Family Tetraclitidae Gruvel, 1903 Eocene–Recent
 Superfamily Balanoidea Leach, 1817 Eocene–Recent
  Family Archaeobalanidae Newman and Ross, 1976 Eocene–Recent
  Family Pyrgomatidae Gray, 1825 Miocene–Recent
  Family Balanidae Leach, 1817 Eocene–Recent

REFERENCES

Alekseev, A. S. 1981. Razvitie usonogikh rakov — toratzid v Mezozeye i Kainozoye. In *Morfogyenyez i Ruti Razvitiya Iskopayemikh Bespozvonochnikh* (V. V. Menner and V. V. Drushits, eds.), pp. 149–64. Izdatelstvo Moskovskogo Universiteta.
Anderson, D. T. 1969. On the embryology of the cirripede crustaceans *Tetraclita rosea*, *T. purpurescens*, *Chthamalus antennatus*, and *Chamaesipho columna* and some considerations of phylogenetic relationships. *Phil. Trans. Roy. Soc. Lond.* (B)**256**:183–235.
Anderson, D. T. 1978. Cirral activity and feeding in the coral-inhabiting barnacle *Boscia anglicum*. *J. Mar. Biol. Assc. U.K.* **58**:607–26.
Anderson, D. T. 1980a. Cirral activity and feeding in the verrucomorph barnacles *Verruca recta* and *V. stroemia*. *J. Mar. Biol. Assc. U.K.* **60**:349–66.
Anderson, D. T. 1980b. Cirral activity and feeding in the lepadomorph barnacle *Lepas pectinata*. *Proc. Linn. Soc. N.S.W.* **104**:147–59.

Anderson, D. T. 1981. Cirral activity and feeding in the barnacle *Balanus perforatus*, with comments on the evolution of feeding mechanisms in thoracican cirripedes. *Phil. Trans. Roy. Soc. Lond.* (B)**291**:411–49.

Anderson, D. T., and J. Buckle. 1983. Cirral activity and feeding in the coronuloid barnacles *Tesseropora rosea* and *Tetraclitella purpurascens*. *Bull. Mar. Sci.* **33**:645–55.

Bainbridge, V., and J. Roskell. 1966. A redescription of the larvae of *Lepas fascicularis* with observations on the distribution of *Lepas* nauplii in the northeastern Atlantic. In *Some Contemporary Studies in Marine Science* (H. Barnes, ed.), pp. 67–82. George Allen & Unwin, London.

Barnes, H. 1955. The hatching process in some barnacles. *Oikos* **6**:114–23.

Barnes, H. 1959. Stomach contents and microfeeding of some common cirripedes. *Can. J. Zool.* **37**:231–26.

Barnes, H. 1962. The composition of the seminal plasm of *Balanus balanus*. *J. Exp. Biol.*, **39**:345–51.

Barnes, H. 1963. Light, temperature, and the breeding of *Balanus balanoides*. *J. Mar. Biol. Assc. U.K.* **43**:717–27.

Barnes, H. 1970. A review of some factors affecting settlement and adhesion in the cyprids of some common barnacles. In *Adhesion in Biological Systems* (R. S. Manley, ed), pp. 89–111. Academic Press, New York.

Barnes, H., and M. Barnes. 1956. The formation of the egg-mass in *Balanus balanoides*. *Arch. Soc. 'Varamo'* **11**:11–16.

Barnes, H., and M. Barnes. 1959. The naupliar stages of *Balanus nubilis*. *Can. J. Zool.* **37**:15–23.

Barnes, H., and J. D. Costlow. 1961. The larval stages of *Balanus balanoides*. *J. Mar. Biol. Assc. U.K.* **41**:59–68.

Barnes, H., and D. J. Crisp. 1956. Evidence of self-fertilization in certain species of barnacles. *J. Mar. Biol. Assc. U.K.* **35**:631–39.

Barnes, H., and E. S. Reese. 1959. Feeding in the pedunculate cirripede *Pollicipes polymerus*. *Proc. Zool. Soc. Lond.* **132**:569–85.

Bassindale, R. 1936. The development of three English barnacles *Balanus balanoides*, *Chthamalus stellatus*, and *Verruca stroemia*. *Proc. Zool. Soc. Lond.* **1936**:57–74.

Briggs, D. E. G. 1983. Affinities and early evolution of the Crustacea: the evidence of the Cambrian fossils. *Crust. Issues* **1**:1–22.

Broch, H. 1927. Cirripedia. In *Handbuch Zoologie*, Vol. 3, part 5 (Kükenthal and Krumbach, eds.), pp. 503–52. de Gruyter, Berlin.

Buckeridge, J. S. 1983. Fossil barnacles of New Zealand and Australia. *N.Z. Geol. Surv. Paleo. Bull.* **50**:1–151, 13 pls.

Burnett, B. R. 1972. Aspects of the circulatory system of *Pollicipes polymerus*. *J. Morph.* **136**:79–108.

Burnett, B. R. 1975. Blood circulation in four species of barnacles. *Trans. San Diego Soc. Nat. Hist.* **17**:293–304.

Burnett, B. R. 1977. Blood circulation in the balanomorph barnacle, *Megabalanus californicus*. *J. Morph.* **153**:299–306.

Calman, W. T. 1909. *Crustacea*. In *A Treatise on Zoology* (E. R.Lankester, ed.). Adam and Charles Black, London.

Chernyshev, B. I. 1931 Cirripeden aus dem Bassin des Donez und von Kuznetsk. *Zool. Anz.* **92**:26–28.

Collier, A., S. Ray, and W. B. Wilson. 1956. Some effects of specific organic compounds on marine organisms. *Science* **124**:220.

Collins, D. H., and D. M. Rudkin. 1981. *Priscansermarinus barnetti*, a probable lepadomorph barnacle from the Middle Cambrian Burgess Shale of British Columbia. *J. Paleo.* **55**:1006–15.

Connell, J. H. 1961. The influence of interspecific competition and other factors on the distribution of the barnacle *Chthamalus stellatus*. *Ecology* **42**:710–23.

Cornwall, I. E. 1953. The central nervous system of barnacles. *J. Fish. Res. Bd. Can.* **10**:76–84.

Crisp, D. J. and A. J. Southward. 1961. Different types of cirral activity of barnacles. *Phil. Trans. Roy. Soc. Lond.* (B)**243**:271–308.

Crisp, D. J., and C. P. Spencer. 1958. The control of the hatching process in barnacles. *Proc. Roy. Soc.* (B)**148**:278–299.

Darwin, C. R. 1851a. A Monograph on the Subclass Cirripedia with Figures of all the Species. The Lepadidae or Pedunculate Barnacles. Ray Society, London.

Darwin, C. R. 1851b. *A Monograph on the Fossil Lepadidae, or, Pedunculate Cirripedes of Great Britain.* Paleontography Society, London.

Darwin, C. R. 1854a. *A Monograph on the Subclass Cirripedia, with Figure of all the Species.* Ray Society, London.

Darwin, C. R. 1854b. A Monograph of the Fossil Balanidae and Verrucidae of Great Britain. Paleontography Society, London.

Dayton, P. K. 1971. Competition, disturbance, and community organization: the provision and subsequent utilization of space in a rocky intertidal community. *Ecol. Monogr.* **41**:351–89.

Dayton, P. K., W. A. Newman, J. Oliver. 1982. The vertical zonation of the deep-sea Antarctic acorn barnacle, *Bathylasma corolliforme*: experimental transplants from the shelf into shallow water. *J. Biogeogr.* **9**:95–109.

Drushits, V. V., and G. B. Zevina. 1969. New Lower Cretaceous Cirripeds from the northern Caucasus. *Paleontol. J.* **2**:73–85.

Elofsson, R. 1966. The nauplius eye and frontal organs of the non-Malacostraca. *Sarsia* **25**:1–128.

Foster, B. A. 1978. The marine fauna of New Zealand: Barnacles. *N.Z. Oceanogr. Inst. Mem.* **69**:1–160.

Foster, B. A. 1983. Complemental males in the barnacle *Bathylasma alearum*. *Aust. Mus. Mem.* **18**:133–39.

Foster B. A., and J. A. Nott. 1969. Sensory structures in the opercula of the barnacle *Elminius modestus*. *Mar. Biol.* **4**:340–44.

Gomez, E. D. 1975. Sex determination in *Balanus galeatus*. *Crustaceana* **28**:105–7.

Hallberg, E., and R. Elofsson. 1983. The larval compound eyes of barnacles. *J. Crust. Biol.* **3**:17–24.

Hines, A. H. 1979. The comparative reproductive ecology of three species of intertidal barnacle. In *Reproductive Ecology of Marine Invertebrates* (S. E. Stancyk, ed.), pp. 213–34. Univ. S. Carolina Press, Columbia.

Kaufmann, R. 1965. Zur Embryonal- und Larvalentwicklung von *Scalpellum scalpellum*, mit einem Beitrag zur Autökologie dieser Art. *Z. Morph. Ökol. Tiere* **55**:161–232.

Lochhead, J. H. 1936. On the feeding mechanism of *Balanus perforatus*. *J. Linn. Soc. Lond.* **39**:429–42.

Moyse, J. 1984. Some observations on the swimming and feeding of the nauplius larvae of *Lepas pectinata*. *Zool. J. Linn. Soc. Lond.* **80**:323–36.

Moyse, J., and E. W. Knight–Jones. 1965. Biology of cirripede larvae. *Proc. Symp. Crust.*, Vol. 2, pp. 595–611. Mar. Biol. Assc. India.

Newman, W. A. 1979a. California transition zone: significance of short-range endemics. In *Historical Biogeography, Plate Tectonics, and the Changing Environment* (J. Gray and A. J. Boucot, eds.), pp. 399–415. Oregon St. Univ. Press, Corvallis.

Newman, W. A. 1979b. On the biogeography of balanomorph barnacles of the southern ocean including new balanid taxa: a subfamily, two genera, and three species. In *Proc. Symp. Mar. Biogeogr. and Evol. South. Hemis.*, Vol. I, pp. 279–306. DSIR Info. Ser. 137, New Zealand.

Newman, W. A. 1979c. A new scalpellid, a Mesozoic relict living near an abyssal hydrothermal spring. *Trans. San Diego Soc. Nat. Hist.* **19**:153–67.

Newman, W. A. 1980. A review of extant *Scillaelepas* including recognition of new species from the North Atlantic, western Indian Ocean and New Zealand. *Tethys* **9**:379–98.

Newman, W. A. 1982. Cirripedia. In *The Biology of Crustacea*, Vol. 1 (L. G. Abele, ed.) pp. 197–220. Academic Press, New York.

Newman, W. A., and D. P. Abbott. 1980. Cirripedia: The Barnacles. In *Intertidal Invertebrates of California*, pp. 504–35, pls. 147–54. Stanford Univ. Press, Stanford.

Newman, W. A., and A. Ross. 1971. Antarctic Cirripedia. *Ant. Res. Ser. 14*, Am. Geophy. Union, Washington.

Newman, W. A., and A. Ross. 1976. Revision of the balanomorph barnacles; including a catalog of the species. *Mem. San Diego Soc. Nat. Hist.* **9**:1–108.

Newman, W. A., and A. Ross. 1977. A living *Tesseropora* from Bermuda and the Azores: first records from the Atlantic since the Oligocene Trans. *San Diego Soc. Nat. Hist.* **18**:207–16.

Newman, W. A., and S. M. Stanley. 1981. Competition wins out overall: reply to Paine. *Paleobiol.* **7**:561–69.

Newman, W. A., V. A. Zullo, and T. H. Withers. 1969. Cirripedia. In *Treatise on Invertebrate Paleontology*, Part R, *Arthropoda* **4**(1), (R. C. Moore, ed.), pp. R206–95. Geol. Soc. Am. and Univ. Kansas Press, Lawrence.

Nott, J. A., and B. A. Foster. 1969. On the structure of the antennular attachment organ of the cypris larva of *Balanus balanoides*. *Phil. Trans. Roy. Soc. Lond.* (B)**256**:115–34.

Ota, K. 1957. Notes on the ecology of *Balanus amphitrite hawaiensis*. *Physiol. Ecol.* **7**:127–30.

Petriconi, V. 1969. Vergleichende anatomische Untersuchungen an Rankenfüssern. *Zool. Anz. Suppl.* **33**:539–46.

Rainbow, P. S., and G. Walker. 1976. The feeding apparatus of the barnacle nauplius larva: a scanning electron microscope study. *J. Mar. Biol. Assc. U.K.* **56**:321–26.

Rainbow, P. S., and G. Walker. 1978. Absorption along the alimentary tract of barnacles. *J. Mar. Biol. Assc. U.K.* **58**:381–86.

Runnström, S. 1925. Zur Biologie und Entwicklung von *Balanus balanoides*. *Bergens Mus. Aarbok* **5**:1–46.

Schram, F. R. 1975. A Pennsylvanian lepadomorph barnacle from the Mazon Creek area, Illinois. *J. Paleo.* **49**:928–30.

Schram, F. R. 1982. The fossil record and evolution of Crustacea. In *The Biology of Crustacea*, Vol. I (L. G. Abele, ed.), pp. 93–147. Academic Press, New York.

Spivey, H. R. 1981. Origin, distribution, and zoogeographic affinities of the Cirripedia of the Gulf of Mexico. *J. Biog.* **8**:153–76.

Stanley, S. M., and W. A. Newman. 1980. Competitive exclusion in evolutionary time: the case of the acorn barnacles. *Paleobiol.* **6**:173–83.

Stone, R. L., and H. Barnes. 1973. The general biology of *Verruca stroemia* I. Geographical and regional distribution, cirral activity, and feeding. *J. Exp. Mar. Biol. Ecol.* **12**:167–85.

Taylor, P. B. 1970. Observations on the function of the frontolateral horns and horn glands of barnacle nauplii. *Biol. Bull.* **138**:211–18.

Thompson, J. Vaughan. 1830. On the cirripedes or barnacles. *Zoological Researches*, Vol. I, Mem. IV, pp. 69–88. Cork.

Walker, G. 1974. The fine structure of the frontal filament complex of barnacle larvae. *Cell. Tiss. Res.* **152**:449–65.

Whyte, M. A. 1976. A Carboniferous pedunculate barnacle. *Proc. York. Geol. Soc.* **41**:1–12.

Wills, L. J. 1963. *Cyprilepas holmi*, a pedunculate cirripede from the Upper Siluvian of Oesel, Estonia. *Palaeontol.* **6**:161–65.

Zevina, G. B. 1981. *Usonogiye raki podotryala Lepadomorpha mirovogo okeana*. Chast I. *Syemyeystvo Scalpellidae*. Nauka Leningradskoye Otdyelyeniye, Leningrad.

Zevina, G. B. 1982. *Usonogiye raki podotryada Lepadomorpha mirovogo okeana*. Nauka Leningradskoye Otdylyeniye, Leningrad.

Zullo, V. A. 1966. Zoogeographic affinities of the Balanomorpha of the eastern Pacific. In *The Galapagos* (R. I. Bowman, ed.), pp. 139–44. Univ. California Press, Berkeley.

41

ACROTHORACICA

DEFINITION   Adults as burrowers generally in shells and calcareous substrates; carapace generally without plates; antennules (of cyprid larva) as organ of initial attachment; burrowing, at least in part by chemical action of the mantle; mandible as serrate blade with palp; trunk limbs as cirri, first pair in vicinity of mouthparts as maxillipedes and distinctly separated from three to five posterior cirri; abdomen absent but caudal appendages usually present in adult; sexes separate, males dwarfed and nonfeeding. Larvae consist of up to four naupliar stages, the last stage metanaupliar in character, nauplii with frontolateral horns and frontal filaments; a post-cyprid and a cyprid stage.

HISTORY   The first acrothoracican, *Alcippe* (= *Trypetesa*) *lampas*, was described by Hancock in 1849 in the shells of gastropods inhabited by hermit crabs. Darwin placed *Trypetesa* in the lepadids, but allied it to his genus *Cryptophialus*. Darwin mistakenly thought the cirri to be abdominal in nature, and so erected an order, Abdominalia, for *Cryptophialus*. *Trypetesa* and other forms were gradually added to the order until Gruvel (1905) recognized Darwin's error and suggested changing the order's name to Acrothoracica. Berndt (1907) recognized two suborders, and taxa have been steadily described since. The most recent review of the group is that of Tomlinson (1969a). Currently there are some 40 species and the group has a worldwide distribution. Though frequently treated by most authors as part of the thoracicans, enough is known about the acrothoracicans to outline their distinctly different character such that a separate status is deemed warranted here.

MORPHOLOGY   Adult acrothoracids occupy small self-excavated cavities in calcareous substrates. The burrow usually has a tapered slitlike opening, a few millimetres in length and less than a millimetre in width. These slits are family-level diagnostic (Fig. 41-1D, E, F). The carapace is developed as a fleshy mantle, envelops the body, is generally without plates, but is covered by chitinous teeth that aid in the excavation of the burrow. In *Weltneria* a calcareous remnant of the rostral plate has been identified (Newman, 1971) that is cemented to the substrate, and from which in turn the body is suspended inside the burrow (Fig. 41-1A). The mantle may be reinforced with lateral bars or rods, that is, chitinous thickenings of the carapace extending from the aperture halfway to the posterior margin (Fig.

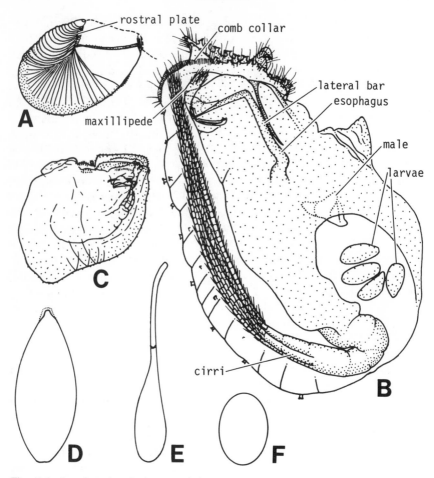

Fig. 41-1. Acrothoracican body types. (A) outer view of *Weltneria hessleri* (from Newman, 1971); (B) *Kochlorine hamata*, body form within the mantle (from Tomlinson, 1969b); (C) *Trypetesa habei*, body form (from Tomlinson, 1969a); (D) aperture opening of lithoglyptids; (E) aperture opening of trypetesid (modified from Tomlinson, 1969b); (F) aperture opening of cryptophialid.

41-1B). The aperture itself is protected with a variety of hairs, spines, hooks, and the comb–collar (a membranous fold with fringed hairs).

The mouthparts are located near the aperture. The labrum is bullate, sometimes almost pointed, and frequently armed with spines or teeth and fine setae. The mandible (Fig. 41-2A) is a serrate blade with four large teeth, the posterior of which can be further subdivided and also has a series of spines on the inferior curvature. The palp is elongate, with stiff setae on the distal and outer margins. (Unlike thoracicans, the palp tends to retain its association with the mandibular gnathobase.)

The maxillules (Fig. 41-2B) are quadrangular in outline and distally armed with robust spines.

The maxillae (Fig. 41-2C) block off the posterior end of the mouth field and are generally simple setose lobes.

The maxillipedes or mouth cirri (Fig. 41-2D) are short and almost club-like. The basal protopod has two elongate segments. The rami are inflated, with short broad segments covered with plumose setae. These limbs may be extremely reduced.

The basis of the posterior cirri (Fig. 41-2E) are separated from the mouthparts by a considerable distance, but extend back up through the mantle cavity toward the aperture. In most acrothoracicans they are extended out through the aperture while feeding. As a result, the cirral

Fig. 41-2. Appendages of *Weltneria hessleri*. (A) mandible; (B) maxillule; (C) maxilla; (D) mouth cirrus; (E) cirrus VI; (F) caudal appendage (from Newman, 1971). (G) male of *Weltneria exargilla* (from Newman, 1974).

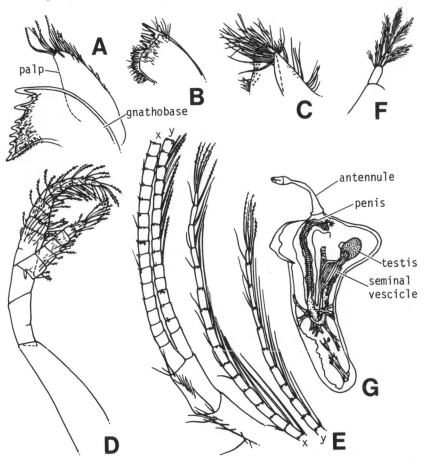

rami are typically rather long and flagelliform. In *Trypetesa* the cirri are short, uniramous, and remain within the mantle (Fig. 41-1C). Caudal appendages (Fig. 41-2F) may or may not be present.

The male is tiny (Fig. 41-2G), about the size of, or even smaller than, a cyprid. They have well-developed antennules, by which they attach to the female or to the burrow, but lack a gut or feeding appendages. They are in effect gonad bags with testes and seminal vesicles but may or may not possess a penis.

The mouth leads to an esophagus, which is situated dorsal to the adductor muscle. The stomach is typically a simple enlarged chamber, except in the cryptophialids, which have a distinct 'cardiac stomach' or 'gizzard' preceding the stomach proper. The stomach is a large, glandular organ, with one or more digestive caeca arising from it. In the pygophorans there is a simple posterior midgut extending to a dorsal anus situated between the terminal cirri. The apygophorans, or trypetesids, have only an esophagus and stomach and lack the posterior gut and anus.

Little is known of the acrothoracican circulatory system. It appears to be lacunar, and no heart has yet been recognized. Respiration appears to be by body diffusion with no specific gill-like structures noted. There are various conical and filamentary body processes and mantle flaps known (Tomlinson, 1969a), and these may serve some accessory respiratory function.

The principal excretory organs are maxillary glands. Utinomi (1960) also noted what he refers to as cephalic nephrocytes, large round cells in scattered masses within the cephalon, and these are similar to isolated cells found in the rest of the body. Since these 'nephrocytes' actively take up carmine dye, it is assumed they have some accessory excretory function.

The ovaries typically lie in the dorsal part of the body and can extend into portions of the mantle. The oviducts exit into an atrium, which in turn opens into the mantle cavity just laterad of the mouth cirri. The male has a testis and seminal vesicle and in a few species is equipped with a long penis.

The nervous system is much reduced, with a single cephalic ganglion and one or two ventral ganglia in the thorax.

NATURAL HISTORY   The only locomotory phase in the acrothoracican life cycle is the dispersive cyprid larva. Only the adult females are capable of feeding.

Details of feeding are somewhat obscure, since we are not sure of what exactly goes on inside the mantle (Tomlinson, 1969a). However, much can be inferred from the anatomy. Cirri in most species are extended from the mantle to sweep in a rhythmic pattern. Because the cirri are clumped near the posterior end of the thorax, they cannot move in a manner designed to progressively move food to the mouth. Rather, it appears that as the cirri are withdrawn, they are cleaned of trapped food particles by the mouth cirri. These in turn are then cleaned by the labrum and mouthparts.

Trypetesa, however, has fewer and less elaborately developed cirri than other acrothoracicans, and these are incapable of being extended beyond the aperture. The gyrations of the body and the beating of the cirri within the mantle in this case set up water currents, which then bring food particles into the mantle chamber where they are entrapped by the mouthparts and mouth cirri. The lack of a complete gut in *Trypetesa* seems to be compensated for by the regurgitation of undigestible particles (Tomlinson, 1969a).

The manner of copulation and exact location of fertilization are not known. However, males are apparently capable of fertilizing several broods, though as they become exhausted they are successively replaced by new males (Turquier, 1972). All nauplii are retained within the mantle; expulsion occurs only when the cyprid stage is achieved.

The cyprid larvae are generally active swimmers and serve as a dispersive stage. However, in the genus *Cryptophialus* the cyprids lack swimming limbs and can only crawl on the substrate after being evicted by the parent. Consequently, such species are noted for their dense infestations. Site selection is achieved by the antennules, and at least the initial excavation of the attachment site is chemically facilitated. Turquier (1968) detected the presence of carbonic anhydrase in the mantle of *Trypetesa nassaroides*. However, whatever the role of chemicals, it seems that by the time metamorphosis occurs the role of mechanical abrasion in the formation and subsequent enlargement of the burrow has become at least equally as important (Tomlinson, 1969b). All acrothoracicans burrow into some form of calcareous hard substrate (Seilacher, 1969), except for *Weltneria exargilla*, which was found in soft, siliceous clay (Newman, 1974).

Tomlinson (1969a) summarized the then current knowledge about acrothoracican zoogeography. Certain generalizations can be made, for example, certain species seem to be circumtropical or circumsubtropical, such as *Lithoglyptes spinatus* and *Kochlorine hamata* (Fig. 41-3). Acrothoracicans seem generally restricted to shallow water in the tropical and temperate zones, except for one species, *Australophialus tomlinsoni*, found south of the Antarctic Convergence (Newman and Ross, 1971).

One could offer comments on the restrictions of certain taxa to particular geographic areas, but the group is still so poorly known to make such statements risky. For example, Tomlinson (1969a) thought the genus *Weltneria* was Indo-Pacific, but Newman (1971, 1974) found two species of the genus in the deep sea from off Bermuda and in the Bay of Biscay. Studies of biogeography are also complicated by incompleteness and uncertainty of our knowledge about the relationships among species, and Tomlinson (1969a) issued caveats concerning difficulties in assessing possible convergences.

DEVELOPMENT The most recent examination of acrothoracican embryology are the studies of Turquier (1967a, b) on *Trypetesa nassaroides*. They

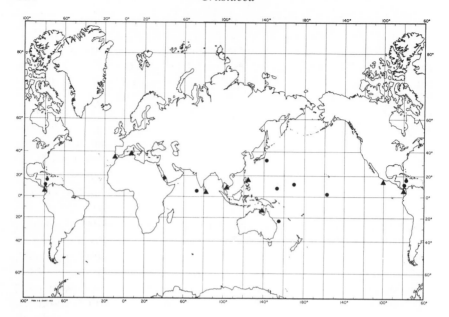

Fig. 41-3. Distribution of the circumtropical–circumsubtropical species *Lithoglyptes spinatus* ● and *Kochlorine hamata* ▲.

reveal a highly modified form of spiral cleavage with a high degree of yolk segregation in the D-cell lineage. Early cleavages (Fig. 41-4A, B) of the oval-shaped egg result in one large yolk cell with a cap of micromeres (Fig. 41-4C). Turquier's numeration of these cells is different from that usually employed and thus difficult to compare to the system employed by Anderson (1969). Nor does it appear to have been possible to determine exact identity of cells in the A–C lineages, which would have allowed the determination of the nature of mesoderm formation. The endoderm is derived from the 3D cell (Fig. 41-4D). Proliferation of the micromere cap proceeds to about 250 cells and eventually achieves gastrulation of the large yolk cell by epiboly (Fig. 41-4D).

Formation of the egg–nauplius (Fig. 41-4E) is then initiated near the anterior part of the embryo, while a caudal papilla is formed in the posterior region. The antennular anlagen are uniramous, those of the antennae and mandibles biramous (Fig. 41-4F).

Subsequent to hatching, four naupliar stages (Fig. 41-4G) are recognized, the last stage being metanaupliar (Fig. 41-4H) in nature. None of these stages feed, and as a result the antennae never develop a naupliar process. The final molt achieves a transformation to a typical cyprid larva (Fig. 41-4I), akin to that seen in thoracicans, including the antennules being modified for probing and attachment (Fig. 41-4J).

Attachment of the cyprid (Fig. 41-5A) results in a molt to a unique pupa (Fig. 41-5B), or postcyprid larva, in which further metamorphosis (Fig. 41-5C) occurs (Turquier, 1970). The larval tissues break down, and the nervous system rotates approximately 180° in relation to the peduncle (Fig. 41-5D). The adult body form slowly appears as the excavation of the burrow begins. Finally the antennules are cast off and the adult stage of the life cycle begins (Fig. 41-5E).

FOSSIL RECORD   A modest fossil record (Newman et al., 1969) exists for the acrothoracicans extending from the Devonian (Rodriguez and Gutschick,

Fig. 41-4. *Trypetesa nassaroides* development. (A) egg; (B) second cleavage; (C) third cleavage forming micromeres and one large macromere; (D) approximately fifth cleavage stage with epiboly beginning around 3D cell; (E) early egg–nauplius; (F) late egg—nauplius; (G) second nauplius; (H) fourth nauplius or metanauplius; (I)cyprid; (J) detail of cyprid antennule. (From Turquier 1967a, b)

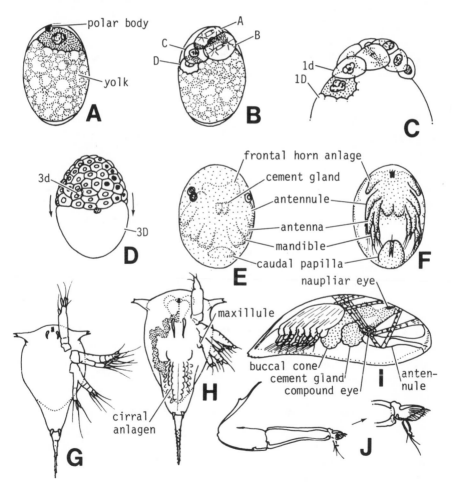

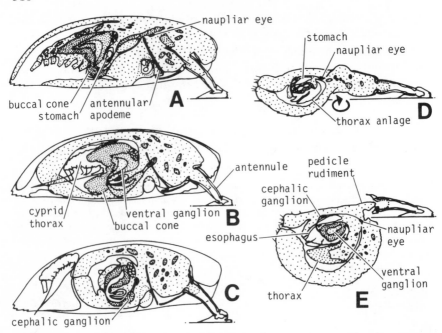

Fig. 41–5. Metamorphosis of *Trypetesa nassaroides*. (A) attached cyprid, tissue breakdown beginning; (B) cyprid undergoing molt; (C) early postcyprid or pupa stage, tissue reorientation advanced; (D) late pupa, tissues rotate in relation to peduncle and burrow excavation begins; (E) juvenile female immediately after the pupal molt. (From Turquier, 1970)

1977) through the Pliocene. These taxa are for the most part trace fossils, consisting typically of burrows of a supposedly distinctive acrothoracican form, though Turner (1973) reported an actual body fossil. Grygier and Newman (1985) cautioned that acrothoracican fossils should be examined closely to determine exactly whether they are just traces or in fact represent remains of the actual body. Tomlinson (1963) described *Trypetesa caveata* from myalinid clams in the Late Pennsylvanian and Early Permian of the central United States. Other more problematic forms were described by Saint-Seine (1951, 1954) and Codez and Saint-Seine (1957) and were generally placed in two families, the Rogerellidae and Zapfellidae, of uncertain affinities, and defined mostly on the basis of burrow depth–length ratio and form of the orifice. However, Tomlinson (1969a) placed these genera within the lithoglyptids.

TAXONOMY   The current higher taxonomy of the group dates from the turn of the century, with some revision of the families made by Tomlinson (1969a). The Pygophora are distinguished by having a complete gut, biramous cirri, and two ventral ganglia. The Apygophora have an incomplete gut, only 3 uniramous cirri, and a single ventral ganglion.

Suborder Pygophora Berndt, 1907 Devonian–Recent
    Family Lithoglyptidae Aurivillius, 1892 Devonian–Recent
    Family Cryptophialidae Gerstaecker, 1866 Recent
Suborder Apygophora Berndt, 1907 (?Carboniferous) Recent
    Family Trypetesidae Stebbing, 1910

REFERENCES

Anderson, D. T. 1969. On the embryology of the cirripede crustaceans *Tetraclita rosea, T. purpurascens, Chthamalus antennatus,* and *Chamaesipho columna* and some considerations of crustacean phylogenetic relationships. *Phil. Trans. Roy. Soc. Lond.* (B)**256**:183–235.

Berndt, W. 1907. Über das Sytem der Acrothoracica. *Arch. Naturgesch.* **73**:287–89.

Codez, J., and R. Sainta-Seine. 1957. Revision des Cirripèdes Acrothoraciques fossiles. *Bull. Soc. Geol. Fr.* (6)**7**:699–719.

Gruvel, A. 1905. *Monographie des Cirrhipèdes ou Thécostraces.* Massonet Cie, Paris.

Grygier, M. J., and W. A. Newman. 1985. Motility and calcareous parts in extant and fossil Acrothoracica, based primarily upon new species burrowing in the deep-sea scleractinian coral *Enallopsammia. Trans. San Diego Soc. Nat. Hist.* **21**:1–22.

Newman, W. A. 1971. A deep-sea burrowing barnacle from Bermuda. *J. Zool.* **165**:423–29.

Newman, W. A. 1974. Two new deep-sea Cirripedia from the Atlantic. *J. Mar. Biol. Assc. U.K.* **54**:437–56.

Newman, W. A., and A. Ross. 1971. Antarctic Cirripedia. *Antarctic Res. Ser.* **14**:1–257. Am. Geophys. Union, Washington.

Newman, W. A., V. A. Zullo, and T. H. Withers. 1969. Cirripedia. In *Treatise on Invertebrate Paleontology,* Part R, *Arthropoda* **4**(1) (R. C. Moore, ed.), pp. R206–95. Geol. Soc. Am. and Univ. Kansas Press, Lawrence.

Rodriguez, J., and R. C. Butschick. 1977. Barnacle borings in live and dead hosts from the Louisiana Limestone of Missouri. *J. Paleo.* **51**:718–24.

Saint-Seine, R. 1951. Un cirripede acrothoracique de Cretace: *Rogerella lecontrei. C. R. Acad. Sci. Paris* **233**:1051–53.

Saint-Seine, R. 1954. Existence de cirripèdes acrothoraciques des la Lias, *Zapfella pattei. Bull. Soc. Geol. Fr.* (6)**4**:447–51.

Seilacher, A. 1969. Paleoecology of boring barnacles. *Am. Zool.* **9**:705–19.

Tomlinson, J. T. 1963. Acrothoracican barnacles in Paleozoic myalinids. *J. Paleo.* **37**:164–66.

Tomlinson, J. T. 1969a. The burrowing barnacles. *Bull. U.S. Nat. Mus.* **296**:1–162.

Tomlinson, J. T. 1969b. Shell-burrowing barnacles. *Am. Zool.* **9**:837–40.

Turner, R. F. 1973. Occurrence and implications of fossilized burrowing barnacles. *Geol. Soc. Am. Abstr. with Programs* **5**:230–31.

Turquier, Y. 1967a. Le développement larvaire de *Trypetesa nassaroides* cirripède acrothoracique. *Arch. Zool. Exp. Gen.* **108**:33–47.

Turquier, Y. 1967b. L'embryogenèse de *Trypetesa nassaroides,* ses rapports avec celle des autres cirripèdes. *Arch. Zool. Exp. Gen.* **108**:111–37.

Turquier, Y. 1968. Recherches sur la biologie des cirripèdes acrothoraciques. *Arch. Zool. Exp. Gen.* **109**:113–22.

Turquier, Y. 1970. Recherches sur la biologie des Cirripèdes Acrothoraciques III. La métamorphose des cypris femelles de *Trypetesa nassasoides* et de *T. lampas. Arch. Zool. Exp. Gen.* **111**:573–628.

Turquier, Y. 1972. Contribution à la connaissance des Cirripèdes Acrothoraciques. *Arch. Zool. Exp. Gen.* **113**:499–551.

Utinomi, H. 1960. Studies on the Cirripedia Acrothoracica II. Internal anatomy of the female *Berndtia purpurea. Publ. Seto Mar. Lab.* **8**:223–79.

42

LARVAL FLOTSAM

HISTORY   Hansen (1899), in a survey of some of the larvae of the Plankton Expedition, recognized several unassignable larval types, designated with Greek or Latin letters. Among them was a series of five particularly distinctive and enigmatic nauplii called Y-larvae. Subsequently, Y-nauplii were recorded by Steuer (1905), McMurrich (1917), and some others, but it was not until Bresciani (1965) that a Y-cypris was recognized. T. Schram (1970a, b, and 1972) and Elofsson (1971) essentially reviewed what is known of these forms. There has never been a consensus as to what Y-larvae are. Hansen (1899) felt they were larvae of apodan cirripedes, though the Apoda are now known to be parasitic isopods. Bresciani (1965) preferred ascothoracidan affinities, a proposition that T. Schram (1970a) rejected but for which he substituted no suggestion of his own except to imply some proximity to cirripedes. Grygier (1983) reviewed the evidence and could conclude nothing beyond suggesting possible maxillopodan affinities, but Grygier (1984) created a separate taxon, Facetotecta, to accommodate them.

Among the Paleozoic fossils, Müller (1983) described and named several species of minute yet exquisitely preserved Cambrian crustaceans. These appear probably, for the most part, to represent metanaupliar stages of some yet unknown animals rather than actual adults.

Y-LARVAE

T. Schram (1970b, 1972) condensed the five Y-nauplii types originally recognized by Hansen into just two, Hansen's type I and IV. The carapace of these nauplii covers the entire body and is marked by sets of plates in a symmetrical pattern. The sutures between the plates are quite strong in the posterior region of the carapace, but in the type I larvae (Fig. 42-1A) the sutures are often absent or poorly developed anteriorly, especially in the smaller or younger stages. The type IV nauplii have more numerous plates, with the sutures clearly marked in all size classes (Fig. 42-1C). The posterior region of the carapace is marked by large spines, three in type I and one in type IV. The single median spine is located in all cases just in front of a curious, slightly raised oval region associated with the dorsocaudal organ inside the larva. The ventral sides of the nauplii are of a typical form (Fig. 42-1B), with a large spatulate labrum and three sets of appendages.

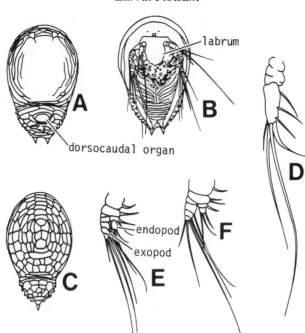

Fig. 42-1. Y-nauplius. (A) dorsal view of type I; (B) ventral view of type I; (C) dorsal view of type IV; (D) antennule; (E) antenna; (F) mandible. (From T. Schram 1970b, 1972)

The antennules (Fig. 42-1D) are uniramous and composed of three joints. The articulation between the first and second joint may be obscure, depending on stage of development, while the distalmost joint bears several large smooth setae.

The biramous antennae (Fig. 42-1E) have a two-segment protopod, a two-segment endopod, and a six-segment exopod. The two basalmost joints of the exopod are only partially divided, and the terminal joint is minute. All elements of the limb bear strong medially and/or ventrally directed spines.

The biramous mandibles (Fig. 42-1F) are similar to the antennae except for a five-jointed exopod and a slightly more robust protopod.

Elofsson (1971) investigated the internal anatomy of the Y-nauplius (Fig. 42-2). The digestive system, though it lacks an anus, is well developed. The labrum contains a large gland whose cells are marked with prominent Golgi bodies and large vacuoles, indicating an active secretory function. The V-shaped esophagus has walls equipped with well-developed circular and longitudinal muscle layers. The midgut is divided into a large anterior portion and a small posterior blind sac. The posterior sac abuts into a deep 'cleft' or invagination that has a T- or Y-shaped cross section. No connection of the midgut into this cleft has been noted (though it is possible this may have pulled away in preparation).

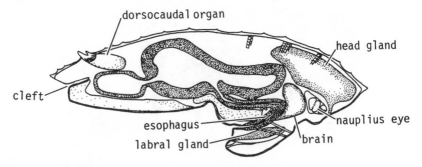

Fig. 42-2. Y-nauplius internal morphology. (From Elofsson, 1971)

In addition to the labral glands, two other glands are noted. The anterior part of the larva is filled with a pair of head glands. These open just to each side of the nauplius eye. These glands seem to resemble similar glands observed in *Pollicipes* and *Trypetesa*. Dorsal to the posterior midgut sac is a large dorsocaudal gland. This organ seems secretory in nature, is closely associated with the median oval plate on the carapace anterior to the posterior median spine, but does not seem to open to the outside by any duct. The function of this unique structure is unknown.

The brain is a typical naupliar type. The anlagen of the compound eyes are present, and the naupliar eye is well developed. The latter has three cups, and Elofsson (1971) compares it closely with the naupliar eye of balanid cirripedes.

McMurrich (1917) was the first to note the Y-cypris. He noted these developing within the cuticle of a type IV Y-nauplius. However, Bresciani (1965) recorded the collection of the first free Y-cypris, and T. Schram (1970a) described the morphology of a slightly different Y-cypris in detail.

The plated carapace is the most distinctive feature of the animal (Fig. 42-3) and extends back over the thorax. The thorax is composed of six segments. The abdomen is divided into an anterior portion of three to five free segments and a posterior portion of apparently fused segments covered with a plated cuticle.

The head is marked by a pair of laterally placed compound eyes beneath the carapace (Fig. 42-3A). The only evident cephalic appendages are the paired antennules (Fig. 42-3A, B). These uniramous limbs are composed of six segments. The second joint is hinged at right angles to the first, forming an 'elbow' and obscuring the proximal joint in ventral view. The large fourth joint has a dorsally hooked spine, the fifth joint bears some small setae, and the bifurcate sixth joint has a small seta and a large spine on one branch with a large aesthetasc on the other. Posterior to the antennules is the oral pyramid, a lobate structure with a wicked-looking set of clawlike hooks on its terminus (Fig. 42-3C). No other appendages, nor any actual mouth opening has been noted on the few Y-cypris specimens known.

The thorax bears six pairs of biramous, paddlelike appendages. The protopods have a distinct bend in their coxae to give the limb pair a lateral spread. The first thoracopod (Fig. 42-3D) has two joints on each ramus; the rest of the limbs (Fig. 42-3E) typically have a three-joint endopod. On some cyprids, the sixth limb varied from the latter pattern with a single protopod joint and a two-jointed endopod (Fig. 42-3F). The thoracopods bear some long, brushlike setae on their termini.

The anterior abdominal segments bear no limbs. The posterior abdominal unit is plated and composed of an undetermined number of fused segments (Fig. 42-3G). This whole unit, termed a 'furca' by T. Schram (1970a) terminates in a median-plated lobe flanked by complex

Figure 42-3. Y-cypris. (A) lateral view of whole larva; (B) ventral view of cephalon; (C) lateral view of oral pyramid; (D) first thoracopod; (E) second thoracopod; (F) sixth thoracopod variant; (G) posterior part of abdomen. (From T. Schram, 1970a)

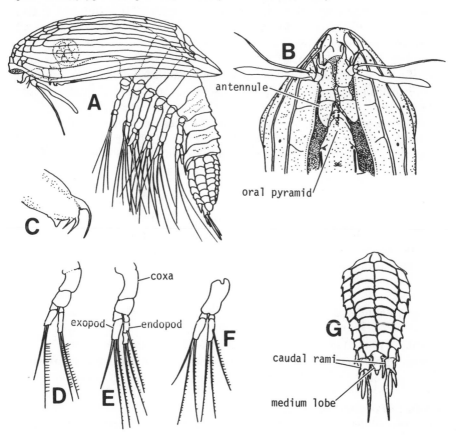

'caudal appendages' or rami. The rami have lobed and spinose proximal portions that bear on their distal ends a set of three lappetlike setae.

NATURAL HISTORY   Our understanding of what Y-larvae do in their environment is hindered by our not knowing what they become when they metamorphose into an adult stage. The glandular tissues of the labral gland and midgut epithelium combined with the muscular esophagus seem to imply active feeding on the part of the Y-nauplius. However, the cephalic morphology of the Y-cypris would imply a shift in that stage to a parasitic mode of feeding. Specific observations of active animals are needed to confirm these suspicions.

There is some confusion in the literature as to the biogeographic distribution of the Y-larvae. Hansen (1899) reported specific localities for each of his five naupliar types: types I and V—in lee of St. Vincent, type II—Canary stream (Cape Verde Islands), type IV—Sargasso Sea and St. Vincent, type IV—around Kiel. At the conclusion of the work he reports these just noted and additional localities in the equatorial Atlantic without any reference to which larval types were found where. Steuer (1905) recorded a type IV nauplius off Trieste in the Adriatic, and McMurrich (1917) found type IV in Passamaquoddy Bay, Canada. Bresciani (1965) mapped all apparent localities known until that time, including his own from the Oresund. T. Schram (1970b, 1972) recorded both types I and IV in and around the Oslofjord, but ignored the equatorial Atlantic localities when summarizing distribution data: type I—Sargasso Sea, Cape Verde Islands, and the Oslofjord; type IV—Baltic, Norwegian coastal waters, north Atlantic, and the Adriatic. In addition to these Atlantic sites, Newman (personal communication) reports Y-larva off the southern California coast (type of nauplius indeterminate), and Ito and Ohtsuka (1984) record Y-cypris in Japan.

T. Schram (1972) also tallied all the collecting records for type IV larvae as to time of year. The nauplii were noted to be present in all months of the year except July, with collecting peaks in February, April, June, and September. It would appear the parent organism, whatever it is, apparently breeds in cycles throughout the year, with the strongest pulses in winter and spring.

ORSTEN LARVAE

Müller (1983) described a series of microscopic crustaceans from the Cambrian of Sweden. The fossils are phosphatized and collected from marls, locally called orsten. The preservation they display is in exquisite detail and allows rather detailed reconstructions to be made not only of body form but of appendages as well.

Though Müller (1983) has chosen to give these forms (Fig. 42-4) indi-

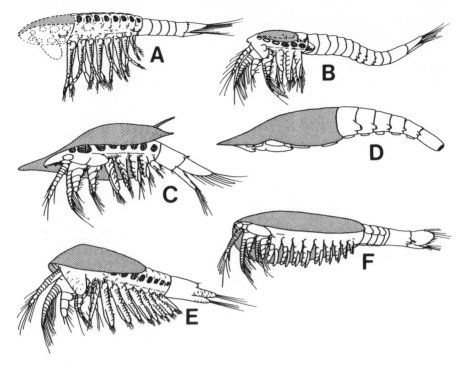

Fig. 42-4. Cambrian orsten larvae reconstructions. (A) *Dala peilertae*; (B) *Skara anulata*; (C) *Walossekia quinquespinosa*; (D) *Oelandocaris oelandica*; (E) *Bredocaris admirabilis*; (F) *Rehbachiella kinnekullensis*. (From Müller, 1983)

vidual generic and specific names, these should probably be viewed as names of convenience, since the fossils themselves in many instances seem to be remains of metanauplii for the most part rather than actual adult organisms. The reasons for so concluding are as follows. The fossils are in the size range for larvae, 800 to 2000 μm. The appendages are of a form characteristic of larvae, that is, joints large relative to the size of the whole limb with only one or a few large, smooth spines or setae associated with each joint. The antennae are pediform, that is, more like the mandibles and posterior appendages than the antennules. The mandibles are pediform and not with especially well-developed gnathobases. And finally, the maxillules and maxillae are not strikingly differentiated; in fact, the maxillae appear not to be developed at all in some of the forms.

One peculiar feature of several of the orsten larvae is the apparent relationship of the carapace to the head and thorax. In some of the species, *Bredocaris admirabilis* (Fig. 42-4E) and *Walossekia quinquespinosa* (Fig. 42-4C), the carapace may arise from the maxillary segment. However, in *Rehbachiella kinnekullensis* (Fig. 42-4F), the carapace apparently arises from three segments posterior to the maxillary somite. If this indeed proves to be so, this might be an argument for interpreting these fossils as ontogenetically very advanced larvae or possibly juveniles.

Though the orsten microcrustaceans will continue to be of tremendous interest, their direct use in phylogenetic or taxonomic analysis of crustaceans may be quite limited unless compared to other larval types.

REFERENCES

Bresciani, J. 1965. Nauplius 'Y', its distribution and relationship with a new cypris larva. *Vidensk. Medd. fra Dansk Naturh. Foren.* **128**:245–58, 1 pl.
Elofsson, R. 1971. Some observations on the internal morphology of Hansen's nauplius Y. *Sarsia* **46**:23–40.
Grygier, M. J. 1983. Ascothoracida and the unity of the Maxillopoda. *Crust. Issues* **1**:73–104.
Grygier, M. J. 1984. Comparative Morphology and Ontogeny of the Ascothoracida, a Step Toward a Phylogeny of the Maxillopoda. Ph.D. dissertation, Scripps Inst. Oceanogr., Univ. California, San Diego.
Hansen, H. J. 1899. Die Cladoceren und Cirripidien der Plankton–Expedition. *Ergeb. Plankton-Exped. Humboldt-Stiftung* **2**:1–58.
Ito, T., and S. Ohtsuka. 1984. Cypris Y from the North Pacific. *Publ. Seto Mar. Biol. Lab.* **29**:179–86.
McMurrich, J. P. 1917. Notes on some crustacean forms occurring in the plankton of Passamaquoddy Bay. *Trans. Roy. Soc. Can.* (3)**11**(4):47–61.
Müller, K. J. 1983. Crustacea with preserved soft parts from the Upper Cambrian of Sweden. *Lethaia* **16**:93–109.
Schram, T. A. 1970a. Marine biological investigations in the Bahamas 14. Cypris Y, a later developmental stage of nauplius Y. *Sarsia* **44**:9–24.
Schram, T. A. 1970b. On the enigmatical larva nauplius Y type I. *Sarsia* **45**:53–68.
Schram, T. A. 1972. Further records of nauplius Y type IV from Scandinavian waters. *Sarsia* **50**:1–24.
Steuer, A. 1905. Über eine neue Cirripedienlarve aus dem Gulf von Trieste. *Arb. Zool. Inst. Wien.* **15**:113–18.

43

PHYLOGENY AND HIGHER CLASSIFICATION

Past discussions of crustacean phylogenetic relationships and classification have been, for the most part, of a rather subjective, evolutionary systematic nature. The use of more objective forms of character analysis has been a recent phenomenon and largely restricted in their scope to particular groups (e.g., spinocalanid copepods, Fleminger, 1983; pagurids, McLaughlin, 1983; tanaidaceans, Sieg, 1983; idoteine isopods, Brusca, 1984; and eumalacostracans, Schram, 1984b). The classic principles of cladistic analysis (Hennig, 1966), tempered with consideration of structural plans (Schram, 1983) and functional morphology (Boxshall, 1983), can prove productive toward a more logical and reasoned understanding of crustacean interrelationships.

The analyses included in this chapter were for the most part performed using Wagner 78 or programs modified therefrom, and utilizing the computer facilities of either the University of Southern California or the California State University system. The programs are designed to produce the most parsimonious arrangement of taxa given the total matrix of characters. This typically will result in the shortest tree, with the highest degree of congruence of characters (high C-index), and the lowest amount of convergence and character reversal (low F-ratio). Wagner programs presuppose that reversals in character expression are possible, and this is a most important consideration! When the data matrices were analyzed with the added restriction in the program that character reversals were to be avoided or minimized, the computer produced longer trees with somewhat lower C-indexes and slightly higher F-ratios than those analyses in which the only restriction was parsimony.

The programs thus occasionally produced some striking character reversals (expressed as minus signs before numbers). For example, Maxillopoda and Phyllopoda (Fig. 43-1) are characterized in part by a lack of mandibular palps. However, this means that the mandibular palp had to reappear in the barnacles, copepods, mystacocarids, and ostracodes (Fig. 43-6). The computer is merely telling us that the total arrangement of characters in which the palp is lost once and then reacquired in some individual groups is more parsimonious than one in which the mandibular palp is independently lost in the ascothoracids, rhizocephalans, branchiurans, tantulocarids, nectiopodans, various phyllopodans, and some

525

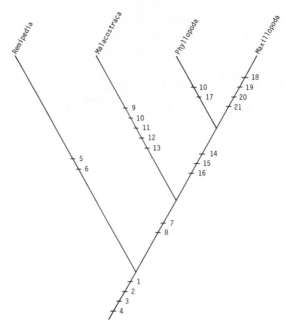

Fig. 43-1. Cladogram depicting the relationships of the four classes of Crustacea herein recognized. Derived characters are: two pairs of antennae (1); biramous antennules and antennae (2); two pairs of maxillae (3); nauplius larva or egg–nauplius stage (4); subrectangular head shield (5); spinose, raptorial posterior mouthparts (6); postcephalic tagmosis (7); typically at most eight thoracic segments (8); malacostracan naupliar eye (9); polyramous limbs (10); thoracic endopod as a stenopod (11); uropods (12); a carapace that covers only, or at least parts of, the thorax and does not envelop the limbs (13); abdomen lacking limbs (14); uniramous antennules (15); generally no mandibular palp (16); leaflike (foliaceous) limbs (17); at most 11 trunk segments (18); no more than six thoracic segments (19); a short bulbous heart (20); and maxillopodan naupliar eye (21).

eumalacostracans. (In a biological context we must assume that flexible developmental programs exist in the animals to allow shifts in character expression. That such is indeed the case is discussed in Chapter 44.) The strength of one hypothesis over the other, however, is testable by the analysis of additional taxa or other characters. The trees presented here are thus working hypotheses but afford a much clearer understanding of crustacean phylogeny than anything proposed heretofore.

Crustaceans can be characterized (Chapter 1) by the possession of (Fig. 43-1) two sets of antennae (1), both of these being biramous (2); two pairs of maxillae (3); and either a nauplius larva or egg–nauplius stage in ontogeny (4). The primitive crustacean condition was probably that of a cephalic feeder with biramous trunk limbs used only for locomotion (see Chapter 44). This latter point is seconded by the initial appearance generally of biramous limb buds in the course of development no matter how complex the ultimate adult limb, and by the retention of biramous limbs posteriorly on the body when more anterior limbs may become specialized to more complex forms. The biramous hypothesis was also supported by analysis of 67 characters for 23 'order' level taxa. When leg characters were scored to treat biramy as primitive, the resulting trees were shorter (109 steps) than when characters were scored to treat polyramy as primitive (113 steps). This means that the assumption of biramous primitive limbs results in more parsimonious arrangement of crustacean orders than assumptions of polyramous primitive limbs.

The interrelationships of the four major classes of crustaceans recognized in this book are presented in the cladogram in Figure 43-1. The Remipedia appear to be the most primitive group and one distinguished by the possession of a great many primitive features: biramous antennules, lack of postcephalic tagmosis with serial homonomy of biramous trunk limbs, lack of a carapace, and segmentation of at least some internal organ systems. Indeed, it is difficult to define shared derived characters uniting the two constituent orders of this group (the living Nectiopoda and extinct Enantiopoda) except for the subrectangular head shield (5) and the specialized, spinose, raptorial mouthparts (6). All other crustacean groups are generally characterized by the possession of postcephalic trunk tagmosis (7) and usually a thorax of eight or fewer segments (8). The preservation on the enantiopod *Tesnusocaris* is not good enough to allow determination of details of the mouthparts. However, another series of characters that might act to unite remipedes is a set of mandibles without palps and these enclosed in a sublabial atrium oris to act as a sort of mandibular mill to process food. More and better material of the Carboniferous fossil some day might allow resolution of these features.

The most primitive of the three 'higher' crustacean classes is the Malacostraca in that they possess appendages on all segments of the body. The taxon Malacostraca is used here *sensu stricto* to include only the Hoplocarida and Eumalacostraca and is characterized by the following derived features: malacostracan naupliar eye (9), polyramous limbs (10), endopod of the thoracopods as a stenopod (11), uropods (12), and a carapace that covers only, or at least parts of, the thorax and does not envelop the limbs (13). The other two crustacean classes are characterized by the abdomen lacking most or all limbs (14), uniramous antennules (15), and the general loss of the mandibular palp (16).

The Phyllopoda are characterized by polyramous limbs (10) with a leaf-like form (17). These would seem to include the Phyllocarida (at least the leptostracans), the cephalocarids *sensu lato* (i.e., Brachypoda and Lipostraca), and the Branchiopoda.

The Maxillopoda are defined by the possession of at most 11 postcephalic trunk segments (18), at most six thoracic segments (19), a small bulbous heart (20), and in general the possession, if present, of a maxillopodan naupliar eye with lens and tapetal cells (21). These include the

Tantulocarida, Mystacocarida, Ostracoda, Branchiura, Copepoda, Rhizocephala, Ascothoracida, and the Cirripedia *sensu stricto*. These characters
and included taxa essentially agree with those of Grygier (1983), and their
grouping as a distinct class may also be reinforced by their lack of frontal
organs and possession of a pair of lateral midgut caeca.

As can be seen from the above, though the form of the thoracopods is
not unimportant (Chapter 44), the main theme of crustacean evolution
appears to be that of tagmosis. From a condition where the trunk is composed of a long series of homonomous units (remipedes) the body is
initially regionalized, and then the tagma become variously shortened or
their development aborted. Consequently, we can recognize and concisely
define four classes of crustaceans rather than the six (e.g., Bowman and
Abele, 1982) or more (e.g., Boxshall, 1983) previously recognized. This
smaller number is more akin to the degree and scope of classes as seen in
other arthropodous phyla (e.g., see Anderson, 1973; Manton, 1977; or
Schram, 1978).

The two currently recognized orders within the Remipedia (Fig. 43-2A)
are characterized by the elongate, subrectangular form of the head shield
(1), and the distinctive development of the posterior mouthparts as specialized, spinose, raptorial limbs (2). The lack of what might be termed
'robustness' in these apparently derived features of the class is related
undoubtedly to the fact that the two known orders of remipedes represent
a rather disparate pair of taxa. These are undoubtedly only part of what
probably was a rather diverse radiation on the primitive crustacean plan.
Certainly, both remipede orders are better characterized by their autapomorphies rather than their shared features with each other. The extinct
Enantiopoda have enlarged antennules (3), reduced antennae (4) and sessile compound eyes (5). The living Nectiopoda have, among other things,
the first trunk segment fused to the cephalon (8), a well-developed maxillipede (7), prehensile or subchelate grappling mouthparts (8), mandibulari-
form proximal maxillulary endites (9), 'internalized' asymmetric mandibles
(10), and a lack of eyes (11). The great number of nectiopodan apomorphies is related to their highly derived and specialized mode of carnivory.

The Malacostraca have traditionally been viewed as being at the apex
of crustacean evolution. Though the trunk is clearly divided into thorax
and abdomen—a derived feature of all higher crustaceans—their retention
of limbs on all the trunk segments is a primitive feature. Two distinct malacostracan subclasses are recognized (Fig. 43-2B), united by possession of
the malacostracan naupliar eye of three cups each with three everse sensory cells (1), polyramous (2) and stenopodous (3) thoracopods, uropods
(4), and a postcephalic carapace structure that does not envelop the abdomen or thoracic limbs (5). The Eumalacostraca, the most diverse of the
two subclasses, is defined by the single segment antennal scale (6), and the
great elaboration of the abdominal musculature (7) to facilitate the caridoid escape reaction. The Hoplocarida are united by their possession of

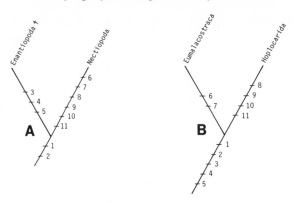

Fig. 43-2. (A) Cladogram depicting the relationships of the two orders of Remipedia. Derived characters are elongate subrectangular head shield (1); spinose, raptorial posterior mouthparts (2); enlarged antennules (3); reduced antennae (4); sessile compound eyes (5); first trunk limb fused to cephalon (6); maxillipede (7); specialized grappling mouthparts (8); mandiblelike basalmost endites of maxillule (9); 'internalized' mandibles (10); loss of compound eyes (11).

(B) Cladogram depicting the relationships of the two subclasses of Malacostraca. Derived characters are: malacostracan naupliar eye (1); polyramous (2); stenopodous (3); thoracopods, uropods (4); thoracic body carapace (5); single-segment antennal scale (6); caridoid abdominal musculature (7); triramous antennules (8); three-segment thoracic protopods (9); four-segment exopod of thoracopods (10); dendrobranchiate gills on the pleopods (11).

triramous antenules (8), three-segmented thoracic protopods (9), four-segmented outer branch of the thoracopods (10), and dendrobranchiate gills on the pleopods (11).

The interrelationships of the Eumalacostraca were reviewed by Schram (1984b). After that paper was submitted for publication, some new taxa were proposed within the eumalocostracans. Felgenhauer and Abele (1983) have recognized the separate status of *Procaris* from all other Eukyphida, which are best left together for now in their own taxon Caridea. Bowman et al. (1985) described a new order, Mictacea, which appears to have a close relationship to the edriophthalmids, that is, the Isopoda and Amphipoda.

Essentially four lines of evolution can be recognized within the Eumalacostraca: the eucarids, belotelsonids, syncarids, and waterstonellids–pancarids–peracarids.

The eucarid lineage (Fig. 43-3) is characterized by possession of a zoea larva (3) and carapace fused to the thorax (4). The Euphausiacea are characterized by possession of a petasma in the males (5), but this is a feature that also developed in the evolution of the Anaspidacea, Dendrobranchiata, and Reptantia. All other eucarids fuse at least the first thoracomere to the cephalon (6). The Amphionidacea do not have an epipodite on their maxillipede (7) and the first pleopod in the females is hypertrophied to

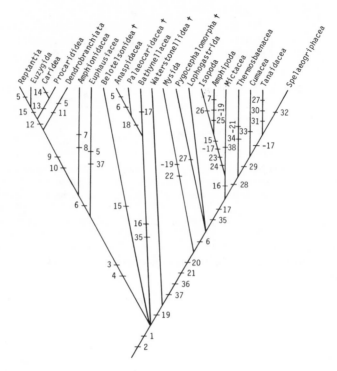

Fig. 43-3. Cladogram depicting the relationships of orders of Eumalacostraca (modified from Schram, 1984b). Derived characters are: single segment antennal scale (1); caridoid musculature in abdomen (2); zoea larva (3); carapace fused to all thoracomeres (4); petasma (5); first thoracomere fused to cephalon (6); loss of maxillipede epipodite (7); first pleopod forming brood pouch under thorax (8); scaphagnathite on maxilla (9); maxillipedes with lamellate protopod with coxal–basal endites directed mediad (10); maxillipede endopod flagelliform (11); eggs brooded on pleopods (12); caridean lobe on maxillipede (13); second pleomere pleuron overlaps that of first and third (14); uniramous thoracopods (15); carapace lost (16); pleopods lost or absent (17); maxillipedes with tendency to form gnathobasic endites, endopod pediform (18); loss of thoracopodal epipodite gills (19); redeveloped thoracopodal epipodite gills (−19), maxillipede with lobate basal endite directed distad (20); oöstegite marsupium (21); loss of oöstegite marsupium (−21); thoracic endopods act as passive filter (22); thoracic coxal plates (23); sessile compound eyes (24); more than one pair of uropods (25); respiratory pleopods (26); male genital cones (27); carapace short (28); maxillipedal epipodite as a cup- or spoonlike respiratory structure (29); maxillipedal epipodite with a tendency to form two to three segments (30); second thoracopods as chelipedes (31); thoracopodal exopods respiratory (32); pseudorostrum and maxillipedal siphons (33); eggs brooded under carapace (34); loss of ventral frontal organ (35); loss of naupliar eye (36); loss of dorsal frontal organ (37); multisegmented uropodal rami (38).

530

form a brood pouch under the thorax (8). The decapods have a scaphagnathite on the maxilla (9) and the maxillipedes are developed with lamellate protopods with medially directed endites (10). The Dendrobranchiata
are characterized by having a petasma (5) and a flagelliform first maxillipedal endopod (11). Gordon (1955) pointed out several apparently derived
features shared by dendrobranchiates and euphausiaceans, namely, abrupt
metamorphosis after each larval stage, the tendency to reduce or lose the
posterior thoracopods, the petasma and spermatophores, thelycum, and
photophores. Any future evaluations of eucarid interrelationships should
reevaluate Gordon's characters, since they could have some significant
implications for realigning eucarid taxonomy. The pleocyemates apparently brood their eggs on their pleopods (12), but whether that in fact is so
in the Procarididea is not known at this time. Eukyphids have a caridean
lobe on the first maxillipede (13), while the only shared feature of the various families of Caridea is the overlapping of the first pleomere pleuron by
the second (14). Reptantia and Euzygida have strictly uniramous thoracopods, and this is a feature that also characterizes the Belotelsonidea as well
as in part the edriophthalmids.

The syncarids lack a carapace (16), a feature that also is shared by the
edriophthalmid–mictacean line. It appears they also lack ventral frontal
organs (35). Bathynellacea have lost and/or reduced the pleopods (17),
while the other syncarids have maxillipedes that apparently had a tendency
to form gnathobasic endites (anaspidaceans at least) and have pediform
endopods (18). The Anaspidacea have a petasma (5), which is not a homolog of the eucarid petasma, and a first thoracomere fused to the cephalon
(6).

The fourth eumalacostracan lineage seems to generally lack thoracic
epipodite gills (19) of which the Waterstonellidea are the most primitive.
Their sister group has distinctive maxillipedes with distally directed lobate
basal endites (20), for the most part has an oöstegite marsupium (21), and
lacks a naupliar eye (36) and dorsal frontal organs (37). The thoracic epipodite gills have reappeared in Mysida (−19), a group that utilizes the
thoracic endopods as passive filters (22). Higher peracarids fuse at least the
first thoracomere to the cephalon (6), of which the Lophogastrida are the
most primitive example, and lack ventral frontal organs (35). The Pygocephalomorpha have genital cones on the males (27). The sister group to
these has a general tendency to reduce or lose the pleopods (17).

The acarideans (i.e., the forms without a carapace, and the use here is
not quite equivalent to the use of this term in Schram, 1981) contain the
more generalized Mictacea with their distinctive multisegmented uropodal
rami (38) and the advanced edriophthalmids. These latter are united in
their possession of uniramous thoracopods (15), well-developed pleopods
(−17), a tendency to form coxal plates on the thoracopods (23), and sessile
compound eyes (24). Amphipods are distinguished by the loss of the epipodite on the maxillipede (7), the reappearance of rather distinctive thoracic

epipodite gills that may not be homologous to other epipodite gills (−19), and more than one pair of uropods (25). The Isopoda are characterized by their development of respiratory pleopods (26). Though the occurrence and degree of development of coxal plates vary among the edriophthalmids, such tendencies or 'underlying synapomorphies' are valid characters (Saether, 1983) since they indicate genetic channeling with the capacity to develop the plates or lose them.

The sister group to the acarideans, the brachycaridans, has a short carapace (28). The Thermosbaenacea have lost the oöstegite brood pouch (−21) and brood their eggs under the carapace (34). The hemicarideans have the distinctive cup- or spoonlike respiratory epipodite on the maxillipede. The Cumacea further specialize the hemicaridean respiratory system with the development of a pseudorostrum and maxillipedal siphons (33). The tanaids and spelaeogriphaceans seem to have redeveloped pleopods (−17). Tanaidacea have male genital cones (27), a tendency to form two to three segments on the maxillipedal epipodite (30), and the second thoracopods developed as chelipedes (31). The Spelaeogriphacea have augmented the respiratory system with the branchial thoracic exopods (32).

The relationships indicated by the above analysis conform only in part to the arrangement of eumalacostracan superorders of Calman (1904, 1909). Problems arise in dealing with the fossil groups, which seem to indicate, especially in regard to the 'peracarids,' that the lineages of eumalocostracans cannot always be concisely defined (Schram, 1984b). However, the above analysis does confirm that in general the *Bauplan* approach of Schram (1981, 1983) finds real expression in the nested sets of relationships in the cladogram of Figure 43-3. However, some of the exact relationships within particular areas in that cladogram remain uncertain. Clarification of the interrelationships among various decapod groups is an area of active study (Burkenroad, 1981; Schram, 1981; Felgenhauer and Abele, 1983 and personal communication). Schram (1984a) has already considered several possible different options in relating syncarids to each other. Finally, the data on the naupliar eye–frontal organ complex of Elofsson (1963, 1965) are quite congruent with the scheme here (characters 35, 36, and 37), and certainly more so than to that of the more traditional schemes of eumalacostracan relationships. A full complement of these sensory structures is likely primitive. Euphausiacea lose the dorsal frontal organ, syncarids apparently lack the ventral frontal organ, 'peracarids' lack at least the naupliar eye and dorsal organ, and acaridans and brachycaridans also lack the ventral frontal organ.

The overall arrangement here within the 'peracarid–pancarid' clade is in stark contrast to that of Siewing (1956) or in part Watling (1981, 1983), particularly in regard to the reassociation of Isopoda and Amphipoda as the edriophthalmids. However, the isopod–amphipod clade is the most resolved of any in the cladogram. The problem is perhaps that in most people's minds when they 'think' amphipod they 'think' a gammaridean, and when they 'think' isopod they 'think' a flabelliferan. In this case 'theory' is predetermining the understanding of 'primitive.' A preliminary analysis of characters within the isopods and amphipods by Dr. Richard Brusca and me indicates that within isopods the anthurideans and phreatoicideans are closer to the basal stock and that within amphipods the ingolfiellideans are near the base. In this respect the juxtaposition of edriophthalmans with the mictaceans is not at all incongruous and does lend some considerable insight into what the ancestral edriophthalman might have looked like. It is also significant that most of these forms occupy cryptic–troglobitic–meiofaunal types of habitats. Such similarities in ecology, rather than be used as an explanation for 'convergence' in form among these groups, might indicate an ancestral habitat type and behavior. From such cryptic infaunal forms, it is not difficult to envision a scenario wherein locomotory and respiratory specializations initiated the separate 'isopodan' and 'amphipodan' radiations on the basic acaridean body plan.

The Hoplocarida seem to have undergone an evolution in which a transition series culminated in a highly specialized raptor (Fig. 43-4). This may be related to the constraint imposed on hoplocaridans by their mandibulogastric anatomy (Kunze 1981, 1983), such that even the generalized aeschronectidans may have been at least scavengers if not pure carnivores. The hoplocaridans are defined by their triflagellate antennules (1), three-segmented thoracic protopods (2), four-segmented exopod of the thoracopods (3), and dendrobranchiate gills on the pleopods (4). Other characters may apply here as well. The kinetic cephalon and movable rostrum are seen only in hoplocaridans and at least some phyllocarids. The unique arrangement of the mandible to the stomach, seen in the living stomatopods, cannot for now be verified in any of the fossil forms. The issues of abdominal fusion raised by Schram (1969) have yet to be adequately reexamined, so we still do not know if stomatopods fuse the anterior of seven abdominal segments to form six or whether they are like eumalacostracans and fuse posterior abdominal segments. Finally, the abdominal musculature is characterized by Hessler (1964, 1983) as 'caridoid' but appears to me to be different from that of eumalacostracans. Hoplocaridans lack the important transverse muscle elements seen in eumalacostracans. The central muscles are not twisted. The ventral longitudinal muscles are much reduced in volume over that of caridians, and the twist of the ventral longitudinal muscles involves three rather than four segments, that is, an ascent of two segments and a descent of one rather than two and two as in eumalacostracans.

The Aeschronectida, a radiation on the primitive form, may be defined by their wide spatulate telsons (5). The rest of the hoplocaridans have subchelate anterior thoracopods (6) and a tendency to reduce (sometimes markedly) the ischiomerus of the subchelipedes (7).

The Palaeostomatopoda are the only hoplocaridans with furcae on the telson (8), which in light of the comments in Chapter 1 are treated as

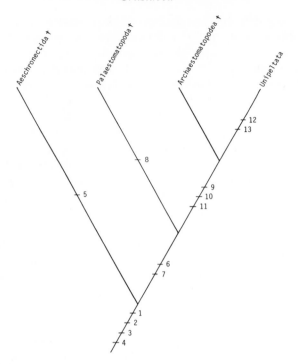

Fig. 43-4. Cladogram depicting the relationships of the orders of Hoplocarida. Derived characters are: triflagellate antennules (1); three-segmented thoracic protopods (2); four-segmented exopod of the thoracopods (3); dendrobranchiate gills on pleopods (4); wide spatulate telson (5); subchelate anterior thoracopods (6); tendency to reduce ischiomerus of subchelipedes (7); furcae on telson (8); reduced carapace (9); subregionalization of thorax (10); shortened carpus on subchelipede thoracopods (11); specialization within subchelipede series (12); rectangular highly armored telson (13).

derived features. The Stomatopoda reduce the carapace so that only the anterior segments are covered (9), have the thorax strongly regionalized with clear divisions within the thoracomere and thoracopod series (10), and possess a shortened thoracopodal carpus (11). The extinct suborder Archaeostomatopodea is a primitive group and a sister group to the living suborder Unipeltata. The latter is characterized by marked specializations within the subchelipede series (12) and a tailfan with a rectangular, highly armored telson, and spinose uropods (13).

The sister group of the malacostracans contains the phyllopodans and maxillopodans, and these are characterized by an abdomen generally lacking limbs, uniramous antennules, and typically a mandible that lacks a palp in the adult. The evolution of these forms in large part involved the exploitation of particular feeding types (Chapter 44). The radiation of phyllopodans has been for the most part directed toward the exploitation of thoracic feeding achieved with the use of polyramous, foliaceous thoracopods. The radiation of maxillopodans was built around a theme of continuing atrophy and modification of the body. The maxillopodans are almost all cephalic feeders and specific groups elaborate on basic variants of maxillary grappling (Chapter 44). However, there is repeated parallel development of parasitism in the class and the evolution of one unique type of thoracic filter feeding in the cirripedes.

Besides the features shared with the maxillopodans, the Phyllopoda are uniquely defined by polyramous (1) and foliaceous (2) thoracic limbs (Fig. 43-5). Another congruent character here may be the distinctive box truss arrangement of the dorsoventral thoracic muscles noted in leptostracans, brachypodans, anostracans, and notostracans. The status of this in any of

Fig. 43–5. Cladogram depicting the relationships of the orders of Phyllopoda. Derived characters are: polyramous (1) and foliaceous (2) thoracopods; all-enveloping (3) generally bivalved (4) carapace; uniramous antennae (5); cephalic kinesis and frequently an articulated rostrum (6); biramous antennules with scalelike second branch (7); caudal rami as paddles (8); redevelopment of pleopods on all but last pleomere (9); reduced antennules (10); pediform maxillae (11); ambulatory thoracic endopods (12); reduction to biramous thoracopods (13); large labrum in at least larvae if not adults (14); large labrum in the adults (15); egg-brooding appendage on females (16); maxillules as claspers in male (17); biramous paddles redeveloped on posterior thoracopods (18); more than eight segments in thorax (19); antennal protopod of nauplius more than half the length of the limb (20); vestigial or absent maxillae (21); anostracan naupliar eye (22); no head shield (23); branchiopod naupliar eye (24); caudal rami as cerci (25); posterior trunk segmentation obscure (26); bilobed carapace (27); caudal rami as abreptors (28); reevolved limbs on all trunk segments (29); loss of trunk tagmosis (30); oligomerized body (31); cephalon exposed, generally separate from carapace (32).

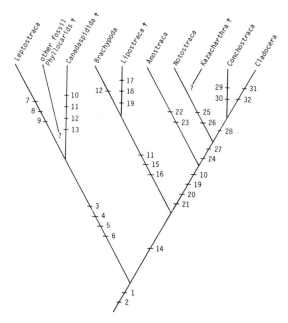

the fossil phyllopodans or in the diplostracan branchiopods is not known. It is a unique system designed to provide a degree of rigidity while allowing the necessary flexibility of the thorax. Three basic groups can be placed within the class: phyllocarids, cephalocarids *sensu lato*, and branchiopods. Placement of the phyllocarids is a problem since only the living leptostracans and the Cambrian canadaspidans are well known enough to allow adequate comparison to other phyllopodans. The other phyllocarids (hymenostracans, archaeostracans, and hoplostracans) are all extinct and their fossils incompletely known, especially in regard to appendage form. Rolfe (1981) offered a cladogram of possible phyllocarid relationships, but he emphasized its tentative nature. For example, when some new information unexpectedly came to light concerning some fossils previously placed within the phyllocarids (Pinna et al., 1982; Briggs and Rolfe, 1983; Secretan, 1983) it provided such insight into the actual anatomy that it is possible the concavicarids and their relatives may not even be crustaceans. The term cephalocarid is used here to include not only the Brachypoda (*Hutchinsoniella* and allies), but also the extinct order Lipostraca.

The Phyllocarida are united in their possession of an all-enveloping (3) generally bivalved (4) carapace, uniramous antennae (5), and apparently a cephalic kinesis and frequently an articulated rostrum (6). Placement of the phyllocarids among the phyllopodans rather than the malacostracans or as a separate taxon by themselves is due to parsimony, the other arrangements require longer trees and/or ones with a higher degree of homoplasy. The Leptostraca display a reappearance of the biramous antennule in the form of a unique scalelike second branch (7), the caudal rami developed as paddles (8), and the apparent reappearance of pleopods on all but the last segment (9). Characters 4, 7, and 9 are really plesiomorphic features, but parsimony suggests 'redevelopment' rather than a continued retention through a series of hypothetical, primitive, intermediate stages. The Canadaspida have reduced antennules (10), pediform maxillae (11)—a feature shared with brachypods along with ambulatory thoracic endopods (12)—and apparent redevelopment of peculiar biramous thoracopods (13). In contrast to the phyllocarids, all other phyllopodans have an enlarged labrum present at least in the larval stages (14), combined with a tendency to lose the carapace (−1).

The Cephalocarida feature pediform maxillae (11), an enlarged labrum even in the adults (15), and an egg–brooding appendage in the female (16). The Brachypoda have ambulatory thoracic endopods (12). The extinct Lipostraca have the maxillules developed as claspers in the males (17), redeveloped biramous posterior thoracopods (18), and have more than eight segments in the thorax (19). Again, these last two features seem best treated as 'reappearances' or character reversals rather than as the retention of a primitive feature.

The Branchiopoda have a reduced antennule (10), more than 8 segments in the thorax (19), a naupliar antennal protopod more than half the length of the limb (20), and vestigial or absent maxillae (21). The Anostraca have a characteristic naupliar eye with three cups each with numerous inverse sensory cells widely separated from a ventral frontal organ (22) and loss of the head shield (23). The calmanostracan branchiopods have developed an all-enveloping bilobed carapace covering not only the thorax but also parts or all of the abdomen and thoracopods (27) and the branchiopod nauplius eye with four cups each with a highly variable number of inverse sensory cells and associated with unique median and distal frontal organs (24).

The Notostraca have the caudal rami developed as cerci (25) and the segmentation in the posterior part of the abdomen obscure (26). Apparently the extinct Kazacharthra with their telson shields bear some kind of relationship to the notostracans, either as a separate group or perhaps as a taxon within the Notostraca proper.

The diplostracans have caudal rami developed as abreptors (28). The Conchostraca have redeveloped limbs on all the trunk segments (29) and even obscured the trunk tagmosis between thorax and abdomen (30). The Cladocera have oligomerized the body (31) and generally have the carapace modified to expose the cephalon and thus only cover the trunk (32).

The Maxillopoda (Fig. 43-6) have at most 11 trunk segments (1), at most six thoracic segments (2), a short bulbous heart (3), and the distinctive maxillopodan nauplius eye containing tapetal and lens cells but lacking associated frontal organs (4). Grygier (1983) pointed out the importance of the location of the gonopores in many of the maxillopodan groups on the seventh trunk somite. This may be another congruent feature useful in recognizing the group, but the opening of the gonopores seems also to be rather variable and consequently has not been incorporated into the cladistic analysis here because of difficulty in polarizing the variant states. Grygier (1981, 1982) originally felt that sperm morphology could be used to define maxillopodans, but this was eventually recognized to be a primitive feature by Grygier (1983). However, he felt that consideration of the arrangement of structures inside the sperm head might still prove useful. Boxshall (1983) also cautioned against relying too heavily on sperm morphology, and it would appear that further study of all crustacean sperm is necessary before any such information can be used in determining crustacean interrelationships.

The Tantulocarida are a separate group whose exact affinities are unresolved. They have several unique features related to their small size and parasitic life-style: lack of cephalic limbs (5), an oral disc (6), less than 11 trunk segments (7), and possibly a lost heart.

Another line seems characterized by a reduction in the number of thoracic segments to less than six (8) and generally a reappearance of mandibular palps (13). In this line the Ostracoda have a highly oligomerized body (9), generally vertically oriented caudal rami (10), and a unique naupliar carapace (11). The branchiurans and mystacocarids both have gonopores

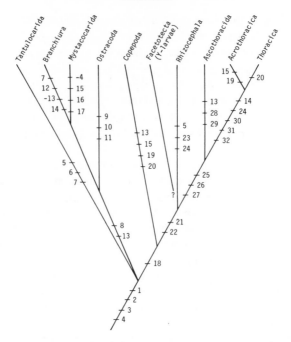

Fig. 43-6. Cladogram depicting the relationships of the major groups within Maxillopoda. Derived characters are: 11 trunk segments at most (1); no more than six thoracic segments (2); a short bulbous heart (3); maxillopodan naupliar eye (4); loss of naupliar eye (−4); lack of cephalic limbs (5); an oral disc (6); less than 11 trunk segments (7); less than six thoracic segments (8); highly oligomerized body (9); vertically oriented caudal rami (10); naupliar carapace (11); flat shieldlike carapace (12); reappearance of mandibular palp (13) (see character 16, Fig. 43-1); loss of palp in adult stage (−13); thoracic limbs as cirri (14); maxillipede (15); divided cephalon (16); caudal rami as pincers (17); at least four or more naupliar stages (18); first thoracomere fused to cephalon (19); six naupliar stages (20); cypris larva (21); antennule as an attachment organ (22); whole adult body reduced to gonad sac (23); frontolateral horns on nauplii (24); saccular carapace or mantle enveloping whole body (25); female genital pores on first thoracomere (26); mandible as a serrate blade (27); naupliar stages reduced from four to two (28); loss of antennae (29); carapace divided into plates (30); uniramous mandibular palp (31); loss of abdomen (32).

located on the fourth trunk segment, but both are better characterized by a series of autapomorphies. The Branchiura have a flat shieldlike carapace (12), a trunk with less than 11 segments (7), no mandibular palp in the adult (−13), and thoracic limbs as cirri (14). The Mystacocarida have a maxillipede (15), a divided cephalon (16), caudal rami as 'pincers' (17), and have lost the naupliar eye (−4).

All higher maxillopodans, that is, copepods and cirripedes *sensu lato*, generally have four or more naupliar stages (18). The Copepoda have a biramous mandibular palp (13), a maxillipede (15), the first thoracomere fused to the cephalon (19), six naupliar stages (20). The cirripedes *sensu lato* all possess a cyprid larval stage (21) and generally utilize the antennule as an attachment organ of some kind (22).

The highly modified parasitic Rhizocephala lack cephalic limbs (5), reduce the adult body to a sac of gonads (23), and have frontolateral horns on the nauplii (24). The members of the rhizocephalan sister group are united in the possession of a saccular all-enveloping carapace or mantle that encloses the body (25), female genital pores on the first thoracomere (26), and the mandible developed as a serrate blade (27). The position of the enigmatic Y-larvae (Facetotecta) is uncertain, though they do have a cypris stage.

The Ascothoracida have in some instances secondarily reduced the number of naupliar stages from four to two (28) and lack antennae (29). The Cirripedia *sensu stricto* have thoracic cirri (14), frontolateral horns on the nauplii (24), the carapace typically divided into plates (30), a uniramous mandibular palp (31), and have lost the abdomen (32). The Acrothoracica have a distinct maxillipede (15) and the first thoracomere fused (or at least closely associated) with the cephalon (19). The Thoracica are characterized by having six naupliar stages (20).

That continued attempts at rigorous character analysis holds great promise for elucidating relationships within groups is indicated by such an analysis (Fig. 43-7) performed on copepods (Boxshall, personal communication). This is all the more enlightening in that the traditional characters (largely involving antennules), used to define copepod orders heretofore, have not proven particularly satisfactory in analyzing interrelationships of these taxa (Boxshall et al., 1984). Copepoda can be recognized by their possession of a maxillipede (1), first thoracomere fused to the cephalon (2), six naupliar stages (3), and to these can be added a geniculate antennule on males (4). The Calanoida develop a flexure point between the sixth thoracic segment and the first abdominal segment (5) and have a single median genital opening (6). All other copepods (traditionally referred to as podopleans) develop a body flexure between the fifth and sixth thoracic segments (7) and lose the maxillulary exites (8). The Misophrioida have a unique carapace fold on the first thoracic or maxillipedal segment (9), while their sister group is characterized by loss of the heart (10) and a ventral fusion between the first and second abdominal segments (11). The Harpacticoida have fused the basis and endopod of the sixth thoracopod (12), while their sister group has lost the endopod of the sixth thoracopod (13) and completely fused (dorsally and ventrally) the first and second abdominal segments (14). The Mormonilloida have lost the geniculate antennules on the males (−4) and also have a single median genital opening (6); their sister group (largely parasitic forms) has no more than two segments on the antennal exopod (15). The Siphonostomatoida have their mouth developed as a siphon (16), while their sister group has lost the antennal exopod (17). The Monstrilloida have lost the antenna altogether (18), and their sister groups have dorsal or dorsolateral genital pores (19).

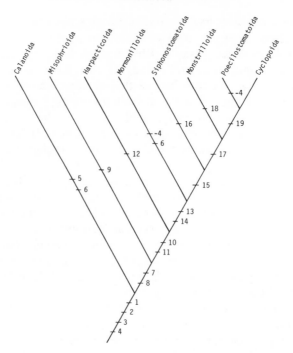

Fig. 43-7. Cladogram depicting the relationships of the orders of Copepoda. Derived characters are: maxillipede (1); first thoracic segment fused to cephalon (2); six naupliar stages (3); male antennules geniculate (4); loss of male geniculate antennules (−4); body flexure between sixth thoracic and first abdominal segments (5); single median genital pore (6); body flexure between fifth and sixth thoracic segments (7); loss of maxillulary exites (8); carapace fold on maxillipede segment (9); loss of heart (10); ventral fusion of first two abdominal segments (11); fusion of basis and endopod of sixth thoracopod (12); loss of endopod of sixth thoracopod (13); complete fusion of first two abdominal segments (14); no more than two segments on antennal exopod (15); siphon (16); loss of antennal exopod (17); loss of antenna (18); dorsal or dorsolateral genital pores. (From Boxshall, personal communication)

The Poecilostomatoida have lost the geniculate antennules in the males. The cyclopoid families cannot be characterized by any unique apomorphies, and one could argue as to whether separate orders for poecilostomes and cyclopoids are warranted since the latter are merely primitive forms of some other taxon characterized by the unique location of the genital pores.

The important role of paedomorphosis related to multiple independent evolutions of parasitism within the maxillopodans makes assessment of relationships rather inexact. The scheme in Figure 43-6 is the most parsimonious one obtained, but choices of slightly different characters or elimination of some others can produce some significant changes in the tree. In addition, some of the characters, such as reductions in numbers of segments in the body and thorax, at the base of the tree are not particularly robust, that is, they seem likely to have occurred several times. Therefore, it is prudent to leave the higher taxonomic categories of Maxillopoda for the most part as they are arranged in Bowman and Abele (1982). The single exception to this is within the cirripedes *sensu lato* where the analysis here coincides in part with some of the important conclusions of Grygier (1984) based on ascothoracids and related groups. Further development of that work will undoubtedly alter our concepts as to how maxillopodans are related.

The taxonomy proposed here is more natural than any proposed heretofore. The level of taxonomic category is determined for the most part by the recognition of *Baupläne* within the clades being considered. It should not be viewed as 'right' and all previous schemes 'wrong.' Rather, in the spirit of Schram (1983) it is merely a working model subject to ongoing modification as more exact character analysis becomes available.

The recognition of only four classes based on degree of tagmosis and appendage form seems a definite advance over past practice. The *Baupläne* revolve around whether or not tagmosis is developed and whether trunk limbs are biramous or polyramous. The four classes can be easily recognized. Such a system abandons past practice of making any newly discovered crustacean group that could not be easily related to any other preexisting group a new class.

The recognition of two subclasses of Malacostraca, originally suggested by Schram (1969) is now general practice (Kunze, 1981, 1983; Bowman and Abele, 1982; Hessler, 1983; McLaughlin, 1980; Schram, 1982). The arrangement of Hessler (1983) merely uses different names for the same taxa recognized here: his Caridoida = my Eumalacostraca and his use of the name Eumalacostraca = my Malacostraca. These differences are related to the removal herein of the phyllocarids as a sister group of hoplocaridans and eumalacostracans (its traditional place) and its reassignment to the phyllopodans. As related above, the classification here of eumalacostracans is based on the recognition of *Baupläne* (Schram 1981, 1983). Uncertainties of relationships of these realigned orders would argue against recognition of any superordinal taxa.

The arrangement of phyllopodan subclasses is based on the recognition of *Baupläne* based on (1) whether or not there is a carapace, (2) if a carapace, whether it is bivalved or bilobed, and (3) if no carapace, whether there is a head shield or not. This arrangement does void the useful taxon Branchiopoda (which appears to be a true clade), but emphasizes the very distinct nature of the Anostraca, separate from the other branchiopods, the Calmanostraca. Indeed, this was a difference long ago recognized by Linder (1945). Some future use of the taxon Diplostraca Gerstaecker, 1866 as a sister group to the Notostraca, for conchostracans and cladocerans, as advocated by Tasch (1969), might eventually prove useful, especially if future character analysis blurs the distinction between those two groups.

The Maxillopoda include *the* most derived of all crustacean forms, such that it is very difficult to determine exact relationships among the groups.

Reduction of body components and adaptation of a parasitic life-style seem to have occurred repeatedly. In addition, certain key anatomical features that proved useful in ordering other crustacean classes and orders have been independently derived several times among maxillopodans: for example maxillipedes in mystacocarids, copepods, and some cirripedes; and carapaces in branchiurans, ostracodes, misophrioid copepods, and at least some thecostracans. The net result is that it seems almost impossible, at least at this time, to recognize discrete *Baupläne* among the maxillopodans. The area of greatest promise seems to be in acquiring a better knowledge of early ontogeny and larval development in all maxillopodans. This has proven of some usefulness in the above analysis and to some extent in Grygier (1984) as well.

The higher taxonomy of crustaceans herein proposed makes both major and minor changes in realigning groups. The disappearance or 'demotion' of traditional 'friends' will undoubtedly upset many. The comment of Gaffney (1979, p.103) is apropos in this regard. 'There is a strong traditional feeling that stability of some sort is important in classification. . . . This is unfortunate because formulation of scientific ideas in a context that imposes stability for traditional purposes breeds the illusion that stability of classifications demonstrates accuracy and depth of understanding. In fact, temporal stability of classifications often reflects ignorance of relationships and lack of work. I hardly advocate change for its own sake, but the maintenance of names for discarded concepts seems useless and misleading.'

The classification derived by me from the above cladograms is as follows ([†] refers to extinct groups; see relevant chapters):

Phylum Crustacea Pennant, 1777

    Class Remipedia Yager, 1981
        Order Enantiopoda[†] Birshtein, 1960
        Order Nectiopoda Schram, 1986

    Class Malacostraca Latreille, 1806
      Subclass Hoplocarida Calman, 1904
        Order Aeschronectida[†] Schram, 1969
        Order Palaeostomatopoda[†] Brooks, 1962
        Order Stomatopoda Latreille, 1817
            Suborder Archaeostomatopodea[†] Schram, 1969
            Suborder Unipeltata Latreille, 1825
      Subclass Eumalacostraca Grobben, 1892
        Order Syncarida Packard, 1885
            Suborder Bathynellacea Chappuis, 1915
            Suborder Anaspidacea Calman, 1904
            Suborder Palaeocaridacea[†] Brooks, 1962

Order Belotelsonidea[†] Schram, 1981
Order Euphausiacea Dana, 1852
Order Amphionidacea Williamson, 1973
Order Decapoda Latreille, 1803
   Suborder Dendrobranchiata Bate, 1888
   Suborder Eukyphida Boas, 1880
      Infraorder Procarididea Felgenhauer and Abele, 1983
      Infraorder Caridea Dana, 1852
   Suborder Euzygida Burkenroad, 1981
      Infraorder Stenopodidea Huxley, 1879
      Infraorder Uncinidea[†] Beurlen, 1930
   Suborder Reptantia Boas, 1880
Order Waterstonellidea[†] Schram, 1981
Order Mysida Boas, 1883
Order Lophogastrida Boas, 1883
Order Pygocephalomorpha[†] Beurlen, 1930
Order Mictacea Bowman et al., 1985
Order Edriophthalma Leach, 1815
   Suborder Isopoda Latreille, 1817
   Suborder Amphipoda Latreille, 1816
Order Thermosbaenacea Monod, 1927
Order Hemicaridea Schram, 1981
   Suborder Cumacea Kröyer, 1846
   Suborder Tanaidacea Dana, 1853
      Infraorder Anthracocaridomorpha[†] Sieg, 1980
      Infraorder Apseudomorpha Sieg, 1980
      Infraorder Neotanoidomorpha Sieg, 1980
      Infraorder Tanaidomorpha Sieg, 1980
   Suborder Spelaeogriphacea Gordon, 1957

Class Phyllopoda Latreille, 1825
   Subclass Phyllocarida Packard, 1879
      Order Archaeostraca[†] Claus, 1888
      Order Canadaspidida[†] Novozhilov, 1960
      Order Hymenostraca[†] Rolfe, 1969
      Order Hoplostraca[†] Schram, 1973
      Order Leptostraca Claus, 1880
   Subclass Cephalocarida Sanders, 1955
      Order Brachypoda Birshtein, 1960
      Order Lipostraca† Scourfield, 1926
   Subclass Sarsostraca Tasch, 1969
      Order Anostraca Sars, 1867
   Subclass Calmanostraca Tasch, 1969
      Order Notostraca Sars, 1867
      Order Kazacharthra[†] Novozhilov, 1957

Order Conchostraca Sars, 1867
Order Cladocera Latreille, 1829

Class Maxillopoda Dahl, 1956
  Subclass Tantulocarida Boxshall and Lincoln, 1983
    Order Tantulocaridida Boxshall and Lincoln, 1983
  Subclass Branchiura Thorell, 1864
    Order Arguloida Rafinesque, 1815
    ? Order Pentastomida Rudolphi, 1819
  Subclass Mystacocarida Pennak and Zinn, 1943
    Order Mystacocaridida Pennak and Zinn, 1943
  Subclass Ostracoda Latreille, 1806
    Order Bradoriida† Matthew, 1902
    Order Phosphatocopida† K. J. Müller, 1964
    Order Leperditicopida† Scott, 1961
    Order Palaeocopida Henningsmoen, 1953
    Order Myodocopida Sars, 1866
    Order Podocopida Sars, 1866
  Subclass Copepoda Milne Edwards, 1840
    Order Calanoida Sars, 1903
    Order Misophrioida Gurney, 1933
    Order Harpacticoida Sars, 1903
    Order Mormonilloida Boxshall, 1979
    Order Siphonostomatoida Thorell, 1859
    Order Monstrilloida Sars, 1903
    Order Cyclopoida Burmeister, 1834
    Order Poecilostomatoida Thorell, 1859
  Subclass Thecostraca Gruvel, 1905
    Order Facetotecta Grygier, 1984
    Order Rhizocephala F. Müller, 1862
    Order Ascothoracida Lacaze-Duthiers, 1880
    Order Cirripedia Burmeister, 1834
      Infraorder Acrothoracica Gruvel, 1905
      Infraorder Thoracica Darwin, 1854

REFERENCES

Anderson, D. T. 1973. *Embryology and Phylogeny in Annelids and Arthropods*. Pergamon Press, Oxford.

Bowman, T. E., and L. G. Abele. 1982. Classification of the Recent Crustacea. In *The Biology of Crustacea*, Vol. 1 (L. G. Abele, ed.), pp. 1–27. Academic Press, New York.

Bowman, T. E., S. P. Garner, R. R. Hessler, T. M. Iliffe, and H. L. Sanders. 1985. Mictacea, a new order of Crustacea Peracarida. *J. Crust. Biol.* **5**:74–78.

Boxshall, G. A. 1983. A comparative functional analysis of the major maxillopodan groups. *Crust. Issues* **1**:121–43.

Boxshall, G. A., F. D. Ferrari, and H. Tiemann. 1984. The ancestral copepod: towards a consensus of opinion at the First International Conference on Copepoda. *Crustaceana Suppl.* **7**:68–84.

Briggs, D. E. G., and W. D. I. Rolfe. 1983. New Concavicarida (new order: ? Crustacea) from the Upper Devonian of Gogo, Western Australia, and the palaeoecology and affinities of the group. *Sp. Pap. Paleo.* **30**:249–76.

Brusca, R. C. 1984. Phylogeny, evolution and biogeography of the marine isopod subfamily Idoteinae. *Trans. San Diego Soc. Nat. Hist.* **20**:99–134.

Burkenroad, M. D. 1981. The higher taxonomy of the Decapoda. *Trans. San Diego Soc. Nat. Hist.* **19**:251–68.

Calman, W. T. 1904. On the classification of the Crustacea Malacostraca. *Ann. Mag. Nat. Hist.* (7)**13**:144–58.

Calman, W. T. 1909. Crustacea. In *A Treatise on Zoology*, Vol. 7 (E. R. Lankester, ed.). Adam and Charles Black, London.

Elofsson, R. 1963. The nauplius eye and frontal organs in Decapoda. *Sarsia* **12**:1–68.

Elofsson, R. 1965. The nauplius eye and frontal organs in Malacostraca. *Sarsia* **19**:1-54.

Felgenhauer, B. E., and L. G. Abele. 1983. Phylogenetic relationships among shrimp-like decapods. *Crust. Issues* **1**:291–311.

Fleminger, A. 1983. Description and phylogeny of *Isaacsicalanus paucisetus*, from an east Pacific hydrothermal vent site. *Proc. Biol. Soc. Wash.* **96**:605–22.

Gaffney, E. S. 1979. An introduction to the logic of phylogeny reconstruction. In *Phylogenetic Analysis and Paleontology* (J. Cracraft and N. Eldredge, eds.), pp. 79-111. Columbia Univ. Press, New York.

Gordon, I. 1955. Importance of larval characters in classification. *Nature* **176**:911–12.

Grygier, M. J. 1981. Sperm of the ascothoracican parasite *Dendrogaster*, the most primitive found in Crustacea. *Int. J. Invert. Reprod.* **3**:65–73.

Grygier, M. J. 1982. Sperm morphology in Ascothoracida: a confirmation of generalized nature and phylogenetic importance. *Int. J. Invert. Reprod.* **4**:323–32.

Grygier, M. J. 1983. Ascothoracida and the unity of Maxillopoda. *Crust. Issues* **1**:73–104.

Grygier, M. J. 1984. Comparative morphology and ontogeny of the Ascothoracida, a step toward a phylogeny of the Maxillopoda. Doctoral dissertation, Univ. California, Scripps Inst. Oceanogr., San Diego.

Hennig, W. 1966. *Phylogenetic Systematics*. Univ. Illinois Press, Urbana.

Hessler, R. R. 1964. The Cephalocarida comparative skeletomusculature. *Mem. Conn. Acad. Arts Sci.* **16**:1–97.

Hessler, R. R. 1983. A defense of the caridoid facies: wherein the early evolution of the Eumalacostraca is discussed. *Crust. Issues* **1**:145–64.

Kunze, J. C. 1981. The functional morphology of stomatopod Crustacea. *Phil. Trans. Roy. Soc. Lond.* (B)**292**:255–328.

Kunze, J. C. 1983. Stomatopoda and the evolution of the Hoplocarida. *Crust. Issues* **1**:165–88.

Linder, F. 1945. Affinities within the Branchiopoda, with notes on some dubious fossils. *Ark. Zool.* **37**(4):1–28.

Manton, S. M. 1977. *The Arthropoda*. Oxford Univ. Press, London.

McLaughlin, P. A. 1980. *Comparative Morphology of Recent Crustacea*. Freeman, San Francisco.

McLaughlin, P. A. 1983. Hermit crabs—are they really polyphyletic. *J. Crust. Biol.* **3**:608–21.

Pinna, G., P. Arduini, C. Pesarini, and G. Teruzzi. 1982. Thylacocephala: una nuova classe di crostacei fossili. *Atti Soc. Ital. Sci. Nat. Museo Civ. Stor. Nat. Milano* **123**:469–82.

Rolfe, W. D. I. 1981. Phyllocarida and the origin of the Malacostraca. *Geobios* **14**:17-27.

Saether, O. A. 1983. The canalized evolutionary potential: inconsistencies in phylogenetic reasoning. *Syst. Zool.* **32**:343–59.

Schram, F. R. 1969. Some Middle Pennsylvanian Hoplocarida and their phylogenetic significance. *Fieldiana: Geol.* **12**:235–89.

Schram, F. R. 1978. Arthropods: a convergent phenomenon. *Fieldiana: Geol.* **39**:61–108.

Schram, F. R. 1981. On the classification of the Eumalacostraca. *J. Crust. Biol.* **1**:1–10.

Schram, F. R. 1982. The fossil record and evolution of Crustacea. In *The Biology of Crustacea* Vol. I (L. G. Abele, ed.), pp. 91–147. Academic Press, New York.

Schram, F. R. 1983. Method and madness in phylogeny. *Crust. Issues* **1**:331–50.

Schram, F. R. 1984a. Fossil Syncarida. *Trans. San Diego Soc. Nat Hist.* **20**:189–246.

Schram, F. R. 1984b. Relationships within eumalacostracan Crustacea. *Trans. San Diego Soc. Nat. Hist.* **20**:301–12.

Secretan, S. 1983. Un groupe énigmatique de crustacés ses représentants du Callovien de la Voulte-sur-Rhone. *Ann. Paléontol.* **6**:59–97.

Sieg, J. 1983. Evolution of Tanaidacea. *Crust. Issues* **1**:229–56.

Siewing, R. 1956. Untersuchungen zur Morphologie der Malacostraca. *Zool. Jahrb. Abt. Anat.* **75**:39–176.

Tasch, P. 1969. Branchiopoda. In *Treatise on Invertebrate Paleontology*, Part R, *Arthropoda* **4**(1) (R. C. Moore, ed.), pp. R128–91. Geol. Soc. Am. and Univ. Kansas Press, Lawrence.

Watling, L. 1981. An alternative phylogeny of peracarid crustaceans. *J. Crust. Biol.* **1**:201–10.

Watling, L. 1983. Peracarid disunity and its bearing on eumalacostracan phylogeny with a redefinition of eumalacostracan superorders. *Crust. Issues* **1**:213–28.

44

EVOLUTIONARY PATTERNS

There are several broad patterns that occur in many, if not all, groups of crustaceans, which indicate that there are recurrent themes in the evolution of the phylum. These are reviewed here and provide some insight into the patterns of relationships discussed in Chapter 43.

PAEDOMORPHOSIS

Gould (1977) and Alberch et al. (1979) have presented a cogent analysis of the role of heterochronic processes in macroevolution. Essentially, acceleration and retardation in the ontogeny of somatic and reproductive anatomy intertwine to produce phylogenetic changes manifested as either recapitulation or paedomorphosis of form.

A shift to an earlier appearance of somatic features results in a speedup of the developmental sequence to produce recapitulation by acceleration. On the other hand, retardation of the timing of the appearance of gonads prolongs the duration and the events of ontogeny to produce recapitulation by hypermorphosis. However, among crustaceans the phenomenon of recapitulation is not common. Nevertheless, Kume and Dan (1968) outlined a sequence of larval types that they felt essentially provided a recapitulatory pattern for eucarid larval evolution.

A shift to an earlier appearance of reproductive organs truncates the developmental sequence and results in progenesis. On the other hand, retardation of the timing of the appearance of somatic characters results in an extension of ontogenetic events to produce neoteny. The phenomenon of progenesis is common among crustaceans. Given that serially homonomous segments and structure are the primitive state, once trunk tagmosis was established the entire thrust of evolution within and between crustacean classes has been to repeatedly abbreviate and truncate the developmental sequence, resulting in the evolution of new body forms with earlier sexual maturation (see, e.g., Newman, 1983). Two whole classes, namely, the phyllopodans and maxillipodans, are essentially progenetic, and the three of the four major lineages within the eumalacostracans repeatedly exhibit progenetic strategies in the course of their evolution. Although the phenomenon of progenesis is widespread, it has received little specific comment in the literature. For example, Baid (1964) observed reproduction in larval *Artemia salina* in shallow Sambar Lake in India that was

547

triggered apparently by the high salinities that develop just before complete desiccation. Parasitic copepods, especially lernaeopids (Wilson, 1911; de Beer, 1958), are noteworthy in that they mature sexually at the first and second copepodid stages.

The endocrine control of the arthropod developmental system easily lends itself to paedomorphic variations in the life cycle. For example, in insects, a lowering of the level of juvenile hormone accelerates the appearance of adult characters; a rise in the level of this hormone extends the time of manifestation of juvenile features. Although there has been extensive laboratory manipulation of this system, Wigglesworth (1954) noted that temperature and photoperiod variations could naturally affect the timing of maturation in *Rhodnius*.

In crustaceans, the Y-organ produces a hormone that serves to suppress the X-organ–sinus gland system, which in turn produces and releases molt-inhibiting hormone. Work is just beginning on the way crustaceans regulate this system, and it has not been subject to the detailed experimental manipulation to which insects have been subjected. However, Laufer et al. (1984) report the detection of juvenile hormone in the spider crab *Libinia emarginata*. McConaugha (1985) and Sanders et al. (1985) have found that specific food requirements and levels of nutrition can affect the timing of molting and metamorphosis. Pressure (Roer et al., 1985), temperature, and photoperiod (Conan, 1985) also have been shown to affect crustacean molting. Most interesting of all, Skinner et al. (1985) postulate that hormones other than the molt inhibitor must operate to bring about ecdysis; for example, an exuvation factor to help shed the old cuticle, a limb growth inhibiting factor, an anecdysial limb autotomy factor that can induce precocious molts, and a proecdysis limb autotomy factor that can interrupt and delay proecdysis. If these factors are confirmed, then study of their operation and control should provide great insight into the genesis of paedomorphosis in Crustacea. Finally, it is also likely that conditions in the environment can override genetic and biochemical controls of the maturation process. In fact, Ra'anan and Cohen (1985) document the role of social interactions in the maturation of male *Macrobrachium rosenbergii*. Such socially controlled maturations will need to be studied from the viewpoint of any potential effects they may exert on paedomorphosis.

Why has paedomorphosis been so important in the evolution of crustaceans? Gould (1977) felt that progenesis redirects natural selection away from selection of anatomical form toward selection for timing of maturation. This implies that morphology in such circumstances need not be perfectly adaptive, merely serviceable in order to get the individual to an early reproductive stage. This idea might receive some support from consideration of the tremendous diversity of form found among crustaceans, more than in any other major group of animals.

Progenesis also can be viewed as a simple pathway to overall biological efficiency. Calow (1983) advanced the idea that it is apparently more efficient to produce gametes rather than somatic tissue. Since levels of nutrition have been demonstrated to have a direct effect on molt and metamorphosis in crustaceans (McConaugha, 1985; Sanders et al., 1985), truncation of the developmental sequence in order to initiate gamete production as soon as possible would ensure the greatest reproductive advantage for the smallest expenditure of energy to procure food. In this regard selection would be directed not so much at the timing of maturation, nor at the details of morphology, but rather at size as an aspect of efficiency. This is a matter of scaling, since achieving reproduction with a small body is more efficient than with a large one. As Gould (1977) pointed out, the retention of a larval or juvenile form is then only a passive consequence of the primary aim to most easily and directly maintain small size. This factor of gamete production efficiency seems to have been operating in the evolution of such groups as mystacocarids (Hessler, 1971), bathynellaceans (Schminke, 1981), and brachypodan cephalocarids (Schram, 1982).

The interplay of paedomorphic with recapitulatory events also affords ready explanation for the 'reappearance' of 'lost' structures in the cladograms of Chapter 43. Acceleration and hypermorphosis can effectively add 'new morphologies' onto a developmental sequence; thus structures may get 'pushed back into' or be 'shifted to' earlier ontogenetic stages. However, as long as they are not completely 'lost,' they can be 'recalled' to an adult or reproductive stage through the agency of progenesis. With progenesis as a principal agency of macroevolution within crustaceans, Dollo's Law cannot be expected to apply rigidly. The interplay of larval and adult characters then is a manifestation merely of subtle shifts in genetic and developmental programs.

SEX

A voluminous literature exists dealing with the origin and adaptive significance of sex (see, e.g., Maynard Smith, 1971, 1978; Thornhill and Alcock, 1983). These explanations essentially have as their nexus the idea that a recombination of characters increases variability in a population and thus imparts potential advantages in the face of natural selection. Though this body of theory has great merit and much mathematical rigor, it need not mask other functions that the variations of modes of sexual reproduction can serve in establishing the life cycle, especially among crustaceans. One effect of varying patterns of sexual reproduction is to dampen phenotypic variation, or at least to minimize the risk that potentially disastrous mutant phenotypes will adversely affect the overall reproductive success of a population.

The dominant feature of the fossil record is the periods of stasis between periods of change. The environment of the planet apparently has remained stable for long intervals. The constancy of faunas in past time periods reflects this (Schram, 1981). Periods of extinction seem to cluster in time around periods of rapid tectonic and climate change (Boucot, 1975). It would thus seem to be advantageous for selection to favor a mode of reproduction that would ensure that variation does not deviate much beyond some 'proven' mean. This is to say that selection is an immediate agent; it operates in context of present suitability, not future fitness.

Those few crustacean groups in which close attention has been paid to reproductive biology do disclose features that would reinforce the interpretation of sex as a means of dampening variation. Notable in this regard are the tanaidaceans (Sieg, 1983). Although species with completely separate sexes are known in the group (see Chapter 15, Fig. 15-4), protogyny is common. Tanaids are not particularly mobile. They live typically in individual burrows or tubes. Males seek out the female lairs. Young are sheltered first in the mother's brood pouch and then on the floor of the maternal burrow. When the young have developed sufficiently, they tunnel out of mother's home to build one of their own nearby. Mother meanwhile might either molt into another copulatory female stage or may metamorphose to a male. Given the highly localized nature of tanaid populations and their lack of great dispersive capabilities, there is a high degree of inbreeding in such populations—males must copulate with siblings, descendents of siblings, or even their own daughters. This mode of reproduction ensures minimal morphologic deviation because of maximum potential for inbreeding, and the preservation of a reproductively successful phenotype (supposedly effectively suited) for local environmental conditions.

Until recently, tanaids were thought to be relatively unique in these regards. Many edriophthalmids are now coming to be recognized as exhibiting serial hermaphroditism. Steele (1967) recorded the stegocephalid amphipod, *Stegocephalus inflatus*, as protandrous. Lowry (personal communication) has identified physically large individuals of the lysianassoid amphipod *Acantiostoma marionis*, which bear both oöstegites and penial processes and concludes that this species appears to be protandrous. Furthermore, Lowry and Stoddart (1983) recognized that the species *Fresnillo fimbriatus* Barnard is actually the secondary male of *Ocosingo borlus* Barnard! So an analogous sexual situation to that seen in tanaidaceans exists at least in some amphipods.

Protandry is also known in some isopods, namely, cryptoniscids (Caullery and Mesnil, 1901), oniscids (Jackson, 1928) and cymothoids (Montalenti, 1948). In addition, protogyny is seen in the anthuridean isopods (Burbanck and Burbanck, 1979; Wägele, 1981; Kensley, 1982).

Even normal modes of sexual reproduction can have interesting effects in terms of dampening phenotypic variation. Some terrestrial isopods at least commonly bear multipaternal broods. Sassaman (1978) noted that 80% of the broods of *Porcellio scaber* are the result of mixing sperm from different males. Johnson (1982) characterized females in such situations as 'living sperm banks,' accepting sperm from a succession of males and storing it. Sperm can be stored of males who have died, and this 'overlap of generations' effectively serves to stabilize the polymorphism of the population.

Thornhill and Alcock (1983), in their discussion of the advantages that sex imparts, point out that the theme common to most evolutionary hypotheses of genetic advantage is that sex is beneficial to creatures in highly variable environments. Sexual offspring are thought to more often than not disperse away from the environment of their parents. However, Ehrlich and Raven (1969) observed that most animals do not move far from the area in which they were born. Nevertheless, Thornhill and Alcock maintain that proof of the 'environmental predictability hypothesis' is afforded by the exception to the rule seen in parthenogenetic populations of species. Parthenogenesis is predicted by the environmental predictability hypothesis to occur in stable environments. In this regard, it is viewed as an aberration that short-circuits the normal purpose of sex, for example, it decreases variation. However, it appears just as logical to maintain that parthenogenesis is not an exception to a rule, but rather another manifestation of sex as a means of ensuring that phenotypic variation does not deviate from some effective form. In this regard, parthenogenesis is merely a special case within a general 'need' that living systems often manifest, that is, to dampen or minimize the effects of mutation and maintain phenotypic stability.

In summary then, among Crustacea the effects of observed sexual and developmental patterns seem to coincide. Progenetic paedomorphosis acts to relieve crustacean species of the burden of selection for morphology. Sexual and asexual parthenogenic reproduction might be viewed as ways to further dampen phenotypic variation, and thus further relieve selection for morphology. The hormonal control of sex observed in crustaceans (Charniaux–Cotton, 1958), rather than genetic, is a consequence of this focus of their evolution.

FEEDING AND EVOLUTION OF FORM

The question of what was the primitive mode of feeding in Crustacea has repeatedly concerned crustacean phylogeneticists. The issue has essentially crystallized around two opposing theories, and the matter of basic feeding type has become linked with attempts to resolve the ancestral form of the crustacean trunk limb.

Borradaile (1917, 1926) advanced the concept that the basic limb type was polyramous and foliaceous, that is, the limb had several branches and/ or lobes (endites, endopod, exopod, and epipods), and was thin and leaf-like. He initially arrived at this idea from a consideration of decapod mouthparts (Borradaile, 1917). This theory received a considerable boost from the discovery of the cephalocarids (Sanders, 1955). These interesting little creatures formed the nexus of a large body of speculation concerning early crustacean evolution (e.g., Sanders, 1957; Hessler, 1964; Hessler and Newman, 1975; Schram, 1982). The primitive mode of feeding was assumed to be that of a thoracic filter feeder, a mode of ingestion supposedly shared by cephalocarids, leptostracan phyllocarids, and most branchiopods. This similarity of thoracophagy suggested recognition of the Thoracopoda as a formal taxon (Hessler, in Hessler and Newman, 1975) more or less uniting the old taxon Phyllopoda with the Malacostraca.

Cannon and Manton (1927) proposed the idea that the basic crustacean limb type was a biramous paddle, that is, the limb had only two branches (endopod and exopod). The paddlelike trunk limbs were used exclusively for locomotion while the mouthparts alone served to gather food. Cannon and Manton felt the primitive mode of feeding was by maxillary filtration, based on a consideration of feeding in many maxillopodan groups and higher malacostracans. This theory has recently received some support from the discovery of the Remipedia (Schram, 1983), a group of Carboniferous and recent creatures with biramous paddlelike limbs on every trunk segment and whose modes of feeding use cephalic appendages.

All of these phylogenetic speculations have focused, almost without exception, on only a few types of animals. There simply has not been a systematic review of feeding in *all* crustaceans. Much information has come to light on this matter in the decades since Cannon and Manton. Some very recent work has provided insights that will bring much to bear on resolving the conflict outlined above. The following is intended as an annotated catalog of what is currently known about crustacean feeding. Details can be found in the relevant chapters of this book.

ENANTIOPODA   This group is known from a single species, *Tesnusocaris goldichi* (Brooks, 1955) from the Carboniferous of Texas. Though this is a remarkably well-preserved fossil, specific details about mouthparts are lacking. The maxillules and maxillae seem to have been spinose, and maxillipedes were lacking. Though the trunk limbs are in a midventral position, the paddle form and distance from the mouthparts would seem to imply that enantiopods were cephalic feeders. Schram (1983) treated these as an extinct sister group to the living nectiopodan remipedes.

NECTIOPODA   The nectiopods are likely have the most unusual set of mouthparts of any crustaceans, though the basic body architecture is quite primitive. The mouthparts are highly specialized to facilitate carnivory. A more detailed consideration of nectiopod feeding is now in preparation, but the essentials seem to be that the maxillules, maxillae, and maxillipedes act to grab and hold victims in a sort of cage while the nectiopod swims on its back. The basal endites of the maxillules act as 'mandibles,' while the actual jaws are hidden within the atrium oris and act as a kind of mechanical mill that shreds and pulverizes food sucked out of the prey by the action of a very muscular esophagus. Thus the nectiopods are specialized cephalic feeding carnivores that grapple their prey with the mouthparts.

STOMATOPODA AND PALAEOSTOMATOPODA   These are two of the few purely carnivorous crustaceans known, and certainly the most highly adapted of any to a raptorial life-style (Dingle and Caldwell, 1978).

AESCHRONECTIDA   Schram (1969) assumed that these Late Paleozoic forms were possibly filter feeders, based on the lack of limb specializations and the very setose thoracic endopods. However, Kunze (1981, 1983) felt that the unique mandible and gastric anatomy of the living stomatopods was so distinctive that the ancestry of such a structural pattern must also have been unique too. Thus, she felt, perhaps with some justification, that all hoplocaridans, aeschronectids included, had predatory habits.

PALAEOCARIDACEA   Nothing is known with certainty about how these Late Paleozoic syncarids may have fed. At least some, such as the family Acanthotelsonidae, may have been raptorial carnivores (Schram, 1984).

ANASPIDACEA   Cannon and Manton (1929) felt that these were maxillary filter feeders and/or raptors. However, they used only preserved material in their study and were perhaps too inclined to draw analogies with *Hemimysis*. Manton (1930) did make some observations on living material that generally confirmed the earlier study.

BATHYNELLACEA   No studies of feeding in living animals have been published for this group. Similarities of form and habitat preference suggest that they are maxillary grazers like mystacocarids.

BELOTELSONIDEA AND WATERSTONELLIDEA   Little can be inferred about the feeding habits of these Paleozoic forms. The belotelsonidans have large maxillulary and maxillary gnathobases, which might indicate raptorial or scavenging habits.

MYSIDA   Cannon and Manton (1927) produced what had become a classic analysis of *Hemimysis lamornae*, which indicated that these were supposedly thoracic filterers feeding from currents produced by the rotary action of the exopods. However, Attramadal (1981) proved that this was not possible. He has gathered some evidence (Atrramadal, personal communication) that mysidans are not even maxillary filterers. Rather, it appears that particles adhere to setose appendages and onto body surfaces and that these limbs and surfaces are groomed and the particles subsequently eaten.

LOPHOGASTRIDA   Manton (1928) studied these forms and found indications that feeding varied in the group. *Gnathophausia* is a maxillary filterer with the maxillae blocking the excurrent filter from the respiratory chamber under the carapace in order to filter out food particles. She did not address how fouling of the respiratory surfaces could be prevented in such circumstances. *Lophogaster*, on the other hand, lacks a food basin behind the mouthfield, so filtration does not seem possible in this form. The massive mandibles might indicate that these were grazers.

AMPHIPODA   Manton (1964) and Watling (1983) anew called attention to the derived and distinctive tranverse action of the mandibles in this group. In this context a carnivorous and/or scavenging mode of feeding is most common, but in such a diverse and successful group many variants are noted. Caine (1979) noted some caprellids use the antennae as filtering devices, when ablation of the so-called 'swimming setae' were found to have little effect on locomotion but profound effect on the ability to filter. Dennell (1933) found that *Haustorius arenarius* typically filtered water with the setose thoracopods; however, they were also seen to use a maxillary filter, using a rotary or grasping–scooping motion of the maxillae to remove particles from the general currents generated around the body by the action of the pleopods (Dahl, 1977). Kaestner (1970) also remarked in passing that *Ampelisia* also filter feeds.

ISOPODA   The transverse mandibular action noted for amphipods (Manton, 1964; Watling, 1983) is also true for isopods (one of several synapomorphies that could serve to unite these two groups). Isopods are generally omnivorous, with several lines specializing in carnivory—scavenging or parasitism. Naylor (1955) documented the basic pattern of mouthpart action for the omnivorous *Idotea*: transverse mandibular biting, with maxillules and maxillae pushing the food to the mandibles and into the mouth. This is a pattern similar to that seen in carnivorous forms (Green, 1957; Jones, 1968).

CUMACEA   Dennell (1934) concluded that *Diastylis bradyi* is a maxillary filterer. The maxillules and maxillae vibrate and suck water into the mouthfield from below and anterior of the head. Water is flushed out laterally over the maxillary exopodal flap and 'mat setae' on the maxillipede. Particles are trapped on the maxillulary setae and are swept into the mouth by the action of the maxillipedal endites.

Other modes of feeding have been noted among cumaceans. Dixon (1944) observed *Cumopsis goodsiri* to pick up sand grains with the first pereiopods, hold them with the maxillipedes, while the mouthparts scrape off adhering organic films. *Diastylis rathkei* has generally been understood to sweep in mud to the mouthfield with the maxillipedes, and then swallow the mass as a whole (e.g., Kaestner, 1970), and *Campylaspis* has styliform molar processes on the mandibles, which suggest carnivorous habits. Clearly, much work on cumacean feeding remains to be done.

TANAIDACEA These animals are generally raptors, using their chelipedes and maxillipedes to shred victims. However, Dennell (1937) noted that *Apseudes talpa* used maxillary filtration. Particles are trapped on the maxillae and are removed and transferred to the mouth by the action of the maxillules and maxillipedes. However, filter feeding is apparently not the prime source of nutrition among tanaids.

SPELAEOGRIPHACEA Formal studies of living animals have yet to be done, but Watling (personal communication) observed individuals of *S. lepidops* to randomly roam over the floor of the pools in which they live, apparently processing detritus.

THERMOSBAENACEA Fryer (1964) found *Monodella argentarii* to be a grazer, using the maxillules to scrape surfaces of their organic films, while the maxillae groom the maxillules and push food into the mouth. The maxillae in turn are cleaned and protected by the maxillipedes.

PYGOCEPHALOMORPHA This extinct Paleozoic order had very robust mouthparts. The large gnathobases on the maxillules and maxillae suggest a possible grazing or scavenging habit.

EUPHAUSIACEA The most recent work on this group (Hamner et al., 1983) indicates that these animals are thoracic endopod filterers. The setose thoracopods form a sweeping filter basket with the mouthparts concentrating the particles to form a bolus. When the bolus is large enough, filtration stops and the bolus is consumed.

AMPHIONIDACEA The adults do not feed and have reduced mouthparts and vestigial guts (Williamson 1973). The larvae have well-developed mouthparts, but how the larvae feed is not known.

DENDROBRANCHIATA No detailed studies of feeding in this economically important group have yet been done. Burkenroad (personal communication) has observed many varieties of feeding. *Penaeus duorarum* is a carnivore, digging trenches on the bottom to trap worms, *P. setiferus* is a detritus feeder, and purely nektonic forms are probably filter feeders.

PROCARIDIDEA Felgenhauer (personal communication) has observed *Procaris* to be a carnivore. Reminiscent in some respects to the nectiopod remipedes, the animals swim on their backs; the thoracopods form a basket to capture and hold prey, while the mandibles munch away until the victim is consumed.

CARIDEA There have been few good studies of feeding in these eukyphidans. They are perceived to be largely carnivores or scavengers. Fryer (1960, 1977) and Felgenhauer and Abele (1983a) noted specialized setose chelae for filter feeding in atyids. However, Felgenhauer and Abele (1983b) recorded such variety of structure in caridean gastric mills that may bespeak many varieties of feeding strategies, which they suggested in turn could be a reflection of possible polyphyletic origins for carideans from among the thalassinidean reptants.

EUZYGIDA These are largely scavengers, with the well-known genus *Stenopus* noted for its penchant for cleaning coral reef fish. Detailed functional morphologic studies of feeding in this group remain to be done.

REPTANTIA This large and diverse group are for the most part carnivores and/or scavengers, with limited filter feeders and some grazing forms noted. Among the carnivores and scavengers, initial food gathering is achieved by the anterior pereiopods, especially the first. The initial processing of food is achieved by the mandible and the third maxillipede, the former being the more passive and the latter the more active (Caine, 1975; Barker and Gibson, 1977). The other mouthparts are used to stuff food chunks into the mouth (definitive processing is achieved by the gastric mill in the foregut). An important component of the reptantian diet can often be detritus (e.g., Schembri, 1982), but here again the first pereiopod typically acts as the initial collector with the maxillipedes and maxillae sorting particles and shoving material through the mouth. Budd et al. (1978) document reptantian filter feeding in juvenile and some adult crayfish. The filter current is produced by the exopods of the first and second maxillipedes, while actual filtration occurs on the setae of the maxillae and first maxillipede with the assistance of some grappling action by the second maxillipedes. In all reptants, whatever the mode of feeding, food collecting is basically achieved by cephalic limbs with the assistance of such anterior thoracic limbs that are incorporated into an 'enlarged cephalon' or cephalothorax.

MYSTACOCARIDA While no detailed study of actual living organisms has yet been done, mystacocarids seem to be maxillary grazers (Buchholz, 1953; Jansson, 1966; Hessler, 1971). A consideration of their anatomy in connection with their habitat seems to indicate they feed on bacterial and fungal films on the sand grains among which they live, that is, the mouthparts scrape organic films off sediment grains into the mouth.

BRANCHIURA These are all ectoparasites. The modified suctorial maxillules and hooked maxillae serve to affix the crustacean to its host, while a suctorial proboscis with enclosed serrate mandibles does the actual feeding.

TANTULOCARIDA This newly discovered group (see, e.g., Boxshall and Lincoln, 1983; Lincoln and Boxshall, 1983) are minute ectoparasites on other crustaceans. A distinctive bulbous organ seems to be everted through what may have been the mouth and appears to be the means of obtaining nutrition from the host.

COPEPODA The work of Cannon (1928b) was one of the earliest works that indicated that the basic mode of feeding in copepods was by maxillary filtration. However, Koehl and Strickler (1981), using high speed cinematography, discovered that copepods are particle grapplers. Their size determines that copepods live in a medium of low Reynolds numbers, where water is a viscous medium with laminar flow (not an inertial medium with turbulent flow as for larger organisms). Water under such circumstances is 'sticky'; setose limbs behave as 'big' solid paddles, and food particles with their adhering water are effectively larger than they really are (Koehl et al., 1984). Rather than trap and comb out particles from a filter, copepods grapple or sweep particles by throwing their limbs back to suck in a particle and its surrounding water, and then bring the limbs back to force the particle toward and into the mouth (see Chapter 37, Fig. 37-9).

Though Koehl and Stickler elucidated the physics of particle feeding under conditions of low Reynolds numbers, the potential importance of grappling in copepod feeding actually had been long recognized (e.g., Gurney, 1931; Lowndes, 1935). Fryer (1957) in fact felt that the grappling or raptorial method of feeding was the only one from which one could most easily evolve the true filterers and grazers as well as parasites. Marcotte (1982) essentially agreed with this, and further observed that true filtration in copepods was probably quite rare.

OSTRACODA Though the ostracodes are numerous and ubiquitous, few papers to my knowledge (e.g., Cannon, 1924) actually recorded observations of living animals. Other papers (Cannon, 1933; Storch, 1933) were more detailed in their analyses, but were made on preserved material only. While there appear to be variations among different ostracodes in the uses to which particular limbs are put, the published consensus among past workers has been that ostracodes are maxillary filterers; that is, the posterior limbs create a current while the anterior limbs filter particles out of that current. However, there seems to be a great deal of unpublished observational data indicating that ostracodes feed by a great variety of means, that is, carnivory, detrital feeding, and grazing. All of these, however, because of the constraints of the ostracode *Bauplan*, are cephalic feeding.

However, the size and morphology of ostracodes is not inconsistent with a Koehl and Strickler (1981) type of analysis. Given their size and the consequent physical regime of low Reynolds numbers in which they must live, ostracodes need to be studied to determine if it is even possible for them to 'filter.'

ASCOTHORACIDA   It has long been widely assumed these forms used their suctorial mouthparts to imbibe body juices from a host. However, Grygier (1981) suggested the possibility in some ascothoracids of filter feeding, though he now discounts this (personal communication). In addition, dendrogastrids live, at least in part, on cells that float in the visceral cavity of the host, which are collected by the parasites' antennules.

RHIZOCEPHALA   These are endoparasites, absorbing nutrition from the host directly into the ramifying body of the parasite.

ACROTHORACICA   These, like their sister group, the thoracicans, are cirral filterers. The trunk limbs or cirri are extended to collect particles. As the cirri are withdrawn back into the mantle cavity, they are cleaned by the mouth cirri (or maxillipedes), which in turn are cleaned by the labrum and mouthparts (Tomlinson, 1969).

THORACICA   These are cirral filterers. Although variations exist concerning the rhythm of the cirral beat and the orientation of the limbs in relation to the barnacle body and passing currents, all are basically thoracic filterers (Anderson, 1978, 1980a, 1980b, 1981; Barnes and Reese, 1959; Crisp and Southward, 1961).

Some modicum of attention has been directed at barnacle naupliar feeding (Lochhead, 1936b; Rainbow and Walker, 1976). These authors termed these larvae maxillary filterers. However, the size of the nauplii and the anatomy and functional morphology of their limbs indicate that they would also fall under the strictures of the observations of Koehl and Strickler (1981) and that in a medium of low Reynolds number they are probably particle grapplers.

LEPTOSTRACA   The work of Cannon (1927) on *Nebalia bipes* revealed these animals to be phyllopodous thoracic filterers. The all-enveloping carapace of these animals forms a perithoracic filter chamber that, with movable rostrum, serves to direct the flow of water around the thoracopods. The functional pattern of the limbs is essentially that seen or assumed for other phyllopods: the exopods and epipods act as lateral valves, interlocking setae on the endopods serve to trap particles, and food is accumulated in a midventral food groove where it is moved forward to the mouth.

LIPOSTRACA   This Devonian fossil, *Lepidocaris rhyniensis* Scourfield, 1926, has the maxilla (see Chapter 27) and first two thoracopods developed as phyllopods, while the posterior limbs are biramous paddles. This probably represents an intermediate stage in the evolution of a complete thoracic filter, to a form like that seen more completely developed in the brachypodan cephalocarids.

BRACHYPODA Despite the past phylogenetic importance of this group, studies of actual feeding in living animals have never been performed (or at least never published). The many remarks in the literature (e.g., Sanders, 1963) that they are thoracic filterers, feeding as they move, are suppositions based on analogy with other phyllopodans and assumptions given apparent locomotory patterns.

FOSSIL PHYLLOCARIDA The Cambrian Hymenostraca and Paleozoic Archaeostraca are so poorly known anatomically that nothing can be inferred about possible feeding habits. The Hoplostraca might have been thoracic filterers; however, Rolfe (1981) saw 'raptorial' antenna on *Kellibrooksia macrogaster* and implied that such structures may indicate a 'carnivorous' habit. However, such 'raptorial' antennae are seen in many different groups of malacostracans and are more typically indicative of fossorial, tube-dwelling habits than carnivory. The Cambrian Canadaspidida are understood in great detail (Briggs, 1978) and, although their limbs are somewhat specialized, the general form and body plan of the organisms seem to indicate that canadaspids were probably perfectly good thoracic filter-feeding phyllopods.

ANOSTRACA The typical pattern of thoracic locomotory filter feeding has been confirmed in this group in several studies (Cannon, 1928a; Cannon and Leak, 1933; and Lowndes, 1933). This last, though confirming Cannon's basic pattern of limb function, suggested that, at least in *Chirocephalus diaphanous*, the creatures were bottom detritus feeders rather than strict nektonic filter feeders. Fryer (1966, 1983) studied the carnivorous *Branchinecta gigas*. Its spinelike setae, which form a food cage to capture and hold prey, are obviously rather specialized. However, Fryer suggested that the primitive anostracan may not have been a filterer, since the basic form associated with predatory habits is close to the assumed primitive type seen in grappling or raptorial copepods.

NOTOSTRACA These appear to be facultative detritus feeders or scavengers, but no detailed work has been done yet on feeding in notostracans.

CONCHOSTRACA Nothing has yet been published concerning the functional morphology of feeding in this group.

CLADOCERA Much attention has been given to feeding in cladocerans. The early studies (Cannon and Leak, 1933; Eriksson, 1935; and Lochhead, 1936a) are remarkably uninformative. The detailed work of Fryer (1963, 1968, 1974) has revealed that there is no common theme. Some forms set up currents with their limbs to suck particles into the carapace chamber; others scrape surfaces of organic films using largely the second thoracopod.

Whatever the mode of feeding, however, there is *no* anteriorly directed current in the food groove, only physical manipulation of food to move it forward to the mouth.

Interesting patterns emerge from the above review. First, it can be seen (Table 44-1) that cephalic feeding is as important, if not more so, than thoracic food gathering among crustaceans overall. A majority of crustacean orders exhibits some development of cephalic feeding, that is, 25 of the 43 ordinal level taxa discussed above.

Second, within the cephalic feeding forms (Table 44-2), some use of maxillary grappling action is of overriding importance, whether it be to grasp and hold prey (as in nectiopodan remipedes), to sweep in suspended food particles (as in copepods, and probably ostracodes and various crustacean larvae), or to scrape and thus graze on surface films (as is probable in mystacocarids and bathynellaceans). It is certainly the most common form of food gathering among crustaceans, and functions on a structural pattern common to all groups that use it. Survey of the literature reveals that the purported importance of maxillary filtration is actually rather questionable, having been demonstrated in only a few groups (cumaceans, one tanaidacean, juvenile crayfish, and perhaps some anaspidaceans and lophogastridans); these cases should probably be carefully reinvestigated, using modern techniques such as high speed cinematography of living animals, in order to confirm whether real filtration actually takes place.

Third, thoracic feeding crustaceans (Table 44-3) do not present the great unity of form and function seen among cephalic feeding types. Many distinct forms occur, such as those with cirri (as in barnacles), phyllopods (e.g., phyllocarids, cephalocarids, and branchiopods), thoracic grapplers (as in at least some anostracans and cladocerans, reptants, carideans, and euzygidans), carnivores of many distinct types (e.g., stomatopods, procaridids, and reptants). Each occurrence of a thoracic feeding type is distinctive anatomically and functionally and is restricted to single groups or small number of closely related groups.

The determination of what might be a primitive or advanced mode of feeding among crustaceans is basically a matter of functional character polarity assessment and, therefore, calls for analysis of sister-group or other outgroup data. Only such outgroup comparisons (Watrous and Wheeler, 1981, Maddison et al., 1984) can provide an acceptable level of probability for achieving some insight into what the ancestral crustacean state may have been. The real problem in this regard is that crustaceans, based on detailed study of form and function as well as ontogeny, do not seem to be closely related to any other known articulates (see, e.g., Anderson, 1973, Manton, 1977; or Schram, 1978)! Given this last point, what then does one use as a sister group or closely related outgroup?

Uniramia, whether herbivores or carnivores, principally utilize modified cephalic appendages. Deviations from this rule, such as occur in the

Table 44–1. Occurrence of major feeding types in various 'orders' of crustaceans.

Order	Cephalophagy	Thoracophagy
Enantiopoda[†]	+	
Nectiopoda	+	
Mystacocarida	+	
Branchiura	+	
Tantulocarida	+	
Ostracoda	+	
Copepoda	+	
Ascothoracida	+	
Rhizocephala	?	
Acrothoracica		+
Thoracica		+
Leptostraca		+
Canadaspidida[†]		+
Lipostraca[†]		+
Brachypoda		+
Anostraca		+
Notostraca		+
Conchostraca		+
Cladocera		+
Stomatopoda		+
Palaeostomatopoda[†]		+
Aeschronectida[†]	+	
Palaeocaridacea[†]	+	+
Anaspidacea	+	
Bathynellacea	+	
Belotelsonidea[†]	+	
Waterstonellidea[†]	?	
Mysida		+
Pygocephalomorpha[†]	+	
Lophogastrida	+	
Amphipoda	+	+
Isopoda	+	
Cumacea	+	
Tanaidacea	+	+
Spelaeogriphacea	?	
Thermosbaenacea	+	
Euphausiacea		+
Amphionidacea	? larval	
Dendrobranchiata	+	+
Procarididea		+
Caridea		+
Euzygida		+
Reptantia	+	+

[†]Extinct taxon.

Table 44–2. Occurrence of subtypes of cephalic feeding among orders that feed in that manner.

Order	Maxillary sweeping	Maxillary filtration	Maxillary scraping	Maxillary carnivory	Suctorial parasitism
Enantiopoda†				?	
Nectiopoda				+	
Mystacocarida			+		
Branchiura					+
Tantulocarida					+
Ostracoda	+				
Copepoda	+		+	+	+
Ascothoracida		+			+
Rhizocephala					+
Aeschronectida†				?	
Palaeocaridacea†		?		+	
Anaspidacea		+		+	
Bathynellacea			+		
Belotelsonidea†				+	
Waterstonellidea†	?	?			
Pygocephalomorpha†				+	
Lophogastrida	+	+			
Amphipoda	+		+	+	(+)
Isopoda	+		+	+	+
Cumacea		+	+	+	
Tanaidacea		+	+	+	
Spelaeogriphacea			?		
Thermosbaenacea			+		
Amphionidacea	? larvae				
Dendrobranchiata		?	+	+	
Reptantia		+			

†Extinct taxon.

carnivorous mantids, with their specialized prothoracic limbs, are obvious specializations.

Cheliceriformes generally use the chelae and sometimes pedipalps for food procurement. The additional use in merostomes of the gnathobasic parts of other opisthosomic limbs seems a specialization unique to that group alone.

Annelida, though they possess a wide array of feeding types, all utilize the head alone and its appendages for food procurement. When special structures like tentacles or palps do occur they are restricted to the head end of the animals.

Trilobita, being extinct, present obvious problems to determining feeding strategies. Cisne (1974, 1975, 1981), based on detailed analyses of

Table 44–3. Occurrence of subtypes of thoracic feeding among orders that feed in that manner.

Order	Phyllopodial filterer	Phyllopodial grappler	Subchela raptors	Thorapling raptors	Grappling raptors	Maxillipede grazers	Thoracopod filterer
Acrothoracica	+						
Thoracica	+						
Leptostraca		+					
Canadaspidida†		+					
Lipostraca†		+					
Brachypoda		+					
Anostraca		+	+				
Notostraca			?				
Conchostraca		?					
Cladocera		+	+				
Stomatopoda				+			
Palaeostomatopoda†				+	+		
Palaeocaridacea†					+		
Anaspidacea					+		
Mysida					+		?
Amphipoda					+		+
Tanaidacea					+		
Euphausiacea					+		+
Dendrobranchiata							+
Procarididea						+	
Caridea					+		+
Euzygida					+		
Reptantia					+		+

†Extinct taxon.

exceptionally well-preserved Ordovician *Triarthrus eatoni*, deduced that they were thoracic feeders and compared them in principle to phyllopods such as cephalocarids. Schram (1978, 1982) cautioned against generalizing about all trilobites based only on *Triarthrus*, since the diverse anatomy of limbs, even in the few trilobites for which limbs are known, would seem to indicate very different modes of feeding among those forms (see Chapter 2). Indeed, Bergstrom and Brassel (1984) described the limbs of *Rhenops*, suggesting that only the cephalic and pygidial appendages may have been involved in feeding (when the animals enrolled with the pygidium opposed to the cephalon). It would seem then that, at least in some trilobites, cephalic feeding may have been a real possibility.

The other so-called trilobitomorphs form a diverse array of articulates, some undoubtedly forming separate articulate phyla (e.g., Briggs 1983). Furthermore, even though we do not know which of these might be closest to the Crustacea, among these strange articulates, cephalic feeding seems to predominate. Thus outgroup analysis of all *potential* sister groups of crustaceans indicate that cephalic feeding is a probable primitive condition for crustaceans, as indeed it appears to be so for *all* articulates.

Demonstrating which mode of cephalic feeding might be primitive within Crustacea is more difficult than establishing cephalic feeding itself as primitive. The unique design of the crustacean head precludes using outgroup analysis, for what should maxillary feeding be compared to?

Although the preferable method of outgroup analysis cannot be effectively used here, the less satisfactory method of ingroup analysis (= commonality) can be utilized. In this case, the striking frequency (Tables 44-1 and 44-2) of some form of maxillary grappling, would seem to indicate that that mode of operation may be the most primitive. Of the 43 ordinal level taxa, 21 are strictly cephalic feeders, 18 strictly thoracic feeders, and four utilize both modes. The importance of cephalic feeding is even somewhat masked in these data in that individual orders of the cephalic feeding ostracodes and copepods are not separately listed in the table. Among the cephalic feeders, the grappling action (whether it be to sweep in particles, to scrape organic films, or to grab and hold prey) is the dominant mode of action.

Furthermore, careful study of the evolution of form and function in such disparate groups as cyclopoid copepods (Fryer, 1957) and branchinectid anostracans (Fryer, 1966) indicates that the grappling mode of cephalic feeding is apparently the only one from which all other more specialized modes of feeding could have easily evolved, whether true maxillary filtration or parasitism. In this context, even the fossil form *Lepidocaris rhyniensis* would seem to indicate how a true thoracic feeding form like that seen in the living phyllopods could have evolved from a maxillary feeder, that is, by gradually integrating more and more trunk limbs into the functional feeding apparatus.

The exact role that maxillary filtration plays in the scheme of things is unclear. Certainly the work of Koehl and Strickler (1981) would seem to indicate that it should be restricted probably to animals with high Reynolds numbers because of the action of the mouthparts.

The two rules for judging primitive limb types, so well summarized by Harding (in Scourfield, 1940), now seem more appropriate than ever. *The rule of ontogeny* tells us that more primitive forms of a limb appear earlier in development than more derived forms. By this rule biramous limbs are to be considered primitive. In groups with polyramous appendages, like the phyllopods and malacostracans, the biramous form of the limb appears first (Fig. 44-1). The fossilized juveniles of *Lepidocaris* clearly display biramous limb buds for the trunk limbs (Scourfield, 1940). The limb buds on *Hutchinsoniella* for the limbs from the maxillules through the anterior trunk limbs in the early naupliar stages all arise as simple biramous structures at their initial appearances (Sanders, 1963). In *Nebalia* the limb anlagen in the embryo and the limb buds after hatching are biramous lobes (Manton, 1934). And the early larvae of both hoplocaridans and eumalacostracans bear biramous trunk limbs (Gurney, 1942).

The rule of functional morphology tells us that the more specialized limbs occur near the anterior end of the adult. *Lepidocaris* has the specialized polyramous limbs near the head, in front of the posterior series of biramous trunk limbs, and in malacostracans the thorax bears the derived polyramous limbs, while the abdomen has biramous limbs.

Combine these observations with the fact that crustaceans that utilize the more primitive cephalic mode of feeding generally have biramous trunk limbs, and it thus follows that the polyramous limbs seen in phyllopods and malacostracans are more likely to be derived from the primitive biramous form rather than vice-versa. This would further serve to argue that thoracic feeding among crustaceans is a highly derived mode of feeding and that cephalic feeding is more primitive.

This analysis also seems to indicate that we should recast the biramous theory. The primitive position of true maxillary filtration now seems in doubt. It is probably best replaced by some form of maxillary grappling, and whether this be a sweeping, scraping, or grabbing action remains to be determined.

It seems idiosyncratic to observe that a more precise solution to the issues raised here will involve more work. This of course is the common caveat of scientific exegesis. However, in this case there is no other way. So much work has been done since the 1920s when the opposing theories were originally advanced. Functional analyses of feeding must be coupled with those of locomotion, basic anatomy, ontogeny, and comparative natural history. Perhaps after another 60 years of such work the vision of the evolution of crustacean feeding should be clearer than that presented here—it may even be quite different!

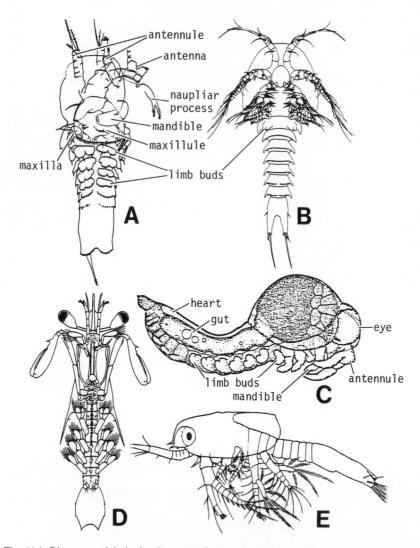

Fig. 44-1. Biramous origin in development of polyramous trunk limbs. (A) juvenile of *Lepidocaris rhyniensis* (from Scourfield, 1940); (B) stage 5 nauplius of *Hutchinsoniella macracantha* (from Sanders, 1963); (C) early hatchling of *Nebalia bipes* (from Manton, 1934); (D) pseudozoea larva of *Squilla mantis* (from Gurney, 1942); (E) protozoea larva of *Peneopsis* sp. (from Gurney, 1942).

In conclusion, crustacean evolution has a wonderfully simple, singular theme to it—the most efficient use of energy resources in order to achieve reproduction. Progenesis of development and cephalic feeding ensure that the trunk serves its locomotory and reproductive functions in the minimal time necessary. To augment this pattern, reproductive strategies are frequently employed to ensure genetic and anatomical uniformity within a wide diversity of body plans.

What beauty there is in such straight forward design. Richard Goldschmidt in the conclusion to his book *The Material Basis of Evolution* (1940) declaimed, ". . . a simplistic attitude is not a flaw but the ideal goal of a theory in science and, therefore, also for a theory of evolution." Our understanding of crustacean evolution can aim at no more worthy an ideal.

REFERENCES

Alberch, P., S. J. Gould, G. F. Oster, D. B. Wake. 1979. Size and shape in ontogeny and phylogeny. *Paleobiol.* **5**:296–317.
Anderson, D. T. 1973. *Embryology and Phylogeny in Annelids and Arthropods*. Pergamon, Oxford.
Anderson, D. T. 1978. Cirral activity and feeding in the coral inhabiting barnacle *Boscia anglicum*. *J. Mar. Biol. Assc. U.K.* **58**:607–26.
Anderson, D. T. 1980a. Cirral activity and feeding in the verrucomorph barnacles *Verruca recta* and *V. stroemia*. *J. Mar. Biol. Assc. U.K.* **60**:349–66.
Anderson, D. T. 1980b. Cirral activity and feeding in the lepadomorph barnacle *Lepas pectinata*. *Proc. Linn. Soc. N.S.W.* **104**:147–59.
Anderson, D. T. 1981. Cirral activity and feeding in the barnacle *Balanus perforatus*, with comments on the evolution of feeding mechanisms in the thoracican cirripedes. *Phil. Trans. Roy. Soc. Lond.* (B)**291**:411–49.
Attramadal, Y. G. 1981. On a non-existent ventral filtration current in *Hemimysis lamornae* and *Praunus flexuosus*. *Sarsia* **66**:283–86.
Baid, I. 1964. Neoteny in the genus *Artemia*. *Acta Zool.* **45**:167–77.
Barker, P. L., and R. Gibson. 1977. Observations on the feeding mechanism, structure of the gut, and digestive physiology of the European lobster *Homarus gammarus*. *J. Exp. Mar. Biol. Ecol.* **26**:297–324.
Barnes, H., and E. S. Reese. 1959. Feeding in the pedunculate cirripede *Pollicipes polymerus*. *Proc. Zool. Soc. Lond.* **132**:569–85.
Bergstrom, J., and G. Brassel. 1984. Legs in the trilobite *Rhenops* from the Lower Devonian Hunsrück Slate. *Lethia* **17**:69–72.
Borradaile, L. A. 1917. On the structure and function of the mouthparts of the palaemonid prawns. *Proc. Zool. Soc. Lond.* **1917**:37–71.
Borradaile, L. A. 1926. Notes upon crustacean limbs. *Ann. Mag. Nat. Hist.* (9)**17**:193–213, pls. 7–10.
Boucot, A. J. 1975. *Evolution and Extinction Rate Controls*. Elsevier, Amsterdam.
Boxshall, G. A., and R. J. Lincoln. 1983. Tantulocarida, a new class of Crustacea ectoparasitic on other crustaceans. *J. Crust. Biol.* **3**:1–16.
Briggs, D. E. G. 1978. The morphology, mode of life, and affinities of *Canadaspis perfecta*, Middle Cambrian, Burgess Shale, British Columbia. *Phil. Trans. Roy. Soc. Lond.* (B)**281**:439–87.

Briggs, D. E. G. 1983. Affinities and early evolution of the Crustacea: the evidence of the Cambrian fossils. *Crust. Issues* **1**:1–22.

Brooks, H. K. 1955. A crustacean from the Tesnus formation of Texas. *J. Paleo.* **29**:852–56.

Buchholz, H. A. 1953. Die Mystacocarida. Eine neue Crustaceenordnung aus dem Lucken-system der Meerssande. *Mikrokosmas* **43**:13–16.

Budd, T. W., J. C. Lewis, and M. L. Tracy. 1978. The filter feeding apparatus in crayfish. *Can. J. Zool.* **56**:695–707.

Burbanck, W. D., and M. P. Burbanck. 1979. *Cyathura*. In *Pollution Ecology of Estuarine Invertebrates* (C. W. Hart and S. L. H. Fuller, eds.), pp. 293–323. Academic Press, New York.

Caine, E. A. 1975. Feeding and masticatory structures of six species of the crayfish genus *Procambarus*. *Form. Funct.* **8**: 49–65.

Caine, E. A. 1979. Functions of swimming setae within caprellid amphipods. *Biol. Bull.* **156**:169–78.

Calow, P. 1983. Energetics of reproduction and its evolutionary implications. *Biol. J. Linn. Soc.* **20**:153–65.

Cannon, H. G. 1924. On the feeding mechanism of a freshwater ostracode, *Pionocypris vidua*. *J. Linn. Soc. Lond. Zool.* **36**:325–35.

Cannon, H. G. 1927. On the feeding mechanism of *Nebalia bipes*. *Trans. Roy. Soc. Edinb.* **55**:355–69.

Cannon, H. G. 1928a. On the feeding mechanism of the fairy shrimp *Chirocephalus diaphanus*. *Trans. Roy. Soc. Edinb.* **55**:807–22.

Cannon, H. G. 1928b. On the feeding mechanism of the copepods *Calanus finmarchicus* and *Diaptomis gracilis*. *Brit. J. Exp. Biol.* **6**:131–44.

Cannon, H. G. 1933. On the feeding mechanism of certain marine ostracodes. *Trans. Roy. Soc. Edinb.* **57**:739–64.

Cannon, H. G., and F. M. Leak. 1933. On the feeding mechanisms of the Branchiopoda. *Phil. Trans. Roy. Soc. Lond.* (B)**222**:267–352.

Cannon, H. G., and S. M. Manton. 1927. On the feeding mechanisms of a mysid crustacean, *Hemimysis lamornae*. *Trans. Roy. Soc. Edinb.* **55**:219–53.

Cannon, H. G., and S. M. Manton. 1929. On the feeding mechanism of the syncarid Crustacea. *Trans. Roy. Soc. Edinb.* **56**:175–89.

Caullery, M., and F. Mesnil. 1901. Recherches sur *l'Hemioniscus balani*, epicaride parasite des Balanes. *Bull. Sci. France Belgique* **34**:316–62.

Charniaux–Cotton, H. 1958. Contrôle hormonal de la différenciation du sexe et de la reproduction chez les Crustacés supérieurs. *Bull. Soc. Zool. France* **83**:314–36.

Cisne, J. L. 1974. Trilobites and the origin of arthropods. *Science* **186**:13–18.

Cisne, J. L. 1975. The anatomy of *Triarthrus* and the relationships of the Trilobita. *Fossils Strata* **4**:45–64.

Cisne, J. L. 1981. *Triarthrus eatoni*: anatomy of its exoskeletal, skeletomuscular, and digestive systems. *Palaeontol. Amer.* **9**:99–142.

Conan, G. Y. 1985. Periodicity and phasing of molting. *Crust. Issues* **3**:73–99.

Crisp, D. J., and A. J. Southward. 1961. Different types of cirral activity of barnacles. *Phil. Trans. Roy. Soc. Lond.* (B)**243**:271–308.

Dahl, E. 1977. The amphipod functional model and its bearing upon systematics and phylogeny. *Zool. Scripta*. **6**:221–28.

de Beer, G. R. 1958. *Embryos and Ancestors*. Clarendon Press, Oxford.

Dennell, R. 1933. The habits and feeding mechanism of *Haustorius arenarius*. *Zool. J. Linn. Soc. Lond.* **38**:363–88.

Dennell, R. 1934. The functional mechanism of the cumacean crustacean *Diastylis bradyi*. *Trans. Roy. Soc. Edinb.* **58**:125–42.

Dennell, R. 1937. On the functional mechanism of *Apseudes talpa* and the evolution of the peracaridan feeding mechanism. *Trans. Roy. Soc. Edinb.* **59**:57–78.

Dingle, H., and R. L. Caldwell. 1978. Ecology and morphology of feeding and agonistic behavior in mudflat stomatopods (Squillidae). *Biol. Bull.* **155**:134–49.

Dixon, A. Y. 1944. Notes on certain aspects of the biology of *Cumopsis goodsiri* and some other cumaceans in relation to their environment. *J. Mar. Biol. Assc. U.K.* **26**:61–71.

Ehrlich, P. R., and P. H. Raven. 1969. Differentiation of populations. *Science* **165**:1228–32.

Eriksson, S. 1935. Studien über die Fangapparate der Branchiopoden. *Zool. Bidr. Uppsala* **15**:23–287.

Felgenhauer, B. E., and L. G. Abele. 1983a. Ultrastructure and functional morphology of feeding and associated appendages in the tropical fresh-water shrimp *Atya innocous* with notes on its ecology. *J. Crust. Biol.* **3**:336–63.

Felgenhauer, B. E., and L. G. Abele. 1983b. Phylogenetic relationships among shrimp-like decapods. *Crust. Issues* **1**:291–311.

Fryer, G. 1957. The feeding mechanism of some freshwater cyclopoid copepods. *Proc. Zool. Soc. Lond.* **1957**:1–25.

Fryer, G. 1960. The feeding mechanism of some atyid prawns of the genus *Caridina. Trans. Roy. Soc. Edinb.* **64**:217–44.

Fryer, G. 1963. The functional morphology and feeding mechanism of the cydorid cladoceran *Eurycercus lamellatus. Trans. Roy. Soc. Edinb.* **65**:335–81.

Fryer, G. 1964. Studies on the functional morphology and feeding mechanism of *Monodella argentarii. Trans. Roy. Soc. Edinb.* **64**:49–90.

Fryer, G. 1966. *Branchinecta gigas*, a non-filter feeding raptatory anostracan, with notes on the feeding habits of certain other anostracans. *Proc. Linn. Soc. Lond.* **177**:19–34.

Fryer, G. 1968. Evolution and adaptive radiation in the Chydoridae, a study in comparative morphology and ecology. *Phil. Trans. Roy. Soc. Lond.* (B)**254**:221–385.

Fryer, G. 1974. Evolution and adaptive radiation in the Macrothricidae, a study in comparative morphology and ecology. *Phil. Trans. Roy. Soc. Lond.* (B)**269**:137–274.

Fryer, G. 1977. Studies on the functional morphology and ecology of the atyid prawns of Dominica. *Phil. Trans. Roy. Soc. Lond.* (B)**277**:57–129.

Fryer, G. 1983. Functional ontogenetic changes in *Branchinecta ferox. Phil. Trans. Roy. Soc. Lond.* (B)**303**:229–343.

Gould, S. J. 1977. *Ontogeny and Phylogeny*. Belknap Press, Cambridge.

Green, J. 1957. The feeding mechanism of *Mesidotea entomon. Proc. Zool. Soc. Lond.* **129**:245–54.

Grygier, M. J. 1981. *Gorgonolaureus muzikae* sp. nov. parasitic on a Hawaiian gorgonian, with special reference to protandric hermaphroditism. *J. Nat. Hist.* **15**:1019–45.

Gurney, R. 1931. *British Freshwater Copepods*, Vol. I. Ray Society, London

Gurney, R. 1942. *Larvae of Decapod Crustacea*. Ray Society, London.

Hamner, W. M., P. P. Hamner, S. W. Strand, and R. W. Gilmer. 1983. Behavior of Antarctic krill, *Euphausia superba*: chemoreception, feeding, schooling, and molting. *Science* **220**:433–35.

Hessler, R. R. 1964. The Cephalocarida comparative skeletomusculature. *Mem. Conn. Acad. Arts Sci.* **16**:1–97.

Hessler, R. R. 1971. Biology of the Mystacocarida: a prospectus. *Smith. Contr. Zool.* **76**:87–90.

Hessler, R. R., and W. A. Newman. 1975. A trilobitomorph orgin for the Crustacea. *Fossils Strata* **4**:437–59.

Jackson, H. G. 1928. Hermaphroditism in *Rhyscotus*, a terrestrial isopod. *Quart. J. Micro. Sci.* **71**:527–39.

Jansson, B. A. 1966. The ecology of *Derocheilocaris remanei. Vie Millieu* **17**:143–86.

Johnson, C. 1982. Multiple insemination and sperm storage in the isopod *Venezillo evergladensis. Crustaceana* **42**:225–32.

Jones, D. A. 1968. The functional morphology of the digestive system in the carnivorous intertidal isopod *Eurydice. J. Zool.* **156**:363–76.

Kaestner, A. 1970. *Invertebrate Zoology*, Vol. II. Interscience, New York.

Kensley, B. 1982. Deep-water Atlantic Anthuridea. *Smith. Contr. Zool.* **346**:1–60.Koehl, M. A. R., A. Y. L. Cheer, and G.-A. Paffenhöfer. 1984. Particle capture by copepods: when does a rake function as a paddle. *Am. Zool.* **24**:105A.

Koehl, M. A. R., and J. R. Strickler. 1981. Copepod feeding currents: food capture at low Reynolds number. *Limnol. Oceanogr.* **26**:1062–73.

Kume, M., and K. Dan. 1968. *Invertebrate Embryology.* U.S. Dept. Health, Educ. Welfare and Nat. Sci. Found., Washington.

Kunze, J. C. 1981. The functional morphology of stomatopod Crustacea. *Phil. Trans. Roy. Soc. Lond.* (B)**292**:255–328.

Kunze, J. C. 1983. Stomatopoda and the evolution of the Hoplocarida. *Crust. Issues* **1**:165–88.

Laufer, H., D. W. Borst, C. Carasco, F. C. Baker, D. A. Schooley. 1984. The detection of juvenile hormone in Crustacea. *Am. Zool.* **24**:33A.

Lincoln, R. J., and G. A. Boxshall. 1983. A new species of *Deoterthron* ectoparasitic on a deep-sea asellote from New Zealand. *J. Nat. Hist.* **17**:881–89.

Lochhead, J. H. 1936a. On the feeding mechanism of a ctenopod cladoceran, *Penilia aviros-tris*. *Proc. Zool. Soc. Lond.* **1936**:335–55.

Lochhead, J. H. 1936b. On the feeding mechanism of *Balanus perforatus*. *J. Linn. Soc. Lond.* **39**:429–42.

Lowndes, A. G. 1933. The feeding mechanism of *Chirocephalus diaphanous*, the fairy shrimp. *Proc. Zool. Soc. Lond.* **1933**:1093–1118, pls. 1–7.

Lowndes, A. G. 1935. The swimming and feeding of certain calanoid copepods. *Proc. Zool. Soc. Lond.* **1935**:687–715.

Lowry, J. K., and H. E. Stoddart. 1983. The shallow-water gammaridean Amphipoda of the subantarctic islands of New Zealand and Australia. Lysianassoidea. *J. Roy. Soc. N.Z.* **13**:279–394.

Maddison, W. P., M. J. Donaghue, and D. R. Maddison. 1984. Outgroup analysis and parsimony. *Syst. Zool.* **33**:83–103.

Manton, S. M. 1928. On some points on the anatomy of the lophogastrid Crustacea. *Trans. Roy. Soc. Edinb.* **56**:103–19.

Manton, S. M. 1930. Notes on the habits and feeding mechanisms of *Anaspides* and *Paranas-pides*. *Proc. Zool. Soc. Lond.* **1930**:791–800.

Manton, S. M. 1934. On the embryology of the crustacean *Nebalia bipes*. *Phil. Trans. Roy. Soc. Lond.* (B)**223**:163–238.

Manton, S. M. 1964. Mandibular mechanisms and the evolution of arthropods. *Phil. Trans. Roy. Soc. Lond.* (B)**247**:1–183.

Manton, S. M. 1977. *The Arthropoda.* Oxford Univ. Press, London.

Marcotte, B. M. 1982. Part 2: Copepods. In *The Biology of Crustacea*, Vol. I (L. G. Abele, ed.), pp. 185–97. Academic Press, New York.

Maynard Smith, J. 1971. What use is sex? *J. Theo. Biol.* **30**:319–35.

Maynard Smith, J. 1978. *The Evolution of Sex.* Cambridge Univ. Press, Cambridge.

McConaugha, J. R. 1985. Nutrition and larval growth. *Crust. Issues* **2**:127–54.

Montalenti, G. 1948. Studi sull'ermafroditismo dei Cirrotoidi. 1. *Emetha audonini* e. *Anilocra physodes*. *Pubb. Staz. Zool. Napoli* **18**:337–94.

Naylor, E. 1955. The diet and feeding mechanism of *Idotea*. *J. Mar. Biol. Assc. U.K.* **34**:347–55.

Newman, W. A. 1983. Origin of the Maxillopoda; urmalacostracan ontogeny and progenesis. *Crust. Issues* **1**:105–19.

Ra'anan, Z., and D. Cohen. 1985. Ontogeny of social structure and population dynamics in the giant freshwater prawn, *Macrobrachium rosenbergii*. *Crust. Issues* **3**:277–311.

Rainbow, P. S., and G. Walker. 1976. The feeding apparatus of the barnacle nauplius larva: a scanning electron microscope study. *J. Mar. Biol. Assc. U.K.* **56**:321–26.

Roer, R. D., R. M. Dillaman, M. G. Shelton, and R. W. Brauer. 1985. Effects of pressure on growth. *Crust. Issues* **3**:251–64.

Rolfe, W. D. I. 1981. Phyllocarida and the origin of the Malacostraca. *Geobios.* **14**:17–27.

Sanders, B., R. B. Lauchlin, J. D. Costlow. 1985. Growth regulation in larvae of the mud crab, *Rithropanopeus harisii. Crust. Issues* **3**:155–61.

Sanders, H. L. 1955. The Cephalocarida, a new subclass of Crustacea from Long Island Sound. *Proc. Natl. Acad. Sci. U.S.A.* **41**:61–66.

Sanders, H. L. 1957. The Cepholocarida and crustacean evolution. *Syst. Zool.* **6**:112–29.

Sanders, H. L. 1963. The Cephalocarida functional morphology, larval development, comparative external anatomy. *Mem. Conn. Acad. Arts Sci.* **15**:1–80.

Sassaman, C. 1978. Mating systems in porcellionid isopods: multiple paternity and sperm mixing in *Porcellio scaber. Heredity* **41**:385–97.

Schembri, P. J. 1982. Feeding behavior of fifteen species of hermit crabs from the Otego region, southeastern New Zealand. *J. Nat. Hist.* **16**:859–78.

Schminke, H. K. 1981. Adaptation of Bathynellacea to life in the interstitial. *Int. Rev. ges. Hydrobiol.* **66**:575–637.

Schram, F. R. 1969. Polyphyly in the Eumalocostraca? *Crustaceana* **16**:243–50.

Schram, F. R. 1978. Arthropods: a convergent phenomenon. *Fieldiana: Geol.* **39**:61–108.

Schram, F. R. 1981. Late Paleozoic crustacean communities. *J. Paleo.* **55**:126–37.

Schram, F. R. 1982. The fossil record and evolution of Crustacea. In *Biology of Crustacea*, Vol. 1 (L. G. Abele, ed.), pp. 93–147. Academic Press, New York.

Schram, F. R. 1983. Remipedia and crustacean phylogeny. *Crust. Issues* **1**:23–28.

Schram, F. R. 1984. Fossil Syncarida. *Trans. San Diego Soc. Nat. Hist.* **20**:189–246.

Scourfield, D. J. 1926. On a new type of crustacean from the Old Red Sandstone (Rhynie Chert Bed, Aberdeenshire)—*Lepidocaris rhyniensis. Phil. Trans. Roy. Soc. Lond.* (B)**214**:153–87.

Scourfield, D. J. 1940. Two new and nearly complete specimens of young stages of the Devonian fossil crustacean *Lepidocaris rhyniensis. Proc. Linn. Soc. Lond.* **152**:290–98.

Sieg, J. 1983. Evolution of Tanaidacea. *Crust. Issues* **1**:229–56.

Skinner, D. M., D. E. Graham, C. A. Holland, D. L. Mykles, C. Soumoff, and L. H. Yamaoka. 1985. Control of molting in Crustacea. *Crust. Issues* **3**:3–14.

Steele, D. H. 1967. The life cycle of the marine amphipod *Stegocephalus inflatus* in the northwest Atlantic. *Can. J. Zool.* **45**:623–28.

Storch, O. 1933. Morphologie und Physiologie des Fangapparates eines Ostrakoden (*Notodromas monacha*). *Biol. Gen.* **9**(1,2):151–98, 355–94; **9**(2,3):299–330.

Thornkill, R., and J. Alcock. 1983. *The Evolution of Insect Mating Systems.* Harvard Univ. Press, Cambridge.

Tomlinson, J. T. 1969. The burrowing barnacles. *Bull. U.S. Natl. Mus.* **296**:1–162.

Wagele, J. W. 1981. Zur Phylogenie der Anthuridea mit Beiträgen zur Lebensweise, Morphologie, Anatomie und Taxonomie. *Zool. Stuttgart* **132**:1–127.

Watling, L. 1983. Peracaridan disunity and its bearing on eumalacostracan phylogeny with a redefinition of eumalocostracan superorders. *Crust. Issues* **1**:213–28.

Watrous, L. E., and Q. D. Wheeler. 1981. The out-group comparison of character analysis. *Syst. Zool.* **30**:1–11.

Wigglesworth, V. B. 1954. *The Physiology of Insect Metamorphosis.* Cambridge Univ. Press, Cambridge.

Williamson, D. I. 1973. *Amphionides reynaudii*, representative of a proposed new order of eucaridan Malacostraca. *Crustaceana* **25**:35–50.

Wilson, C. B. 1911. North American parasitic copepods. *Proc. U.S. Natl. Mus.* **39**:1–189.

TAXONOMIC INDEX

SUBJECT INDEX

595